Protein Biochemistry

Protein Biochemistry

Editor: Anton Torres

RCallisto Reference

www.callistoreference.com

Callisto Reference,
118-35 Queens Blvd., Suite 400,
Forest Hills, NY 11375, USA

Visit us on the World Wide Web at:
www.callistoreference.com

ISBN: 978-1-63239-803-1 (Hardback)

Cataloging-in-publication Data

Protein biochemistry / edited by Anton Torres.
    p. cm.
Includes bibliographical references and index.
ISBN 978-1-63239-803-1
1. Proteins. 2. Biochemistry. 3. Proteins--Metabolism. 4. Metabolism. I. Torres, Anton.
QP551 .P76 2017
612.01575--dc23

Table of Contents

Preface

This book on protein biochemistry explains themes related to protein bonding, protein folding and other similar protein dynamics that occur at a molecular level. The study of proteins is very important for the understanding of DNA molecules as they help in the synthesis of protein. Biochemical study of protein deals with protein structures that perform a variety of roles such as the processing of food and muscle movement which is extensively studied through kinematic systems and informatics analysis. This book unravels the recent studies in the field of protein biochemistry. It strives to provide a fair understanding about the discipline and to help better understand the latest advances within this field. This book is appropriate for students seeking detailed information in this area as well as for experts. It is a vital tool for all researching and studying this field.

The world is advancing at a fast pace like never before. Therefore, the need is to keep up with the latest developments. This book was an idea that came to fruition when the specialists in the area realized the need to coordinate together and document essential themes in the subject. That's when I was requested to be the editor. Editing this book has been an honour as it brings together diverse authors researching on different streams of the field. The book collates essential materials contributed by veterans in the area which can be utilized by students and researchers alike.

Each chapter is a sole-standing publication that reflects each author´s interpretation. Thus, the book displays a multi-facetted picture of our current understanding of applications and diverse aspects of the field. I would like to thank the contributors of this book and my family for their endless support.

Editor

Design and Analysis of a Petri Net Model of the Von Hippel-Lindau (VHL) Tumor Suppressor Interaction Network

Giovanni Minervini[1], Elisabetta Panizzoni[1], Manuel Giollo[1,2], Alessandro Masiero[1], Carlo Ferrari[2], Silvio C. E. Tosatto[1]*

1 Dept. of Biomedical Sciences, University of Padua, Padua, Italy, **2** Dept. of Information Engineering, University of Padua, Padua, Italy

Abstract

Von Hippel-Lindau (VHL) syndrome is a hereditary condition predisposing to the development of different cancer forms, related to germline inactivation of the homonymous tumor suppressor pVHL. The best characterized function of pVHL is the ubiquitination dependent degradation of Hypoxia Inducible Factor (HIF) via the proteasome. It is also involved in several cellular pathways acting as a molecular hub and interacting with more than 200 different proteins. Molecular details of pVHL plasticity remain in large part unknown. Here, we present a novel manually curated Petri Net (PN) model of the main pVHL functional pathways. The model was built using functional information derived from the literature. It includes all major pVHL functions and is able to credibly reproduce VHL syndrome at the molecular level. The reliability of the PN model also allowed in silico knockout experiments, driven by previous model analysis. Interestingly, PN analysis suggests that the variability of different VHL manifestations is correlated with the concomitant inactivation of different metabolic pathways.

Editor: Manuela Helmer-Citterich, University of Rome Tor Vergata, Italy

Funding: This work was supported by Associazione Italiana per la Ricerca sul Cancro(AIRC) Grant MFAG 12740 to ST. GM is an AIRC research fellow (http://www.airc.it/). The funders had no role in study design, data collection and analysis, decision to publish, or preparation of the manuscript.

* E-mail: silvio.tosatto@unipd.it

Introduction

Pathological deregulation of cellular pathways often results in a family of complex and correlated diseases commonly termed cancer [1]. Cancer is a multi factorial disease where different causes contribute to its development. Several computational methods have been developed to explore the functional pathways involved in tumorigenesis. Some of them focus on differential gene expression between healthy and pathologic tissues [2–3], on protein-protein interaction network analysis [4–5] or on molecular dynamics simulations [6]. Other methods approach the disease through discretization of pathological components that result in tumor [7]. All of these approaches are very powerful when the variables related to the disease, although complex, are well known and studied. A multi-factorial disease can be approached by means of mathematical theory, building a theoretical model where cell components are connected with each other. In biology, several problems were dealt with network theory [8–9]. A network is a group of objects strongly inter-connected with each other (e.g. proteins and enzymes of a pathway or animals belonging to interacting populations). Their construction and subsequent simulation is made via mathematical analysis of the connections between nodes found in the system and their time-dependent behavior [10]. A biological network is generally composed of proteins, nucleic acids and cofactors connected by biological reactions such as protein complex formation or enzyme activity regulation [10]. Von Hippel-Lindau syndrome (VHL) [11] is a

good study case to test the network theory applied to cancer due to the similar medical history and pathological phenotype that patients share. While hereditary cancers represent only a small part of all human tumors, their investigation represents a challenge to understand the pathway leading to tumor formation. In 2010, Heiner et al. first approached VHL using the so-called Petri Net (PN) simulation networks [12]. Their work, inspired by a previous theoretical model of cellular oxygen-related pathways [13] [14], was a preliminary investigation of the core oxygen sensing system and its connection with VHL onset. Heiner and coworkers proposed three different functional modules responsible for hypoxia network control and for HIF-1α degradation [12]. In other words, they theorized that hereditary forms of cancer, such as different manifestations of VHL, are the result of different and concomitantly compromised metabolic pathways.

Von Hippel-Lindau Disease

Von Hippel-Lindau protein (pVHL) is the product of the von Hippel-Lindau gene, located in the short arm of 3rd chromosome, and constantly transcribed in both fetal and adult tissues [15]. Mutations of pVHL are related to a pathological outcome termed VHL syndrome, an inherited form of cancer [16]. VHL syndrome is characterized by cysts and tumors growing in specific parts of the organism [16–17]. It is considered a severe autosomal dominant genetic condition with inheritance of one person in over 35,000 [18]. The tumor injuries, which can be either benign or malign, are usually located in the retina, adrenal glands, epididymis, central nervous system, kidneys and pancreas [19]. As a genetic disorder, VHL syndrome follows Knudson's two hit principle. A copy of the gene is mutated in the germ line, but the other gene copy still produces a functional protein. Complete protein inactivation appears during life due to somatic inactivation of the remaining functional copy [20]. On the contrary, mutations occurring during early fetal formation result in unsuccessful development [21]. The pVHL gene has 11,213 base pairs including three exons [18] and the final transcript is a protein commonly present in two isoforms: pVHL30 and pVHL19, of 213 and 160 residues respectively. Neither isoform contains a known enzymatic domain, but rather appears to serve as a multipurpose adapter protein engaging in multiple protein-protein interactions [22]. pVHL structure is organized in an α- and β-domain and its stability was demonstrated to be ensured by direct interaction with other proteins such as Elongins B and C [23]. Both Elongin B and C are also required for the best characterized function of pVHL, the ubiquitination dependent degradation of Hypoxia Inducible Factor (HIF) via the proteasome [24]. However, pVHL is considered a multipurpose protein due to its high number of known interactors. At the time of writing, the IntAct database [25] presents more than 200 different interaction partners, with some of them competing for the same Elongin binding site. Indeed, pVHL was found in different cellular compartments and seems to be involved in many different cellular processes such as apoptosis, cell proliferation, survival and motility [26]. Considering the huge number of interactors and multiple cellular localizations, many different functions have been described or hypothesized, such as regulation of cytoplasmic microtubules during mitosis [27] and endothelial extracellular matrix deposition [28]. On the other hand, considering the huge number of players involved in VHL syndrome and the lack of reliable kinetic data, a PN based approach may be a preferable option for an entire VHL pathway simulation.

Petri Net for Interaction Pathways

Since their invention, by Carl Adam Petri in the early sixties, PNs were mostly used to describe technical systems, but later the utility in describing biological and biochemical functions has also been demonstrated [29]. PNs were successfully used in many studies to describe biological networks [30], such as the regulation and etiopathology in human Duchenne Muscular Dystrophy [31] and the hypoxia response network [12]. PNs are qualitative mathematical models that can graphically represent many object types, not only metabolites but also different protein states and are useful to simulate networks where not only metabolites are involved. Indeed, PNs can be a powerful tool to study all concurrent interactions in a specific pathway, even if the proteins or kinetics are not well-known. Due to the large number of different pVHL functions involved in VHL disease progression, we decided to extend the PN based analysis of [12] increasing the number of considered protein-protein interactions. We generated a novel manually curated PN model of the entire VHL regulation system collecting data from the literature and including the signaling pathways and glucidic metabolism. In order to build a realistic network, literature from both biochemical experiments and *in silico* predictions were used as source. It was decided to build a PN with only confirmed pVHL interactions whose function was also known. The resulting PN was validated using an analysis of specific properties as suggested by previous studies using the same method [29]. After validating the PN structure, *in silico* knock outs of specific proteins were done in order to observe the different network behaviors and the resulting biological effect.

Methods

The network was designed in the Snoopy PN framework (version 2, revision 1.13) [32], respecting the mathematical PN formalism as described in [12–33]. PN were demonstrated to be useful in describing discrete and concurrent processes in a simple graphical representation [30] and have been used to describe biomedical processes due to their capacity of representing sequential steps in a process. PN modeling methods are actively used to describe, simulate, analyze, and predict the behavior of biological systems. The Snoopy PN framework provides an extensible multi-platform framework to design, animate, and simulate Petri nets [32]. We chose Snoopy to facilitate future extensions of the VHL pathway presented here. Among different available PN types a standard PN was chosen to limit the number of variables. Both Charlie and PInA analyzers were used for PN analysis and validation (34). Further, *in silico* knock out experiments were used to test the biological reliability of the model. Structural model validation was made by analysis of the T-invariants to demonstrate whether the system was covered by T-invariants and to confirm the biological meaning of each invariant. The use of T- and P-invariants is given by their own properties: they are a set (of transitions or places, respectively) that allow the reproduction of the same state after n transformations. A P-invariant represents a set of places where the number of tokens is constant and independent on the firing rate. A T-invariant instead represents a set of transitions that cyclically comes back to show the same initial set. Biologically a P invariant can represent the process of regulating a protein, whereas T invariants can represent cyclical biochemical transformations such as metabolic reactions. To this end, the computed invariants were grouped in Maximal Common Transition Sets (MCTS) and Clusters, the former based on occurrence of specific sets of transition inside the various T-invariants, and the latter based on similarities between T-invariants. Different numbers of clusters will be defined depending on the resulting square matrix. Where MCTS create disjunctive nets, Clusters merge together similar T-invariants. Behavioral validation was made by selectively deleting tokens inside the model, imitating possible biological disruptions such as disease-causing mutations. The resulting network behavior was compared to what is reported in the literature. Total runtime for invariants computation were less than ten seconds on a mainstream Linux x86 workstation. Literature sources used to build the model are reported in Table S1. The Snoopy framework for PN construction, Charlie and PInA tools for analysis are available at the website (URL: http://www-dssz.informatik.tu-cottbus.de/DSSZ/Software). Finally, the model was used to simulate the network behavior through visual inspection of both token movement and accumulation in specific parts of the network. For a visual explanation of token movement in a PN refer to Video S1.

Model Availability

The resulting VHL disease PN model is available in File S1.

Results

Notations and Assumptions

The PN built here focuses on pVHL interactions that were already proven by biochemical experiments and reported in the literature. We chose to model a realistic VHL disease pathways based on confirmed literature data, including all known VHL functions, VHL related signal pathway and glucidic metabolism. All bibliographic sources used to design the model are presented in Table S1. The final PN is composed of 323 places and 238 transitions, connected by 801 arcs. Tables S1 and S2 show all places and transitions and the related biological correspondence. Places are mainly proteins and enzymes, while some represent DNA or small molecular substrates such as glucose and cofactors (e.g. ATP). Notation for both pre- and post-places and their biological meaning are explained in Table S1. In a few cases, places are used to represent a whole group of changes generated by DNA transcription, (e.g. p_32 and p_33 or Et_eff1 and Et_eff2). Transitions instead symbolize complex formation between two proteins or post-translational modifications. Output transitions stand for degradation or movement to other parts of the cell or organism to complete their functions (e.g. degrad_1 and degrad_2) whereas input transitions show the generation of a substrate or protein. In order to simplify the design of such a large network, we decided to use macro nodes to group reactions representing complex molecular pathways such as signaling pathways or secondary signal cascades. The whole process is merged into a single node with a given name to allow visual inspection only in case of need. From the top level all transitions can still be found in a hierarchical lower layout level. Logic nodes were used for places participating in many reactions throughout the network such as ATP and ADP (7 logical copies each) or NAD and NADH (4 logical copies each). A total nesting depth of two was chosen to model macro nodes. Special arcs were not used while we chose to model the permanent presence of some objects using double arcs (e.g. for elob, eloc and places standing for enzymatic activity). In case of proteins which are actively degraded, it was preferred to create an input transition simulating constant production (or synthesis) and an output for consumption. This is the case for pkcz2, Jade1, pVHL and HIF-1α. As can be seen from Figures 1 to 3, which represent the entire model, two major nodes can be immediately identified: pVHL and vcb, the complex made by pVHL and the two elongins. Another relevant part is the glucidic metabolism, modeled due to its hypoxia induced regulation. It is represented in detail in Figure 2.

HIF-1α Transcription Activity

The HIF-1α transcription factor stimulates proliferation of endothelial cells to create new blood vessels during localized or broad hypoxia. In human, it is present as three different paralogs: HIF-1α, HIF-2α and HIF-3α. The sequence is quite conserved between the former two, whereas the latter is slightly shorter and seems to have completely different functions compared to the other two [33–34]. Both HIF-1α and -2α stimulate DNA transcription but the exact products of this activity are still poorly understood. In our model, only HIF-1α *in vivo* activity was considered. It cannot be excluded that other biological effects depend on the second paralog. Indeed, both have a pro-angiogenetic function and are degraded by pVHL via proline-directed hydroxylation. HIF is a heterodimer of HIF-1α and HIF-1β, the latter being also termed Aryl hydrocarbon Receptor Nuclear Translocator (ARNT). We started from the transcription activity of HIF due to its regulation is the most studied pVHL function. Our model, as expected from literature data, shows that HIF-1α enters the nucleus when not degraded by pVHL. It subsequently binds HIF-1β to form the HIF heterocomplex which interacts with DNA. Our model correctly simulates the increased affinity of HIF towards DNA. Transcription is enhanced by some co-factors binding both subunits of HIF and other proteins such as p300, Creb and cjun. This takes place in a specific DNA promoter sequence termed Hypoxia Response Element (HRE). Furthermore, during transcription some pro-angiogenic factors are produced: Vascular Endothelial Growth Factor (VEGF), Endo-

thelin (ET) and Erythropoietin (EPO). All described pathways are in agreement with previous observations reported in [35].

Metabolic Processes

HIF-1α transcription activity includes some proteins which are dependent on oxygen but involved in other pathways (e.g. oxidative metabolism) or completely independent (e.g. metallo-proteinase MT1MMP). Further, HIF-1α stimulates production of proteins involved in the glucidic pathway. The final product of the metabolism is adenosine triphosphate (ATP), a molecular form of energy, composed by adenosine, an adenine ring connected to a ribose sugar, and three phosphate moieties. When a phosphate moiety is hydrolyzed it releases energy, used by cells for enzymatic reactions. The glucidic metabolism is composed of glycolysis, Krebs cycle, glycogen formation and respiratory chain with ATP synthesis. Glucose is absorbed in cells by enzymatic glucose transporters (GLUT), which carry the molecule to the location inside the cell where the metabolism takes place [36]. There are many isoforms of these transporters: GLUT1 is present in all cells and in particular in erythrocytic membranes, neurons and glia [37]. GLUT2, located in both liver and pancreatic beta cells, is characterized by low affinity for glucose, hence it requires a higher glucose concentration to be activated [38]. Right after eating, glucose concentration increases, thereby quickly activating them. GLUT2 stimulates production of insulin, a hormone regulating the plasmatic glucose concentration. Glucose plasmatic concentration can also increase due to an opposite pathway, originating from liver glycogen being decomposed into glucose and reaching systemic circulation. GLUT3 is mostly present in neurons, whereas GLUT4 is the insulin activated transporter located in myocytes, adipocytes and cardiomyocytes [36–39]. In our model, we chose to exclude GLUT3 due to its specific role in neuronal cells. Glycolysis occurs in the cytoplasm and during this process each glucose molecule is phosphorylated, consuming two molecules of ATP, then divided into two smaller molecules. Further modifications of these two molecules result in new ATP production. The molecule obtained at the end of glycolysis is pyruvate, which can be again modified through three different pathways. It can be decarboxylated and linked to Co-enzyme A to form acetyl-Co-enzyme A. It can then be carboxylated to obtain oxalacetate, or transformed through lactate dehydrogenase into lactic acid. Pyruvate can also be generated by other metabolic pathways, like protein or fatty acid disruption and amino-acid modifications. Acetyl-CoA and oxalacetate are the molecules used in the following glucidic metabolism process, the Krebs cycle, taking place in the mitochondrial matrix. The Krebs cycle starts with acetyl-CoA and oxalacetate merging to create citric acid, which continues undergoing modifications until oxalacetate is formed again. During the process some co-enzymes are modified. Decarboxylation of pyruvate to form acetyl-CoA already transforms a NAD+ (Nicotinamide Adenine Dinucleotide) in NADH (reduced form), afterwards obtaining one more of ATP, GTP, $FADH_2$ (Flavin Adenine Dinucleotide) and three more NADH per pyruvate molecule entering the Krebs cycle. The redox co-enzymes are considered electron transporters. During metabolic reactions they reduce themselves and get electrons (and protons) to oxidize the substrate of the enzymatic reaction. Electrons taken during the glucose metabolism are then used in the respiratory chain taking place in the internal mitochondrial membrane. The respiratory chain consists in transporting electrons through enzymes called cytochromes and others co-enzymes, characterized by the capability to receive and donate electrons. NADH ($FADH_2$) is oxidized again by cytochromes going back to the form of NAD (or FAD). Electrons gained through oxidation are used to reduce

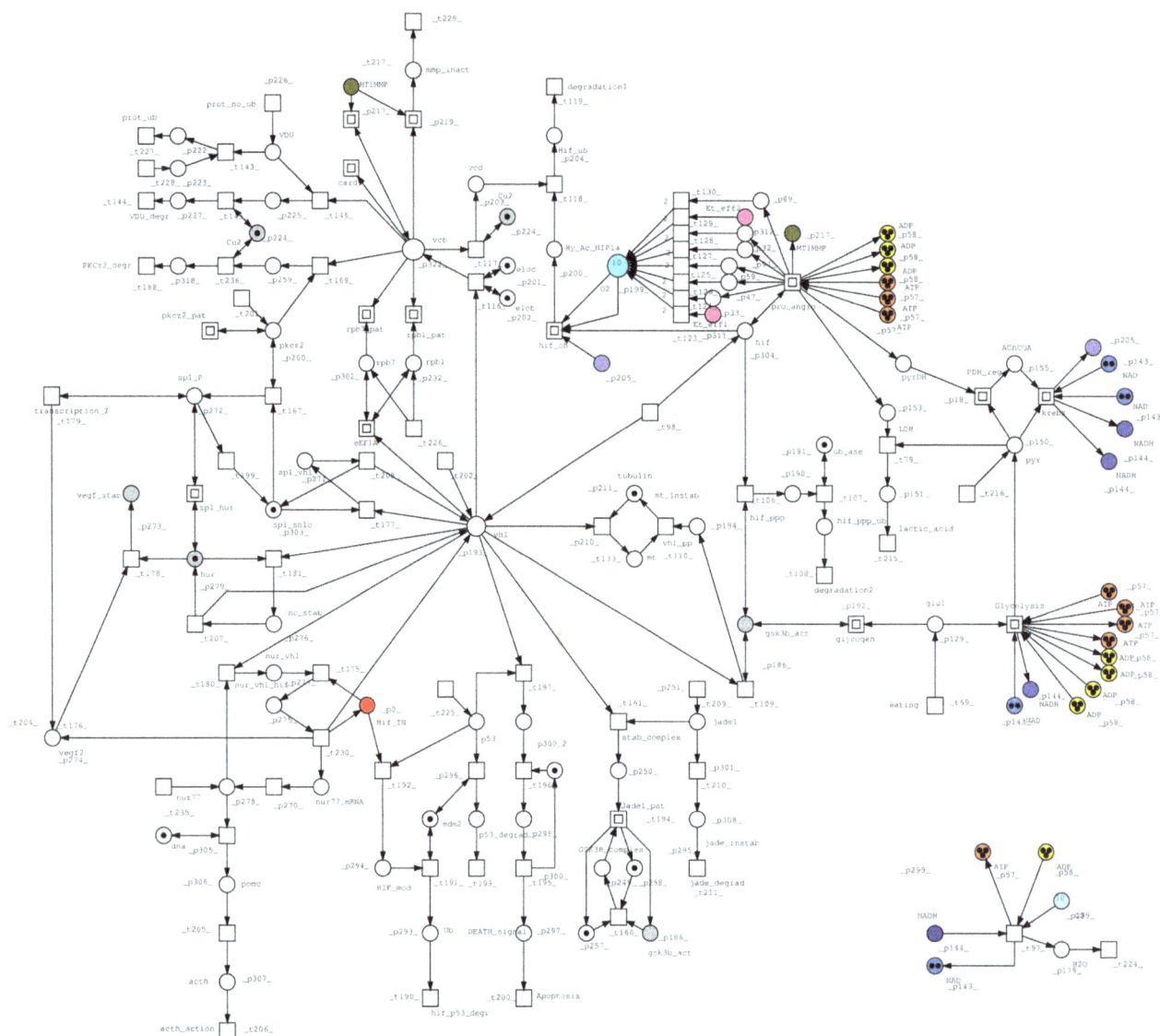

Figure 1. Top level model. The colors of some tokens were arbitrarily chosen to give a clearer identification of the central nodes (ATP, Vcb and oxygen) or for nodes involved in more reactions such as GSK3β. The group of nodes in the bottom left is not disconnected from the central body of the network thanks to the presence of logic nodes for ATP synthesis (t_97).

half a molecule of oxygen into water, releasing more energy. The $FADH_2$ and NADH redox chain establishes a chemical potential causing the push of protons outside the internal membrane towards the inter-membrane space, which stays between the mitochondrial inner and outer membrane. This also causes a higher concentration of protons outside the inner membrane. The resulting gradient causes the tendency of protons to enter the cell. The final step is ATP-synthetase, formed by a channel that allows protons to enter, pushed by the gradient, allowing the enzyme to change conformation and make its reaction. This kinetic energy is converted into ATP. In our model, the glycolytic and Krebs cycles were described in detail, represented at the hierarchical second level by the coarse transition Glycolysis. The respiratory chain was instead merged into a single node (t_97). We chose to represent creation and consumption of ATP in order to show the effects of lower and higher oxygen concentration on the network. On the other hand, oxygen consumption for ATP synthesis during the respiratory chain creates a flow of oxygen in the model. Oxygen is

not the only connection between glucidic metabolism and hypoxia. Indeed, HIF-1α transcription activity enhances the transcription of many GLUT isoforms (such as 1, 3 and 9) and the pyruvate dehydrogenase kinase, which determines the pyruvate dehydrogenase (PyrDH) inactivation and consequent Acetyl-CoA formation from pyruvate. Finally, Lactate dehydrogenase is also produced, to ensure an alternative compound, creating energy needed for cell survival [36–40].

pVHL-dependent Processes

Some interactors can bind pVHL in regions interacting with Elongin C. These are HuR, Nur77, p53 and Jade1. Nur77 has a complex function and its role in pVHL tumor suppressor activity is still not entirely clear. Nur77 can bind pVHL, inhibiting Elongin binding while allowing HIF-1α binding. Its transcription is stimulated by HIF-1α itself, and pVHL-HIF-1α-Nur77 complex formation stabilizes the transcription activity of HIF-1α by inhibiting the pVHL-dependent degradation [41]. Another

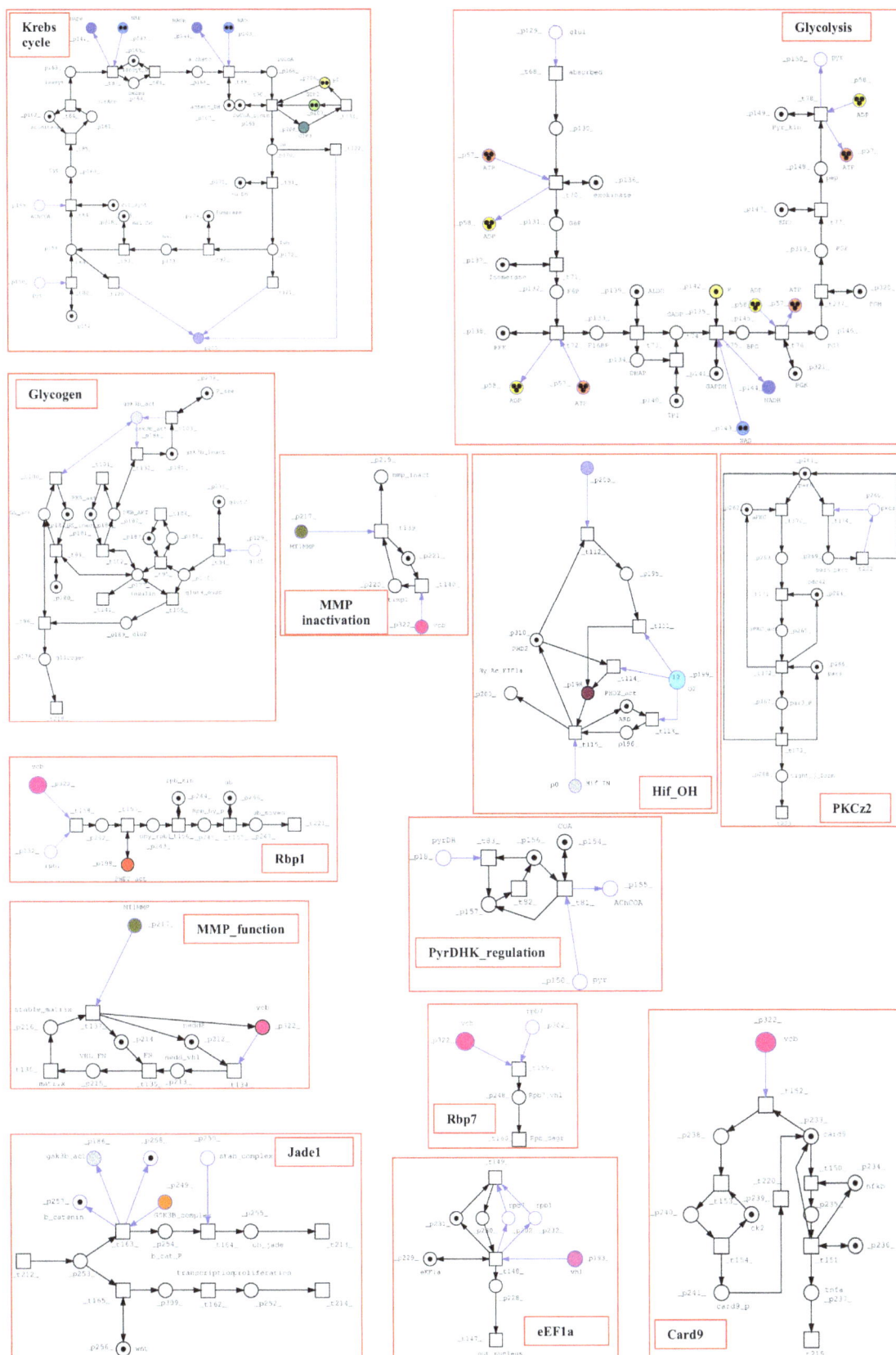

Figure 2. Lower hierarchical PN levels. Pathways from the top level are grouped in macro-nodes (functional subordinated layer), in particular glucidic metabolism and various VHL functions.

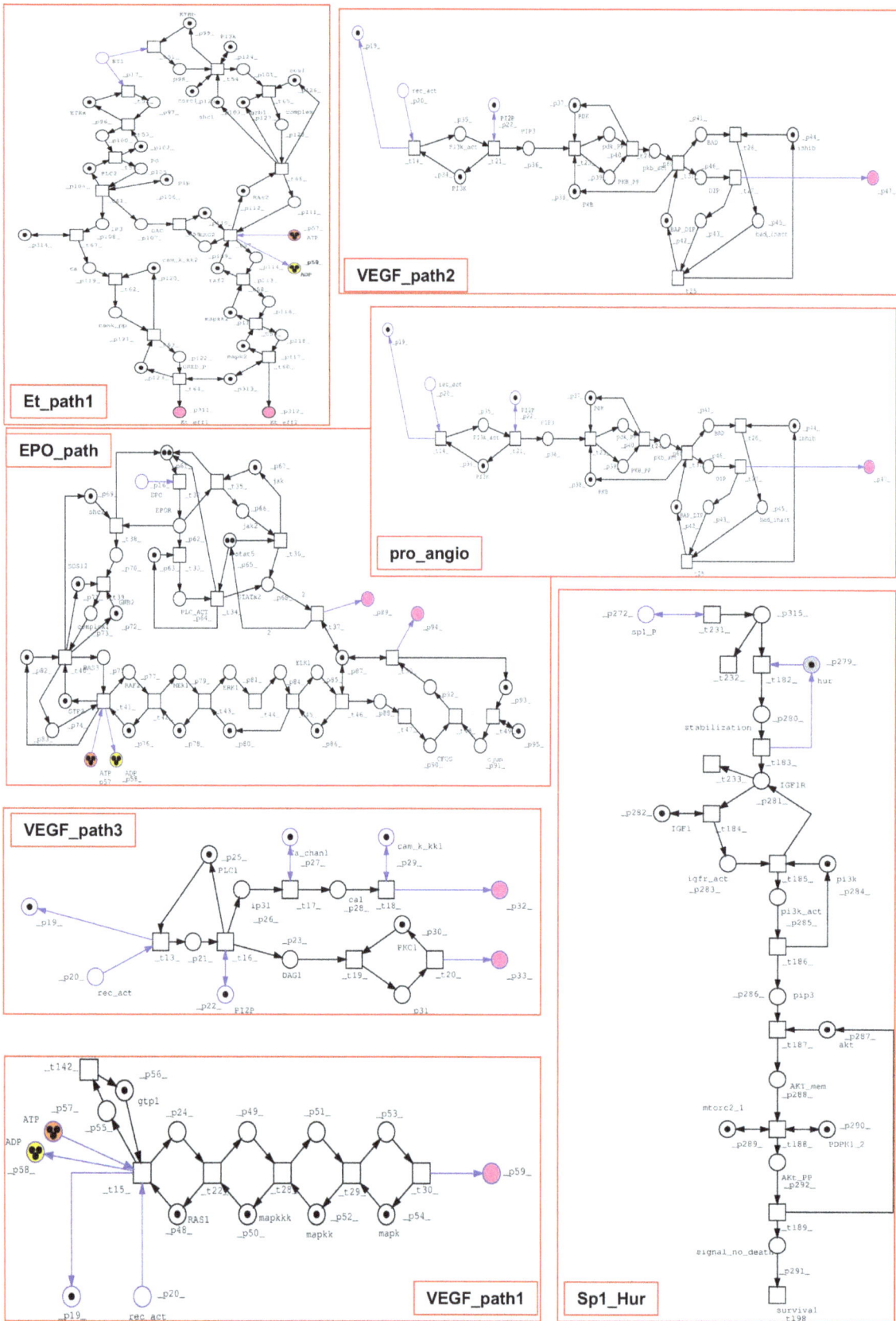

Figure 3. Lower hierarchical PN levels, in particular HIF-1α regulation and HIF-1α-dependent pro-angiogenic signaling. VEGF and EPO pathways are at a lower hierarchical level than the pro_angio macro-node.

Nur77 function is the stimulation of proopiomelanocortin (POMC) transcription, which is a precursor for adrenocorticotropic hormone (ACTH) formation. This hormone has an important stress response function, stimulating cortisol production and other neurotransmitters from the adrenal glands, to enhance the organism reaction to danger and stress *stimuli* e.g. increase of gluconeogenesis and muscle mass. An excess of this hormone can cause desensitization of its receptors for feedback down-regulation and thus muscular weakness, tiredness, hyperglycemia and osteoporosis [42]. p53 can bind to pVHL avoiding the degradation of this tumor suppressor. Instead, it stimulates the apoptotic signal cascade via the p300 co-activator, which stimulates production of proteins enhancing the cell programmed death. If p53 cannot bind pVHL, two more mechanisms are described in the model. One is its modification and degradation by Mdm2 and the other is the pVHL-independent degradation of HIF-1α. Interaction with Mdm2 is needed in both cases [43–44]. Jade1 is a short-lived protein whose main function is to stimulate the phosphorylation-dependent degradation of β-catenin. This is a subunit of the cadherin protein complex acting as an intracellular signal transducer in the Wnt signaling pathway. It seems that β-catenin is able to stop cell division via a contact-dependent inhibition signal, whereas in Wnt signaling it is also involved in proliferative transcription. When Wnt is not present, β-catenin can be phosphorylated by Glycogen Synthetase Kinase, type 3β (GSK3β) in complex with APC (Adenomatous Polyposis Coli) and Axin. β-catenin can interact with Jade1 and be only successfully degraded after this interaction [45]. Related functions are represented in the macro node Jade1_pat. GSK3β seems to be a protein involved in many different pathways. GSK3β is involved in Glycogen Synthetase deactivation and can even phosphorylate pVHL and HIF-1α. In the case of HIF-1α, it generates a pVHL-independent degradation pathway, where phosphorylation allows ubiquitination, whereas in the case of pVHL, it inhibits pVHL stabilization of microtubules [46].

Structural Model Analysis

Based also on previous observations of Heiner et al., [47], in 2008 Grunwald et al., demonstrated that PN can be used to describe large and complex metabolic pathways [31]. They postulated the following set of minimal rules that a PN should satisfy to be considered biologically reliable: (1) the network should be entirely connected, (2) the network should be covered by T-invariants, and (3) each T-invariant and P-invariant should have a biological meaning. The model described here was tested with respect to what previously done by Grunwald and co-workers [31] and resulted to be covered by T-invariants, connected, homogeneous and each place has a pre-transition and a post-transition. Transitions without pre- or post-places were used to simulate the system interface to the surroundings. The network is alive, in other words, it continues to work forever, with all transitions contributing to the net behavior forever, and no dead transitions. The MCTS and Cluster analysis were used due to the large number of T- and P-invariants included in the model. Both methods are used in PN theory to reduce the complexity connected with such a large network and to reduce the errors connected with manual investigation. From the 238 transitions present at the beginning in the model, 393 T-invariants were computed without considering 10 trivial invariants. The latter consist in a pair of transitions that usually represent a forward and backward reaction, such as the active and inactive state of a protein. Trivial invariants could be erased to reduce the dimension of the network without disturbing the overall system when the interest is focused on the steady state behavior [12]. T-invariants were grouped into 44

Clusters using the Tanimoto coefficient with similarity threshold of 65%, as described in [31]. Only 11 of these 44 comprised more than one T-invariant. The three biggest Clusters are C9, composed of 144 T-invariants, C8 of 72 and C11 of 64 T-invariants. Separation into clusters allows easier analysis of networks pathways represented by each T-invariant, since they are grouped by similarity, specifically the common transitions by which they are composed. T-invariants named in the text are shown in Table S3, while T-invariants grouped in C8, C9, C10, C11 are explained in Table S4 and described as follows.

Cluster C8

Cluster C8 groups all transitions included in HIF-1α pathways, including transcription, signaling cascades, degradation via pVHL, p53 and GSK3β, and eventually the Krebs cycle. For the EPO signaling pathway, two transitions (t_35 and t_36) are not included which cause Jak activation and consequent Stat5 activation to stimulate DNA transcription. Matrix stability regulation is also part of the cluster due to the destabilization induced by HIF-1α transcription of metallo-proteinase (MMP), transitions from t_134 to t_140. The largest T-invariant in C8 is Inv_280 (93 transitions) while the smallest is Inv_377 (81 transitions). The differences between T-invariants show the possibility of alternative pathways inside the model. For example, the VEGF dependent signal cascade can proceed in three different ways: t_13, t_14 and t_15, which lead to the pathways being merged in the coarse nodes Vegf_path3, Vegf_path2 and Vegf_path1, respectively. The occurrence rate in C8 is 24 transitions for each path. The Endothelin, VEGF and Erythropoietin pathways are not in conflict and occurring together. Disaggregation of the matrix via MMPs is present in 18 T-invariants, whereas inhibition of these proteins, i.e. matrix stabilization, is present in the remaining 54 transitions. Regarding the Krebs cycle, 47 T-invariants have t_91, of which only 24 reach t_92 and t_93, representing the last three steps of the cycle: succinate to fumarate, fumarate to malate, and malate to oxalacetate. All the malate being produced is used to regenerate oxalacetate. Degradation of HIF-1α occurs in any T-invariant of the cluster. The pVHL-dependent degradation of HIF-1α is always present (transitions t_116 to t_119). In 19 T-invariants degradation takes place via p53 (t_191 to t_193) or, alternatively, via phosphorylation by GSK3β in another 17 T-invariants. Two of the three pathways can be present in the same T-invariant, as in Inv_227, where degradation via pVHL and degradation via p53 are both present. This was considered as the HIF-1α dependence on the lack of degradation by these proteins. All three degradation pathways never appear in the same T-invariant. The p53 and GSK3β paths are never present together but each of them is accompanied by pVHL-dependent proteasomal degradation. Inv_377 lacks the EPO signaling pathway but is the only one in this cluster to have t_34, t_33 and t_37. These invariants have all input and output transitions. For example, t_202 the second input for pVHL, is present in only 18 invariants. Other inputs are t_98, always present, leading to formation of HIF-1α and pVHL, t_192, producing p53 and t_216, representing other pyruvate generating metabolic pathways. The latter is also present in each invariant allowing formation of the pyruvate needed for Krebs cycle progression.

Cluster C9

Cluster C9 is the largest cluster in our model and includes 144 T-invariants. It is characterized by complete EPO pathway abrogation which goes through formation of the Shc-Grb-Sos complex and the consequent mapk-dependent phosphorylation cascade. Transition t_127, representing EPO effects on oxygen

Table 1. List of Trivial T-invariants excluded from calculation with their associated biological meaning.

Trivial T-Invariants	ID transitions	Biological Meaning
TInv_1	t_99, t_100	Glycongen Synthase regulation
TInv_2	t_101, t_102	Pkb regulation
TInv_3	t_103, t_132	GSK3β active-inactive state
TInv_4	t_174, t_222	Par6 inactivation via aPKCζ2
TInv_5	t_177, t_208	VHL binding to Sp1
TInv_6	t_167, t_199	Sp1 phosphorylation and dephosphorylation
TInv_7	t_0, t_2	Hif transport in and out of nucleus
TInv_8	t_0, t_234	Hif inhibition via FIH
TInv_9	t_181, t_207	Hur inhibition via VHL
TInv_10	t_231, t_232	IGFR mRNA production and destruction

production, is absent. In its place, t_35 and t_36 are considered, which are present in 72 T-invariants. In cluster C9, the largest T-invariants are Inv_278 and Inv_279 (74 transitions) while the shortest ones are Inv_101, Inv_105, Inv_144 and Inv_148 with 65 transitions each.

Cluster C10

Cluster C10, composed of 52 T-invariants, is characterized by the presence of glycolysis between many transitions grouped in the cluster. This is also the cluster containing the most populated T-invariant of all computed 393 non-trivial T-invariants. This is Inv_245, including 101 transitions and covering almost half of the whole model. Cluster C10 also includes Inv_125, the shortest invariant of this model, composed by 85 transitions due to lack of the Krebs cycle. Another difference with the other three major clusters is that here both EPO paths are present, specifically, the Jak pathway belongs to 4 T-invariants and Shc-Grb-Sos is observed throughout the cluster. Vegf_path1 seems to be more common in this cluster, being present in 36 T-invariants, whereas the other two are present 12 times each. This time they are present even in the same invariant, as for Inv_60, Inv_129, Inv_172 and Inv_215, with both t_13 and t_15, and Inv_142, Inv_185 and Inv_228 with t_14 and t_15 and all subsequent signaling appearing at the same time. Despite glycolysis being present in all cluster invariants, the Krebs cycle appears only in 11 cases. p53-dependent degradation of HIF-1α occurs in 11 cases while the phosphorylation-dependent one appears in 13. An input transition has been added with respect to the other major clusters so far analyzed (i.e. t_69_eating) without which glycolysis could never take place.

Cluster C11

Cluster C11 is composed of 64 T-invariants. Only part of the EPO pathway is described here, with the major difference that the Krebs cycle is completely abrogated while Prolyl Hydroxylase type 2 (PHD2) regulation by oxalacetate is included. HIF-1α interaction with Nur77 and transcription of VEGF by Sp1 are also present. t_80 (transformation of pyruvate in oxalacetate) is not present in the first 42 cluster T-invariants. Nur77 interaction with HIF-1α is present only in 8 T-invariants, specifically Inv_87 to Inv_94. Sp1 transcription activity is appearing in twice the amount, including the same 8 invariants just mentioned. VEGF transcription via Sp1 activity is aPKCζ2 phosphorylation dependent, which does however not appear in the cluster. When VEGF

is synthesized, it is subsequently stabilized by Hur, followed by t_178 and Hur is recreated to allow other functions. Indeed, it is one of the few places without input transition but with a token that goes forward and backward again. Compared to the other clusters, C11 also shows one less transition in the coarse PHD regulation node, specifically t_81, which shows the transformation of pyruvate by pyrDH into acetyl-Coenzyme A, needed for the Krebs cycle. The four clusters C-8 to C12 are very similar to each other, as can be seen from the distance tree in Figure 4. They all contain the HIF-1α transcription activity and signaling pathways caused by EPO, VEGF and the HIF-1α degradation options. They include the effects of other transcription activity products, like metallo-proteinase and pyruvate dehydrogenase kinase, which regulate the activation state of PyrDH. All include part of the glucidic metabolism but not Glycogen formation itself. Other five clusters from C12 to C16 have a smaller number of T-invariants and fewer transitions present in each invariant. They do not include transcription activity but are only formed by the VEGF and glycolytic pathways. The information contents of these clusters turned out to be uninformative and their analysis was not included. The same applies to clusters composed by 1–3 T-invariants. Finally, some transitions are not present in the clusters and not listed in the T-invariants because trivial invariants were excluded from cluster analysis. These transitions are shown in Table 1 with their respective biological meaning.

MCTS Analysis

Another way to group invariants is by the amount of single transitions present in them. Maximal common transition set (MCTS) analysis provides a PN decomposition into non-overlapping subnets, sharing parts of the same T-invariants [29]. In a biochemical network, MCTS could be interpreted as enzyme subsets operating together under steady state conditions, computed based on the support of a T-invariant. MCTS computation does not consider stoichiometric relations, describing exclusively sets of reactions present in a maximal number of T-invariants resulting shared by different signaling pathways [48]. A total of 40 non-trivial MCTS were identified, with results and related biological means shown in Table 2. Some transitions do not belong to a non-trivial MCTS, because their occurrence has no similarity with other transitions and they create separate MCTS (specifically: t_69, t_82, t_91, t_94, t_98, t_114, t_116, t_120, t_121, t_122, t_179, t_202, t_209, t_212, t_216, t_225 and t_229). MCTS define transitions that always take place together, but are not necessarily connected, thus representing disjunct building

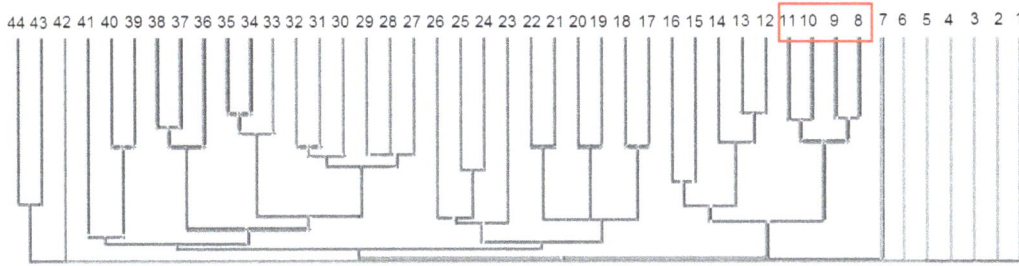

Figure 4. PinA Distance Matrix clustering, using Tanimoto coefficient and 65% threshold of. The numbers indicates clusters. In C8, C9, C10, C11 are highlighted a red square.

blocks constituting the network. Considering both analyses, a table was automatically built in PInA [29] showing a correlation between clusters and MCTS. Transitions (t) or MCTS (M) are compared to evaluate how many T-invariants clusters cover the selected M or t (if the transition is not already part of the MCTS, as listed above). The more covered a transition or set is, the more central it could be considered for the network behavior. Recently, a network coarsening method based on abstract dependent transition sets (ADT) was presented [49]. It is formulated without the requirement of pre-computation of the T-invariants and is a tool commonly used for the decomposition of large biochemical networks into smaller subnets. Due to the manually designed nature of our model, we preferred to maintain a logic hierarchy based on metabolic pathways in order to maintain the network centered on pVHL and its interaction. The MCTS calculation results shows that the most covered set by cluster T-invariants is M20 with 358 T-invariants covering all transitions in the set, indicating that this MCTS corresponds to more T-invariants than the others. All transition sets are an important link to the others, as tokens pass through these transitions more often. A transition not present in any set but most covered by T-invariants is t_98, which is also the most frequently occurring transition, see Figure 5. The 10 most occurring transitions are listed in the Table 3.

P-invariant Analysis

Although the network is not covered by P-invariants, it has 130 P-invariants. 47 of these are trivial P-invariants, comprising a single place, connected with double arcs to imitate an activator arc function. Another object represented with double arcs is the enzymatic activity catalyzing a reaction and immediately going back to the steady state. P-invariants show places or sets of places where token numbers always remain equal and do not move outside the subnetwork induced by the P-invariant in the initial marking. In other words, they do not grow nor diminish. The remaining P-invariants are mostly located in signal transduction pathways, such as situations in which a protein is sequestered from its function and then goes back after a second reactivation mechanism. This scenario is present in p_41, p_42 and p_45 located in invariant P_58. It is important to notice that ATP and ADP, as well as NAD and NADH, are modeled as P-invariants. P_90, P_91 and t_97 are able to transform ATP and ADP. More in general, all energy consuming transitions are considered to be backward transitions of invariants. Invariants not related to signal transduction are places located in the Hur system, where Hur is removed from its function by pVHL. This is a good approximation for sequential modifications that momentarily activate proteins. Afterwards, Hur can go back and stabilize VEGF to increase its transcription activity.

In Silico Knock Out Experiments

The previously described clustering and MCTS analysis for T-invariants allowed us to identify the most common transitions and to understand which transitions can be depleted in our knock out experiments in order to get the most important biological effect. The knock out experiments were performed erasing selected transitions or tokens and observing which transitions or MCTS become inactivated. Considering our results and the literature, we decided to knock out the following pathway elements: (i) pVHL, (ii) HIF1α alone and with Sp1, (iii) t_98, (iv) PHD2, (v) MCTS1, (vi) t_97 and (vii) GSK3β. In the following, we describe the effect of each knock out scenario on our model.

(i) pVHL knock out. Degradation of HIF-1α is not completely depleted due to presence of both p53- and GSK3β-dependent alternative degradation pathways. All other processes usually inhibited by pVHL take place in an uncontrolled way, including creation of VEGF via Sp1 transcription activity and increased matrix regulation due to lack of fibronectin crosslinking. Hur resulted constantly activated and nur77 can stimulate synthesis of Proopiomelanocortin, precursor for the Adrenocorticotropic hormone. Card9 increases release of tumor necrosis factor, and NF-kB when not inhibited by pVHL. Instead, Jade1 is unable to survive long enough to inhibit β−catenin, generating a proliferation signal with Wnt. Lactic acid is also not produced due to LDH enzyme production being HIF-1α transcription activity dependent.

(ii) HIF-1α knock out. VEGF is still created thanks to Sp1, thus oxygen is still generated even if in lower proportion. If HIF-1α and Sp1 are both knocked out at the same time, oxygen is quickly consumed and the metabolism is soon unable to proceed. Lactic acid is not produced due to LDH enzyme production being HIF-1α transcription activity dependent. Glycolysis and glycogen are produced normally and the metabolism is not inhibited by PyrDH negative regulation and lactic acid formation. Since pVHL is present, other tumor suppressor activities are enabled, except for proteasomal degradation of HIF-1α due to the substrate being non-existent.

(iii) HIF-1α and pVHL double knock out. This generates a situation where the metabolism is normal but oxygen regeneration is less productive, with only Sp1 acting for transcription. Due to absence of pVHL, all proliferation-stimulating processes are active, causing an unbalanced consumption of resources. Our model shows that this condition is compatible with cell growth and multiplication, but new blood vessel generation is consistently slower and glucidic metabolism appears principally based on the glycolysis reaction. Similar activity reduction applies to both tight junction and cellular external matrix (ECM) pathway regulation. It cannot be excluded that some observed effects could be mitigated by both HIF-2α and HIF-3α activity *in vivo*.

Table 2. List of MCTS and transitions from PInA.

MCTS	ID Transitions
MCTS 1 (M1)	t_0, t_190, t_191, t_192;
MCTS 2 (M2)	t_1, t_3, t_4, t_5, t_6, t_7, t_8, t_9, t_10, t_11, t_12, t_32, t_51, t_52, t_53, t_54, t_55, t_56, t_57, t_58, t_59, t_60, t_61, t_62, t_63, t_64, t_65, t_66, t_67, t_79, t_83, t_123, t_129, t_138, t_215;
MCTS 3 (M3)	t_2, t_99, t_100, t_101, t_102, t_103, t_132, t_167, t_174, t_181, t_199,_177, t_207, t_208, t_222, t_232, t_234;
MCTS 4 (M4)	t_13, t_16, t_17, t_18, t_19, t_20, t_124, t_128;
MCTS 5 (M5)	t_14, t_21, t_23, t_24, t_25, t_26, t_27, t_31, t_126;
MCTS 6 (M6)	t_15, t_22, t_28, t_29, t_30, t_125, t_142;
MCTS 7 (M7)	t_33, t_34;
MCTS 8 (M8)	t_35, t_36;
MCTS 9 (M9)	t_37, t_130;
MCTS 10 (M10)	t_38, t_39, t_40, t_41, t_42, t_43, t_44, t_45, t_46, t_47, t_48, t_49, t_50, t_127;
MCTS 11 (M11)	t_68, t_70, t_71, t_72, t_73, t_74, t_75, t_76, t_77, t_78, t_237;
MCTS 12 (M12)	t_80, t_111, t_112;
MCTS 13 (M13)	t_81, t_84, t_85, t_86, t_87, t_88, t_89, t_90, t_131;
MCTS 14 (M14)	t_92, t_93;
MCTS 15 (M15)	t_95, t_104, t_141;
MCTS 16 (M16)	t_96, t_105, t_218;
MCTS 17 (M17)	t_97, t_224;
MCTS 18 (M18)	t_106, t_107, t_108;
MCTS 19 (M19)	t_109, t_110, t_133;
MCTS 20 (M20)	t_113, t_115, t_117, t_118, t_119;
MCTS 21 (M21)	t_134, t_135, t_136, t_137;
MCTS 22 (M22)	t_139, t_140, t_217;
MCTS 23 (M23)	t_143, t_227, t_228;
MCTS 24 (M24)	t_144, t_145, t_146;
MCTS 25 (M25)	t_147, t_148, t_149;
MCTS 26 (M26)	t_150, t_151, t_219;
MCTS 27 (M27)	t_152, t_153, t_154, t_220;
MCTS 28 (M28)	t_155, t_156, t_157, t_158, t_159, t_160, t_221, t_226;
MCTS 29 (M29)	t_161, t_163, t_164, t_166, t_213;
MCTS 30 (M30)	t_162, t_165, t_214;
MCTS 31 (M31)	t_168, t_169, t_201, t_236;
MCTS 32 (M32)	t_170, t_171, t_172, t_173, t_223;
MCTS 33 (M33)	t_175, t_176, t_180, t_230;
MCTS 34 (M34)	t_178, t_203;
MCTS 35 (M35)	t_182, t_183, t_231, t_233;
MCTS 36 (M36)	t_184, t_185, t_186, t_187, t_188, t_189, t_198;
MCTS 37 (M37)	t_193, t_194;
MCTS 38 (M38)	t_195, t_196, t_197, t_200;
MCTS 39 (M39)	t_204, t_205, t_206, t_235;
MCTS 40 (M40)	t_210, t_211;

(iv) PHD2 knock out. The protein is involved in pVHL mediated and oxygen dependent degradation of HIF-1α. Further, PHD2 is involved in hydroxylation of the RNA polymerase II subunit Rpb1 to allow its translocation to less chromatin-concentrated areas of the nucleus. When it is knocked out, HIF-1α degradation can continue via alternative pathways as seen in the pVHL knock out experiment and there is more RNA polymerase II activity, even if rpb7 can still be inactivated by pVHL.

(v) MCTS1 knock out. MCTS1 groups some reactions involved in the HIF-1α p53-dependent degradation pathway (Table 2). To perform this knock out, we erased the necessary token in mdm2, making the precondition insufficient to enable the MCTS transitions. p53 is not degraded and can continue its proapoptotic signal. On the other hand, a HIF-1α degradation

Table 3. Ranking of the 10 most occurring transitions with biological meaning and percentage of occurrence.

Rank	Transitions	Biological meaning	Occurrence %
1	t_98	Input transition for Hif and VHL	95.165
2	t_116	Interaction of VHL with Elongin B and C	94.148
3	t_113	Activation by oxygen of ARD	94.094
4	t_115	Acetylation and hydroxilation of Hif	94.094
5	t_117	Interaction of complex Vcb with Cu2	94.094
6	t_118	Interaction of complex Vcb with modified Hif	94.094
7	t_119	Degradation VHL dependent of Hif	94.094
8	t_97	ATP formation	89.059
9	t_224	Water Output transition	89.059
10	t_82	Pyruvate Dehydrogenase inactivation	88.041

mechanism is also knocked out resulting in an increased HIF-1α transcription activity.

(vi) t_97 knock out. This is the ATPase transition, allowing the model to imitate oxygen consumption for ATP synthesis. If this transition is inactive, oxygen accumulates infinitely and ATP is not regenerated after few simulation steps. At the beginning, ATP is formed during the first step of glycolysis but afterwards it is consumed again. At some point, these reactions do not have any ATP available to allow the system to re-balance the consumed ATP. After few simulation steps, oxygen reaches a high level due to slower consumption in the PHD2 regulation process. Biologically, this means that the metabolism stops and the cell is not able to create energy to survive. There is no accumulation other than glucose in the model. A few oxygen creation processes are blocked as well due to absence of ATP, e.g. t_15, t_41 and t_57.

(vii) GSK3β knock out. This enzyme is involved in negative glycogen synthetase (GS) regulation and is inactivated when phosphorylated. When GSK3β is knocked out, glycogen is continuously produced due to the enzyme remaining in an active state. In a real organism there are alternative forms of GSK3β which can inactivate GS, hence the effect will be less sharp. GSK3β is also involved in the degradation of HIF-1α, causing its phosphorylation and following ubiquitination. It is also involved in the degradation of β-catenin, where it is responsible for primary phosphorylation. If knocked out, even if Jade1 can be stabilized by pVHL, the effect will be similar to a knock out of Jade1, where β-catenin is free to continue proliferation stimulating transcription activity.

Discussion

We started from a core model of hypoxia response [12] and extended the original network with functional data derived from the literature in order to represent a complete description of the pVHL interaction pathway according to current knowledge. VHL syndrome is characterized by the formation of tumors and cysts affecting different organism districts and tissues. Indeed, pVHL is a tumor suppressor whose functions are connected to inhibition of proliferation and survival, growth and stability of extracellular matrix and microtubules, as well as cell polarity and migration. The IntAct database reports more than 200 suspected pVHL interactors and for most of them interaction and function details remain largely unknown. We chose to model the pVHL interactions in a credible cellular context with many protein activities occurring at the same time. The main idea was to create a novel manually curated PN description of the entire VHL disease pathway, including glucidic metabolism and signaling pathways. The model was designed as a standard PN and is composed of 238 transitions and 323 places, connected by 801 edges. A biologically realistic PN model needs to be covered by T-invariants, meaning each transition in the model has to be included in a T-invariant, and each invariant needs to have a biological meaning [31–47]. We used the T-invariant analysis to validate the reliability of the model. We computed a total of 393 T-invariants, plus 10 trivial invariants, which were excluded from analysis. These were grouped into 44 Clusters and, through use of T-invariants, transitions were grouped into 40 MCTS. The model

Occurrence of transitions in t-invariants

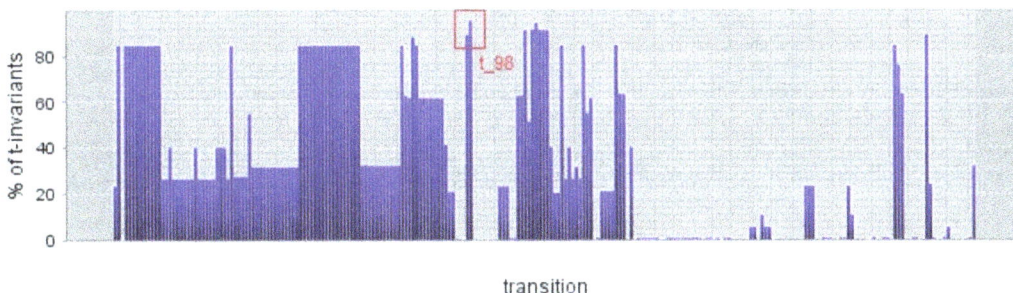

Figure 5. Transitions occurrence T-invariants. Transitions are ordered by name and t_98 is highlighted in red.

obtained is connected, covered by T-invariants with each invariant holding a biological meaning. MCTS analysis was used to identify the most frequent crucial transitions occurring in the model. This specific subset was further used to plan *in silico* knock out experiments and for the model validation and analysis of expected biological behavior. The model was then used to perform *in silico* knock out experiments inactivating specific transitions during qualitative network analysis. Our results showed that the model is able to represent important transitions reflecting real biological outcomes, i.e. transitions involving species such as oxygen or ATP are correctly inactivated under certain circumstances as expected from the bibliographic data. Biological energy-related reactions (e.g. ATP production from ADP) were modeled as P-invariants. Although the network is intentionally not covered by P-invariants, P-invariant analysis was used to verify all modeled energy consuming transitions. Both the ATP and NADH balances appeared constant during the simulation, with irrelevant P-invariants located in the Hur system. This approximation was used to verify the Hur-dependent regulation of VEGF, with results in accordance with [50]. The specific pVHL knock out suggests that this protein alone is not sufficient for complete HIF-1α inactivation. Indeed, other concurrent HIF-1α degradation pathways promote a sort of cell cycle regulation backup. On the contrary, simple deletion of pVHL turned out to be sufficient to increase all its other inhibitory functions, showing similar effects to pathological VHL symptoms. Indeed, ECM destabilization increases cell migration to other areas, promoting metastasis outbreak in case of tumor cells. Further, pVHL-dependent inhibition of tight junction formation by aPKCζII participates in an easier cellular detachment. The interactions of Nur77 could be considered a good example for pathological effects. It is a stimulator of Proopiomelanocortin production, a precursor for the Adrenocorticotropic hormone. If excessively released, it promotes an overproduction of adrenergic neurotransmitters by adrenal glands. Coming at clinical condition known as Cushing syndrome. On the very long term, Nur77 deregulation is known to cause tumors of the pituitary and adrenal glands [51], [52]. This happens in pheochromocytoma, which is one of the main VHL disease manifestations. We speculate that continuous VEGF transcription, even in situations where HIF-1α (but not Sp1) is knocked out, could be the explanation for clinical studies where VEGF-targeting drugs have turned out to be effective in kidney cancer treatment as reported in [53]. Although we used only confirmed data from the literature, Nur77 may be involved in

other regulation systems which were not considered in our model. The transitions for pVHL fibronectin stabilization show a behaviour which is coherent with biochemical experiments, illustrating a complete abrogation of ECM stabilization and an increased matrix metallo-proteinase action. Although the results are encouraging, the presented model will need further improvements since standard PNs do neither allow a complete transition control nor enzymatic activity modulation. Nevertheless, thanks to its manual curation our model can be used to plan new *in vitro* and *in vivo* experiments. The results are convincing enough to suggest our model as a comprehensive pathway model to simulate the main pVHL functions.

Supporting Information

Table S1   List of model transitions. The sequential number, name, biological functions and bibliographic source are listed.

Table S2   List of all model places. The progressive ID number, name and biological meaning are shown.

Table S3   List of T-invariants named in the text and their composition.

Table S4   Composition in terms of invariants of the clusters C8, C9, C10, C11.

File S1   VHL disease network model in spped format.

Video S1   Snoopy PN framework running demo.

Acknowledgments

The authors are grateful to Stefano Moro and to members of the BioComputing UP lab for insightful discussions.

Author Contributions

Conceived and designed the experiments: GM CF ST. Performed the experiments: GM EP MG. Analyzed the data: EP MG AM. Contributed reagents/materials/analysis tools: ST. Wrote the paper: GM EP ST.

References

1. Soga T (2013) Cancer metabolism: Key players in metabolic reprogramming. Cancer Sci 104: 275–281.
2. Golub TR, Slonim DK, Tamayo P, Huard C, Gaasenbeek M, et al. (1999) Molecular classification of cancer: class discovery and class prediction by gene expression monitoring. Science 286: 531–537.
3. Thomas JG, Olson JM, Tapscott SJ, Zhao LP (2001) An efficient and robust statistical modeling approach to discover differentially expressed genes using genomic expression profiles. Genome Res 11: 1227–1236. doi:10.1101/gr.165101.
4. Leonardi E, Murgia A, Tosatto SCE (2009) Adding structural information to the von Hippel-Lindau (VHL) tumor suppressor interaction network. FEBS Lett 583: 3704–3710.
5. Leonardi E, Martella M, Tosatto SCE, Murgia A (2011) Identification and in silico analysis of novel von Hippel-Lindau (VHL) gene variants from a large population. Ann Hum Genet 75: 483–496.
6. Minervini G, Masiero A, Moro S, Tosatto SCE (2013) In silico investigation of PHD-3 specific HIF1-α proline 567 hydroxylation: A new player in the VHL/HIF-1α interaction pathway? FEBS Lett. 587(18): 2996–3001.
7. Wu M, Liu L, Hijazi H, Chan C (2013) A multi-layer inference approach to reconstruct condition-specific genes and their regulation. Bioinforma Oxf Engl: 1–12. doi:10.1093/bioinformatics/btt186.
8. Almaas E (2007) Biological impacts and context of network theory. J Exp Biol 210: 1548–1558.
9. Weitz JS, Benfey PN, Wingreen NS (2007) Evolution, interactions, and biological networks. PLoS Biol 5: e11. doi:10.1371/journal.pbio.0050011.
10. Proulx SR, Promislow DEL, Phillips PC (2005) Network thinking in ecology and evolution. Trends Ecol Evol 20: 345–353.
11. Mahon PC, Hirota K, Semenza GL (2001) FIH-1: a novel protein that interacts with HIF-1alpha and VHL to mediate repression of HIF-1 transcriptional activity. Genes Dev 15: 2675–2686.
12. Heiner M, Sriram K (2010) Structural analysis to determine the core of hypoxia response network. PLoS One 5: e8600. doi:10.1371/journal.pone.0008600.
13. Kohn KW, Riss J, Aprelikova O, Weinstein JN, Pommier Y, et al. (2004) Properties of switch-like bioregulatory networks studied by simulation of the hypoxia response control system. Mol Biol Cell 15: 3042–3052.
14. Yu Y, Wang G, Simha R, Peng W, Turano F, et al. (2007) Pathway switching explains the sharp response characteristic of hypoxia response network. PLoS Comput Biol 3: e171. doi:10.1371/journal.pcbi.0030171.
15. Stolle C, Glenn G, Zbar B, Humphrey JS, Choyke P, et al. (1998) Improved detection of germline mutations in the von Hippel-Lindau disease tumor suppressor gene. Hum Mutat 12: 417–423.
16. Gnarra JR, Tory K, Weng Y, Schmidt L, Wei MH, et al. (1994) Mutations of the VHL tumour suppressor gene in renal carcinoma. Nat Genet 7: 85–90.
17. Latif F, Tory K, Gnarra J, Yao M, Duh FM, et al. (1993) Identification of the von Hippel-Lindau disease tumor suppressor gene. Science 260: 1317–1320.

18. Kim WY, Kaelin WG (2004) Role of VHL gene mutation in human cancer. J Clin Oncol Off J Am Soc Clin Oncol 22: 4991–5004.

19. Vortmeyer AO, Huang SC, Pack SD, Koch CA, Lubensky IA, et al. (2002) Somatic point mutation of the wild-type allele detected in tumors of patients with VHL germline deletion. Oncogene. 14; 21(8): 1167–70.

20. Knudson AG Jr (1971) Mutation and cancer: statistical study of retinoblastoma. Proc Natl Acad Sci U S A 68: 820–823.

21. Gnarra JR, Ward JM, Porter FD, Wagner JR, Devor DE, et al. (1997) Defective placental vasculogenesis causes embryonic lethality in VHL-deficient mice. Proc Natl Acad Sci U S A 94: 9102–9107.

22. Frew IJ, Krek W (2008) pVHL: a multipurpose adaptor protein. Sci Signal 1: pe30. doi:10.1126/scisignal.124pe30.

23. Schoenfeld AR, Davidowitz EJ, Burk RD (2000) Elongin BC complex prevents degradation of von Hippel-Lindau tumor suppressor gene products. Proc Natl Acad Sci U S A 97: 8507–8512.

24. Semenza GL (1999) Regulation of mammalian O2 homeostasis by hypoxia-inducible factor 1. Annu Rev Cell Dev Biol 15: 551–578.

25. Kerrien S, Aranda B, Breuza L, Bridge A, Broackes-Carter F, et al. (2012) The IntAct molecular interaction database in 2012. Nucleic Acids Res 40: D841–846.

26. Frew IJ, Krek W (2007) Multitasking by pVHL in tumour suppression. Curr Opin Cell Biol 19: 685–690.

27. Thoma CR, Toso A, Gutbrodt KL, Reggi SP, Frew IJ, et al. (2009) VHL loss causes spindle misorientation and chromosome instability. Nat Cell Biol 11: 994–1001.

28. Tang N, Mack F, Haase V (2006) pVHL function is essential for endothelial extracellular matrix deposition. Mol Cell 26. doi:10.1128/MCB.26.7.2519.

29. Sackmann A, Heiner M, Koch I (2006) Application of Petri net based analysis techniques to signal transduction pathways. BMC Bioinformatics 7: 482.

30. Chaouiya C (2007) Petri net modelling of biological networks. Brief Bioinform 8: 210–219.

31. Grunwald S, Speer A, Ackermann J, Koch I (2008) Petri net modelling of gene regulation of the Duchenne muscular dystrophy. Biosystems 92: 189–205.

32. Rohr C, Marwan W, Heiner M (2010) Snoopy–a unifying Petri net framework to investigate biomolecular networks. Bioinforma Oxf Engl 26: 974–975.

33. Maynard MA, Evans AJ, Hosomi T, Hara S, Jewett MAS, et al. (2005) Human HIF-3alpha4 is a dominant-negative regulator of HIF-1 and is down-regulated in renal cell carcinoma. FASEB J Off Publ Fed Am Soc Exp Biol 19: 1396–1406.

34. Li QF, Wang XR, Yang YW, Lin H (2006) Hypoxia upregulates hypoxia inducible factor (HIF)-3alpha expression in lung epithelial cells: characterization and comparison with HIF-1alpha. Cell Res 16: 548–558.

35. Smith TG, Robbins PA, Ratcliffe PJ (2008) The human side of hypoxia-inducible factor. Br J Haematol 141: 325–334.

36. Richardson SM, Knowles R, Tyler J, Mobasheri A, Hoyland JA (2008) Expression of glucose transporters GLUT-1, GLUT-3, GLUT-9 and HIF-1alpha in normal and degenerate human intervertebral disc. Histochem Cell Biol 129: 503–511.

37. Vannucci SJ, Maher F, Simpson IA (1997) Glucose transporter proteins in brain: Delivery of glucose to neurons and glia. Glia 21: 2–21.

38. Takeda J, Kayano T, Fukomoto H, Bell GI (1993) Organization of the human GLUT2 (pancreatic beta-cell and hepatocyte) glucose transporter gene. Diabetes 42: 773–777.

39. Heather LC, Pates KM, Atherton HJ, Cole MA, Ball DR, et al. (2013) Differential Translocation of FAT/CD36 and GLUT4 Coordinates Changes in Cardiac Substrate Metabolism During Ischemia and Reperfusion. Circ Heart Fail. doi:10.1161/CIRCHEARTFAILURE.112.000342.

40. Kim J, Tchernyshyov I, Semenza GL, Dang CV (2006) HIF-1-mediated expression of pyruvate dehydrogenase kinase: a metabolic switch required for cellular adaptation to hypoxia. Cell Metab 3: 177–185.

41. Kim BY, Kim H, Cho EJ, Youn HD (2008) Nur77 upregulates HIF-alpha by inhibiting pVHL-mediated degradation. Exp Mol Med 40: 71–83.

42. Choi J-W, Park SC, Kang GH, Liu JO, Youn H-D (2004) Nur77 activated by hypoxia-inducible factor-1alpha overproduces proopiomelanocortin in von Hippel-Lindau-mutated renal cell carcinoma. Cancer Res 64: 35–39.

43. Roe J-S, Youn H-D (2006) The positive regulation of p53 by the tumor suppressor VHL. Cell Cycle Georget Tex 5: 2054–2056.

44. Fels DR, Koumenis C (2005) HIF-1alpha and p53: the ODD couple? Trends Biochem Sci 30: 426–429.

45. Berndt JD, Moon RT, Major MB (2009) Beta-catenin gets jaded and von Hippel-Lindau is to blame. Trends Biochem Sci 34: 101–104.

46. Hergovich A, Lisztwan J, Thoma CR, Wirbelauer C, Barry RE, et al. (2006) Priming-Dependent Phosphorylation and Regulation of the Tumor Suppressor pVHL by Glycogen Synthase Kinase 3. Mol Cell Biol 26: 5784–5796.

47. Heiner M, Koch I, Will J (2004) Model validation of biological pathways using Petri nets–demonstrated for apoptosis. Biosystems 75: 15–28.

48. Bortfeldt RH, Schuster S, Koch I (2011) Exhaustive analysis of the modular structure of the spliceosomal assembly network: a petri net approach. Stud Health Technol Inform 162: 244–278.

49. Heiner M (2009) Understanding Network Behavior by Structured Representations of Transition Invariants. Algorithmic Bioprocesses. Natural Computing Series. Springer Berlin Heidelberg. 367–389. Available: http://link.springer.com/chapter/10.1007/978-3-540-88869-7_19.

50. Zhu JL, Kaytor EN, Pao CI, Meng XP, Phillips LS (2000) Involvement of Sp1 in the transcriptional regulation of the rat insulin-like growth factor-1 gene. Mol Cell Endocrinol 164: 205–218.

51. Okabe T (1998) [Functional role of nur77 family in T-cell apoptosis and stress response]. Nihon Rinsho Jpn J Clin Med 56: 1734–1738.

52. Murphy EP, Conneely OM (1997) Neuroendocrine regulation of the hypothalamic pituitary adrenal axis by the nurr1/nur77 subfamily of nuclear receptors. Mol Endocrinol Baltim Md 11: 39–47.

53. Tabernero J (2007) The Role of VEGF and EGFR Inhibition: Implications for Combining Anti-VEGF and Anti-EGFR Agents. Mol Cancer Res 5: 203–220.

2

IIS – Integrated Interactome System: A Web-Based Platform for the Annotation, Analysis and Visualization of Protein-Metabolite-Gene-Drug Interactions by Integrating a Variety of Data Sources and Tools

Marcelo Falsarella Carazzolle[1,2], **Lucas Miguel de Carvalho**[1], **Hugo Henrique Slepicka**[3], **Ramon Oliveira Vidal**[2], **Gonçalo Amarante Guimarães Pereira**[2], **Jörg Kobarg**[1], **Gabriela Vaz Meirelles**[1]*

1 Laboratório Nacional de Biociências, Centro Nacional de Pesquisa em Energia e Materiais, Campinas, São Paulo, Brazil, **2** Laboratório de Genômica e Expressão, Departamento de Genética e Evolução, Instituto de Biologia, Unicamp, Campinas, São Paulo, Brazil, **3** Laboratório Nacional de Luz Síncrotron, Centro Nacional de Pesquisa em Energia e Materiais, Campinas, São Paulo, Brazil

Abstract

Background: High-throughput screening of physical, genetic and chemical-genetic interactions brings important perspectives in the Systems Biology field, as the analysis of these interactions provides new insights into protein/gene function, cellular metabolic variations and the validation of therapeutic targets and drug design. However, such analysis depends on a pipeline connecting different tools that can automatically integrate data from diverse sources and result in a more comprehensive dataset that can be properly interpreted.

Results: We describe here the Integrated Interactome System (IIS), an integrative platform with a web-based interface for the annotation, analysis and visualization of the interaction profiles of proteins/genes, metabolites and drugs of interest. IIS works in four connected modules: (i) Submission module, which receives raw data derived from Sanger sequencing (e.g. two-hybrid system); (ii) Search module, which enables the user to search for the processed reads to be assembled into contigs/singlets, or for lists of proteins/genes, metabolites and drugs of interest, and add them to the project; (iii) Annotation module, which assigns annotations from several databases for the contigs/singlets or lists of proteins/genes, generating tables with automatic annotation that can be manually curated; and (iv) Interactome module, which maps the contigs/singlets or the uploaded lists to entries in our integrated database, building networks that gather novel identified interactions, protein and metabolite expression/concentration levels, subcellular localization and computed topological metrics, GO biological processes and KEGG pathways enrichment. This module generates a XGMML file that can be imported into Cytoscape or be visualized directly on the web.

Conclusions: We have developed IIS by the integration of diverse databases following the need of appropriate tools for a systematic analysis of physical, genetic and chemical-genetic interactions. IIS was validated with yeast two-hybrid, proteomics and metabolomics datasets, but it is also extendable to other datasets. IIS is freely available online at: http://www.lge.ibi.unicamp.br/lnbio/IIS/.

Editor: Frederique Lisacek, Swiss Institute of Bioinformatics, Switzerland

Funding: This work was supported by Fundação de Amparo à Pesquisa do Estado São Paulo (FAPESP), Conselho Nacional de Pesquisa e Desenvolvimento (CNPq), Brazilian Biosciences National Laboratory (LNBio) and Center for Computational Engineering and Sciences at UNICAMP/Brazil (FAPESP/CEPID project #2013/08293-7). The funders had no role in study design, data collection and analysis, decision to publish, or preparation of the manuscript.

Competing Interests: The authors have declared that no competing interests exist.

* Email: gabriela.meirelles@lnbio.cnpem.br

Introduction

High-throughput screening of physical, genetic and chemical-genetic interactions brings new important perspectives in the Systems Biology field, as the analysis of these interactions provides new insights into protein/gene function, help to unravel how cellular networks are organized and facilitates the validation of therapeutic targets and drug design.

Recently, many experimental procedures have been developed to help elucidate the intricate networks of proteins, genes and drugs interactions, ranging from high-throughput experiments based on genomic scale analyses [1–6] to molecular biology approaches on a specific key pathway [7,8]. Molecular interactions data related to human and model organisms are currently being integrated in diverse databases, such as BioGRID [9], Intact [10], DIP [11], STRING [12], MINT [13], HPRD [14], DrugBank [15], ChemBL [16], HMDB [17], YMDB [18], ECMDB [19], as well as KEGG [20] and Reactome [21]. However, the integration of different datasets is not a trivial task, since they vary widely in coverage, data quality and annotation. Moreover, the information available can be derived from diverse experimental methods, such as yeast two-hybrid (Y2H), mass spectrometry (MS), immunoprecipitation (IP), or fluorescence resonance energy transfer (FRET) assays to demonstrate protein interactions and, in some cases, interaction networks are determined solely by bioinformatics tools [22,23], which rarely consider the subcellular localization of the interactors.

A major fraction of protein-protein interactions (PPIs) deposited in these public databases is generated by the yeast two-hybrid technology. Indeed, Y2H allows high-throughput screening of direct physical PPIs at a proteome scale, but requires the sequencing of hundreds to thousands of cellular preys per experiment. Moreover, the analyses of sequences derived from such interaction assays are difficult to proceed without an appropriate pipeline connecting different tools that can automatically integrate data derived from diverse sources and result in a more comprehensive and organized dataset that can be properly visualized and interpreted.

In response, several software projects became available to offer computer-assisted data and software integration. Notable among these are G2N [24], GeneMANIA [25], STRING [12], Ingenuity [26], and pISTiL [27] softwares. Though, most of them show some limitations. pISTil works well on chromatograms processing and partial annotation, but lacks the connection to visualization and analysis of interaction networks. The other software work well on the integration of a variety of bioinformatic tools with focus on the interaction networks, but lack the chromatograms processing feature or are restricted to a small number of model organisms and types of molecules.

Here we present the Integrated Interactome System (IIS), a new platform integrating a variety of tools and data sources used in systems biology analyses. It comprises a pipeline that receives raw sequence data from screening methods based on Sanger sequencing, like yeast two-hybrid system, or lists of proteins/genes, metabolites and drugs of interest, which are automatically processed, annotated and linked to interaction networks that can be filtered by the scoring system proposed by mathematical approaches, and evaluated according to expression/concentration fold change values and to the enriched biological processes and pathways in the network.

As major advantages over other systems, IIS supports the entire data analysis of experiments such as two-hybrid assays, besides other omics approaches, from the sequencing all the way to generating publication-ready interaction networks and annotation tables. In the process, all the challenges related to this type of experiment are addressed: processing/assembling reads, mapping them to the correct gene, automatically retrieving annotations from multiple resources and interactors from nine public databases, assigning annotations and interactions via orthologs if required, and building networks that gather novel identified interactions, protein and metabolite expression/concentration levels, subcellular localization, topological metrics and enriched biological processes and pathways. Each one of those tasks being very time-consuming and hard to manually integrate using separate different tools.

We also describe the construction of the Global Protein-Metabolite-Gene-Drug Interaction Database (GPMGDID) and discuss the workflow of IIS website. We then validate IIS's ability to perform the proposed tasks with three case studies: (i) human Nek6 yeast two-hybrid screening [28], (ii) *Saccharomyces cerevisae* encapsulated cells proteome [29] and (iii) primary and metastatic human ovarian cancer metabolome [30], on which we evaluate the benefits of using IIS to interpret the interaction profiles of a variety of conditions (e.g. interactions of specific genes or based on the omics data from different cell types or treatments).

Methods

The Integrated Interactome System (IIS) is an integrative platform with a web-based interface, which integrates four different modules for processing, annotation, analysis and visualization of the interaction profiles of proteins/genes, metabolites and/or drugs of interest. IIS organizes the analysis in a project context and the user can create several projects protected by password. The project is a structure inside the system where researchers can develop and organize their thematic studies, choosing between two types: (i) chromatogram project or (ii) genes/metabolites/drugs project.

Submission Module

The submission module is divided in nomenclature edition and chromatogram submission. The nomenclature edition allows the user to manage the description of the experiment, considering the laboratory, organism, cDNA library, strategy, project, sequencing plate and sequencing orientation. The chromatogram submission was developed to input the chromatograms (originated from Sanger sequencing derived from Y2H experiments, transcriptome, etc.) into the system. The chromatograms need to be organized in ZIP files and named according to the position in the 96 well plates used in the sequencing process (e.g. A01 to H12). The system receives the uploaded chromatograms file in a ZIP format (each file containing up to 96 chromatograms), checks the ZIP file and the individual chromatograms integrity after uncompressing, organizes the uncompressed chromatograms in a directory structure and runs PHRED base calling and quality scoring [31], generating reads sequences (FASTA and QUAL files) (Figure 1.1). The reads are then submitted to quality analysis and identification of vector and adaptor sequences, using the BDTrimmer program [32], and a report is sent by email to the user summarizing the information about the chromatograms processing (see also Methods S1 for more details). At the end of this module, the resulting processed reads are then aligned against a protein sequence database (GenBank/NR) by using the BLASTx alignment tool [33] with e-value threshold of 1e-10 to partially annotate them.

Search Module

In the second module, the partially annotated reads from the SUBMISSION MODULE are available to be checked, added to the user's project (chromatogram project type) and assembled into clusters (contigs and singlets) using CAP3 program [34], in order to eliminate redundant reads typically generated by Y2H and transcriptome assays (Figure 1.2).

In the genes/metabolites/drugs project type, lists of genes/proteins (UniProt Accession, RefSeq or gene symbol), metabolites (HMDB, YMDB or ECMD IDs) and/or drugs (DrugBank ID or CAS number) can also be uploaded by the user as a single column TXT file and added to the project (Figure 1.2). Because of the gene symbols redundancies and the presence of aliases in the databases, searching for gene symbols in the selected organism is first performed on Swiss-Prot database and in the case of unreviewed proteins it is extended to TrEMBL database. It is also possible to upload a two-column TXT file containing UniProt Accession, RefSeq or gene symbol and fold change values, respectively, the second one representing expression/concentration levels.

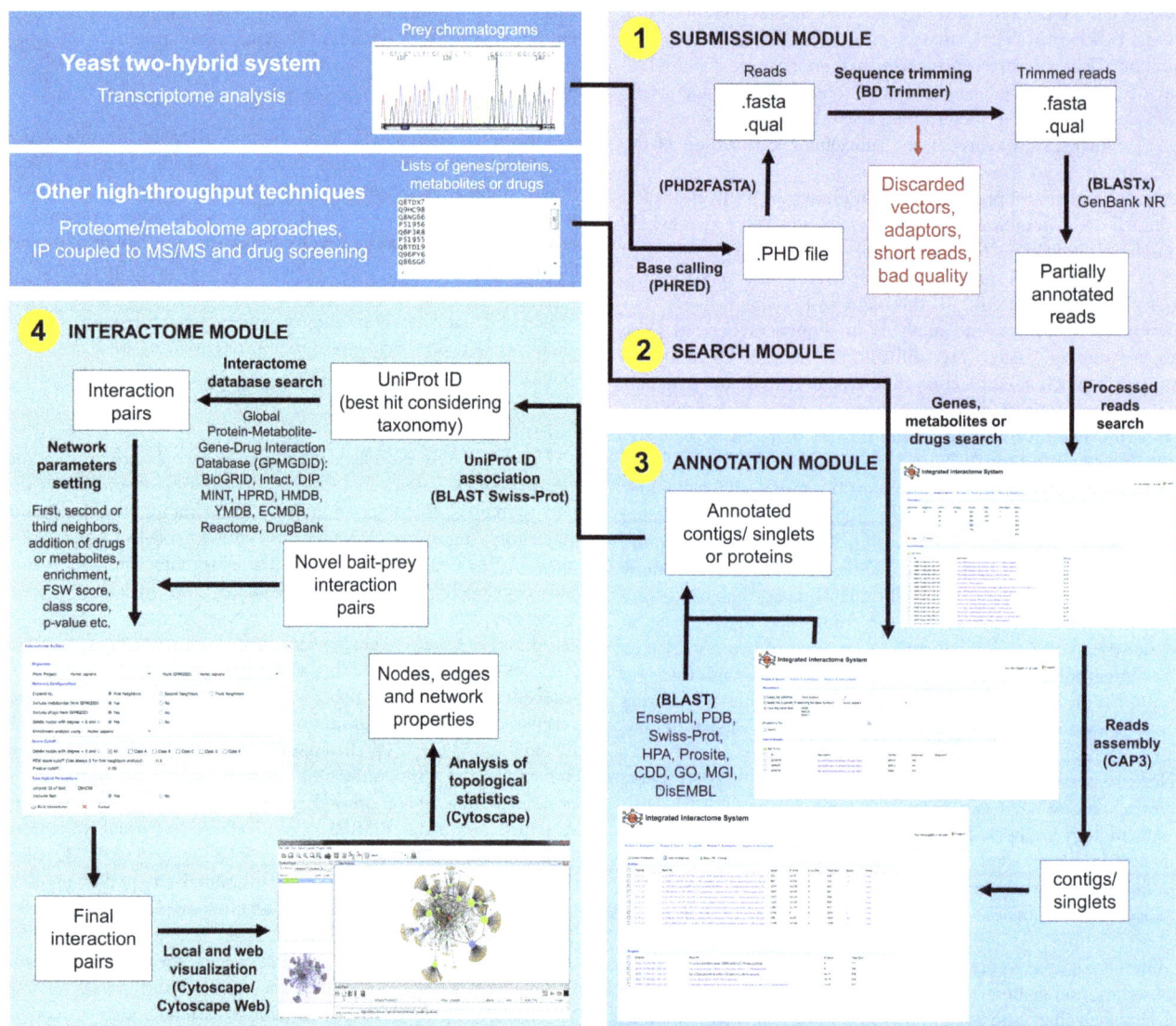

Figure 1. Workflow used in IIS, showing the integration of the (1) SUBMISSION, (2) ANNOTATION, (3) SEARCH and (4) INTERACTOME MODULES for data analysis. All steps are indicated by arrows alongside a term, out or in parentheses (both in black and bold font) that correspond to a sequence of actions (the term in parentheses meaning the tool/database used in that step).

Annotation Module

In the third module, the partially annotated contigs and singlets, or the lists of proteins/genes uploaded by the user, are searched against nine databases (Gene Ontology [35], HPA [36], CDD [37], MGI [38], PDB [39], DisEMBL [40], Prosite [41], Ensembl [42] and Swiss-Prot [43], all of them queried monthly for updates) in order to generate tables with automatic annotation that can be exported to other software (e.g. Excel) for editing/formatting purposes (Figure 1.3). The lists of proteins/genes are searched by their respective UniProt Accession numbers, and the contigs/singlets are first blasted against Swiss-Prot database and their best hits used to make an association between their sequences from the selected organism defined by the user and their respective UniProt Accession numbers [44].

This module was designed in order to allow users to export publication-ready tables with a more complete annotation of their data (for an example and more details see Table S1 and Methods S1). Users can also create their annotation tables containing only some desired fields by selecting the databases of interest. By doing this, thematic annotation tables can be created, e.g. structural annotation tables (by selecting the CDD, PDB, DisEMBL and Prosite databases), functional annotation tables (by selecting the Gene Ontology, HPA and MGI databases), etc., according to the user needs.

Interactome Module

The fourth module is used to blast the input contigs/singlets against the Swiss-Prot database to retrieve the corresponding UniProt Accession numbers of the organism of interest, or search the input lists of proteins/genes, metabolites and drugs that are already linked to their own or related UniProt Accession numbers, and use them as queries in our Global Protein-Metabolite-Gene-Drug Interaction Database (GPMGDID) to build the networks (Figure 1.4). The latter was constructed in a MySQL structure by grouping more than 1 million interactions from nine public available databases: BioGRID [9], Intact [10], DIP [11], MINT

[13], HPRD [14], DrugBank [15], HMDB [17], YMDB [18], and ECMDB [19], all of them queried monthly for updates. There are five parameters classes to select in this module: the organism, the network configuration, the score cutoff, the two-hybrid parameters and the expression analysis. IIS works with diverse organism datasets that can be chosen independently for the input dataset (project) and the GPMGDID, enabling also the construction of networks with interactions between different organisms (e.g. host-pathogen interactions) or using ortholog relationship. The network configuration parameter considers the interaction level of expansion from first to third neighbors, the addition or not of metabolites and drugs from GPMGDID in the network expansion, the deletion of nodes with connectivity degree of 0 and 1 (yielding a more connected network), and the selection of the background organism for the enrichment analysis. The score cutoff parameters can be used to filter the network for more confident interactions by three types of score: the Class score, the FSW score and the p-value, which are described in more details in the following sections. The order considered in the algorithm to reduce the network size by filters is: (i) Class score, (ii) p-value, (iii) deletion of nodes with connectivity degree of 0 and 1, and (iv) FSW score. In the two-hybrid parameters, if the user is working with two-hybrid or immunoprecipitation techniques and has a bait of interest to connect with the identified novel preys, it can be done using this option. Finally, in the expression analysis parameters, if working with omics datasets, the user can set cutoff values to color the input nodes as up- or down-regulated and change the node sizes according to their fold change in expression/concentration levels. Regarding the enrichment analysis, the program calculates the enrichment for the GO biological processes and KEGG pathways in the generated network using the hypergeometric distribution [45]. The exact and approximated hypergeometric distributions were implemented in the interactome algorithm using gamma and log-gamma function, respectively, to calculate factorial number. The second one was necessary to avoid stack overflow related to large factorial numbers [46] (the empirical tests showed that the transition from exact to approximated function occurs for GO term or KEGG pathway with more than 1,800 related proteins in the GPMGDID database).

This module generates a XGMML file containing all annotations and metrics described below that can be directly visualized on the website using Cytoscape web [47] from our web server (Figure 1.4) or can be imported into Cytoscape platform [48]. The Cytoscape platform is an open source software that enables the visualization of all interactions (or defined subgroups of interactions) and the analysis and correlation of node and edge properties with topological network statistics using a set of core modules and external plugins. The information available in the XGMML file has been standardized in order to communicate with these plugins.

Construction of the GPMGDID database. The Global Protein-Metabolite-Gene-Drug Interaction Database (GPMGDID) is a non-redundant database which integrates all protein-metabolite-gene-drug interactions described in several public databases, divided by organism, where the interaction pairs are classified by data type (experimental or predicted), methodology (e.g. two-hybrid, pull down, genetic interference, etc.), organism and source (PubMed ID of the paper that published the interaction), while the proteins/genes involved in the interactions are characterized by biological process, molecular function and cellular component allowing the enrichment and compartmentalization analysis performed by the INTERACTOME MODULE.

The publicly available interaction databases have non-standard protein identifications, file formats and are not uniquely indexed and annotated, which compromises the development of a single algorithm to integrate all datasets. Therefore, the UniProt Accession was chosen as the reference ID for the unification of the different datasets, generating the following possible interaction pairs: UniProtID1_UniProtID2, UniProtID1_HMDBID1, UniProtID1_YMDBID1, UniProtID1_ECMDBID1 or UniProtID1_DrugBankID1. A large amount of interaction redundancies generated because the same information is described in different interaction databases was also eliminated by concatenating interaction pairs with the source from which they were described (PubMed IDs), producing an interaction pair ID given by UniProtID1_UniProtID2_PubMedID. Figure 2 summarizes the pipeline applied to construct GPMGDID database (see also Methods S1 for more details).

Filtering high-confidence interactions by mathematical approaches. The interacting pairs constructed by the method described above may be error prone and must undergo a validation step. In order to achieve a more reliable result, some facts should be considered: proteins that actually interact are expected to share the same cellular compartment and have common interaction partners. It has been shown that a pair of genuine interacting proteins is generally expected to have a common cellular role and proteins that have common interaction partners have a higher chance of sharing a common function. Moreover, even if two proteins are consistently predicted to interact they must be located at the same cell compartment and at the same time [49–53]. Therefore, three validation approaches were considered to verify the quality of interaction pairs in networks constructed from GPMGDID database: Class score, Functional Similarity Weight score (FSW score) and p-value. These mathematical approaches are further described and can be used as filters in the INTERACTOME MODULE to reduce the network size for more reliable interactions.

Class score. The interactions in the GPMGDID present a Class score similar to the cellular compartment classification (C^3) described by Brandão et al. [54], and it is based on three characteristics: type of interaction (experimental or predicted), number of papers describing the interaction in PubMed (PubMed ID), and cellular component (CC) described for the interacting nodes in the Gene Ontology database [35]. The CC used by IIS corresponds to a concise list of the main selected subcellular compartments from GO and are depicted in bold in Table S2 (GO CC children terms were grouped for each selected main ancestral CC term, considering only terms annotated for ≥10 genes). This classification divides the interactions into four classes according to their evidence and subcellular localization. Class score value attributed for the type of interaction is +4 if it is based on experimental data, and 0 if there is no experimental data available (predicted); for co-localization we attribute score +1, otherwise we display score 0; if the interaction is described in more than one PubMed ID considering at least one paper not related to high-throughput experiments we score +4, if the interaction is described in more than one PubMed ID we score +3, if it is described in only one PubMed ID we score +1, and 0 if not published. We consider high-throughput experiment papers those describing more than 500 interactions. The Class scores are used in IIS to depict different edge widths to the generated networks, in order to visually assign interactions confidence (Table 1).

Functional Similarity Weight score (FSW score). In GPMGDID, due to its integrative profile, the reliability index for a reported interaction can be postulated in terms of the proportion of interaction partners that two proteins have in common. A mathematical approach called Functional Similarity Weight (FSWeight) [51] has been proposed to assess the reliability of protein interaction data based on the number of common

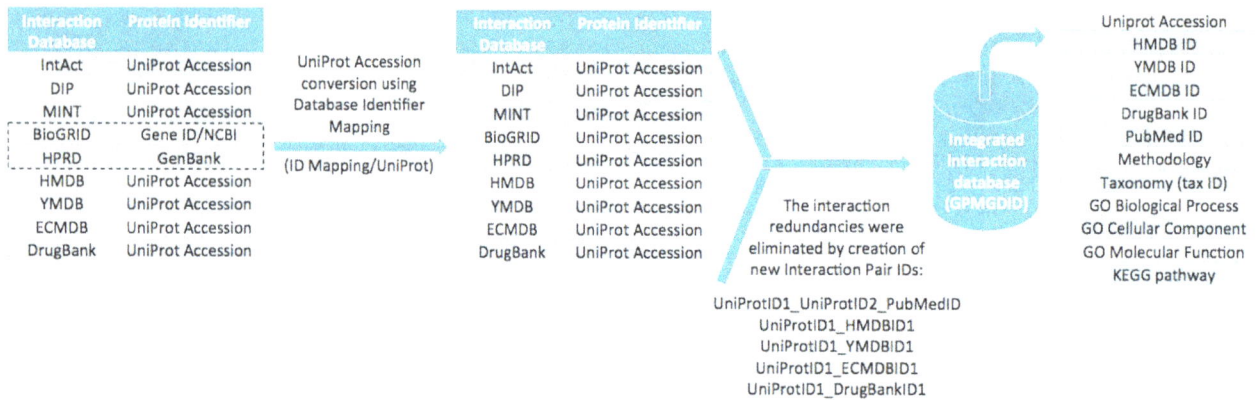

Figure 2. Global Protein-Metabolite-Gene-Drug Interaction Database (GPMGDID) construction. The UniProt Accession was chosen as the reference ID for the unification of the nine different databases used to construct GPMGDID: IntAct, DIP, MINT, BioGRID, HPRD, HMDB, YMDB, ECMDB and DrugBank databases. The interaction redundancies were eliminated by concatenating pairs of interactions with the source (PubMed IDs), generating an interaction pair ID given by UniProtID1_UniProtID2_PubMedID. The resultant database integrates several protein-metabolite-gene-drug interactions classified by source, methodology and organism.

neighbors of two proteins. The FSWeight approach was initially designed to predict protein functions, and lately has shown a good performance in evaluating the reliability of protein interactions [53]. The interaction pairs of proteins that are classified with high score by this method are likely to be true positives. On the other hand, the pairs of proteins that are classified with low scores are likely to be false positives. The most interesting feature of the FSWeight is that it is able to rank the reliability of an interaction between a pair of proteins using only the topology of the interactions between that pair of proteins and their neighbors within a short radius in a graph network [49,50].

Therefore, we implemented in GPMGDID the Functional Similarity Weight score (FSW score) calculation originally proposed by Chua et al. [51], and described by Brandão et al. [54], for all first, second and third level interactions present in our database. The effect of FSW score threshold in the network is exemplified and discussed in the Results and Discussion section.

P-value. Finally, a statistical hypothesis testing was implemented to avoid random interaction pairs generated during network expansion using GPMGDID database. Every time the user builds a new subnetwork from the GPMGDID, p-values are calculated for each protein in the generated subnetwork, in order to assign confidence. The p-value is calculated based on the work by Berger et al. [24]. First, the z-score value is calculated for each protein using a binomial proportion test that depends on the total

of interactions of the protein in the subnetwork, the total of interactions of the protein in the GPMGDID filtered by a specific organism, the total of interaction pairs in the subnetwork and the total of interaction pairs in the GPMGDID filtered by a specific organism. Next, a normal distribution that depends on the variance and average of the values already calculated was used for converting the z-scores to p-values.

Web interface. IIS web interface was built in JavaScript, JSON and PHP, and locally hosted on a Linux server at http://www.lge.ibi.unicamp.br/lnbio/IIS/. The web interface allows the user to work in the thematic project, protected by password, organizing and updating the set of proteins/genes, metabolites and drugs of interest and their respective annotations and networks.

Results and Discussion

We have validated IIS's ability to perform the analysis of interaction profiles for both specific genes or omics data originated from different cell types or conditions with three case studies: (i) an yeast two-hybrid screening [28], (ii) an yeast proteome [29] and (iii) a human cancer metabolome [30].

First case study: hNek6 yeast two-hybrid screening

The human NIMA-related kinase 6 (hNek6) was chosen based on a previous work by our group [28] in which the PPI network of hNek6 was manually generated, annotated and visually analyzed

Table 1. Interactions confidence measured by Class scores used to represent different edge widths in the networks.

Class	Score	Edge width	Parameters[1]
A	+9	2.5	Experimental/PubMed ID >1 (at least one not HT)/same CC
B	+7/+8	2.0	Experimental/PubMed ID >1 or Experimental/PubMed ID >1/same CC
C	+6	1.5	Experimental/PubMed ID = 1/same CC
D	+5	1.0	Experimental/PubMed ID = 1
E	+4	0.5	Experimental/PubMed ID = 0

[1]Parameters used to calculate the Class scores: interaction described as experimental (not predicted) (+4); interaction described in more than one paper (PubMed ID >1) and at least one paper not describing high-throughput (HT) experiments (+4); interaction described in more than one paper (PubMed ID >1) (+3); interaction described in only one paper (PubMed ID = 1) (+1); interacting nodes described in the same cellular component (CC) (+1). For novel interactions not described in any paper (PubMed ID = 0), even if the interacting nodes are described in the same CC, it will be assigned Class score E.

by Osprey software [55] using the BioGRID database [9]. Here we used IIS to perform all the steps from the chromatograms processing to the annotation and interactome construction and analysis using a standardized pipeline executed in a significantly shorter period of time. First, hNek6 prey cDNAs were sequenced and their chromatograms were organized into files to be submitted to IIS in a new chromatogram project. After submission, the chromatograms were immediately processed into reads, assembled into contigs and singlets and blasted against GenBank/NR for a partial annotation. The complete annotation table (Table S1) against diverse databases was then generated by selecting all the contigs and singlets in the "Module 3: Annotation" tab inside the project and using the "Create Annotation" button. The same selection was done in the "Module 4: Interactome" tab, and the "Create Interactome" button was used to build the hNek6 networks, from the first to the third neighbors levels of interactions (Figure 3A, B and D). All the networks were visualized both on the website using the Cytoscape web [47], and locally using the Cytoscape software [48], which was also used to manipulate and analyze the networks. The example chromatogram files available for the user on the IIS website correspond to fifteen hNek6 interactions confirmed by *in vitro* and *in vivo* assays (described in Table 5 by Meirelles et al. [28]).

Figure 3 shows the hNek6 interaction networks generated by IIS and visualized by Cytoscape. Our new automatic analysis using IIS made it possible to verify, as described before [28] that this kinase is a hub (node with several connections in the network) involved in several biological processes through its interaction with diverse types of proteins in different cellular compartments (Table S3), possibly at different time points during the cell cycle. In our previous work [28], we manually curate from the literature the hNek6 putative cellular roles, considering all the novel interacting partners retrieved by the yeast two-hybrid screening, which were as follows: cell cycle, cytoskeleton organization, DNA repair, NF-kappaB and Notch signalings and cancer-related interactions. Using our new approach, by building a network for the hNek6 interactions confirmed by *in vitro* and *in vivo* assays, considering only the top enriched biological processes (p≤0.05) and, particularly, the second neighbors expansion, we were able to identify mostly the same processes but also new ones, e.g. apoptotic process (GO enrichment p-value of 2.2e-48), cell division (1.1e-41), epidermal growth factor receptor signaling pathway (3.3e-38), transcription, DNA-dependent (6.8e-34), cell proliferation (1.0e-31), DNA repair (1.5e-22), I-kappaB kinase/NF-kappaB cascade (1.7e-16) and others (Table S3). Beyond cell cycle and DNA repair biological processes/pathways, which have been more extensively explored for Neks [56], the NF-kappaB cascade kept our attention, since NEK6 gene was described among others to activate the NF-kappaB signaling pathway, in a large-scale screening [57]. However, there is no explanation of how hNek6 activates this pathway and the first possible links to that question were addressed by our yeast two-hybrid results that showed hNek6 interactions with Transcription factor RelB (RELB), Prx-III (PRDX3) and TRIP-4 (TRIP4) [28]. The first neighbors expansion of our network was not able to show enrichment in I-kappaB kinase/NF-kappaB cascade, but in apoptotic process and transcription, where these proteins were found as most enriched. Though, from the second neighbors expansion, we could observe I-kappaB kinase/NF-kappaB cascade enrichment, forming a cluster of five proteins: Protein-tyrosine phosphatase-like A domain-containing protein 1 (PTPLAD1), NF-kappa-B inhibitor-interacting Ras-like protein 1 (NKIRAS1), E3 ubiquitin-protein ligase parkin (PARK2), GTPase RhebL1 (RHEBL1) and Ubiquitin D (UBD) (Figure 3C). Interestingly, the first neighbors of this cluster have hNek6 as a

component forming another smaller cluster of proteins also annotated to be involved in NF-kappaB cascade (by analyzing all their enriched biological processes depicted in Table S3): hNek6 (NEK6), Beta-arrestin-1 (ARRB1), Estrogen receptor (ESR1) and Sequestosome-1 (SQSTM1). Moreover, five hNek6 protein partners identified by the yeast two-hybrid system (40S ribosomal protein S7, Cell division control protein 42 homolog, E3 ubiquitin-protein ligase RBBP6, Prx-III and TRIP-4), including two of the three interactors described above also interact with two other proteins from this hNek6 cluster (Beta-arrestin-1 and Estrogen receptor), both of which negatively regulate NF-kappaB cascade [58,59] (Figure 3C, red edges). Therefore, our hypothesis is that hNek6 may interact directly with any of those two-hybrid interactors, possibly regulating them by phosphorylation, which could regulate their interaction with Beta-arrestin-1 and/or Estrogen receptor, finally inhibiting these proteins and activating the pathway. This analysis adds novel possible clues on how hNek6 activates NF-kappaB cascade. Although the Transcription factor RelB was found to interact only with hNek6 from the referred cluster, it is already a direct link to the NF-kappaB cascade activation, since it is a component of the NF-kappa-B RelB-p50/p52 complex. Nek6 is also directly linked to Protein-tyrosine phosphatase-like A domain-containing protein 1 (PTPLAD1), enriched in the I-kappaB kinase/NF-kappaB cascade cluster. Altogether, these findings may suggest a novel non-mitotic function for hNek6 through this pathway.

Second case study: *S. cerevisiae* encapsulated cells proteome

As an example of a proteomics study, we chose the *S. cerevisiae* proteome of encapsulated cells in liquid core alginate-chitosan capsules in comparison with cells grown freely in suspension described by Westman et al. [29]. In the context of bioethanol production, encapsulation of yeast cells has been shown to improve the fermentative performance in toxic lignocellulosic hydrolysates [60] and to increase thermotolerance [61]. It has been shown that the yeast metabolism changed significantly upon encapsulation [29], so we used IIS to build a network for the 116 up- and 95 down-regulated proteins in yeasts growing in capsules (described in Table S1 by Westman et al. [29]) to comparatively analyze how encapsulation affects the cells on a more integrated molecular level. First, we uploaded a single two-column TXT file containing both the up- and down-regulated proteins, available as UniProt Accession numbers and respective fold change values, in the "Module 2: Search" tab inside the project. Then the retrieved proteins were selected and added to the project, annotated in the "Module 3: Annotation" tab, and used as queries to build a network in the "Module 4: Interactome" tab, setting expression analysis parameters to consider fold change ≥1.3 as up-regulated and fold change ≤−1.3 as down-regulated proteins. The network was visualized and manipulated using the Cytoscape software.

Figure 4 shows the interactome of encapsulated *S. cerevisiae* built from the proteome data. Our new analysis using IIS showed the same and other functional categories enriched among the up- and down-regulated proteins as described before [29], but using the GO database instead and with one considerable advantage: together with Cytoscape it enabled the visualization of the (i) distribution of the biological processes among the identified proteins, (ii) the number, identity and type of each protein (up- or down-regulated and interactors from database) in each process, (iii) the relative fold change levels of each protein and (iv) their interactions, all resultant data integrated in the same network. It was also possible to analyze the network according to the enriched KEGG pathways and GO cellular components, since these

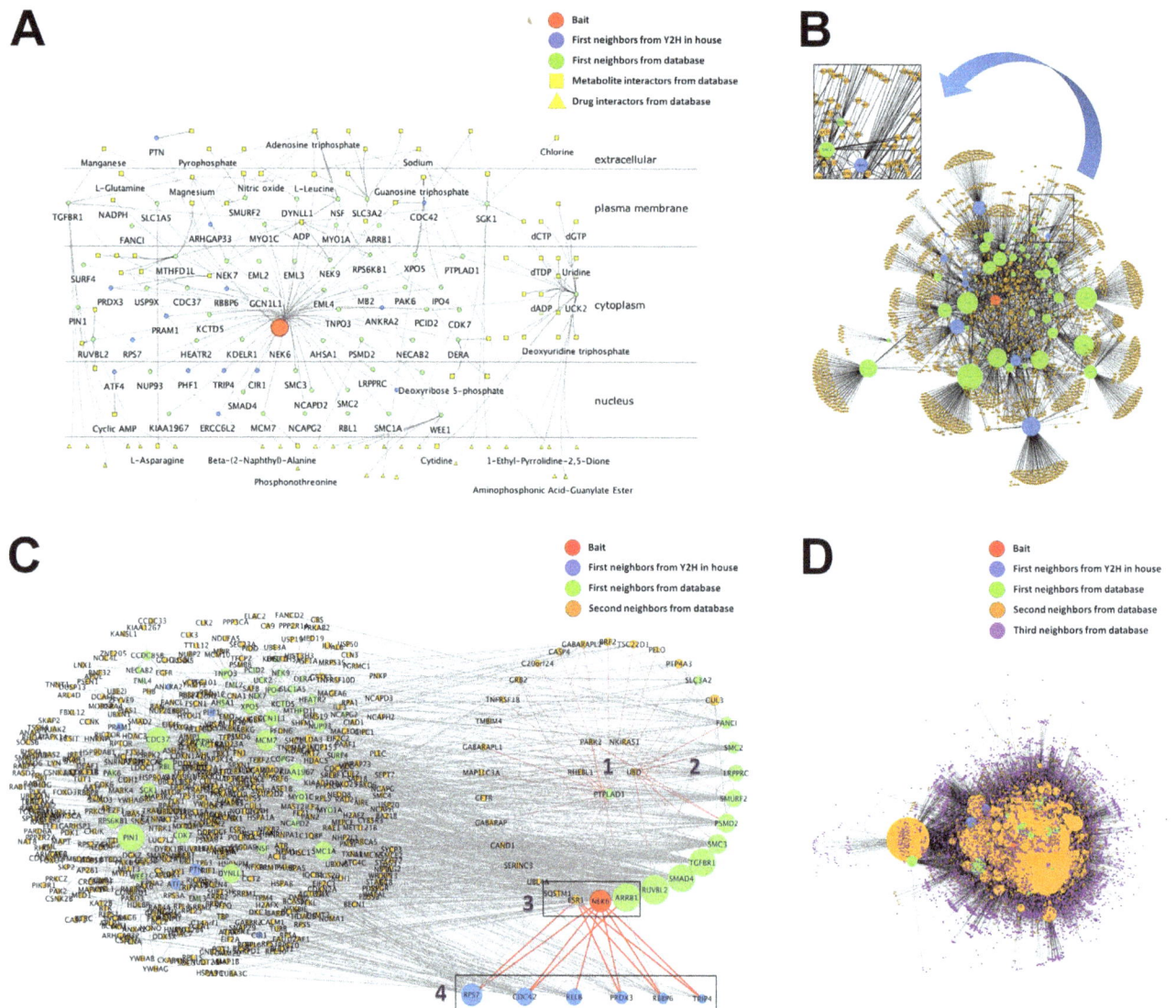

Figure 3. Human Nek6 interactome built from yeast two-hybrid data. (A) hNek6 first neighbors network, showing the bait hNek6 in red, the Y2H first neighbors in blue, the first neighbors described in the GPMGDID database in green, and the metabolites/drugs interactors described in the GPMGDID database in yellow and in different shapes: squares for metabolites and triangles for drugs. The proteins were localized according to their cellular components (GO) described in the "Selected CC" node attribute field by using the Cerebral Cytoscape plugin. (B) hNek6 second neighbors network, showing the second neighbors in orange. The proteins were distributed according to the organic layout. The insertion is depicting the different edge widths, according to our confidence Class scores. (C) hNek6 second neighbors network showing the following protein clusters: 1. top enriched NF-kappaB cascade, 2. first neighbors of cluster 1, 3. enriched NF-kappaB cascade subset of cluster 2, and 4. hNek6 yeast two-hybrid interactors. The proteins were distributed according to the organic and degree-sorted circle layouts, and proteins with degree 0 and 1 were deleted from the network. (D) hNek6 third neighbors network, showing the expansion from the first to the third level of interaction with the third neighbors in purple. The proteins were distributed according to the organic layout. The networks were visualized using Cytoscape v2.8.3.

information were also computed and available in the generated network (data not shown).

In a more global perspective, it was of immediate observation that the majority of up-regulated proteins was involved in cellular metabolic processes (eg. heme biosynthetic process, glycolysis, NADH oxidation, fatty acid metabolic process, ergosterol biosynthetic process and glycogen biosynthetic process), unlike the down-regulated proteins, mostly involved in RNA processing (comprising the most down-regulated protein Drs1p), translation and cellular component organization or biogenesis (Figure 4A). Regarding the metabolic process clusters in the network, as also emphasized by Westman et al. [29], the glycolytic pathway enzyme Tdh1p was found in a significantly higher level in the

encapsulated yeast (Figure 4B), and the high affinity hexose transporters Hxt6p and Hxt7p, although not clustered together, were visually identified as the most up-regulated proteins. Moreover, our analysis was able to identify many proteins in the glycogen biosynthetic process cluster (eg. Gsy1p, Gsy2p, Pgm2p, Glc3p, Ugp1p and Gdb1p) (Figure 4C), and proteins involved in NADH oxidation (the alcohol dehydrogenases Adh1p and Adh5p, which reduce acetaldehyde to ethanol) (Figure 4B), which were all up-regulated. These findings strongly indicate a carbon limitation inside the capsules, but an accumulation of glycogen as the capsules filled up with cells, considering its importance as a storage carbohydrate in slowly growing or starved yeast, and, more relevant, an increase in ethanol yields. Notably, proteins involved

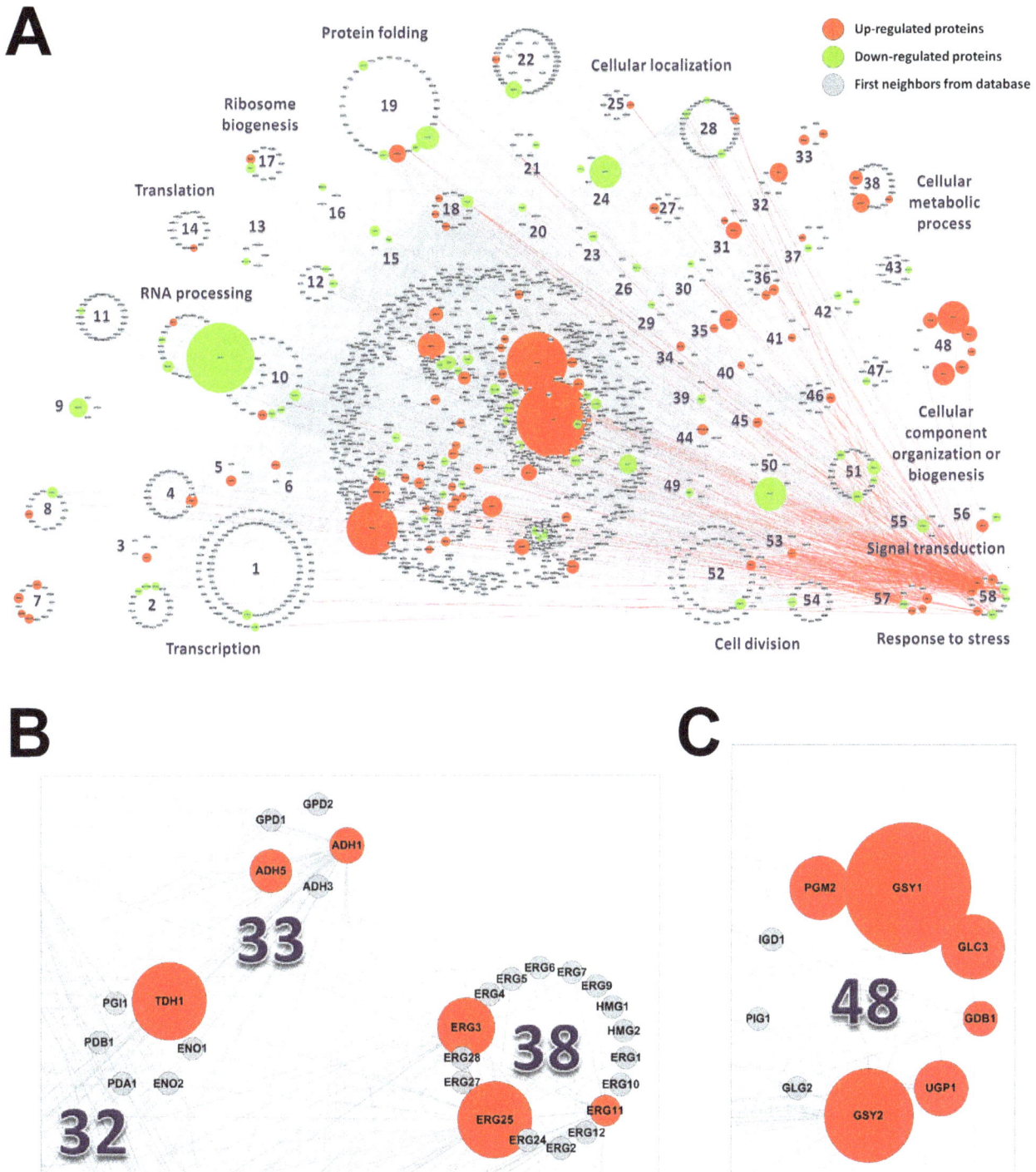

Figure 4. Interactome of *S. cerevisiae* encapsulated in liquid core alginate-chitosan capsules vs. cells grown freely in suspension, built from proteome data. (A) The enriched GO biological processes (p≤0.05) among the up-regulated proteins (red), the down-regulated proteins (green) and the background intermediary proteins (grey) from GPMGDID are depicted in the network by clustering the proteins involved in each of the biological processes with a circle layout. Clusters were assigned only to biological processes containing more than three proteins with at least one from the proteome data; proteins belonging to more than one biological process were assigned to clusters with the best enrichment p-values. More specific biological processes are shown only for proteins with more specific annotation in GO database. The nodes sizes of up- and down-regulated proteins are depicted proportional to their fold change (FC ≥1.3, FDR p≤0.05, as described by Westman et al.) [29]. (B) Network zoom showing the glycolysis (GO enrichment p-value of 1.7e-02), NADH oxidation (2.1e-04) and ergosterol biosynthetic process (4.3e-15) clusters. (C) Network zoom showing the glycogen biosynthetic process (2.5e-06) cluster. The network was built using first neighbors expansion, deletion of nodes with degree 0 and 1, addition of different colors and sizes to proteins according to their fold change, and was filtered by Class scores A to C. The network was visualized using Cytoscape v2.8.3 and the proteins were distributed according to selected enriched biological processes (GO) from the "Top Enriched BP" node attribute field by using the group attributes layout. The following enriched biological processes clusters are shown in the network: 1. transcription, DNA-dependent (3.8e-25), 2. chromatin silencing at telomere (6.1e-15), 3. positive regulation of RNA elongation from RNA polymerase II promoter (5.0e-10), 4.positive regulation of transcription from RNA polymerase II promoter (4.2e-19), 5. negative regulation of

transcription, DNA-dependent (7.1e-03), 6. positive regulation of transcriptional preinitiation complex assembly (2.5e-05), 7. vacuolar acidification (1.1e-10), 8. replicative cell aging (3.9e-12), 9. pseudohyphal growth (3.6e-08), 10. rRNA processing (4.8e-16), 11. maturation of SSU-rRNA from tricistronic rRNA transcript (1.2e-15), 12. regulation of translation (3.2e-10), 13. regulation of translational fidelity (2.7e-05), 14. mitochondrial translation (9.9e-04), 15. mature ribosome assembly (5.2e-04), 16. ribosomal small subunit assembly and maintenance (1.1e-05), 17. ribosomal large subunit biogenesis and assembly (3.9e-12), 18. protein refolding (1.8e-11), 19. protein folding (7.7e-09), 20. mRNA transport (9.7e-08), 21. poly(A)+ mRNA export from nucleus (5.2e-09), 22. protein transport (1.9e-11), 23. ribosomal small subunit export from nucleus (1.3e-08), 24. protein localization (4.9e-07), 25. protein import into nucleus (6.9e-11), 26. protein targeting to ER (5.2e-04), 27. ER to Golgi vesicle-mediated transport (1.1e-07), 28. endocytosis (9.6e-19), 29. lysine biosynthetic process via aminoadipic acid (7.1e-03), 30. pantothenate biosynthetic process (4.5e-04), 31. heme biosynthetic process (1.3e-03), 32. glycolysis (1.7e-02), 33. NADH oxidation (2.1e-04), 34. phospholipid biosynthetic process (1.1e-02), 35. fatty acid metabolic process (8.9e-04), 36. fatty acid biosynthetic process (2.7e-05), 37. protein amino acid N-linked glycosylation (2.1e-03), 38. ergosterol biosynthetic process (4.3e-15), 39. branched chain family amino acid catabolic process (2.4e-04), 40. pentose-phosphate shunt (2.7e-02), 41. 2-oxoglutarate metabolic process (2.4e-03), 42. one-carbon compound metabolic process (1.2e-02), 43. DNA recombination (5.8e-03), 44. metabolic process (3.0e-03), 45. deoxyribonucleotide biosynthetic process (6.6e-06), 46. protein deubiquitination (1.8e-09), 47. aerobic respiration (3.3e-04), 48. glycogen biosynthetic process (2.5e-06), 49. actin cytoskeleton organization and biogenesis (3.0e-05), 50. actin filament organization (5.1e-12), 51. chitin- and beta-glucan-containing cell wall organization and biogenesis (1.0e-12), 52. cell division (2.8e-21), 53. mitosis (1.5e-17), 54. establishment of cell polarity (2.8e-13), 55. TOR signaling pathway (8.3e-09), 56. Ras protein signal transduction (1.2e-07), 57. response to osmotic stress (1.1e-10), 58. response to stress (3.3e-06).

in the ergosterol biosynthetic process cluster (eg. Erg25p, Erg3p and Erg11p) were also visually identified as greatly up-regulated (Figure 4B), although not discussed in the previous report by Westman et al. [29]. Since ergosterol is the major sterol of the fungal plasma membrane, important for the fluidity and integrity of the membrane and for the proper function of many membrane-bound enzymes, with its biosynthetic pathway consisting in a pivotal target of antifungal drugs [62], these findings may also explain the differences between encapsulated and free growing yeast cells. Indeed, a more intact membrane supports higher concentrations of ethanol. Furthermore, among the stress response proteins, comprising both up- and down-regulated proteins, it was suggested by Westman et al. [29] that a more plausible explanation for the apparent osmotic stress response is a cross-talk between nutrient starvation and other environmental stress responses. In our network analysis, this hypothesis could be visualized by the broad spectrum of connections among the stress response clusters with other clusters in the network (Figure 4A, red edges).

Third case study: primary and metastatic human ovarian cancer metabolome

For the metabolomics analysis, we used as an example the work by Fong et al. [30], which described the metabolome of the human normal ovary and its transformation in primary epithelial ovarian cancer (EOC) and metastatic ovarian cancer (MOC). In the context of oncogenesis and the importance of a comprehensive metabolic analysis of solid tumors to reveal possible biomarkers for early diagnosis and monitoring of cancer progression and recurrence, IIS was used to build two comparative networks: one for the up- and down-regulated metabolites in EOC and the other one for the up- and down-regulated metabolites in MOC (described in Table S2 by Fong et al. [30]). First, we converted the metabolite names to HMDB IDs and uploaded a single two-column TXT file containing both the up- and down-regulated metabolites for each condition (EOC and MOC), as a list of HMDB IDs and respective fold change values, in the "Module 2: Search" tab inside each project (EOC and MOC). Then the retrieved metabolites were selected and added to the project, and used as queries to build the networks in the "Module 4: Interactome" tab, setting expression analysis parameters to consider fold change ≥ 1.2 as up-regulated and fold change ≤ -1.2 as down-regulated metabolites, as described by Fong et al. [30]. The network was visualized and manipulated using the Cytoscape software.

Figure 5 shows the interactomes of (A) EOC and (B) MOC built from the metabolome data. Our new analysis using IIS showed similar metabolic pathways as described before [30], and also other signaling and metabolic pathways enriched among the up-

and down-regulated metabolites. We analyzed the data based on the KEGG database enrichment performed by IIS, which was able to retrieve 5 of each list of 15 enriched pathways for EOC and MOC identified by the Ingenuity Pathway Analysis (IPA): Aminoacyl-tRNA biosynthesis (KEGG enrichment p-value of 1.3e-26), Urea cycle and metabolism of amino groups (1.4e-21), Glycine, serine and threonine metabolism (1.1e-10), Methionine metabolism (2.5e-09), and Phenylalanine, tyrosine and tryptophan biosynthesis (1.4e-07) for EOC; and Alanine and aspartate metabolism (1.8e-16), Purine metabolism (8.3e-12), Arginine and proline metabolism (2.3e-11), Glutamate metabolism (2.0e-10), and Pyrimidine metabolism (1.2e-10) for MOC. Though, IIS also retrieved other 25 and 16 significant enriched pathways for EOC and MOC, respectively, including signaling and metabolic pathways (Figure 5). Notable among these are the Glycan structures degradation (1.3e-04 in EOC; 3.0e-09 in MOC) and Fatty acid metabolism (1.3e-11 in EOC; 2.1e-14 in MOC) pathways enriched in both EOC and MOC, which could explain the increase in fucose (2.75 fold in EOC; 1.81 fold in MOC) and carnitine (1.79 fold in EOC; 1.88 fold in MOC) levels. The enriched Pyruvate metabolism (5.6e-17) and Glycolysis/Gluco-neogenesis (5.5e-32) pathways in EOC and MOC, respectively, could also explain the increase in lactate levels when compared to normal ovarian tissue (1.46 fold in EOC; 1.37 fold in MOC).

In order to reduce complexity, Figure 5 shows the metabolites in only a few metabolic pathway clusters, since they are the ones containing interacting proteins with the best enrichment p-values, although the metabolites are also connected to the other clusters by interactions with different proteins, e.g. carnitine is connected to the Purine metabolism (7.9e-36) cluster in EOC by its interaction with Xanthine dehydrogenase/oxidase (XDH), and also connected to the Fatty acid metabolism (1.3e-11) cluster by its interaction with Carnitine palmitoyltransferase 1A (CPT1A) (Figure 5A, red edges). Clusters composed of at least one first neighbor interactor represent probably the most confident pathways, since they group direct interactors of metabolites. As in the proteomics approach, IIS metabolomics analysis connected to Cytoscape enabled the visualization of all resultant data integrated in the same network, making it easier to interpret the whole dataset and its relations, since they can bring together information concerning: the (i) distribution of the pathways among the identified metabolites, (ii) the number, identity and type of each metabolite (up- or down-regulated) in each process, (iii) the relative fold change levels of each metabolite and (iv) their interactions.

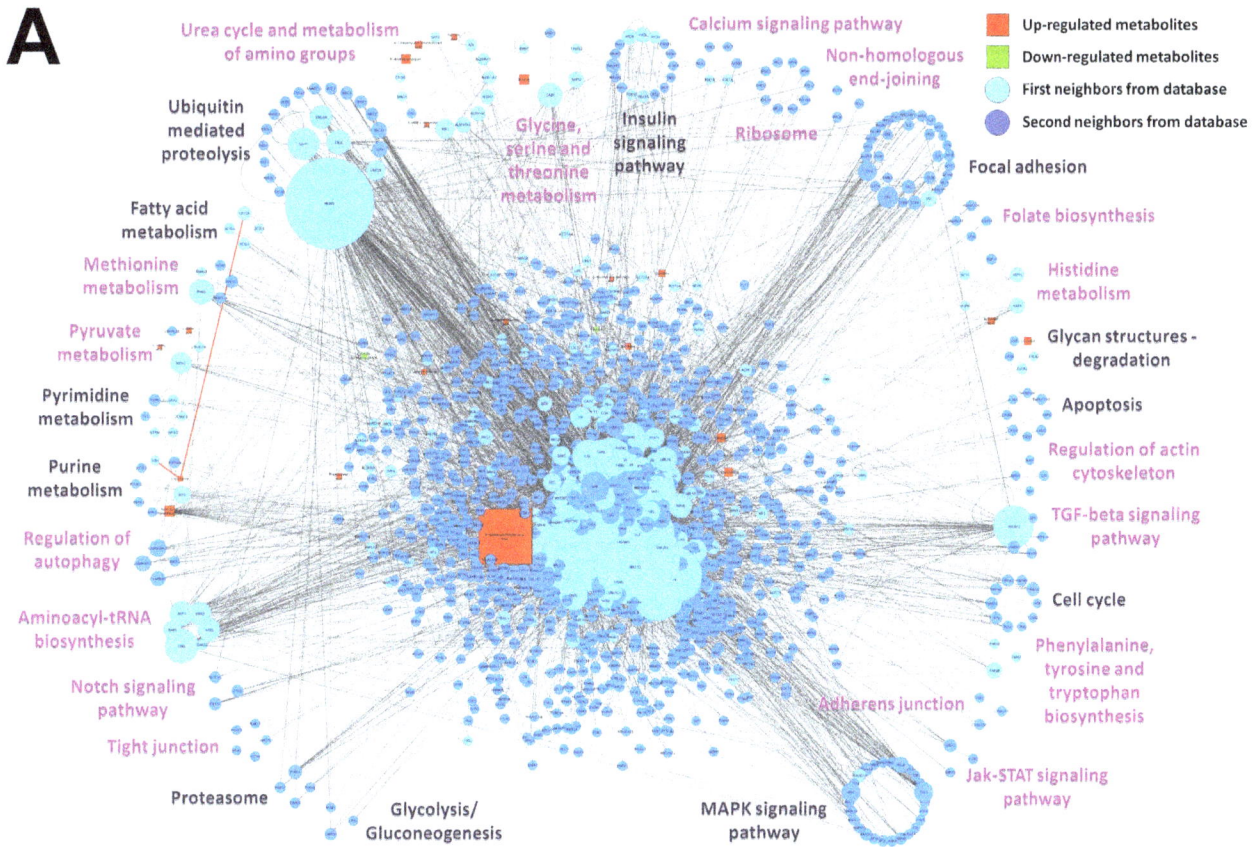

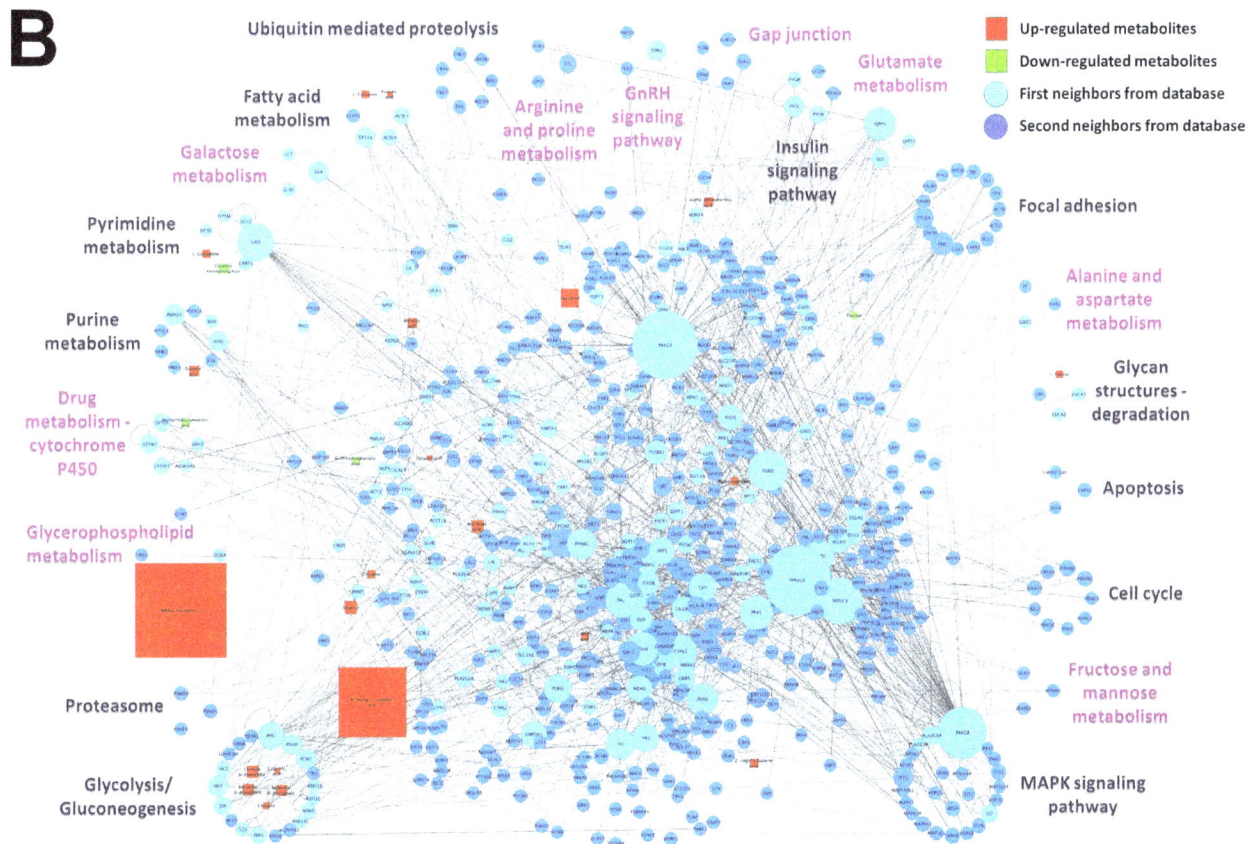

Figure 5. Comparison between the interactomes of (A) primary human epithelial ovarian cancer and (B) metastatic ovarian cancer vs. normal human ovary, built from metabolome data. The enriched KEGG pathways (p≤0.05) among the up-regulated metabolites (red squares), the down-regulated metabolites (green squares) and the background intermediary proteins (light blue circles for first neighbors and dark blue circles for second neighbors) from GPMGDID are depicted in the networks by clustering the proteins involved in each of the pathways with a circle layout. Enriched KEGG pathways specifically for each network (A) or (B) are depicted in purple and the ones in common are depicted in black. Clusters were assigned only to pathways containing more than three proteins (disease pathways or pathways specific for defined cell types were not considered), and metabolites were assigned only to metabolic pathway clusters containing interacting proteins with the best enrichment p-values. The nodes sizes of up- and down-regulated metabolites are depicted proportional to their fold change (FC≥1.2, p≤0.05, as described by Fong et al. [30]) and the nodes sizes of the background intermediary proteins are depicted proportional to their connectivity degree. The networks were built using second neighbors expansion, deletion of nodes with degree 0 and 1 and addition of different colors and sizes to proteins according to their fold change. The networks were visualized using Cytoscape v2.8.3 and the proteins were distributed according to selected enriched pathways (KEGG) from the "Top Enriched KEGG" node attribute field by using the group attributes layout.

Network attributes and parameters

It is important to point out that the network construction by IIS considers the degree of each node in the network, showing a gradient of node sizes, which makes it easy to distinguish the hubs. The generated network also brings the cellular components and the enriched biological processes and pathways of each node, which can be used to easily separate the nodes into cell compartments (e.g. by using different layouts or the Cerebral Cytoscape plugin [63], as shown in Figure 3A), or cluster the nodes into functional modules. It also considers each type of node

(proteins, metabolites and drugs) as different entities, which can be distinguished by their different node shapes (Figure 3A), and depicts different edge widths according to the interaction confidence Class score, described above (Figure 3B). Other confidence interaction measures, such as the FSW score and p-value, or different types of interactions, can be accessed as node and edge attributes (Methods S1). Besides, all of these parameters can be changed by the users according to their specific needs.

As a metric of how many neighbors a pair of proteins share, the FSW score was implemented so that it can also be used as a filter

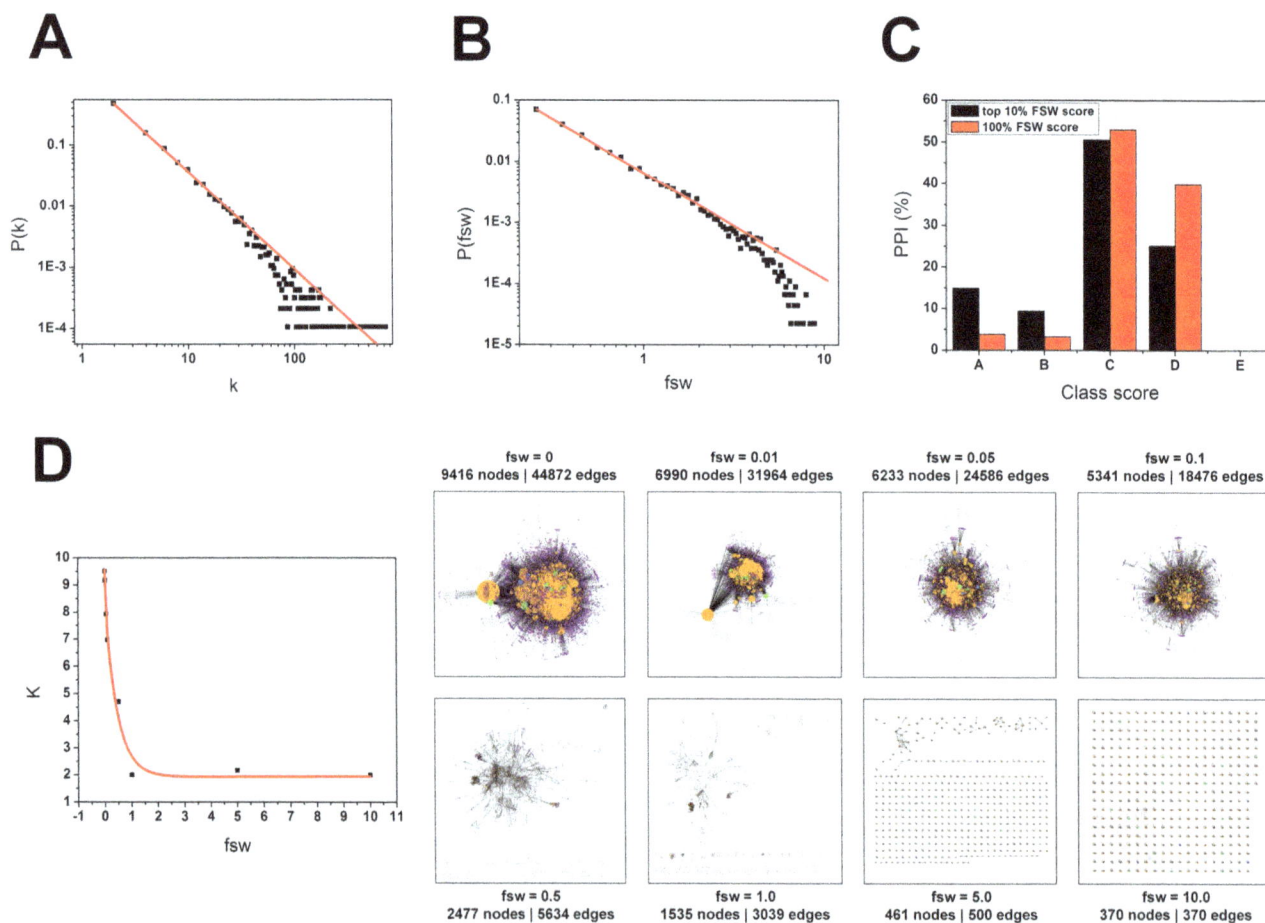

Figure 6. Comparison between FSW score, degree and Class score. (A) Degree distribution of hNek6 third neighbors network ($\gamma = -1.59$). (B) FSW score distribution of hNek6 third neighbors network ($\gamma = -1.72$). (C) Percentage of PPIs characterized by the best FSW score and Class score in hNek6 third neighbors network. (D) Correlation between the average degree and the FSW score of hNek6 third neighbors network from FSW score 0 to 10. Both the degree distribution and the FSW score distribution approximate a power-law and are scale-free in topology. The slopes (γ) were determined by linear fitting where $P(k)$ approximates a power-law: $P(k) \approx k^{-\gamma}$ (k: total number of links; K: average degree; γ: slope of the distribution on the log-log plot; fsw: functional similarity weight; PPI: protein-protein interaction).

of hubs when building the networks. To statistically compare the FSW distribution to the degree distribution in the networks generated by IIS, the degree distribution $P(k)$ and the average degree (K) of each network were calculated as described by Stelzl et al. [4]. We found that the FSW score distribution is similar to the degree distribution and is also scale-free in topology (Figure 6A and B). Therefore, the FSW score could be used as a parameter to filter hubs from the networks, as shown by using score values 0.01 to 1.0, where the average degree of the network is greatly reduced and most of the hubs fall outside the network (Figure 6D).

Furthermore, the effectiveness of using FSW score as a PPI reliability index was demonstrated before [49,50,54]. Here we ranked the top 10% of protein interactions in the hNek6 third neighbors network by the FSW score and compared to the Class score. We found that the top 10% of PPIs with the best FSW scores were also enriched with the best Class scores A and B: 15.0% were characterized by Class score A and 9.4% by Class score B, compared to 3.8% and 3.3%, respectively, considering the total PPIs in the network (Figure 6C).

IIS annotates nodes and edges using diverse databases and metrics, and offers a variety of filters to build the networks, which can be used depending on the type and amount of data to be analyzed. Though, in general, a few steps may be considered: if working with (i) large datasets or organisms with huge interaction databases (Table S4), the network size can be reduced by using the Class score or FSW score filters; (ii) small datasets, the network can give more information when expanded to second or third neighbors; (iii) organisms for which only a few interactions were described, the network can be built by using the "ortholog relationship" option selecting a phylogenetically close model organism; (iv) transcriptome or proteome datasets, the network can be more coherent and concise by expanding it only to first neighbors and using the "delete nodes with degree 0 and 1" option; (v) metabolome datasets, an expansion to second neighbors may be more interesting, since it will probably allow clusters of metabolites and first neighbors to connect with each other; and (vi) drugs datasets, the same as for metabolome datasets.

Therefore, from the analyses presented above, IIS comes as a platform to perform an integrative analysis of omics data focused on interaction networks, mainly visualized via web or by Cytoscape software, in a more complete and easy-to-interpret way, in order to give a first overview of all the components, their emergent properties and relations and assist researchers to direct further relevant experiments and take important insights of their data. IIS is freely available online at: http://www.lge.ibi.unicamp.br/lnbio/IIS/. IIS code and database can be downloaded at: http://bioinfo03.ibi.unicamp.br/lnbio/IIS2/download.php.

Supporting Information

Table S1 Automatic annotation table generated from the Annotation Module.

Table S2 Cellular component (GO) used in our database. Children terms were grouped for each selected ancestral cellular component, considering only terms annotated for ≥ 10 genes.

Table S3 Node attributes from hNek6 second neighbors network. All attributes are described in details in Methods S1. Enriched GO Biological Processes (BP) and KEGG Pathways are depicted with a p-value in parentheses for each protein in the network. Only enriched terms with p≤0.05 were considered in the network analyses. hNek6 interactors retrieved by yeast two-hybrid are depicted in bold. Nodes with degree 0 and 1 were deleted from the network.

Table S4 Statistics from GPMGDID.

Methods S1 GPMGDID construction, IIS pipeline and XGMML file generation.

Acknowledgments

We thank Dr. Marcelo M. Brandão, Dr. Paulo S. L. de Oliveira and Dr. Tiago J. P. Sobreira for significant discussions and ideas.

Author Contributions

Conceived and designed the experiments: MFC GVM. Performed the experiments: MFC LMC HHS ROV. Analyzed the data: JK GVM. Contributed reagents/materials/analysis tools: GAGP. Wrote the paper: MFC JK GVM.

References

1. Ho Y, Gruhler A, Heilbut A, Bader GD, Moore L, et al. (2002) Systematic identification of protein complexes in Saccharomyces cerevisiae by mass spectrometry. Nature 415 (6868): 180–3.
2. Giot L, Bader JS, Brouwer C, Chaudhuri A, Kuang B, et al. (2003) A protein interaction map of Drosophila melanogaster. Science 302: 1727–1736.
3. Li S, Armstrong CM, Bertin N, Ge H, Milstein S, et al. (2004) A map of the interactome network of the metazoan C. elegans. Science 303: 540–543.
4. Stelzl U, Worm U, Lalowski M, Haenig C, Brembeck FH, et al. (2005) A human protein-protein interaction network: a resource for annotating the proteome. Cell 122: 957–968.
5. Tong AH, Evangelista M, Parsons AB, Xu H, Bader GD, et al. (2001) Systematic genetic analysis with ordered arrays of yeast deletion mutants. Science 294 (5550): 2364–8.
6. Zhu F, Shi Z, Qin C, Tao L, Liu X, et al. (2012) Therapeutic target database update 2012: a resource for facilitating target-oriented drug discovery. Nucleic Acids Res 40 (Database issue): D1128–36.
7. Kang HG, Klessig DF (2008) The involvement of the Arabidopsis CRT1 ATPase family in disease resistance protein-mediated signaling. Plant Signal Behav 3: 689–690.
8. Kormish JD, Sinner D, Zorn AM (2009) Interactions between SOX factors and Wnt/beta-catenin signaling in development and disease. Dev Dyn 239 (1): 56–68.
9. Stark C, Breitkreutz BJ, Reguly T, Boucher L, Breitkreutz A, et al. (2006) BioGRID: a general repository for interaction datasets. Nucleic Acids Res 34: D535–9.
10. Hermjakob H, Montecchi-Palazzi L, Lewington C, Mudali S, Kerrien S, et al. (2004) IntAct: an open source molecular interaction database. Nucleic Acids Res 32 (Database issue): D452–5.
11. Xenarios I, Rice DW, Salwinski L, Baron MK, Marcotte EM, et al. (2000) DIP: the database of interacting proteins. Nucleic Acids Res 28 (1): 289–91.
12. Szklarczyk D, Franceschini A, Kuhn M, Simonovic M, Roth A, et al. (2011) The STRING database in 2011: functional interaction networks of proteins, globally integrated and scored. Nucleic Acids Res 39 (Database issue): D561–8.
13. Zanzoni A, Montecchi-Palazzi L, Quondam M, Ausiello G, Helmer-Citterich M, et al. (2002) MINT: a Molecular INTeraction database. FEBS Lett 513 (1): 135–40.
14. Keshava Prasad TS, Goel R, Kandasamy K, Keerthikumar S, Kumar S, et al. (2009) Human Protein Reference Database—2009 update. Nucleic Acids Res 37 (Database issue): D767–72.
15. Knox C, Law V, Jewison T, Liu P, Ly S, et al. (2011) DrugBank 3.0: a comprehensive resource for 'omics' research on drugs. Nucleic Acids Res 39 (Database issue): D1035–41.
16. Gaulton A, Bellis LJ, Bento AP, Chambers J, Davies M, et al. (2012) ChEMBL: a large-scale bioactivity database for drug discovery. Nucleic Acids Res 40 (Database issue): D1100–7.

17. Wishart DS, Knox C, Guo AC, Eisner R, Young N, et al. (2009) HMDB: a knowledgebase for the human metabolome. Nucleic Acids Res 37 (Database issue): D603–610.

18. Jewison T, Neveu V, Lee J, Knox C, Liu P, et al. (2012) YMDB: The Yeast Metabolome Database. Nucleic Acids Res 40 (Database assue): D815–20.

19. Guo AC, Jewison T, Wilson M, Liu Y, Knox C, et al. (2013) ECMDB: the E. coli Metabolome Database. Nucleic Acids Res 41(Database issue): D625-30.

20. Kanehisa M, Goto S, Kawashima S, Nakaya A (2002) The KEGG databases at GenomeNet. Nucleic Acids Res 30 (1): 42–6.

21. Joshi-Tope G, Gillespie M, Vastrik I, D'Eustachio P, Schmidt E, et al. (2005) Reactome: a knowledgebase of biological pathways. Nucleic Acids Res 33(Database issue): D428–32.

22. De Bodt S, Proost S, Vandepoele K, Rouze P, Peer Y (2009) Predicting protein-protein interactions in Arabidopsis thaliana through integration of orthology, gene ontology and co-expression. BMC Genomics 10: 288.

23. Lin M, Hu B, Chen L, Sun P, Fan Y, et al. (2009) Computational identification of potential molecular interactions in Arabidopsis. Plant Physiol 151: 34–46.

24. Berger SI, Posner JM, Ma'ayan A (2007) Genes2Networks: connecting lists of gene symbols using mammalian protein interactions databases. BMC Bioinformatics 8: 372.

25. Mostafavi S, Ray D, Warde-Farley D, Grouios C, Morris Q (2008) GeneMANIA: a real-time multiple association network integration algorithm for predicting gene function. Genome Biol (Suppl 1): S4.

26. Ingenuity Pathway Analysis (IPA) website. Available: http://www.ingenuity.com/products/pathways_analysis.html. Accessed 2014 Feb 25.

27. Pellet J, Meyniel L, Vidalain PO, de Chassey B, Tafforeau L, et al. (2009) pISTil: a pipeline for yeast two-hybrid Interaction Sequence Tags identification and analysis. BMC Res Notes 2: 220.

28. Meirelles GV, Lanza DCF, Silva JC, Bernachi JS, Leme AP, et al. (2010) Characterization of hNek6 interactome reveals an important role for its short N-terminal domain and colocalizations with proteins at the centrosome. J Proteome Res 9(12): 6298–316.

29. Westman JO, Taherzadeh MJ, Franzén CJ (2012) Proteomic analysis of the increased stress tolerance of saccharomyces cerevisiae encapsulated in liquid core alginate-chitosan capsules. PLoS One 7 (11): e49335.

30. Fong MY, McDunn J, Kakar SS (2011) Identification of metabolites in the normal ovary and their transformation in primary and metastatic ovarian cancer. PLoS One 6 (5): e19963.

31. Ewing B, Green P (1998) Base-calling of automated sequencer traces using Phred. II. Error probabilities. Genome Res 8: 186–194.

32. Baudet C, Dias Z (2006) Analysis of slipped sequences in EST projects. Genet Mol Res 5 (1): 169–181.

33. Altschul SF, Madden TL, Schaffer AA, Zhang J, Zhang Z, et al. (1997) Gapped BLAST and PSI-BLAST: a new generation of protein database search programs. Nucleic Acids Res 25: 3389–3402.

34. Huang X, Madan A (1999) CAP3: a DNA sequence assembly program. Genome Res 9: 868–877.

35. Ashburner M, Ball CA, Blake JA, Botstein D, Butler H, et al. (2000) Gene ontology: tool for the unification of biology. The Gene Ontology Consortium. Nat Genet 25 (1): 25–9.

36. Uhlen M, Oksvold P, Fagerberg L, Lundberg E, Jonasson K, et al. (2010) Towards a knowledge-based Human Protein Atlas. Nat Biotechnol 28 (12): 1248–50.

37. Marchler-Bauer A, Lu S, Anderson JB, Chitsaz F, Derbyshire MK, et al. (2011) CDD: a Conserved Domain Database for the functional annotation of proteins. Nucleic Acids Res 39(Database issue): D225–9.

38. Mouse Genome Informatics (MGI). Available: http://www.informatics.jax.org/phenotypes.shtml. Accessed 2014 May 29.

39. Berman HM, Westbrook J, Feng Z, Gilliland G, Bhat TN, et al. (2000) The Protein Data Bank. Nucleic Acids Res 28 (1): 235–42.

40. Linding R, Jensen LJ, Diella F, Bork P, Gibson TJ, et al. (2003) Protein disorder prediction: implications for structural proteomics. Structure 11 (11): 1453–9.

41. Sigrist CJ, Cerutti L, de Castro E, Langendijk-Genevaux PS, Bulliard V, et al. (2010) PROSITE, a protein domain database for functional characterization and annotation. Nucleic Acids Res 38: D161–6.

42. Hubbard T, Barker D, Birney E, Cameron G, Chen Y, et al. (2002) The Ensembl genome database project. Nucleic Acids Research 30 (1): 38–41.

43. Boeckmann B, Bairoch A, Apweiler R, Blatter MC, Estreicher A, et al. (2003) The Swiss-Prot protein knowledgebase and its supplement TrEMBL in 2003. Nucleic Acids Research 31: 365–370.

44. Magrane M, Consortium U (2011) UniProt Knowledgebase: a hub of integrated protein data. Database (Oxford) 2011: bar009.

45. Boyle EI, Weng S, Gollub J, Jin H, Botstein D, et al. (2004) GO::TermFinder—open source software for accessing Gene Ontology information and finding significantly enriched Gene Ontology terms associated with a list of genes. Bioinformatics 20 (18): 3710–5.

46. Press WA, Teukolsky SA, Vetterling WT, Flannery BP (2007) Numerical Recipes: The Art of Scientic Computing. Cambridge University Press.

47. Lopes CT, Franz M, Kazi F, Donaldson SL, Morris Q, et al. (2010) Cytoscape Web: an interactive web-based network browser. Bioinformatics 26 (18): 2347–8.

48. Shannon P, Markiel A, Ozier O, Baliga NS, Wang JT, et al. (2003) Cytoscape: a software environment for integrated models of biomolecular interaction networks. Genome Res 13 (11): 2498–504.

49. Chen J, Hsu W, Lee ML, Ng SK (2006) Increasing confidence of protein interactomes using network topological metrics. Bioinformatics 22: 1998–2004.

50. Chen J, Chua HN, Hsu W, Lee M-L, Ng S-K, et al. (2006) Increasing confidence of protein-protein interactomes. In 17th International Conference on Genome Informatics. Yokohama, Japan. 2006: 284–297.

51. Chua HN, Sung WK, Wong L (2006) Exploiting indirect neighbours and topological weight to predict protein function from protein-protein interactions. Bioinformatics 22: 1623–1630.

52. Gerstein M, Lan N, Jansen R (2002) Proteomics. Integrating interactomes. Science 295: 284–287.

53. Liu G, Wong L, Chua HN (2009) Complex discovery from weighted PPI networks. Bioinformatics 25: 1891–1897.

54. Brandão MM, Dantas LL, Silva-Filho MC (2009) AtPIN: Arabidopsis thaliana protein interaction network. BMC Bioinformatics 10: 454.

55. Breitkreutz BJ, Stark C, Tyers M (2003) Osprey: a network visualization system. Genome Biol 4 (3): R22.

56. Meirelles GV, Perez AM, Souza EE, Basei FL, Papa PF, et al. (2014) "Stop Ne(c)king around": How systems biology can help to characterize the functions of Nek family kinases from cell cycle regulation to DNA damage response. World J Biol Chem 5 (2): 141–160.

57. Matsuda A, Suzuki Y, Honda G, Muramatsu S, Matsuzaki O, et al. (2003) Large-scale identification and characterization of human genes that activate NF-kappaB and MAPK signaling pathways. Oncogene 22 (21): 3307–3318.

58. Wang Y, Tang Y, Teng L, Wu Y, Zhao X, et al. (2006) Association of beta-arrestin and TRAF6 negatively regulates Toll-like receptor-interleukin 1 receptor signaling. Nat Immunol 7 (2): 139–47.

59. Liu H, Liu K, Bodenner DL (2005) Estrogen receptor inhibits interleukin-6 gene expression by disruption of nuclear factor kappaB transactivation. Cytokine 31 (4): 251–7.

60. Talebnia F, Taherzadeh MJ (2006) In situ detoxification and continuous cultivation of dilute-acid hydrolyzate to ethanol by encapsulated S. cerevisiae. J Biotechnol 125: 377–384.

61. Westman JO, Manikondu RB, Franzen CJ, Taherzadeh MJ (2012) Encapsulation-induced stress helps Saccharomyces cerevisiae resist convertible lignocellulose derived inhibitors. Int J Mol Sci 13: 11881–11894.

62. Lupetti A, Danesi R, Campa M, Del Tacca M, Kelly S (2002) Molecular basis of resistance to azole antifungals. Trends Mol Med. 8 (2): 76–81.

63. Barsky A, Gardy JL, Hancock RE, Munzner T (2007) Cerebral: a Cytoscape plugin for layout of and interaction with biological networks using subcellular localization annotation. Bioinformatics 23 (8): 1040–2.

3

Long-Term Supranutritional Supplementation with Selenate Decreases Hyperglycemia and Promotes Fatty Liver Degeneration by Inducing Hyperinsulinemia in Diabetic *db/db* Mice

Chaoqun Wang[1,2], Shulin Yang[2], Ningbo Zhang[2], Yulian Mu[2], Hongyan Ren[2], Yefu Wang[1]*, Kui Li[2]*

1 College of Life Sciences, Wuhan University, Wuhan, Hubei, P. R. China, 2 State Key Laboratory for Animal Nutrition, Institute of Animal Science, Chinese Academy of Agricultural Sciences, Beijing, P. R. China

Abstract

There are conflicting reports on the link between the micronutrient selenium and the prevalence of diabetes. To investigate the possibility that selenium acts as a "double-edged sword" in diabetes, cDNA microarray profiling and two-dimensional differential gel electrophoresis coupled with mass spectrometry were used to determine changes in mRNA and protein expression in pancreatic and liver tissues of diabetic *db/db* mice in response to dietary selenate supplementation. Fasting blood glucose levels increased continuously in *db/db* mice administered placebo (DMCtrl), but decreased gradually in selenate-supplemented *db/db* mice (DMSe) and approached normal levels after termination of the experiment. Pancreatic islet size was increased in DMSe mice compared with DMCtrl mice, resulting in a clear increase in insulin production and a doubling of plasma insulin concentration. Genes that encode proteins involved in key pancreatic β-cell functions, including regulation of β-cell proliferation and differentiation and insulin synthesis, were found to be specifically upregulated in DMSe mice. In contrast, apoptosis-associated genes were downregulated, indicating that islet function was protected by selenate treatment. Conversely, liver fat accumulation increased in DMSe mice together with significant upregulation of lipogenic and inflammatory genes. Genes related to detoxification were downregulated and antioxidant enzymatic activity was reduced, indicating an unexpected reduction in antioxidant defense capacity and exacerbation of fatty liver degeneration. Moreover, proteomic analysis of the liver showed differential expression of proteins involved in glucolipid metabolism and the endoplasmic reticulum assembly pathway. Taken together, these results suggest that dietary selenate supplementation in *db/db* mice decreased hyperglycemia by increasing insulin production and secretion; however, long-term hyperinsulinemia eventually led to reduced antioxidant defense capacity, which exacerbated fatty liver degeneration.

Editor: Silvia C. Sookoian, Institute of Medical Research A Lanari-IDIM, University of Buenos Aires-National Council of Scientific and Technological Research (CONICET), Argentina

Funding: This research was supported by grants from the national high technology research and development program 863 (2012AA020603), the State Key Laboratory of Animal Nutrition of China, Independent Research Project (2004DA125184G1109) and the Agricultural Science and Technology Innovation Program (ASTIP-IAS05). The funders had no role in study design, data collection and analysis, decision to publish, or preparation of the manuscript.

Competing Interests: The authors have declared that no competing interests exist.

* Email: wangyefu@whu.edu.cn (YW); likui@caas.cn (KL)

Introduction

Selenium is a necessary trace element in the body that is currently used as a nutritional supplement for humans and animals, although whether it is beneficial has remained a subject of controversy. Initially considered a toxin, during the mid-20th century researchers discovered that selenium exerts positive effects on human and animal health and can thus be either beneficial or detrimental [1,2], acting as a "double-edged sword". Further studies have demonstrated that the safe range of selenium intake is very narrow and that the effects of selenium on health follow a U-shaped risk curve [2,3]. While the negative health consequences [3,4] of dietary selenium deficiency (e.g., Keshan disease and Kashin-Beck disease) and selenium excess (e.g., hair loss, brittle,

thickened and stratified nails) have been noted, the effects of intermediate levels of selenium are less certain.

There are conflicting reports on the link between selenium micronutrient status and the prevalence of type 2 diabetes [5]. On the one hand, selenium can act as an antioxidant nutrient in different cell types via incorporation of selenocysteine into selenoproteins through a complex genetic mechanism encoded by the UGA codon [6], and thereby contribute to the prevention of cardiovascular disease, cancer, and diabetes [7,8]. Furthermore, selenium has insulin-like properties and could been qualified as a potential antidiabetic agent [9]. Many studies have demonstrated a protective effect of selenium against type 1 and type 2 diabetes [8], and the use of appropriate selenium supplements may improve glucose metabolism by alleviating hyperglycemia, regulating glycolysis and gluconeogenesis, and activating key components of the insulin signaling cascade [10,11]. On the other hand, more recent findings from large-scale human studies [12–14] and animal experiments [15–17] have shown that high selenium status or intake is positively correlated with an increased risk of type 2 diabetes. Thus, it may be important to examine the effect of selenium supplementation on the development of type 2 diabetes. In patients with type 2 diabetes, selenium causes adverse effects on blood glucose homeostasis, even when the plasma selenium concentration is raised from "deficient" levels to the optimal concentration for antioxidant activity [18]. Thus, regarding the pathologies involved in type 2 diabetes, selenium may act as a "double-edged sword", and therefore the detailed molecular mechanism that underlie how selenium promotes or prevents the development of type 2 diabetes require further investigation.

Sodium selenite and sodium selenate are the most commonly used inorganic selenium compounds for dietary selenium supplementation. Sodium selenate is a more effective insulin mimetic than either sodium selenite or organoselenium compounds such as selenomethionine [19–21]. In the present study, we designed experiments to determine whether selenium acts as a "double-edged sword" in type 2 diabetes. For this purpose, we administered daily oral sodium selenate at a moderate dose for 9 weeks to *db/db* mice, which are a model system for the development of spontaneous type 2 diabetes. Genetic microarray and proteomic analyses were used to determine the effects of selenate on transcription and translation, as well as to decipher the possible mechanisms underlying the effects of long-term supranutritional selenate supplementation.

Methods and Materials

Animals and sodium selenate supplementation

Seven-week-old male C57BL/KsJ-*lepr^{db}*/*lepr^{db}* diabetic (*db/db*) mice were purchased from the Model Animal Research Center of Nanjing University (Nanjing, China) and were housed in a standard specific-pathogen free animal feeding room and provided free access to food and water. After habituation, mice were randomly assigned to two groups as follows: the control group (DMCtrl, $n = 8$) and the selenate supplementation group (DMSe, $n = 8$). All the animals were fed a standard mouse chow which met the basic nutritional requirements for mice, while diet made with sodium selenite contained 0.2 mg selenium per kilogram. Sodium selenate was purchased from Sigma-Aldrich (Shanghai, China) and dissolved in sterile water. The mice in the DMSe group were administered 0.8 mg sodium selenate per kilogram body weight (BW) via daily tube feeding, and the control mice were given an equal volume of sterile water. BWs were measured once per week to allow for dose adjustment. Blood was obtained by tail incision and used to determine glucose concentrations using a glucometer (OneTouch Ultra, LifeScan, Milpitas, California) once every 2 weeks after an overnight fast.

After 9 weeks, the mice in all experimental groups were anaesthetized by intraperitoneal injection of sodium pentobarbital (Sigma-Aldrich, Shanghai, China) at a dose of 50 mg/kg BW and subsequently decapitated. Blood samples were collected and pancreatic and liver tissues were immediately removed, weighed, rinsed with cold physiological saline, frozen in liquid nitrogen, and stored at −80 °C. Small pieces of these tissues were fixed in 4% paraformaldehyde for histopathological studies. Liver samples used for hepatic glycogen analysis were fixed in 85% ethanol.

The Institutional Animal Care and Use Committee at Wuhan University approved this study, which was conducted in accordance with the guidelines of the National Institutes of Health (Bethesda, Maryland, USA) for animal care.

Biochemical assays

Insulin levels in plasma and pancreatic tissue homogenates were determined using an Enzyme Linked Immunosorbent Assay (ELISA) kit (Nanjing Jiancheng Bioengineering Institute, Nanjing, China). Analysis of glycosylated hemoglobin (HbA1c) content was performed on 100 μl of fresh total blood using a turbidimetric immunoassay performed with a Hitachi 7080 (Tokyo, Japan) automatic biochemistry analyzer. Plasma triglyceride (TG), total cholesterol (TC), and low-density lipoprotein (HDL) levels were measured using the Hitachi 7080 system with the cognate kits.

Kits purchased from Nanjing Jiancheng Bioengineering Institute were used to measure malondialdehyde (MDA) levels, catalase (CAT) activity, superoxide dismutase (SOD) activity, glutathione peroxidase (GSH-Px) activity, and reduced glutathione (GSH) levels in plasma or liver tissue homogenates. An anthrone-sulfuric acid colorimetric method [22] was used to determine liver glycogen content.

Histopathological analysis

Liver and pancreatic tissues were dehydrated, embedded in paraffin, and cut into 6-μm thick sections. Liver samples were subjected to conventional hematoxylin and eosin (H&E) staining and periodic acid-Schiff (PAS) staining for glycogen, and pancreatic samples were stained using the conventional H&E method. Immunohistochemical analysis of pancreatic islets was performed using an antibody against insulin and proinsulin (ab8403) purchased from Abcam (New Territories, Hong Kong). Specimens were observed using a visible-light microscope DP72 (Olympus, Tokyo, Japan). Pancreatic islet size was measured and calculated using Image-Pro Plus software v. 6.0 (Media Cybernetics, Washington, USA) with more than 100 islets randomly selected from five fields for every pancreatic slice from each experimental group. Specimens from healthy wild-type C57BL/J mice (WT) of the same age and genetic background were used for reference.

Gene expression analysis

Total RNA was isolated from the liver and pancreas using Trizol reagent (Invitrogen, Shanghai, China) and an RNApure High-purity Total RNA Rapid Extraction Kit (spin column) (BioTeke, Beijing, China). RNA integrity and concentration were evaluated using a UVP GDS-7600 (California, USA) imaging instrument and a NanoDrop ND-1000 (Wilmington, USA) spectrophotometer, respectively. Roche NimbleGen Mouse Gene Expression 12×135 K arrays were used for gene expression microarray profiling analysis. All subsequent technical procedures and quality control analyses were performed at CapitalBio Corporation in Beijing, China [23]. Three independent replicates from each of the DMSe and DMCtrl groups were performed.

Differentially expressed genes (DEGs) were identified using the significance analysis of microarrays (SAM) method with SAMR software version 3.02 [24] with a selection threshold of the false discovery rate (q value (%)≤5). Differences in expression levels≥ 1.5 for up-regulation or<0.667 for down-regulation were considered statistically significant. Evaluation of DEGs according to Gene Ontology (GO) terms and Kyoto Encyclopedia of Genes and Genomes (KEGG) pathways were performed using the CapitalBio Molecule Annotation System (MAS) platform (http://bioinfo. capitalbio.com/mas).

Proteomic analysis of liver samples

Liver samples were placed in liquid nitrogen and ground to a very fine powder using a mortar and pestle. The powder (approximately 100 mg per sample) was transferred to sterile tubes containing 1 ml ice-cold lysis buffer (30 mM Tris-HCl [pH 8.5], 7 M urea, 2 M thiourea, 4% CHAPS, and 20 µl/ml protease-inhibitor cocktail [Roche, Mannheim, Germany]) for protein extraction via ultrasonication (400 W for 5 min) in an ice bath. The homogenate was centrifuged at 4,000×g for 1 h at 4 °C, and the supernatant was collected to remove the fat layer and stored at −80 °C. A Micro BCA Protein Assay Kit (Thermo Scientific, Rockford, USA) was used to determine protein concentrations, and samples were adjusted to a final concentration of 5 mg/ml.

Three biological replicates from each of the two groups to be compared (50 µg of each protein sample) were labeled with Cy3 or Cy5 dye (GE Healthcare, Washington, USA). Additionally, a pooled sample was generated using equal amounts of all test samples and was labeled with Cy2 as an internal control. Labeled samples were analyzed using a Bio-Rad two-dimensional differential gel electrophoresis (2D-DIGE) system (Bio-Rad, Hercules, USA). Briefly, labeled samples were loaded onto 24-cm immobilized pH gradient strips (pH 3–10) and incubated at room temperature for 30–60 min. Isoelectric focusing was performed at 17 °C. The focused strips were equilibrated for 15 min in 6 M urea, 20% glycerol, 2% SDS, 375 mM Tris-HCl (pH 8.8), and 1% DTT, and then for an additional 15 min in the same buffer with DTT replaced by 4% iodoacetamide. After equilibration, proteins were separated on a 12% SDS polyacrylamide gel at 20 mA/gel overnight. The 2D gels were scanned using a Typhoon 9410 Scanner (GE Healthcare, Washington, USA), and imaging analysis was performed using DeCyder software version 5.02 (GE Healthcare, Washington, USA). Statistical significance was assessed using the Student t test; P-values≤0.05 were considered significant. Selected protein spots were cut out from the gels using a scalpel and digested with trypsin (Promega, Madison, WI) for protein peptide extraction. A MALDI-TOF 4800 mass spectrometer (Applied Biosystems, Foster City, CA) was used to identify differentially expressed proteins.

Quantitative real-time polymerase chain reaction and western blot analyses

To validate the differential expression of genes and proteins detected using microarray analysis and 2D-DIGE, respectively, quantitative real-time polymerase chain reaction (qRT-PCR) assays and western blot analyses were performed. The cDNA was reverse-transcribed from total RNA using a First Strand cDNA Synthesis Kit (Fermentas, Beijing, China), and qRT-PCR assays were conducted using primers designed with NCBI Primer-BLAST (Table 1), a SYBR Green kit (Takara, Dalian, China), and an ABI-7500 Real-Time PCR system (Applied Biosystems, Foster City, CA). Gene expression levels were calculated according to the $2^{-\Delta\Delta Ct}$ method [25]. Antibodies against PAX6 (ab5790), NEU-ROD1 (ab60704), FBP2 (ab131253), PDX1 (ab47267), SCD1 (ab19862), GLUL (ab64613), and ALDOB (ab75751) were purchased from Abcam (New Territories, Hong Kong), and an antibody against GSTA1/2 (sc-323939) was purchased from Santa Cruz Biotechnology, Inc. (Santa Cruz, CA, USA). An anti-GAPDH antibody purchased from CST China (Shanghai, China) served as an internal reference. Protein samples were separated using SDS-PAGE and transferred to PVDF membranes. The blots were visualized using the SuperSignal West Pico Chemiluminescent Substrate (Thermo Scientific, Shanghai, China) and the intensity of protein bands was determined by densitometry using Gel-Pro Analyzer software v. 4.1 (Media Cybernetics, Washington, USA). The statistical significance of the differences between groups was estimated using the Student t test, with $P<0.05$ considered significant.

Results

Insulin levels and hepatic oxidative stress capacity in diabetic mice treated with selenate

To investigate whether selenate supplementation affects animal health and the onset of diabetes, we monitored changes in food intake, weight gain, and fasting blood glucose levels in all experimental mice during 9 weeks of supplementation. All mice were in good condition, and there was no difference in the changes in BW between groups (Figure 1A). Fasting blood glucose levels increased continuously in DMCtrl mice, but decreased gradually in DMSe mice and had returned to near-normal levels by the end of the supplementation period (Figure 1B). In DMCtrl mice, the glycosylated hemoglobin content at time of sacrifice was markedly higher than during pretreatment, and this was accompanied by an increase in glucose level. In contrast, the level of glycosylated hemoglobin in DMSe mice did not change. The plasma insulin concentration of DMSe mice was nearly twice that of controls, and there was a 30% increase in insulin protein levels in pancreatic tissue homogenate. Compared with the control group, the DMSe mice had increased total plasma cholesterol (132%) and low-density lipoprotein (116%) levels, but decreased levels of plasma TG (51.7%) and MDA (73.3%, $P=0.06$) (Table 2).

The activity of antioxidant enzymes such as GSH-Px, CAT and SOD were measured to estimate the oxidative stress capacity of *db/db* mice after selenate supplementation. There was a slight decrease in hepatic GSH-Px activity, but a small increase in hepatic SOD activity. However, no significant difference in hepatic CAT activity was found between the two groups, and the global levels of GSH-Px activity, SOD activity and GSH content were also unchanged (Table 3).

Increase in pancreatic islet size and accumulation of liver fat in mice receiving selenate supplementation

H&E staining was performed in order to analyze pathological changes in the pancreas and liver histologically, while PAS staining was used to determine hepatic glycogen storage capacity (Figure 2). Compared to WT mice, where the area of the islets was mostly less than 10000 µm^2 (approximately 74.33%), we observed a dramatic increase in pancreatic islet size in DMCtrl mice, especially the percentage between 10000 µm^2 and 20000 µm^2 in area, which increased from 14.86% to 27.12%. A further increase in islet area occurred in DMSe mice after selenate supplementation, with 20.69% of the islets greater than 20000 µm^2 (Figure 2J). Although the islets were irregularly shaped and loosely arranged in diabetic *db/db* mice, there was no significant difference in islet cell density between WT mice and *db/db* mice with or without selenate treatment (Figure 2K). Occasional capillary congestion and islet cell necrosis was observed in both DMCtrl and DMSe mice.

Limited instances of fatty degeneration and vacuolization were observed in the livers of DMCtrl mice; however, selenate treatment exacerbated these lesions, which featured large numbers of swollen hepatocytes, lipid vacuoles, steatosis, and hepatic cord congestion. Moreover, hepatic glycogen storage capacity was reduced in DMSe mice.

In addition, immunohistochemical analysis was performed to measure the synthesis and secretion of insulin. Although islet size was increased in DMCtrl mice compared with WT mice, the relative positive rate of insulin expression decreased by 15.14%. Selenate treatment relieved the effect, resulting in a decline of only 4.22% in DMSe mice compared to WT mice (Figure 3).

Table 1. Nucleotide sequence of primers used for qRT-PCR.

Gene Name	Forward (5′→3′)	Reverse (5′→3′)
Pax6	TTCCCGAATTCTGCAGACCC	TCTTGGCTTACTCCCTCCGA
Neurod1	CCCTACTCCTACCAGTCCCC	GAGGGGTCCGTCAAAGGAAG
Fbp2	CGCACCTTGGTCTATGGAGG	CCTCCTGCTTGCTCGATGAT
Pdx1	CCTTTCCCGAATGGAACCGA	TTCCGCTGTGTAAGCACCTC
Scd1	CACCTGCCTCTTCGGGATTT	TCTGAGAACTTGTGGTGGGC
Pltp	CGCAAAGGGCCACTTTTACTAC	GCCCCCATCATATAAGAACCAGT
Aldob	CCGCTTGCAGGAACAAACAA	ACGCCACTTCCCAAAGTCAA
Glul	TTATGGGAACAGACGGCCAC	TAACCTCCGCATTTGTCCCC
Gsta1	AGCCCGTGCTTCACTACTTC	CAATCTCCACCATGGGCACT
InsR	TCAAGACCAGACCCGAAGATTT	TCTCGAAGATAACCAGGGCATAG
Gapdh	CATGTTCCAGTATGACTCCACTC	GGCCTCACCCCATTTGATGT

Analysis of expression levels of genes associated with pancreatic β-cell function

Among the differentially expressed genes in the pancreas, 81 were upregulated and 34 were downregulated in DMSe mice (Table S1). A wide range of genes were closely associated with either key pancreatic β-cell functions or the development of diabetes (Figure 4A). Of these, there were increases in the levels of mRNAs encoding transcription factors that regulate insulin synthesis and secretion, such as paired box gene 6 (*Pax6*), neurogenic differentiation 1 (*Neurod1*), and activin A receptor type 1c (*Acvr1c*). We detected elevated levels of mRNAs encoding components of the insulin signal transduction pathway, such as insulin (*Ins1* and *Ins2*), ribosomal protein S6 kinase polypeptide 6 (*Rps6ka6*), and doublecortin-like kinase 2 (*Dclk2*). In contrast, the level of *c-Jun* expression decreased. The transcription of genes encoding proteins involved in glucolipid metabolism, including glucose-6-phosphatase catalytic subunit 2 (*G6pc2*), transforming growth factor beta 1 induced transcript 1 (*Tgfb1i1*), solute carrier family 2 member 5 (*Slc2a5*), ATP-binding cassette sub-family G

member 8 (*Abcg8*), and ATP-binding cassette sub-family C member 8 (*Abcc8*), was induced. However, the levels of fructose bisphosphatase 2 (*Fbp2*), elongation of very long chain fatty acids (*Elovl2*), and acyl-CoA thioesterase 1 (*Acot1*) were suppressed.

Along with the increase in pancreatic islet size, as determined by histological analysis in *db/db* mice treated with selenate (DMSe), the expression of cell-cycle components was also elevated. Transcription of genes encoding proteins known to stimulate cell proliferation, such as *Spc25*, *Cdc27* and *Rprm*, increased while the transcription of genes that encode proteins that inhibit cell cycle progression, such as *Jun*, *Junb*, and *Fosl2*, was decreased. The expression levels of apoptosis-associated genes, including growth arrest and DNA-damage-inducible 45 beta (*Gadd45b*), nuclear receptor (*Nr4a1*), and B cell translocation gene 2, anti-proliferative (*Btg2*), decreased. Moreover, transcription levels of genes encoding proteins that mediate selenide metabolism, such as selenophosphate synthetase 1 (*Sephs1*) and selenocysteine lyase (*Scly*), also increased, whereas the expression of genes encoding major selenoproteins, such as glutathione peroxidase 1 (*GPx1*), seleno-

A.

B.

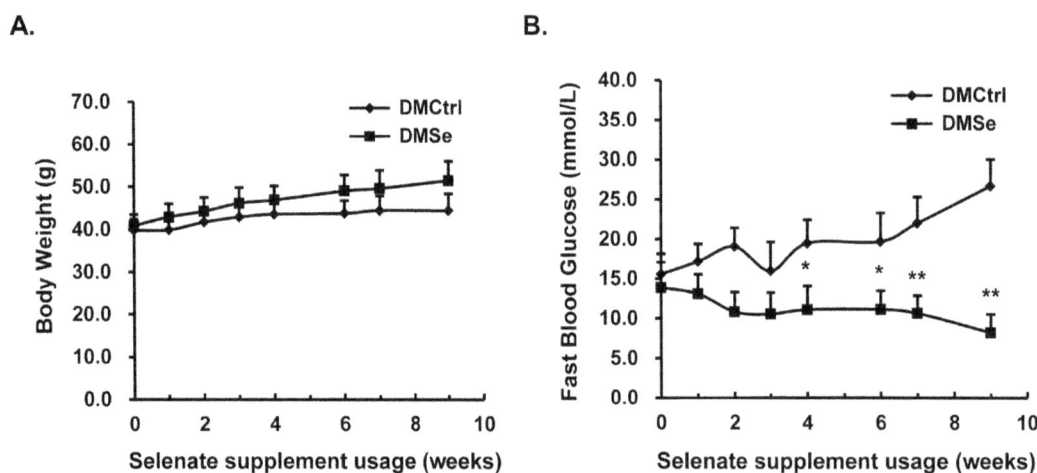

Figure 1. Phenotypic changes during selenate supplementation. (A) There were no differences in the changes in body weight between DMCtrl and DMSe mice. (B) Fasting blood glucose levels increased continuously in placebo-supplemented mice (DMCtrl), but decreased gradually in selenate-supplemented mice (DMSe) and returned to normal when supplementation ceased. Error bars were calculated for eight animals in the DMCtrl group and eight animals in the DMSe group. The asterisk (*) represents values found to be significantly different from the reference group (DMCtrl) using the Student *t* test (*$P<0.005$, **$P<0.0001$).

Table 2. Biochemical parameters (Mean±SD) #of the *db/db* mice with or without dietary selenate supplementation.

Parameter	DMCtrl	DMSe	P Value
Blood			
Pretreatment: HbA1c [%(mmol/mol)]	4.2±0.3(21±2)	4.0±0.5(20±6)	
At sacrifice: HbA1c [%(mmol/mol)]	5.9±0.6(41±6)	4.4±0.5(25±5)**	0.00516 (0.00331)
Plasma			
glucose, mmol/L	29.26±6.52	15.99±5.21***	0.00097
Insulin, pmol/L	11.78±1.67	21.44±8.16*	0.01640
HOMA-IR index	16.56±3.12	13.56±1.91NS	0.09395
triglycerides, mmol/L	1.18±0.39	0.61±0.22*	0.04155
cholesterol, mmol/L	2.74±0.22	3.61±0.60*	0.04061
low-density lipoprotein, mmol/L	1.91±0.08	2.21±0.23*	0.01206
malondialdehyde, mmol/L	3.71±0.73	2.72±0.73NS	0.06013
Pancreatic tissue homogenate			
Insulin, pmol/mg protein	282.20±57.24	366.25±58.48*	0.01668

#Mean±SD were calculated for five animals in the DMCtrl group and eight animals in the DMSe group.
*Values significantly different from DMCtrl mice, by the Student t test (*P<0.05, **P<0.01, *** P<0.001).
NSindicates the values were not significantly different from DMCtrl mice by the Student t test (P>0.05).

protein P (*Sepp1*), and 15 kD selenoprotein (*15-Sep*), did not change significantly. These findings indicate that supplementation with dietary selenate did not affect the overall expression of selenoproteins in the transcriptome.

Differential expression of the genes of interest (*Pax6*, *Neurod1*, *Fbp2*, and *Pdx1*) and their encoded proteins (PAX6, NEUROD1, FBP2 and PDX1) was validated using qRT-PCR (Figure 4B) and western blot analyses (Figure 4C, 4D), and the results showed coordinated changes in mRNA and protein levels.

Analysis of the expression of genes involved in lipogenesis and inflammation

DNA microarray analysis of liver samples showed that the expression levels of 582 transcripts had changed by a factor of≥1.5 in response to selenate supplementation (Table S2). The levels of 301 genes were elevated while 281 genes were decreased (Figure 5A). Functional annotation and pathway analysis showed that a large proportion of these differentially expressed genes were

involved in regulating lipid metabolism, including genes involved in fatty acid synthesis through arachidonic acid metabolism and the PPARα signaling pathway. The other differentially expressed genes were mainly associated with detoxification, inflammatory response, and cell cycle regulation.

We found that genes encoding key enzymes in fatty acid synthesis, such as prostaglandin D2 synthase (*Ptgds*), stearoyl-coenzyme A desaturase 1 (*Scd1*), acetyl-coenzyme A carboxylase beta (*Acacb*), NADP-dependent malic enzyme 1 (*Me1*), and elongation of very long chain fatty acids (*Elovl5*), were upregulated in the DMSe mice. Conversely, the gene encoding acyl-CoA synthetase medium-chain family member 5 (*Acsm5*) was downregulated in these animals. Carnitine palmitoyltransferase 1b (*Cpt1b*), hydroxyacid oxidase 2 (*Hao2*), and adiponectin receptor 2 (*Adipor2*), are involved in fatty acid oxidation and were induced by selenate supplementation. The transcription of phospholipid transfer protein (*Pltp*) and 3-hydroxy-3-methylglutaryl-Coenzyme A synthase 1 (*Hmgcs1*), which function in ketogenesis, was

Table 3. Oxidative stress parameters (Mean±SD) #of the *db/db* mice with or without dietary selenate supplementation.

Parameter	DMCtrl	DMSe	P Value
Plasma			
GSH, μmol/L	31.42±8.67	33.85±9.93	NS
GSH-Px activity, U/ml	85.54±11.66	79.81±13.06	NS
SOD activity, U/ml	124.38±41.67	133.76±31.54	NS
Hepatic			
glycogen, mg/g tissue	31.79±8.10	26.26±7.91	NS
CAT activity, U/mg protein	10.93±2.54	12.515±2.10	NS
SOD activity, U/mg protein	38.37±10.74	55.35±10.99**	0.00712
GSH-Px activity, U/mg protein	1183.54±98.74	1037.93±117.18*	0.02508

#Mean±SD were calculated for eight animals in the DMCtrl group and eight animals in the DMSe group.
*Values significantly different from DMCtrl mice, by the Student t test (*P<0.05, **P<0.01).
NS means no significance (P>0.05) between the DMCtrl group and DMSe group.

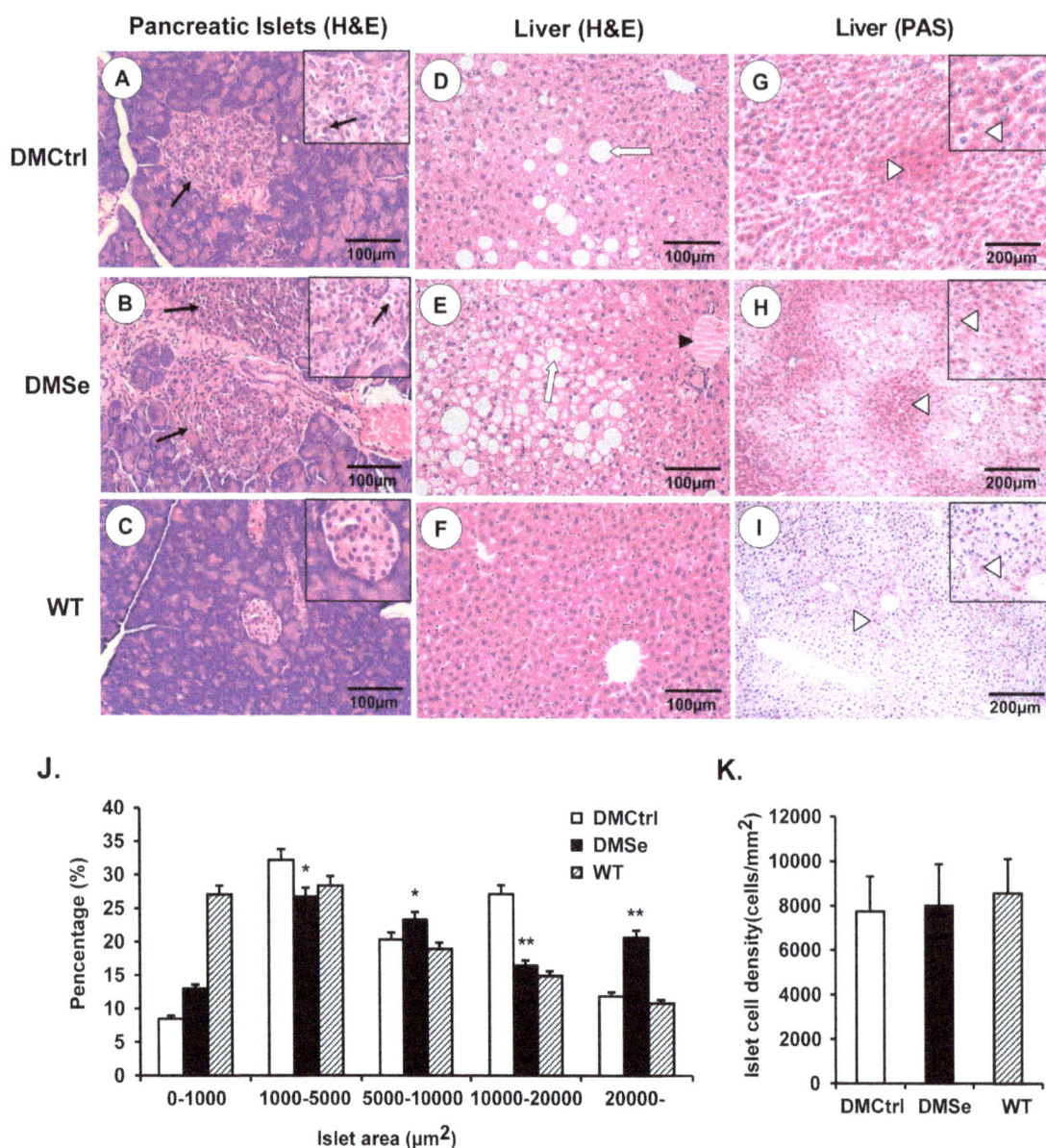

Figure 2. Histological analysis of pathological changes in the pancreas and liver. Compared with the results from the DMCtrl group (A), the pancreatic islets of selenate-treated DMSe mice (B) increased in size. Occasional capillary congestion and islet cell necrosis (black arrows) was observed. (C) Pancreatic islets of WT mice were shaped regularly and arranged evenly. (D) Modest fatty liver degeneration and vacuolization was observed in the DMCtrl control group, but (E) this was exacerbated after selenate treatment and was accompanied by a large number of swollen hepatocytes and lipid vacuoles (white arrows). Steatosis and hepatic cord congestion (black triangles) were also present. (F) There were no abnormal lesions in the liver tissue of WT mice. (H) PAS staining for hepatic glycogen (purple-reddish granules, white triangles) revealed a reduced capacity for glycogen storage, along with increased fatty liver damage in DMSe mice compared with (G) that in DMCtrl mice. (I) There were occasional purple-reddish glycogen granules in the liver tissue of WT mice. Changes in islet area (J) and islet cell density (K) were measured and calculated using Image-Pro Plus software v. 6.0 (Media Cybernetics, Washington, USA) from more than 100 randomly selected islets in five fields from each pancreatic slice from each experimental group. The asterisk (*) represents significant differences (*$P<0.05$, **$P<0.01$) between the DMSe group and DMCtrl group.

increased in response to selenate treatment. Moreover, the transcription of genes ncoding multiple cytochrome P450 isoforms, including *Cyp2a4*, *Cyp2b10*, *Cyp2c44*, *Cyp4a12b*, and *Cyp7b1*, was significantly downregulated in response to selenate treatment. The expression of other genes, such as hydroxysteroid 11-beta dehydrogenase 1 (*Hsd11b1*), lysosomal acid lipase A (*Lipa*), and phospholipase (*Plcg1*), was also downregulated. Most of the differentially expressed genes listed above encode components of the PPAR-α signaling pathway that function to induce the synthesis of fatty acids and ketone bodies in the liver.

Genes that encode proteins involved in detoxification or the response to oxidative stress, including serine (or cysteine) peptidase inhibitor subfamily members (*SerpinA1a*, *SerpinA1b*, *SerpinA1d*, *SerpinA3f*, *SerpinA3k* and *SerpinA3m*), were expressed at lower levels in the DMSe group than that in the DMCtrl group. Expression of the glutathione *S*-transferase (GST) family genes *Gsta1* and *Gsta2* was downregulated in the DMSe group, while *Gstm3* expression was upregulated.

Consistent with our histological data indicating exacerbated inflammation and necrosis in DMSe mice, our expression profiling

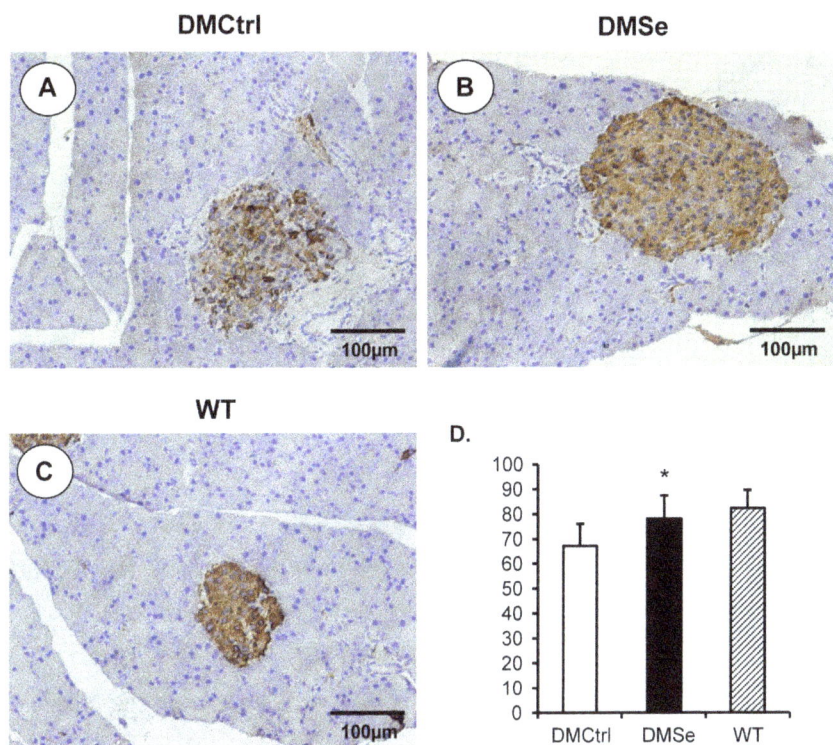

Figure 3. Immunohistochemical analysis to measure the production of insulin. Positive cells displayed brownish yellow granules in the cytoplasm. (A) Although islet size was increased in DMCtrl mice, the relative positive rate of insulin expression was decreased as compared with WT mice (C). (B) Selenate treatment relieved the decline and maintained a similar levels of insulin expression in both DMSe mice and WT mice. (D) The insulin positive ratios of cells were measured and calculated using Image-Pro Plus software v. 6.0 (Media Cybernetics, Washington, USA) from more than 100 islets randomly selected in five fields for every pancreatic slice of each experimental group. The asterisk (*) represents significant differences (*$P<0.01$) between the DMSe group and DMCtrl group.

data indicated the activation of inflammatory stress pathways. In the DMSe group, we detected increased levels of expression of genes encoding chemokines involved in the inflammatory response (*Flt1* and *Ccl19*), whereas expression of *Cxcl1* and *Cx3cl1* decreased. Similarly, the expression levels of certain apoptosis-related genes were significantly altered by selenate treatment. There was increased expression of genes encoding the apoptosis-inducing factors *Aifm1* and *Aifm3*, cell death-inducing DNA fragmentation factor alpha subunit-like effector A and C (*Cidea, Cidec*), death associated protein-like 1 (*Dapl1*), MAP-kinase activating death domain (*Madd*), DNA damage-inducible transcript 4 (*Ddit4*), and gap junction protein beta 6 (*Gjb6*).

All of the data as well as detailed information from the microarray analysis were deposited in the NCBI-GEO database, with the accession number GSE55636.

Proteomic analysis of liver using 2D-DIGE and MALDI-TOF/MS-MS identified five proteins whose expression were significantly changed in DMSe mice compared with DMCtrl mice, although the differences were modest (Table 4). There was no direct correlation between the results of the cDNA microarray analysis and the proteomic analysis, with the exception of aldolase B fructose-bisphosphate (ALDOB). Analysis of these proteins using the online biological software DAVID (http://david.abcc.ncifcrf.gov/) revealed that they represented two functions: (i) proteins involved in glucolipid metabolism, such as aldolase B (ALDOB), glutamate-ammonia ligase (GLUL), and isocitrate dehydrogenase 1 (NADP+) soluble (IDH1); and (ii) proteins involved in the endoplasmic reticulum assembly pathway, such as protein disulfide isomerase associated 3 (PDIA3) and Calreticulin (CALR).

The differential expression of genes such as *Scd1, Pltp, Aldob, Glul, Gsta1* and *InsR* was validated by qRT-PCR. The results of cDNA microarray and qRT-PCR analyses were essentially consistent (Figure 5B). The results of western blot analysis for representative proteins such as GLUL, ALDOB, SCD1 and GSTA1/2 showed changes in protein levels between the two treatment groups (Figure 5C, D), and these differences were similar to those found by cDNA microarray analysis and 2D-DIGE coupled with mass spectrometry analysis.

Discussion

In vivo and *in vitro* studies have demonstrated the insulin-mimetic properties of selenium and have indicated that appropriate dietary supplementation with selenium can prevent diabetes. However, different forms of selenium have differing effects on insulin-regulated carbohydrate metabolism, indicating various functional mechanisms of selenium compounds. For example, selenate may suppress the increase in fasting plasma glucose concentrations in diabetes models by increasing insulin sensitivity and by acting as an insulin-mimetic in liver and adipose tissue [11,20,26], while selenite has been found to stimulate insulin production and secretion from islets and further enhance carbohydrate efficiency via high insulin levels [27]. It has even been reported that selenite may counteract insulin-induced signaling [28]. In the present study, after oral selenate administration for 9 weeks the increase in fasting blood glucose in diabetic *db/db* mice was suppressed, while plasma insulin concentrations increased significantly. The pancreatic islets were enlarged and the expression of insulin protein in the

A.

Gene Symbol	DMSe	DMCtrl	Relative Signal Strength (DMCtrl)	Fold Change (DMSe/DMCtrl)
†Acot1			18970	0.536
†Elovl2			850	0.532
†Gadd45b			9563	0.567
†Nr4a3			1505	0.040
†Fos			7452	0.162
†Cyr61			3449	0.145
†Junb			6886	0.279
†Fosl2			1511	0.517
Arl4d			3349	0.304
Egr2			1143	0.099
Egr4			1087	0.099
†Fosb			938	0.051
†Jun			3364	0.527
†Nr4a1			3487	0.264
†Fbp2			2120	0.669
†Tbc1d1			849	0.375
†Abcg8			189	3.432
Amy2b			3293	3.428
†Defb1			144	4.851
†Aldh1a3			445	2.589
†G6pc2			3113	2.747
†Bace2			581	2.361
†Ins2			2637	2.749
†Ucn3			319	3.664
†Acvr1c			471	1.521
Ank1			437	2.366
Cxx1a			308	2.359
Eme2			1103	1.534
†Slc2a5			666	1.902
†Rps6ka6			395	1.812
Sephs1			340	1.894
Ece2			246	2.103
†Ins1			11810	2.610
†Isl1			383	2.123
†Tbc1d7			965	1.632
Hes5			406	1.995
†Tpcn1			600	1.470
†Neurod1			701	1.825
†Pax6			211	2.799
Tgfb1i1			221	2.170
†Abcc8			581	1.985
Ifi27l2b			467	1.865
Pappa2			666	2.876
†Tmem27			2873	1.796
†Mgat4a			407	1.522
†Pde5a			239	2.171
†Scg3			7825	1.515
†Scgn			654	2.884
†Vil1			180	3.439
†Scly			437	1.432

B.

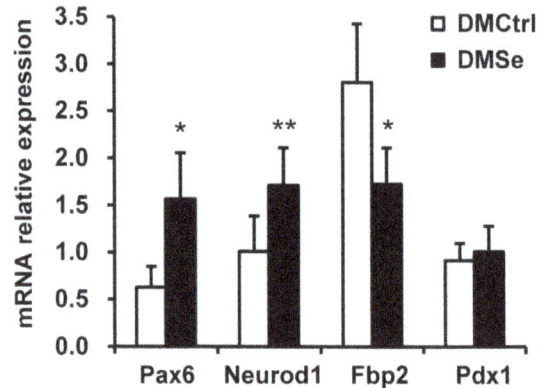

C.

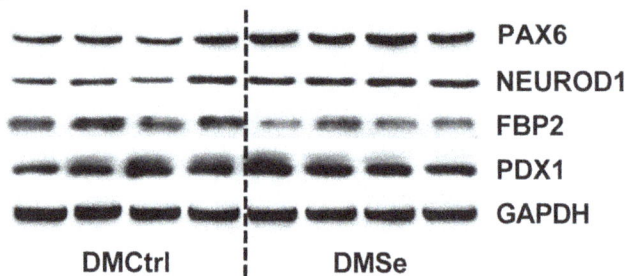

D.

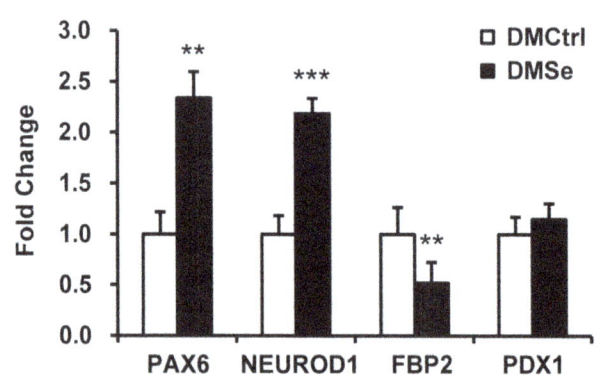

Figure 4. Pancreatic-specific genes and proteins differentially expressed in *db/db* mice. (A) Heat map showing differences in the mRNA expression levels (green, lower; red, higher between DMCtrl and DMSe groups). The (†) indicates genes that are closely linked to key pancreatic β-cell functions or the development of diabetes. The relative signal strength represents the mean gene expression value in the DMCtrl group derived from Robust Multi-Array analysis. (B) qRT-PCR analysis of *Pax6*, *Neurod1*, *Fbp2*, and *Pdx1* expression was used to validate the cDNA microarray data from the pancreas. There was strong consistency between the two assays. Three replicates were performed for each mouse (eight mice per group). *Gapdh* was used as an internal reference gene. The results are presented as the relative changes in mRNA expression levels in DMSe vs. DMCtrl mice. (C) Representative western blots and (D) quantitation of the levels of PAX6, NEUROD1, FBP2 and PDX1. GAPDH was used as an internal reference. Protein band intensity was quantified using densitometry using Gel-Pro Analyzer software v. 4.1 (Media Cybernetics, Washington, USA), and the results are presented as the ratio of densitometric values from DMSe to DMCtrl. The asterisk (*) represents values shown to be significantly different from DMCtrl by the Student *t* test (*$P<0.05$, **$P<0.01$, ***$P<0.001$).

pancreas was also upregulated. Additionally, the insulin transcription factors Pax6 and Neurod1, which are important for the regulation of insulin synthesis and secretion [29,30], were also upregulated at the mRNA and protein expression levels. Thus, dietary selenate supplementation has a similar effect on enhancing islet function. Although a previous report [27] indicated that selenate stimulates the *Ipf1* (insulin promoter factor 1) gene promoter to a lesser extent than selenite, the authors concluded

that this was due to the presence of specific selenate reductase activity in the pancreatic extract. This blocked the reduction of selenate to selenite, which is readily reduced to selenide and subsequently assimilated into selenoproteins. Another study found that free SeIV compounds, the cellular metabolic intermediates of selenate, act as inhibitors of protein tyrosine phosphatases (PTPs), which is why increasing selenate concentration directly *in vitro* effected no inhibition of PTPs [20]. Taking this into consideration,

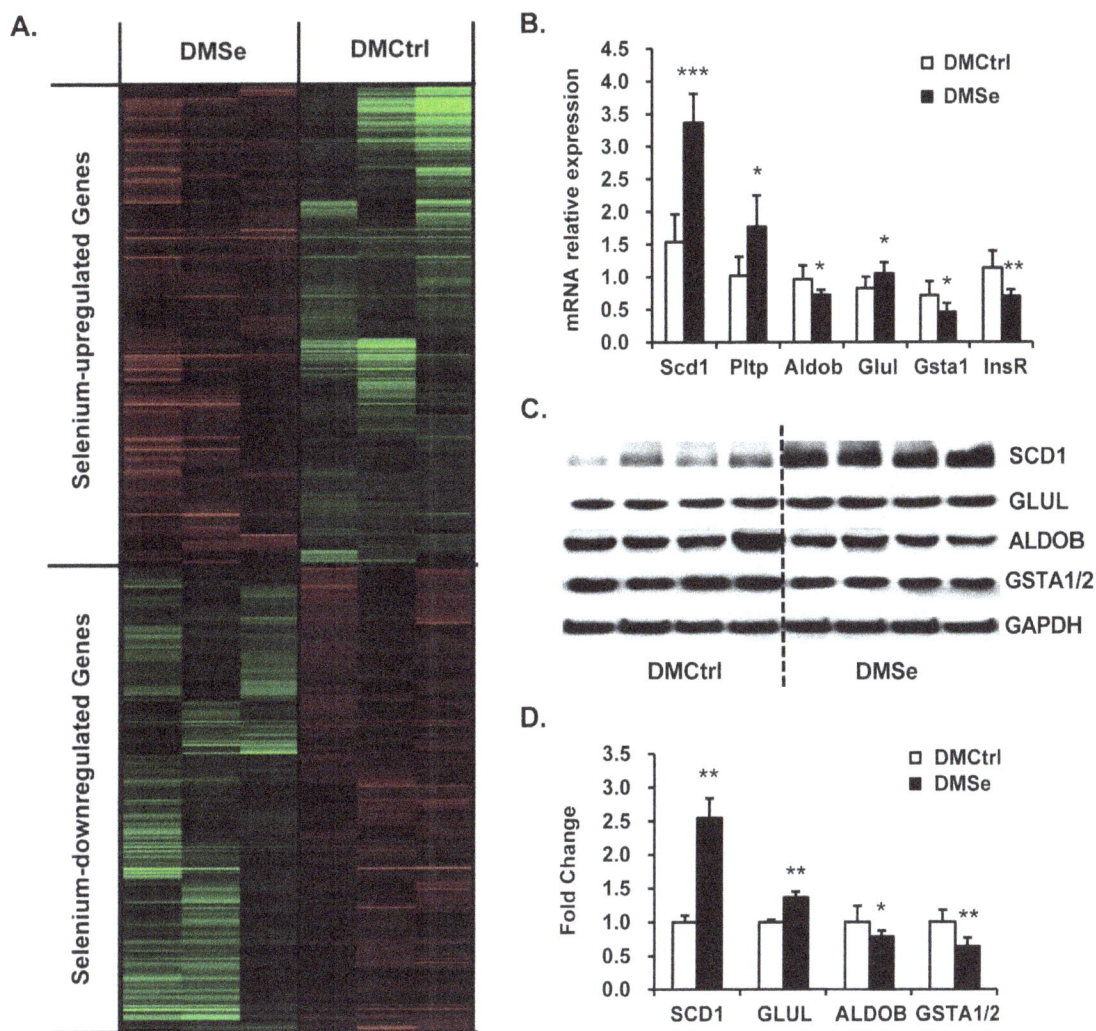

Figure 5. Hepatic genes and proteins differentially expressed between DMCtrl and DMSe groups. (A) The heat map depicts differences in mRNA expression levels according to brightness (green, downregulated; red, upregulated compared with DMCtrl mice). (B) qRT-PCR analysis of *Scd1*, *Pltp*, *Aldob*, *Glul*, *Gsta1* and *InsR*, which were selected to validate the cDNA microarray data from the liver. The data indicate excellent consistency between the two techniques. Three replicates were performed for each mouse (eight mice per group). *Gapdh* was used as an internal reference gene. The results are presented as the relative changes in mRNA expression levels in DMSe vs. DMCtrl mice. (C) Representative western blots and (D) quantitation of selected proteins. There was agreement in the expression levels of GLUL and ALDOB as detected using 2D-DIGE and western blot analyses. SCD1 and GSTA1/2 displayed the same differences in expression level seen in the microarray data. GAPDH was used as an internal reference. The intensity of protein bands was quantified by densitometry using Gel-Pro Analyzer software v. 4.1 (Media Cybernetics, Washington, USA), and the results are presented as the ratio of densitometric values from DMSe to DMCtrl. The asterisk (*) represents values shown to be significantly different from DMCtrl by the Student t test (*$P<0.05$, **$P<0.01$, ***$P<0.001$).

Table 4. Liver proteins differentially expressed in *db/db* mice with or without dietary selenate supplementation.

Protein name	Accession number[#]	Symbols	Fold change DMSe vs. DMCtrl	P. Value
Protein disulfide-isomerase A3 precursor	gi|112293264	Pdia3	1.36	0.0410
Glutamate-ammonia ligase (glutamine synthetase)	gi|31982332	Glul	1.27	0.0035
Calreticulin precursor	gi|6680836	Calr	1.25	0.0400
Isocitrate dehydrogenase 1 (NADP⁺), soluble	gi|57242927	Idh1	1.22	0.0330
Aldolase B, fructose-bisphosphate	gi|21707669	Aldob	−1.35	0.0490

[#]NCBI database accession number.

we propose that the mechanism by which selenate stimulates insulin production may also be associated with a metabolic intermediate product of selenate. Unfortunately, our results failed to prove this hypothesis, which remains to be confirmed by future research.

Unexpectedly, hepatopathy observed in the DMSe mice was more severe than that in the DMCtrl mice and was accompanied by increased hepatocyte lipid vacuolation and hepatic cord congestion. Levels of plasma cholesterol and low-density lipoprotein were increased while triglyceride levels decreased, and cDNA microarray analysis detected upregulation of lipogenic genes, which may account for this fat accumulation [31]. Of particular interest are genes involved in fatty acid oxidation, which could cause an increase in reactive oxygen species (ROS) and oxidative stress [32]. The HOMA-IR index, calculated based on the levels of fasting plasma glucose and fasting plasma insulin, declined in DMSe mice compared with DMCtrl mice, although the effect was not significant. However, there was no difference in expression of genes or proteins involved in the insulin signaling cascade, including PTP1B, which can be suppressed by selenate metabolic intermediates [20].

We observed that the expression of a group of genes that encode proteins with oxidoreductase activity in the liver was suppressed. These included cytochrome P450 family genes related to arachidonic acid metabolism [33], which generate metabolites such as eicosatetraenoic acids that activate the MAPK and PI3K/ AKT signaling pathways [34], enhance insulin sensitivity, and inhibit hepatic inflammation [35] as well as apoptosis [36]. Further, several enzymes of the GST family known to be involved in ROS detoxification [37] were down-regulated, indicating a possible reduction in antioxidant defense capability. Moreover, there was no difference in the expression of any selenoproteins between the DMSe and DMCtrl groups, although hepatic GSH-Px activity decreased. Thus, we conclude that oral selenate administration did not increase insulin sensitivity, but instead reduced antioxidant defense capacity and exacerbated fatty liver degeneration in db/db mice, in contrast to what has been reported in previous studies [26,38].

The proteomic data presented here provide novel insight on the effects of selenate supplementation in diabetic db/db mice. Aldolase B (ALDOB) is an important enzyme in glycolysis and gluconeogenesis that catalyzes the dissimilation of fructose 1, 6-bisphosphate (FBP) or fructose 1-phosphate (F1P) [39]. Mutations leading to defects in aldolase B result in a condition known as hereditary fructose intolerance [40], where the lack of functional aldolase B leads to the accumulation of F1P in bodily tissues. This accumulation damages tissues and traps phosphate in an unusable form that does not return to the general phosphate pool, eventually depleting phosphate and ATP stores [41]. The lack of readily available phosphate terminates glycogenolysis in the liver, which causes hypoglycemia [41]. The accumulation of F1P also inhibits glycogenolysis and further reduces blood glucose levels [41]. ALDOB was downregulated in DMSe mice in this study, indicating decreased functional aldolase B levels that may account for the reduction in blood glucose levels.

Our liver proteome analysis also found that, hepatic glutamine ligase (GLUL) and NADP+-dependent isocitrate dehydrogenase 1 (IDH1) were upregulated in DMSe mice. GLUL (glutamine synthetase, GS) catalyzes the ATP-dependent reaction of glutamate with ammonia that yields glutamine, and also plays an important role in the regulation of nitrogen metabolism [42,43]. Recent studies have found that selenium affects nitrogen and glutamate metabolism via regulation of GS expression in lettuce

plants [43], although the specific mechanism involved is unknown. Isocitrate dehydrogenases catalyze the oxidative decarboxylation of isocitrate to 2-oxoglutarate, generating NADPH. IDH1 plays a vital role in the regulation of lipogenesis and metabolism, and the phenotypes of transgenic mice that overexpress IDH1 include fatty liver, hyperlipidemia, and obesity [44,45]. Protein disulfide isomerase associated 3 (PDIA3) and Calreticulin (CALR) were also upregulated after selenate treatment in our study. PDIA3, also known as ERp57, is a thiol oxidoreductase with protein disulfide isomerase activity and is a rate-limiting enzyme for protein folding in the endoplasmic reticulum [46]. Calreticulin, an endoplasmic reticulum chaperone and calcium regulator, plays an important role in regulating collagen expression and fibrosis [47]. PDIA3 and CALR act together in the major histocompatibility complex class I assembly pathway [46,48]. The increased expression of these two endoplasmic reticulum proteins in DMSe mice indicate increased endoplasmic reticulum stress and the progression of hepatic steatosis to fibrosis.

Based on the above-mentioned phenotypes, it is possible that the doses applied in the present study resulted in chronic selenium toxicity. A standard mouse chow diet made with sodium selenite contains 0.2 mg Se/kg, and our supplementation dose was 0.8 mg sodium selenate/kg body weight, which would correspond to approximately 0.336 mg Se/kg body weight. We assume the consumption of diet per day is one-tenth of the body weight and that a 50 g mouse would consume 5 g diet per day. Thus, the supplemented amount is approximately 16.8-fold the recommended dietary amount (0.2 mg Se/kg diet). It has been previously reported that [20] selenium concentrations up to 20-fold over the recommended amount (4 mgSe/kg diet) do not affect animal health or mortality after 2 years (survival rate > 90%). However, after careful consideration of the phenotypes mentioned in this manuscript, we concluded that chronic selenium toxicity was a distinct possibility, especially in liver tissue, which is the most sensitive organ to selenium toxicity.

In conclusion, our findings suggest that nutritional supplementation of db/db mice with selenate does not improve or prevent the symptoms of type 2 diabetes, although hyperglycemia was decreased as a result of increasing insulin production and secretion. Moreover, long-term hyperinsulinemia eventually led to reduced antioxidant defense capacity and exacerbated fatty liver degeneration.

Supporting Information

Table S1 Expression information of all the differential genes with a change factor of ≥ 1.5 in pancreas.

Table S2 Expression information of all the differential genes with a change factor of ≥ 1.5 in liver.

Acknowledgments

The authors thank Dr. Yanxin Hu of the College of Biological Science, China Agricultural University and Dr. Bintao Qiu of the Beijing Proteome Research Center for their skillful technical assistance.

Author Contributions

Conceived and designed the experiments: KL CW. Performed the experiments: CW NZ. Analyzed the data: CW SY HR. Contributed reagents/materials/analysis tools: KL YW YM. Contributed to the writing of the manuscript: CW KL.

References

1. Oldfield JE (1987) The two faces of selenium. J Nutr 117: 2002–2008.
2. Rayman MP (2012) Selenium and human health. Lancet 379: 1256–1268.
3. Fairweather-Tait SJ, Bao Y, Broadley MR, Collings R, Ford D, et al. (2011) Selenium in human health and disease. Antioxid Redox Signal 14: 1337–1383.
4. Rayman MP (2008) Food-chain selenium and human health: emphasis on intake. Br J Nutr 100: 254–268.
5. Rayman MP, Stranges S (2013) Epidemiology of selenium and type 2 diabetes: can we make sense of it? Free Radic Biol Med 65: 1557–1564.
6. Papp LV, Lu J, Holmgren A, Khanna KK (2007) From selenium to selenoproteins: synthesis, identity, and their role in human health. Antioxid Redox Signal 9: 775–806.
7. Burk RF (2002) Selenium, an antioxidant nutrient. Nutr Clin Care 5: 75–79.
8. Rocourt CR, Cheng WH (2013) Selenium supranutrition: are the potential benefits of chemoprevention outweighed by the promotion of diabetes and insulin resistance? Nutrients 5: 1349–1365.
9. Stapleton SR (2000) Selenium: an insulin-mimetic. Cell Mol Life Sci 57: 1874–1879.
10. Becker DJ, Reul B, Ozcelikay AT, Buchet JP, Henquin JC, et al. (1996) Oral selenate improves glucose homeostasis and partly reverses abnormal expression of liver glycolytic and gluconeogenic enzymes in diabetic rats. Diabetologia 39: 3–11.
11. Iizuka Y, Ueda Y, Yagi Y, Sakurai E (2010) Significant improvement of insulin resistance of GK rats by treatment with sodium selenate. Biol Trace Elem Res 138: 265–271.
12. Bleys J, Navas-Acien A, Guallar E (2007) Serum selenium and diabetes in U.S. adults. Diabetes Care 30: 829–834.
13. Laclaustra M, Navas-Acien A, Stranges S, Ordovas JM, Guallar E (2009) Serum selenium concentrations and diabetes in U.S. adults: National Health and Nutrition Examination Survey (NHANES) 2003–2004. Environ Health Perspect 117: 1409–1413.
14. Stranges S, Sieri S, Vinceti M, Grioni S, Guallar E, et al. (2010) A prospective study of dietary selenium intake and risk of type 2 diabetes. BMC Public Health 10: 564.
15. Labunskyy VM, Lee BC, Handy DE, Loscalzo J, Hatfield DL, et al. (2011) Both maximal expression of selenoproteins and selenoprotein deficiency can promote development of type 2 diabetes-like phenotype in mice. Antioxid Redox Signal 14: 2327–2336.
16. Zeng MS, Li X, Liu Y, Zhao H, Zhou JC, et al. (2012) A high-selenium diet induces insulin resistance in gestating rats and their offspring. Free Radic Biol Med 52: 1335–1342.
17. Pinto A, Juniper DT, Sanil M, Morgan L, Clark L, et al. (2012) Supranutritional selenium induces alterations in molecular targets related to energy metabolism in skeletal muscle and visceral adipose tissue of pigs. J Inorg Biochem 114: 47–54.
18. Faghihi T, Radfar M, Barmal M, Amini P, Qorbani M, et al. (2013) A Randomized, Placebo-Controlled Trial of Selenium Supplementation in Patients With Type 2 Diabetes: Effects on Glucose Homeostasis, Oxidative Stress, and Lipid Profile. Am J Ther.
19. Mueller AS, Pallauf J, Rafael J (2003) The chemical form of selenium affects insulinomimetic properties of the trace element: investigations in type II diabetic dbdb mice. J Nutr Biochem 14: 637–647.
20. Muller AS, Most E, Pallauf J (2005) Effects of a supranutritional dose of selenate compared with selenite on insulin sensitivity in type II diabetic dbdb mice. J Anim Physiol Anim Nutr (Berl) 89: 94–104.
21. Mueller AS, Mueller K, Wolf NM, Pallauf J (2009) Selenium and diabetes: an enigma? Free Radic Res 43: 1029–1059.
22. Leyva A, Quintana A, Sanchez M, Rodriguez EN, Cremata J, et al. (2008) Rapid and sensitive anthrone-sulfuric acid assay in microplate format to quantify carbohydrate in biopharmaceutical products: method development and validation. Biologicals 36: 134–141.
23. Li R, Sun Q, Jia Y, Cong R, Ni Y, et al. (2012) Coordinated miRNA/mRNA expression profiles for understanding breed-specific metabolic characters of liver between Erhualian and large white pigs. PLoS One 7: e38716.
24. Storey JD, Tibshirani R (2003) Statistical significance for genomewide studies. Proc Natl Acad Sci U S A 100: 9440–9445.
25. Livak KJ, Schmittgen TD (2001) Analysis of relative gene expression data using real-time quantitative PCR and the 2(-Delta Delta C(T)) Method. Methods 25: 402–408.
26. Mueller AS, Pallauf J (2006) Compendium of the antidiabetic effects of supranutritional selenate doses. In vivo and in vitro investigations with type II diabetic db/db mice. J Nutr Biochem 17: 548–560.
27. Campbell SC, Aldibbiat A, Marriott CE, Landy C, Ali T, et al. (2008) Selenium stimulates pancreatic beta-cell gene expression and enhances islet function. FEBS Lett 582: 2333–2337.
28. Pinto A, Speckmann B, Heisler M, Sies H, Steinbrenner H (2011) Delaying of insulin signal transduction in skeletal muscle cells by selenium compounds. J Inorg Biochem 105: 812–820.
29. Gosmain Y, Katz LS, Masson MH, Cheyssac C, Poisson C, et al. (2012) Pax6 is crucial for beta-cell function, insulin biosynthesis, and glucose-induced insulin secretion. Mol Endocrinol 26: 696–709.
30. Fu D, Gilbert ER, Liu D (2013) Regulation of insulin synthesis and secretion and pancreatic Beta-cell dysfunction in diabetes. Curr Diabetes Rev 9: 25–53.
31. Perfield JW 2nd, Ortinau LC, Pickering RT, Ruebel ML, Meers GM, et al. (2013) Altered hepatic lipid metabolism contributes to nonalcoholic fatty liver disease in leptin-deficient Ob/Ob mice. J Obes 2013: 296537.
32. Sanyal AJ, Campbell-Sargent C, Mirshahi F, Rizzo WB, Contos MJ, et al. (2001) Nonalcoholic steatohepatitis: association of insulin resistance and mitochondrial abnormalities. Gastroenterology 120: 1183–1192.
33. Spector AA (2009) Arachidonic acid cytochrome P450 epoxygenase pathway. J Lipid Res 50 Suppl: S52–56.
34. Karkoulias G, Mastrogianni O, Lymperopoulos A, Paris H, Flordellis C (2006) alpha(2)-Adrenergic receptors activate MAPK and Akt through a pathway involving arachidonic acid metabolism by cytochrome P450-dependent epoxygenase, matrix metalloproteinase activation and subtype-specific transactivation of EGFR. Cell Signal 18: 729–739.
35. Bystrom J, Wray JA, Sugden MC, Holness MJ, Swales KE, et al. (2011) Endogenous epoxygenases are modulators of monocyte/macrophage activity. PLoS One 6: e26591.
36. Zhao G, Wang J, Xu X, Jing Y, Tu L, et al. (2012) Epoxyeicosatrienoic acids protect rat hearts against tumor necrosis factor-alpha-induced injury. J Lipid Res 53: 456–466.
37. Aleksunes LM, Manautou JE (2007) Emerging role of Nrf2 in protecting against hepatic and gastrointestinal disease. Toxicol Pathol 35: 459–473.
38. Faure P, Ramon O, Favier A, Halimi S (2004) Selenium supplementation decreases nuclear factor-kappa B activity in peripheral blood mononuclear cells from type 2 diabetic patients. Eur J Clin Invest 34: 475–481.
39. Dalby AR, Tolan DR, Littlechild JA (2001) The structure of human liver fructose-1,6-bisphosphate aldolase. Acta Crystallogr D Biol Crystallogr 57: 1526–1533.
40. Coffee EM, Tolan DR (2010) Mutations in the promoter region of the aldolase B gene that cause hereditary fructose intolerance. J Inherit Metab Dis 33: 715–725.
41. Bouteldja N, Timson DJ (2010) The biochemical basis of hereditary fructose intolerance. J Inherit Metab Dis 33: 105–112.
42. Liaw SH, Kuo I, Eisenberg D (1995) Discovery of the ammonium substrate site on glutamine synthetase, a third cation binding site. Protein Sci 4: 2358–2365.
43. Rios JJ, Blasco B, Rosales MA, Sanchez-Rodriguez E, Leyva R, et al. (2010) Response of nitrogen metabolism in lettuce plants subjected to different doses and forms of selenium. J Sci Food Agric 90: 1914–1919.
44. Shechter I, Dai P, Huo L, Guan G (2003) IDH1 gene transcription is sterol regulated and activated by SREBP-1a and SREBP-2 in human hepatoma HepG2 cells: evidence that IDH1 may regulate lipogenesis in hepatic cells. J Lipid Res 44: 2169–2180.
45. Koh HJ, Lee SM, Son BG, Lee SH, Ryoo ZY, et al. (2004) Cytosolic NADP+-dependent isocitrate dehydrogenase plays a key role in lipid metabolism. J Biol Chem 279: 39968–39974.
46. Zhang Y, Kozlov G, Pocanschi CL, Brockmeier U, Ireland BS, et al. (2009) ERp57 does not require interactions with calnexin and calreticulin to promote assembly of class I histocompatibility molecules, and it enhances peptide loading independently of its redox activity. J Biol Chem 284: 10160–10173.
47. Zimmerman KA, Graham LV, Pallero MA, Murphy-Ullrich JE (2013) Calreticulin regulates transforming growth factor-beta-stimulated extracellular matrix production. J Biol Chem 288: 14584–14598.
48. Del Cid N, Jeffery E, Rizvi SM, Stamper E, Peters LR, et al. (2010) Modes of calreticulin recruitment to the major histocompatibility complex class I assembly pathway. J Biol Chem 285: 4520–4535.

4

Integrative Analyses of Hepatic Differentially Expressed Genes and Blood Biomarkers during the Peripartal Period between Dairy Cows Overfed or Restricted-Fed Energy Prepartum

Khuram Shahzad[1,2], Massimo Bionaz[3]*, Erminio Trevisi[4], Giuseppe Bertoni[4], Sandra L. Rodriguez-Zas[1,2,5], Juan J. Loor[1,2]*

1 Department of Animal Sciences and Division of Nutritional Sciences, University of Illinois at Urbana-Champaign, Urbana, Illinois, United States of America, 2 Illinois Informatics Institute, University of Illinois at Urbana-Champaign, Urbana, Illinois, United States of America, 3 Department of Animal and Rangeland Sciences, Oregon State University, Corvallis, Oregon, United States of America, 4 Istituto di Zootecnica and Centro di ricerca sulla nutrigenomica, Universitá Cattolica del Sacro Cuore, Piacenza, Italy, 5 The Institute for Genomic Biology, University of Illinois at Urbana-Champaign, Urbana, Illinois, United States of America

Abstract

Using published dairy cattle liver transcriptomics dataset along with novel blood biomarkers of liver function, metabolism, and inflammation we have attempted an integrative systems biology approach applying the classical functional enrichment analysis using DAVID, a newly-developed Dynamic Impact Approach (DIA), and an upstream gene network analysis using Ingenuity Pathway Analysis (IPA). Transcriptome data was generated from experiments evaluating the impact of prepartal plane of energy intake [overfed (OF) or restricted (RE)] on liver of dairy cows during the peripartal period. Blood biomarkers uncovered that RE vs. OF led to greater prepartal liver distress accompanied by a low-grade inflammation and larger proteolysis (i.e., higher haptoglobin, bilirubin, and creatinine). Post-partum the greater bilirubinaemia and lipid accumulation in OF vs. RE indicated a large degree of liver distress. The re-analysis of microarray data revealed that expression of >4,000 genes was affected by diet × time. The bioinformatics analysis indicated that RE vs. OF cows had a liver with a greater lipid and amino acid catabolic capacity both pre- and post-partum while OF vs. RE cows had a greater activation of pathways/functions related to triglyceride synthesis. Furthermore, RE vs. OF cows had a larger (or higher capacity to cope with) ER stress likely associated with greater protein synthesis/processing, and a higher activation of inflammatory-related functions. Liver in OF vs. RE cows had a larger cell proliferation and cell-to-cell communication likely as a response to the greater lipid accumulation. Analysis of upstream regulators indicated a pivotal role of several lipid-related transcription factors (e.g., PPARs, SREBPs, and NFE2L2) in priming the liver of RE cows to better face the early postpartal metabolic and inflammatory challenges. An all-encompassing dynamic model was proposed based on the findings.

Editor: Lars Kaderali, Technische Universität Dresden, Medical Faculty, Germany

Funding: The authors have no support or funding to report.

Competing Interests: The authors have declared that no competing interests exist.

* E-mail: massimo.bionaz@oregonstate.edu (MB); jjloor@illinois.edu (JJL)

Introduction

The liver performs essential functions in mammals. These include, but are not limited to, gluconeogenesis and glycogen synthesis, synthesis of several plasma proteins encompassing clotting factors and acute phase proteins (APP) (e.g., haptoglobin, albumin, and fibrinogen), metabolism of amino acids and lipids, and detoxification including ammonia removal [1,2]. During the period around parturition in dairy cattle, also known as the peripartal or "transition period" [2], the importance of the liver becomes even more critical due to the greater metabolic demands imposed by the onset of lactation, particularly the need to increase gluconeogenesis, fatty acid metabolism, and to control the inflammatory response [3,4].

The level of dietary energy fed prepartum can alter the physiological adaptations to the transition period both in dairy [5]

and beef cattle [6,7]. Transcriptome profiling studies of peripartal cattle also demonstrated molecular adaptations in this organ some of which could alter its function, e.g. immune response and lipid metabolism [8–10]. The limited bioinformatics analyses performed in previous studies [9,10] indicated that moderate overfeeding of energy pre-partum (OF) results in transcriptional changes predisposing cows to fatty liver and potentially compromising liver health early postpartum. The aggregation of the transcriptomics dataset from cows fed recommended [9] or different levels of energy prepartum [10] revealed an extremely large effect of OF or restricting energy (RE) prepartum on the transcriptome adaptations during the transition period with a very modest effect observed when cows were fed recommended levels of energy [11].

Due to the more pronounced effect on the liver transcriptome of prepartal OF or RE relative to feeding to requirements, in the present work we took advantage of the advancements in bioinformatics and statistical tools to re-analyze microarray data from the liver of OF and RE cows from the previously published study from Loor et al. [10]. The functional analysis of DEG between the two treatments at each time point during the transition period was performed using both the Dynamic Impact Approach (**DIA**), a novel bioinformatics approach developed by Bionaz et al. [12], and the classical enrichment analysis approach by means of Database for Annotation, Visualization and Integrated Discovery (**DAVID**) [13] coupled with previously published [10] and new blood biomarkers. In addition, we used Ingenuity Pathway Analysis (**IPA**) to study the upstream regulators of transcriptomics differences. The primary aim of the present study was to propose an all-encompassing dynamic model to explain the main effects of prepartal dietary management approaches on the physiological adaptations of transition dairy cattle.

Results and Discussion

Blood profiling and overall metabolism

Summary of previous data. We reported previously that prepartal high dietary energy (OF) vs. feed restriction (RE) markedly increased prepartal insulin concentration and early postpartum concentration of NEFA, BHBA, total protein, and liver TAG. The metabolism of palmitate in liver tissue slices also was more pronounced in OF vs. RE prepartum with no significant differences postpartum [10]. Those data clearly indicated that the higher energy prepartum had a strong metabolic effect during dietary treatment (i.e., pre-partum) and a carry-over effect during early post-partum, considering that in the postpartum the diet was the same for the two groups.

New blood biomarkers. To further understand the differential effects on metabolism and inflammation status between the two groups and to better interpret the transcriptomics differences in liver, we have performed analysis of an additional 17 blood biomarkers plus a re-analysis of total protein concentration (see Materials and Methods for details). Among the additional blood biomarkers reported in the present study those with an overall statistically significant ($p<0.05$) effect due to diet or diet × time or with tendency ($p<0.10$) are included in Figure 1. Results of other biomarkers that were unaffected by dietary treatment, are included in Figure S1. Among those was total protein, for which the statistical effect of the diet × time differed from the previous analysis [10]. This is likely due to the different assay used and the larger number of data points in the previous analysis [10] (Figure S1). However, the pattern of protein concentration over time was overall similar between the two analyses.

The results indicated that the concentration of phosphorus was overall lower and zinc overall greater in OF vs. RE. Calcium, gamma-glutamyl transpeptidase (GGT), bilirubin, and creatinine concentrations had a significant diet × time interaction while vitamin A and vitamin E concentrations tended to have a significant interaction. The concentration of haptoglobin tended to be overall greater in RE vs. OF.

As indicated by the numerically higher GGT and significantly higher bilirubin, the data suggest that during the prepartal period the liver from cows in RE experienced a more pronounced state of distress. This might be partly due to the inflammatory-like conditions in RE vs. OF, as indicated by the greater concentration of haptoglobin and lower concentration of zinc [14]. However, in our study the inflammatory-like conditions did not seem to be pronounced as indicated by the lack of differences in the plasma concentration of indices of negative APP [14] such as albumin,

paraoxonase, and cholesterol [14,15] (Figure S1). The moderately higher inflammatory-like conditions prepartum in RE vs. OF cows might have been a consequence of a dietary protein deficiency because the cows in this group were only allowed to consume feed to meet 80% of the overall dietary requirements including protein. This conclusion is partly supported by a previous study in rats, where an acute protein deficiency induced a low-grade inflammation [16]. In contrast, during the early postpartum period the blood biomarkers indicated that liver from cows in OF experienced a more pronounced state of distress, i.e., numerically greater GGT at 28 d and a larger increase soon after parturition, and larger bilirubin [14,15] (Figure 1). The lower concentration of haptoglobin in OF vs. RE postpartum and the lack of difference in negative APP suggests that the observed stress response postpartum in OF vs. RE was not a consequence of higher inflammation [14,15] but potentially a consequence of greater TAG accumulation in liver [10].

The numerical difference between the two groups in prepartum concentrations of vitamin A, and partly, vitamin E, are likely related with the dry matter intake which was higher in OF vs. RE prepartum (see [10] for feed intake). The concentration of vitamin E tended to be lower after +14 days relative to parturition (**d**) in OF vs. RE despite the lack of difference in feed intake postpartum [10] (Figure 1).

The markedly greater plasma creatinine in RE vs. OF prepartum was striking (Figure 1). Creatinine is considered an index of muscle mass and/or renal function (i.e., used for the estimation of glomerular filtration rate) [17,18]. In our case the greater concentration of creatinine in RE cows prepartum was not due to greater muscle mass because it decreased in proportion to the body weight when cows were energy-restricted [5]. An increase of plasma creatinine due to feed restriction have been observed previously [19–21] but not always [22,23]. A greater creatinine concentration in blood despite a lack of increase in muscle mass has been explained by a decrease in renal filtration and/or by an increase in muscle proteolysis [19]. In our study it is likely that RE cows had an increase in proteolysis due to a reduction in body weight [5]. This is supported by previous data from dairy cows where an increase of plasma creatinine was observed immediately before and after calving when proteolysis normally increases [24]. Creatinine also can be produced by the liver [18]. In our microarray we measured the expression of genes coding for 2 out of 3 enzymes involved in the formation of creatine phosphate from glycine and arginine, the guanidinoacetate N-methyltransferase (*GAMT*) and the creatine kinase, mitochondrial 1B (*CKMT1B*). The former was numerically and the latter was significantly more expressed in RE vs. OF cows (File S1). This might indicate an additional source of creatinine in RE vs. OF cows.

Differentially expressed genes (DEG) between OF and RE

A mixed model ANOVA with FDR correction resulted in 4,790 cDNA array ID with a time × diet interaction (FDR≤0.05). Out of these cDNA ID, 4,111 were annotated with bovine Entrez gene ID (3,460 unique Entrez gene ID). Among these only the genes with a significant difference ($p≤0.05$) between OF and RE at each time point were used for the analysis (Figure 2). The number of DEG in Figure 2 indicated that there was a marked difference in expression of genes due to dietary energy level prepartum. A large number of DEG was observed between −14 to +14 d. It is noteworthy that the number of DEG between OF and RE did not change (or even decreased) during the first 35 days of the dietary treatment but changed tremendously afterwards. The large number of DEG observed up to 2 weeks after the treatments

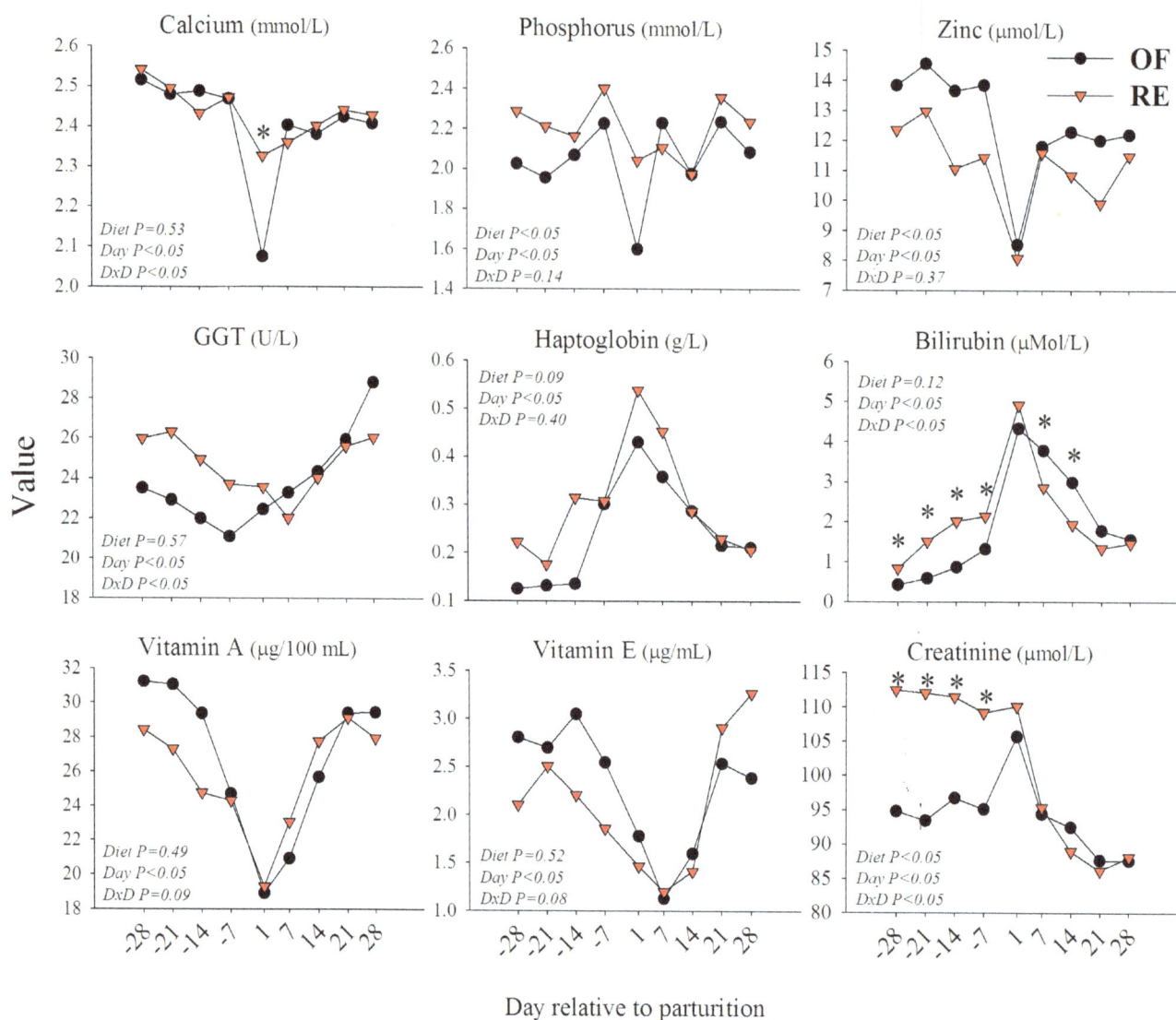

Figure 1. Results of plasma parameters significantly affected (*P*<0.05 or tendency *P*<0.10) by prepartal dietary energy level (OF = overfed energy and RE = restricted-fed energy prepartum) or diet × time interaction. *Difference between treatments at each time point. The GGT and Haptoglobin were log$_2$-transformed while Bilirubin and Vitamin E were square root transformed before statistical analysis and back-transformed for plotting due to non-normal distribution.

ceased is indicative of a brief, but significant, carry-over effect, i.e. an effect lasting through a stage when the same diet was fed to both groups of cows [10].

From the above data, it is obvious that at +1 d and, more pronounced, at +14 d, there was a greater number of DEG more expressed in OF vs. RE compared with the DEG more expressed in RE vs. OF (Figure 2A), indicating a larger transcriptional sensitivity of the OF group compared with the RE group due to stimuli from parturition/initiation of lactation. By applying the > 2-fold ratio expression threshold, the number of DEG was greater in RE vs. OF prepartum, but there were more DEG with a higher expression of >2-fold in OF vs. RE postpartum, particularly on day 1 post-partum (OF [n = 99] vs. RE [n = 21]) Figure 2B. The greater gene expression and fold difference at +1 d in OF vs. RE indicates an overall larger liver transcriptional activity due to higher dietary energy prepartum.

Summary view of KEGG pathway analysis

The "Impact" value in the DIA analysis represents an estimate of the perturbation in a biological pathway, while the "Direction of the Impact" (or flux) value represents the overall direction of the perturbation (i.e., a pathway is either activated or inhibited) [12]. The DIA also provides a summary of the KEGG pathways in the form of categories and sub-categories (Figure 3), besides providing the details of each pathway (File S2: sheet "Details of pathways").

In accordance with the number of DEG (Figure 2) all the KEGG pathway categories were more impacted from −14 d to + 14 d in parallel with the other time comparisons. The category 'Metabolism' followed by the 'Genetic Information Processing' was the most-impacted (Figure 3). With exception of the subcategories of pathways within 'Metabolism of Terpenoids and Polyketides', 'Transcription', 'Replication and Repair', 'Cell Growth and Dead', and 'Environmental Adaptation', all other categories and sub-categories of pathways were more induced in RE vs. OF at −14 d.

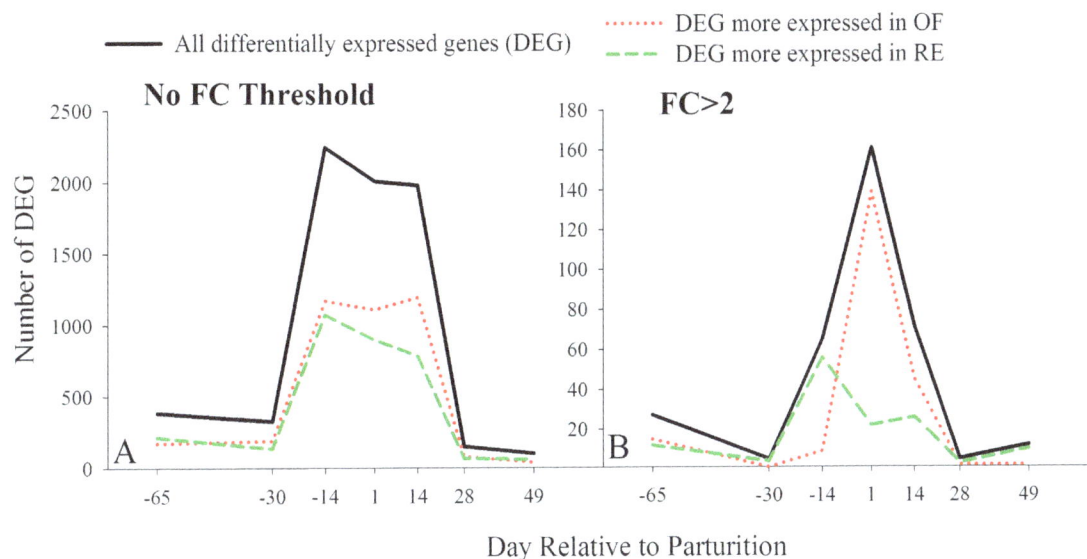

Figure 2. Number of differentially expressed genes (DEG) in liver of dairy cows fed restricted (RE) energy or receiving a higher energy diet (OF) prepaprtum. A). DEG with no fold change (FC) threshold. A larger number of DEG between the two diets were observed from two weeks prepartum to two weeks postpartum with a peak at −14 d. Except at 14 d, there was an overall similar number of genes with greater expression in OF vs. RE compared with those with greater expression in RE vs. OF. At 14 d there was an increase in the number of genes that were more expressed in OF vs. RE compared with RE vs. OF. B). DEG with 2-fold change (FC) cutoff. The total number of DEG between the two diets with > 2-fold difference in expression was <7% of total DEG. Most of the differences in gene expression between OF *vs.* RE occurred during −14 d to 14 d. The larger number of DEG more expressed in RE vs. OF was observed at −14 d while the larger number of DEG more expressed in OF vs. RE was observed at +1 d.

Few of the subcategories of pathways were or remained more induced in RE vs. OF until 2 weeks postpartum (Figure 3). This was evident particularly for 'Energy Metabolism', 'Amino Acid Metabolism', 'Metabolism of Other Amino Acids', 'Xenobiotics Biodegradation and Metabolism', 'Membrane Transport' and 'Circulatory System'. Almost all the other subcategories of pathways had a larger induction in OF vs. RE early postpartum which was more evident in non-metabolic related sub-categories of pathways with the exception of 'Glycan Biosynthesis and Metabolism' and 'Metabolism of Cofactors and Vitamins'. A very large induction in OF vs. RE postpartum was observed for subcategories of pathways related to 'Transcription', 'Translation', cell proliferation (e.g., 'Cell Growth and Death' and 'Replication and Repair'), signaling and cell-to-cell communication (e.g., 'Signaling Molecules and Interaction' and 'Cell Communication') (Figure 3).

The summary of the KEGG pathways indicated that after ca. 50 days of dietary treatment prepartum (coinciding with 2 weeks prepartum) the RE cows had an overall larger induction of the pathways, particularly metabolic-related. With the exception of few pathways, the postpartum carry-over effect was characterized by an overall large induction of non-metabolic related pathways in OF cows (Figure 3). In addition, the larger induction of transcription-related pathways in OF vs. RE is supportive of the greater number of DEG in OF vs. RE compared with RE vs. OF from +1 d to +14 d (Figure 2). Those results are suggestive of increased cell proliferation (i.e., larger induction of 'Replication and Repair' and 'Cell Growth and Dead' sub-categories of pathways) in OF vs. RE, as also discussed previously [11]. The 'Human Disease' KEGG pathways category was not included in the results and discussion section because of its low biological significance for the present study (however, the results can be found in File S2).

Biological interpretation of KEGG pathways in combination with Gene Ontology and enrichment analyses

Here we provide an integral view of the pathways combining the DIA results (File S2) with the visual results of several pathways obtained via KegArray (File S3) plus DIA results of Gene Ontology (**GO**) biological process (Files S2 and S4) and enrichment analysis performed using DAVID (Figure 4 and File S5 for details). In order to simplify the interpretation of data, the discussion was separated between metabolic-related and non-metabolic related pathways. The pathways related to metabolism were discussed in combination with the blood profiling reported in Figure 1 and the data reported previously [10].

Dietary energy and protein restriction prepartum might prime the liver to face high metabolic demands. It has been clearly established that the metabolism of carbohydrate [25], energy, lipid [26–28], and nitrogen [29] is critical in the liver of transition dairy cows. Liver has a great flexibility to adapt to the metabolic changes occurring during the transition period [30]. Our data indicated that pathways related to the above metabolism categories were highly-impacted by the prepartal level of energy and protein intake in the diet (Figure 3). The details of the pathways (File S2) revealed a large effect of increased dietary energy and protein intake prepartum (i.e., high impact) on 'Carbohydrate Metabolism' (Figure 3). Considering the most-impacted pathways (Figure 5), the OF cows had a larger induction of several carbohydrate-related pathways between −14 d and + 14 d, including 'Glycolysis/Gluconeogenesis'. Considering only the 'Glycolysis/Gluconeogenesis' pathway it is difficult to ascertain if the glucose metabolism was more towards synthesis or utilization of glucose (File S3). In a previous experiment, liver from cows fed ad-libitum vs. restricted before parturition had only numerically higher glycogen concentration [5]. The larger plasma insulin

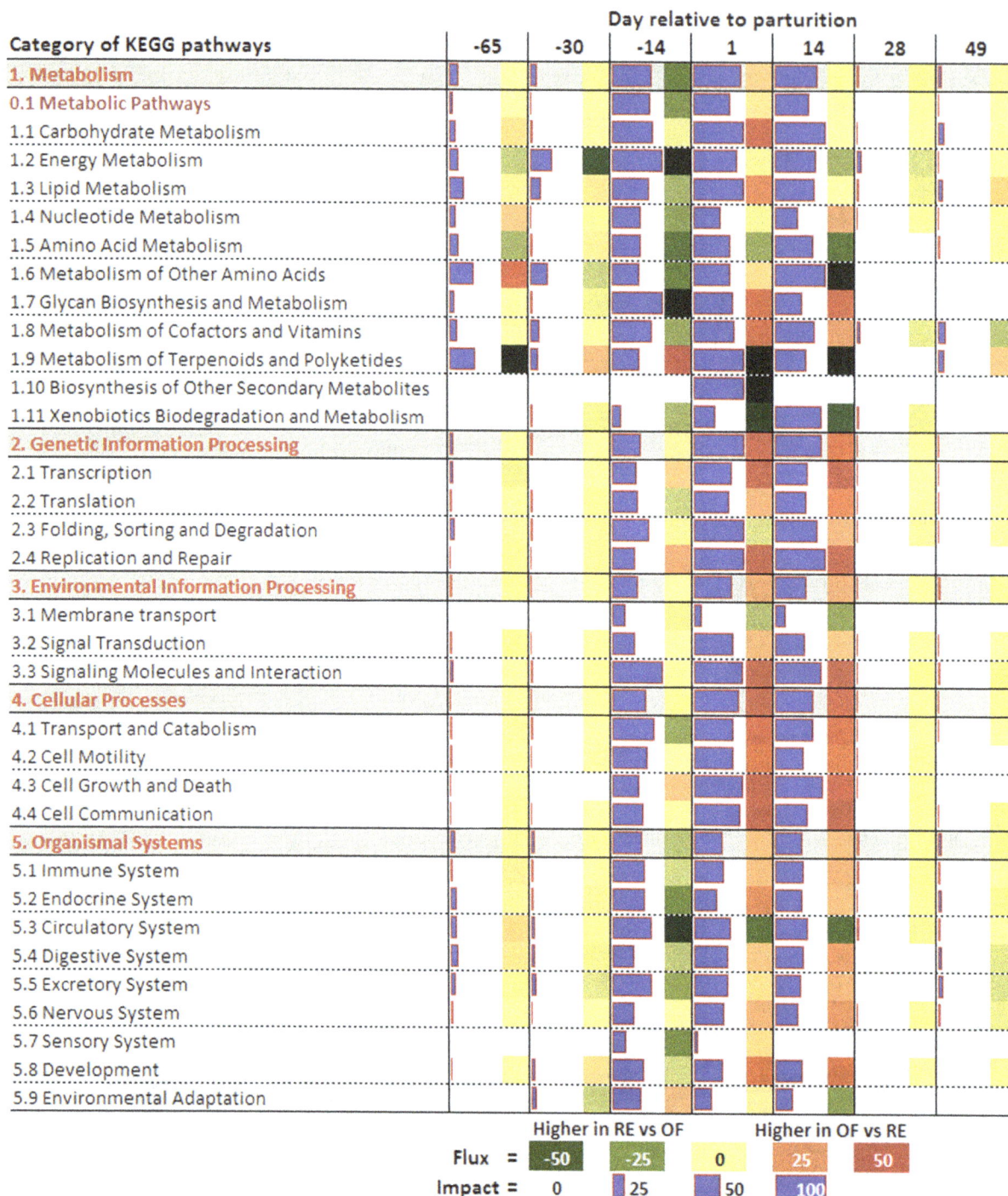

Figure 3. The summary of KEGG pathways encompassing categories and sub-categories of pathways as provided by the Dynamic Impact Approach (DIA). The 'Impact' is represented by the horizontal blue bars (larger the bar larger the impact) and the 'Direction of the Impact' (Flux) is represented by green (more induced in RE vs. OF) to red (more induced in OF vs. RE) rectangles.

prepartum in OF vs. RE [10] might have played a role in determining the expression of genes related to carbohydrate metabolism, and might have induce glycogenesis [31]. However, the higher glucose and insulin prepartum in OF vs. RE are indicative of a decrease in insulin sensitivity; in rats, insulin resistance has been associated with decreased liver glycogen content [32]. Therefore, the cause for the response in gene

expression observed might be due to the contrasting effect of higher insulin but also insulin resistance.

The liver of RE cows had a higher activation both pre- and post-partum of energy production through the 'Oxidative phosphorylation' pathway (Figure 5). The data also indicated a higher induction of 'Citrate cycle (TCA)' and degradation of amino acids (AA; Files S2 and S3) in RE vs. OF during the peripartal period, which support the apparently greater proteolysis suggested by the

	-14 d		1 d		14 d	
	RE	OF	RE	OF	RE	OF
KEGG pathway						
bta03010:Ribosome	▉		▉			
bta00190:Oxidative phosphorylation					▉	
bta03050:Proteasome						
Gene Ontology term						
GO:0003735~structural constituent of ribosome	▉					
GO:0003677~DNA binding				▉		▉
GO:0031966~mitochondrial membrane			▉			
GO:0005783~endoplasmic reticulum			▉			
GO:0045454~cell redox homeostasis			▉			
GO:0070003~threonine-type peptidase activity			▉			
GO:0044271~nitrogen compound biosynthetic process					▉	
GO:0009165~nucleotide biosynthetic process					▉	

Legend: **FDR<** | 0.05 | 0.01 | 0.001 |

Figure 4. Summary of GO and KEGG results from the DAVID analysis for DEG in liver between cows fed restricted (RE) or receiving a higher energy diet (OF) prepaprtum. Reported are the unique significantly-enriched terms with a Benjamini FDR<0.05 for at the least one comparison at −14, 1, and/or 14 d. Purple shades denote enrichment with a FDR<0.05.

higher creatinine (Figure 1). The AA can play an important role during early lactation to meet the glucose requirements through hepatic gluconeogenesis [33]. In addition certain AA such as alanine, aspartate and glutamate play a significant role in hepatic gluconeogenesis during starvation or nutrient restriction [34]. Overall, the AA metabolism was more induced in RE vs. OF (File S2, Figure 3). The data indicated a greater utilization of AA for gluconeogenesis in RE vs. OF, as is the case for 'Alanine, aspartate and glutamate metabolism' that was more induced at −14 d in RE vs. OF cows (File S3), likely due to greater glucose synthesis in this group as consequence of restricted energy and protein intake [35].

Among the energy-related pathways, as for the 'Oxidative phosphorylation', also the 'Sulfur metabolism' was more activated in RE vs. OF, particularly post-partum (File S2). The role of sulfur metabolism in the liver of periparturient dairy cows has not yet been investigated thoroughly. However, because of the anionic property of the sulfur compounds, this pathway appears essential in order to balance the cation-anion concentrations in the liver [36]. It may also be involved in the synthesis of sulfur containing AA [37]. The metabolism of sulfur in sulfur-containing compounds (e.g., enzymes, hormones and xenobiotics) can play an important role in the regulation of different cellular and metabolic processes in the liver such as the sulfate-conjugation of xenobiotics and steroid hormones which are needed for their metabolism, bioactivation and detoxification often resulting in a decrease in biological activity and an increase in their urinary excretion [38].

The pathways involved in lipid synthesis, especially 'Glycerolipid metabolism', were evidently more induced postpartum in OF vs. RE (Figure 5), which is consistent with a greater degree of esterification of fatty acids (FA) observed *in vitro* [10]. This mechanism also was supported by the GO BP analysis with DIA, which uncovered a higher activation in OF vs. RE of terms related to TAG synthesis and storage (File S4). The data also suggested that during the last month of being on diets (i.e. end of pregnancy) the RE vs. OF cows had a lower degree of sterol synthesis, which is consistent with the observed inhibition of cholesterol synthesis in cows feed-restricted both during mid-lactation [39] or in early postpartum [40]. However, in our experiment, the blood

biomarker analyses did not reveal differences in blood cholesterol between the two groups. The data from DIA indicated that the RE cows compared with OF had an overall higher induction of sterol synthesis early postpartum (File S2) which also was supported by the DIA analysis of GO BP (File S4). At least in humans, cholesterol is essential for the synthesis of very-low density lipoproteins [41], and this is the only means for liver to export TAG to peripheral tissues. The above data indicate that the higher liver TAG observed in the OF vs. RE group might be due to a concomitant higher synthesis of TAG and lower VLDL formation (also due to lower cholesterol availability).

Despite having a greater NEFA concentration postpartum [10], the transcriptomics data suggest that cows in OF vs. RE had a lower degree of lipid catabolism (Figure 5). As previously proposed [11], a greater NEFA concentration rather than a change in gene expression appears more important in terms of a flux increment towards oxidation leading to ketone body production. This is supported by the lower overall induction of synthesis of ketone bodies in OF vs. RE during the first two weeks postpartum (Figure 5) despite the greater plasma BHBA [10]. However, the apparently greater fatty acid catabolism in RE vs. OF is not entirely supported by the *in vitro* oxidation of palmitate, which did not differ between groups [10]. Taken together, the modest induction of genes related to synthesis of cholesterol and lower fatty acid catabolism coupled with the higher induction of genes involved in synthesis of TAG in OF vs. RE might have contributed to the greater degree of liver TAG accumulation observed in OF vs. RE [10].

The observations highlighted above indicate that RE vs. OF induced a higher degree of degradation of molecules in the liver in order to obtain energy during the feed restriction period but, paradoxically, also when the cow was allowed to consume ad-libitum energy- and protein during the postpartal period. Together, the data indicate that the dietary treatment prepartum "primed" the liver of RE cows for higher metabolic capacity. In support of this in RE vs. OF cows there also was a greater induction of 'Fatty acid metabolism' and other energy-related pathways such as 'Sulfur metabolism' and 'Oxidative phosphorylation'. The latter also was enriched in DAVID analysis together

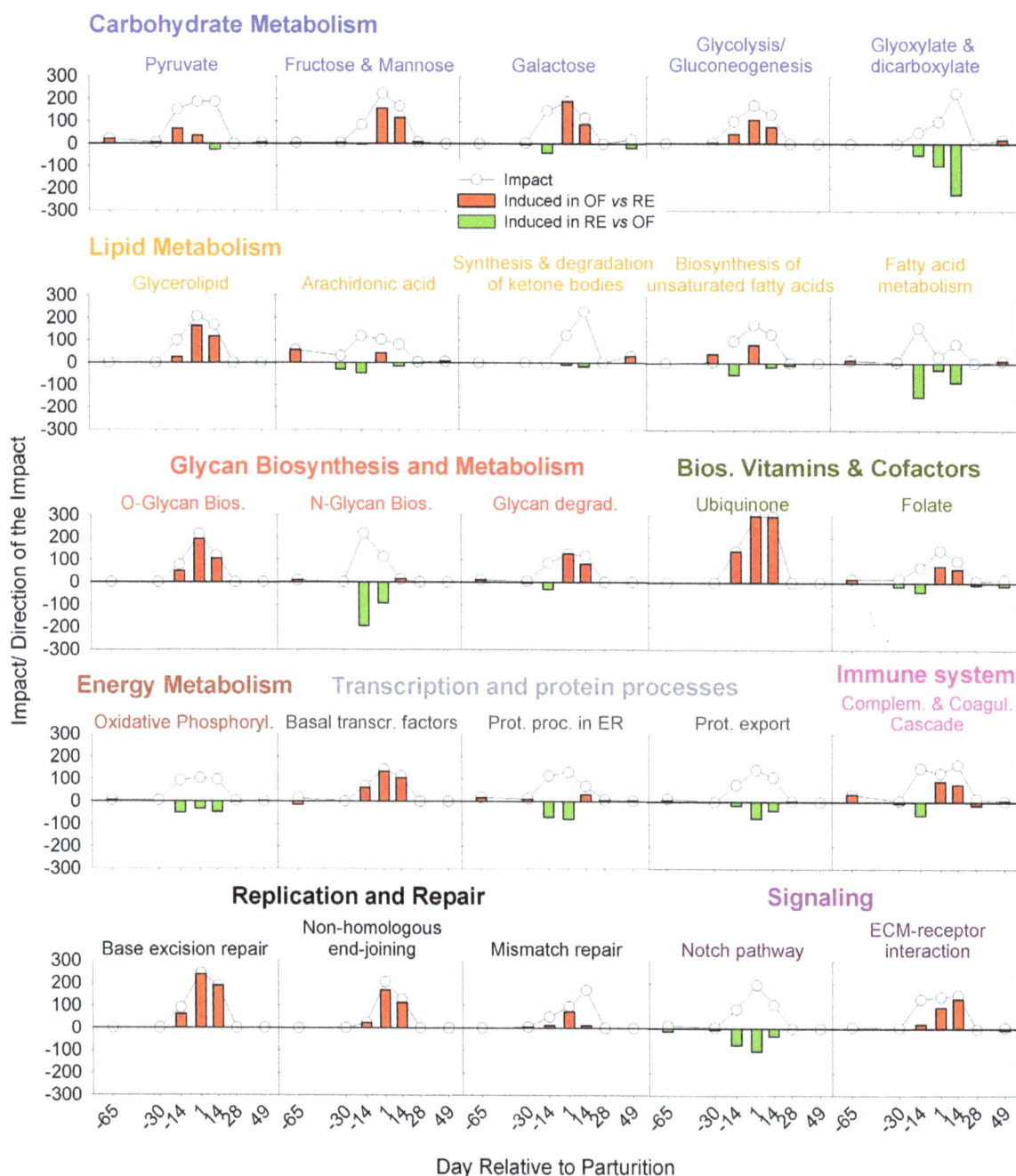

Figure 5. Dynamic Impact Approach (DIA) results (Impact and Direction of the Impact) for the 25 most impacted KEGG pathways grouped in sub-categories of pathways.

with enrichment of mitochondria and oxidative related terms (Figure 4 and File S5). Terms related to mitochondria and oxidation were also among the most-impacted GO BP terms in DIA, at least at −14 d (File S4).

Glycan biosynthesis and ER stress. Glycans are carbohydrate molecules that are linked with lipids and protein moieties to form glycolipids and glycoproteins and also act as signaling molecules [42]. The most-impacted glycan pathways in our experiment included 'O-Glycan biosynthesis', 'Other glycan degradation', and 'N-Glycan Biosynthesis' (Figure 5 and File S2). Except the latter, all were more induced in OF vs. RE (Figure 5). The potential role of N-Glycan biosynthesis in the liver

is to handle the misfolded proteins in the endoplasmic reticulum (ER) during stress conditions [43]. During protein synthesis, the function of the ER is affected by both intracellular and extracellular stimuli leading to ER stress, which results in accumulation of misfolded proteins in the ER lumen [44]. It is interesting that the 'Protein processing in the ER' pathway was among the most impacted and had a similar induction as 'N-Glycan Biosynthesis' in RE vs. OF (Figure 5). This might be indicative of a greater degree of ER stress, likely related to the higher oxidative conditions in RE vs. OF cows during the last month of dietary treatments; however, our inference needs to be corroborated by quantitative measurements of ER stress prepartum. The data also could be indicative of an increased capacity of the liver prepartum in RE cows in order to handle ER stress postpartum. The ER stress response in RE cows appears to be partly under control of the transcription factor *XBP1* (X-box binding protein 1), as previously discussed [45].

Dietary energy and protein restriction prepartum might prime the liver for a better response to the inflammatory challenge of early lactation. Among the non-metabolic related pathways the most-impacted were related to transcription, translation, cell cycle and cell signaling/communication (Figure 5 and File S2). Overall, during the postpartum these categories of pathways were more induced in OF vs. RE (Figure 3). The details of these pathways (Files S2 and S3) clearly support a larger induction of transcription, replication and repair of DNA, and cell-to-cell interaction/communication in OF vs. RE postpartum; however, the 'Ribosome' pathway (i.e., protein synthesis machinery components) and pathways related to protein synthesis (e.g., 'Protein export' and 'Proteasome') were overall more induced in RE vs. OF (File S2). This also was supported by the significant enrichment of pathways and GO terms related to ribosome within the DEG related to greater expression in RE *vs.* OF at −14 d and +1 d (Figure 4). However, the "Proteasome" pathway was more induced in RE vs. OF at +14 d which was also supported by the enrichment analysis (Figure 4). These data indicate that, on the one hand, the overall amount of RNA (and/or number of different transcripts) was greater in OF vs. RE, as supported by the number of DEG (Figure 1), but, on the other hand, the overall degree of protein synthesis and export, particularly in the ER (File S2), was greater in RE vs. OF (Figure 5), as also discussed in the previous section.

The higher protein synthesis and protein export in liver of RE vs. OF cows may be associated with a greater capacity to produce and export proteins such as signaling molecules (e.g., IGFBPs and interleukins, see File S1) and positive APP. This is supported by the higher concentration of plasma haptoglobin in RE vs. OF cows (Figure 1) but also by the greater expression of some of the known positive APP measured by the microarray (Figure S2 and File S2). A greater production of positive APP is normally associated with a decrease of negative APP [14]. This phenomenon is associated with an impaired capacity of the liver to face metabolic challenges, as indicated by the detrimental effect of inflammation on the peroxisome proliferator-activated receptor (PPAR)[46], which (at least in non-ruminants) is involve in assuring the normal functions carried out by the liver [14,15]. It is noteworthy that both the blood biomarker (Figure 1 and Figure S1) and gene expression analyses (Figure S2) indicated a more pronounced inflammatory-like conditions around parturition in RE vs. OF cows (Figure 1), which did not appear to elicit a detrimental effect on the "normal" liver function. There was no detectable lower plasma concentration of negative APP in RE vs. OF (Figure S1) or in their mRNA abundance in liver (Figure S2).

A higher capacity for or sensitivity to an inflammatory response in RE vs. OF also was suggested by the observed changes in the 'Arachidonic acid metabolism' pathway (Figure 5), particularly at the end of the dietary treatment phase (i.e. parturition). Arachidonic acid is a long-chain polyunsaturated fatty acid [47], and the larger induction of its metabolism may be indicative of an increased rate of inflammation via the production of pro-inflammatory lipids such as prostaglandins [48].

Overall, the immune-related pathways suggested a higher immune/inflammatory response in RE vs. OF by the end of pregnancy/dietary treatment phase but with a greater induction of the same pathways postpartum in OF vs. RE (Figure 3), especially for the 'Complement and coagulation cascades' (Figure 5). This

pathway is part of the innate immune response and links the inflammatory response (i.e., complement) with coagulation [49]. The detailed visualization of the pathway (File S3) uncovered that the induction in OF vs. RE was exclusively due to the coagulation pathway.

The higher immune/inflammatory response in RE vs. OF also was suggested by the greater induction of 'NOD-like receptor signaling pathway' in RE vs. OF cows. This innate immune response pathway has been previously reported to be among the most-activated in several studies involving mammary gland bacterial infection [50]. Innate immune response-related pathways including 'Toll-like receptor' (File S2), but especially the pathways related to the immune cell migration or activity (e.g., 'Leukocyte transendothelial migration', and 'T cell receptor signaling pathway'), were more induced in OF vs. RE, particularly at + 14 d. The DIA analysis of GO BP indicated a larger activation of macrophage in OF vs. RE but, contrary to the KEGG pathway results, indicated a more induced chemotaxis of some of the immune cells in RE vs. OF at +14 d (File S4).

Overall, the data suggest that the liver in RE cows was likely more responsive to inflammatory-like conditions and also better able to handle them. Despite the apparently higher inflammatory response the liver did not have a decrease of negative APP or inhibition of metabolic or detoxification pathways. The greater detoxification capacity of liver of RE vs. OF cows was substantiated by the lower plasma bilirubin (e.g., higher clearance capacity) in early lactation (Figure 1) and inferred by the higher activation of 'Drug metabolism – other enzymes' and 'ABC transporters' (File S2). The ATP binding cassette (ABC) transporters play an important role in removal of xenobiotic compounds through their excretion in bile salts [51,52].

The data suggest that the OF vs. RE had a higher activation of immune cells but there is some discordance about the capacity of the immune cells to perform chemotaxis. The suggested higher activation of the immune cells in the present analysis appears to be similar to what is observed in monogastrics during the occurrence of fatty liver [53]. It is noteworthy that our data suggest a positive role of moderate inflammatory-like conditions in the peripartal period to prime the liver for a better inflammatory and metabolic response post-partum. Those findings seem to support recent findings where inhibition of inflammatory-like conditions using salicylate in early postpartal cows had a negative effect on the homeorhetic adaptations of the liver [54].

It is important to stress the word "moderate" in the context of inflammation. Previous studies from some of the authors have clearly established that substantial and prolonged inflammatory-like conditions (i.e., haptoglobin >0.3 g/L and a significant decrease of negative APP for the first two weeks postpartum) negatively affect productive and reproductive performance and increase the likelihood of developing health disorders in dairy cows early postpartum [14,15]. In addition, several studies have observed a positive effect of preventing/decreasing inflammation in peripartal cows [39,56]. Overall, the present data support a positive role of moderate and likely diet-derived-stress-related response inflammatory-like conditions on the liver of transition dairy cows, especially post-partum.

Higher cell signaling and cell-to-cell communication in OF vs. RE: a response to liver fat accumulation?. The proliferation of hepatic cells appeared to be greater in OF vs. RE, as indicated by an overall higher induction of KEGG pathways and GO BP terms related to cell cycle (Figure 3 and Files S2 and S4), including the 'DNA replication and repair' pathway (Figure 5). Reduced proliferation due to an acute energy restriction has been observed previously [55]. It is noteworthy that the higher induction of pathways related to proliferation in liver of OF vs. RE was observed postpartum (i.e., carryover effect). This could have been due to the greater degree of lipid accumulation. Ethanol-induced steatosis in liver of rats increased hepatocyte proliferation as a mechanism to reduce the injury due to the large degree of lipid infiltration [56].

The potentially greater degree of liver proliferation also was accompanied by an overall greater degree of cell-to-cell communication in OF vs. RE, as suggested by a larger induction of signaling and cell communication-related pathways (Figure 3) such as 'ECM [Extracellular matrix] receptor interaction' (Figure 5) and GO terms related to ECM disassembly (File S4). The activation of the extracellular matrix is a common finding in non-ruminant liver steatosis [57] and it is determined by the interplay between several cell types but apparently initiated by stellate cells. In line with this previous observation, our data suggest that macrophages cells were more activated in OF vs. RE (e.g., more induced macrophage activation, see File S4; and several immune-related KEGG pathway, File S2).

The change in ECM deposition is an essential step for the onset of fibrosis, which is one marker of liver damage. It is challenging to conclude that the liver in OF cows was damaged compared with RE cows because a histological analysis was not performed; however, functional analysis of transcriptome differences between the two groups of cows, the numerically greater amount of GGT at 28 d, the higher increase in plasma bilirubin compared with pre-partum (although not pathological), and the greater hepato-cellular TAG [10] could be considered reasonable indicators of a more injured liver in OF compared with RE.

With the exception of 'Notch signaling' and 'Hedgehog signaling' all the pathways involved in signaling communication were more induced in OF vs. RE (Figure 3) (Figure 5 and File S2). The inhibition of 'Notch signaling' in OF vs. RE is likely a consequence of the greater lipid accumulation. It has been clearly demonstrated that steatosis in mice inhibits Notch signaling [58]. It is noteworthy that the pathway was already more inhibited in OF vs. RE at −14 d, well-before the greater TAG concentration was observed (Figure 5 and [10]). This indicate that the larger inhibition of the Notch signaling pathway in OF vs. RE prior to the peak of NEFA might have allowed for a greater lipid accumulation postpartum. This idea is supported by the higher induction of 'MAPK signaling' in OF vs. RE particularly during early lactation (File S2). This pathway is mainly involved in the control of cellular growth and proliferation but it also is essential for the induction of liver steatosis in mouse by cross-talk with the peroxisome proliferator-activated receptor gamma (PPARγ; [59]). Although our data seem to support such a role, the expression of PPARγ in bovine liver is nearly non detectable relative to the alpha and delta isotypes [60].

The PPAR, particularly PPARα, is considered important in the whole economy of lipid metabolism in liver of mammals, including dairy cows during the transition period [60]. Our analysis indicated that the 'PPAR signaling' pathway (which includes all three PPAR subtypes) played a role in regulating aspects of lipid metabolism in response to different levels of dietary energy fed prepartum. As commonly observed in monogastrics [61], the 'PPAR signaling' pathway was more induced in liver of RE compared with OF cows (File S2). It is noteworthy that the larger activation of 'PPAR signaling' in RE vs. OF also was observed at +14 d, namely due to the higher expression of genes encoding for proteins involved in FA oxidation (File S3). These data support a role of PPAR in controlling lipid catabolism in response to prepartal dietary energy level. However, the higher activation of this pathway in RE vs. OF in spite of a lower concentration of

plasma NEFA [10], which should increase its activation [60], also supports the idea of a liver in RE vs. OF that was better "primed" to respond to the increased NEFA flux postpartum potentially by having a higher metabolic capacity.

Up-stream regulators

The analysis of up-stream regulators among the DEG in the comparison of RE vs OF at −14, +1, and +14 d surprisingly revealed few gene targets overall and a higher number in the comparison on −14 d (Figure 6) compared with other time points (Figures S3 and S4). Among up-stream regulators at −14 d, several are related to lipid metabolism including PPARA, SREBF1, SREBF2, and SCAP, all more activated in OF vs. RE. The enrichment analysis revealed significant associations of down-stream genes with functions related to lipid metabolism and inflammation (Figure 6), among these 'Hepatic steatosis'. Remarkably, an obvious increase in lipid accumulation in liver of OF cows was not evident until ca. 28 days later (i.e., +14 d) [10]. The transcription factor NFE2L2 was also uncovered to be an important up-stream regulator of the DEG between OF and RE at −14 d (Figure 6). The NFE2L2 is crucial regulator of the proper oxidative stress response and detoxification in liver [62]. This transcription factor was among the most important up-stream regulators also among DEG at +1 and +14 d (Figures S3 and S4). Interestingly, our data suggested that the higher liver capacity to respond to oxidative stress (including ER stress) and detoxification in RE vs. OF cows was more pronounced early post-partum compared to pre-partum (Figures 3 and 5). These data further support the idea that dietary energy prepartum "primed" the liver for a postpartum metabolic response.

Surprisingly, few up-stream regulators were uncovered by IPA at +1 d, with enrichment of lipid metabolism and proliferation related down-stream genes (Figure S3). At +14 d even fewer predicted up-stream regulators (and down-stream genes) were uncovered by IPA but there was an apparent prominent role and activation in OF vs. RE of osteopontin (SPP1) and metallopepti-dase inhibitor 1 (TIMP1) both of which are related to induction of ECM formation [63,64].

Conclusions

The use of additional blood biomarkers and novel bioinfor-matics tools for the analysis of published transcriptomics data provided a more holistic understanding of the role of plane of nutrition during late-pregnancy on hepatic adaptations around the period of parturition. An all-encompassing dynamic model using these data is reported in Figure 7. Overfeeding energy prepartum enhances body fat deposition, partly in response to chronic hyperinsulinemia, which leads to more pronounced and sustained increase in blood NEFA postpartum and greater TAG accumu-lation in liver at least in part by reducing lipid catabolism and partly due to "dampened" PPARα activation. Despite such response, in overfed cows there was an attempt to counterbalance these negative effects by reducing Notch signaling and activating other cellular pathways of which cell cycle and ECM receptor interaction would likely help the liver repair from cellular damage (suggested by higher blood bilirubin and, numerically, GGT). On the contrary, although cows fed restricted energy appeared to catabolize substantially more muscle mass prepartum, their liver was able to adapt to the higher postpartal metabolic state well-ahead of parturition. This adaptation was likely driven by molecular processes partly controlled by transcription regulators such as PPARA and NFE2L2, of importance in fatty acid oxidation and cellular stress. As a result, restricted-fed cows had

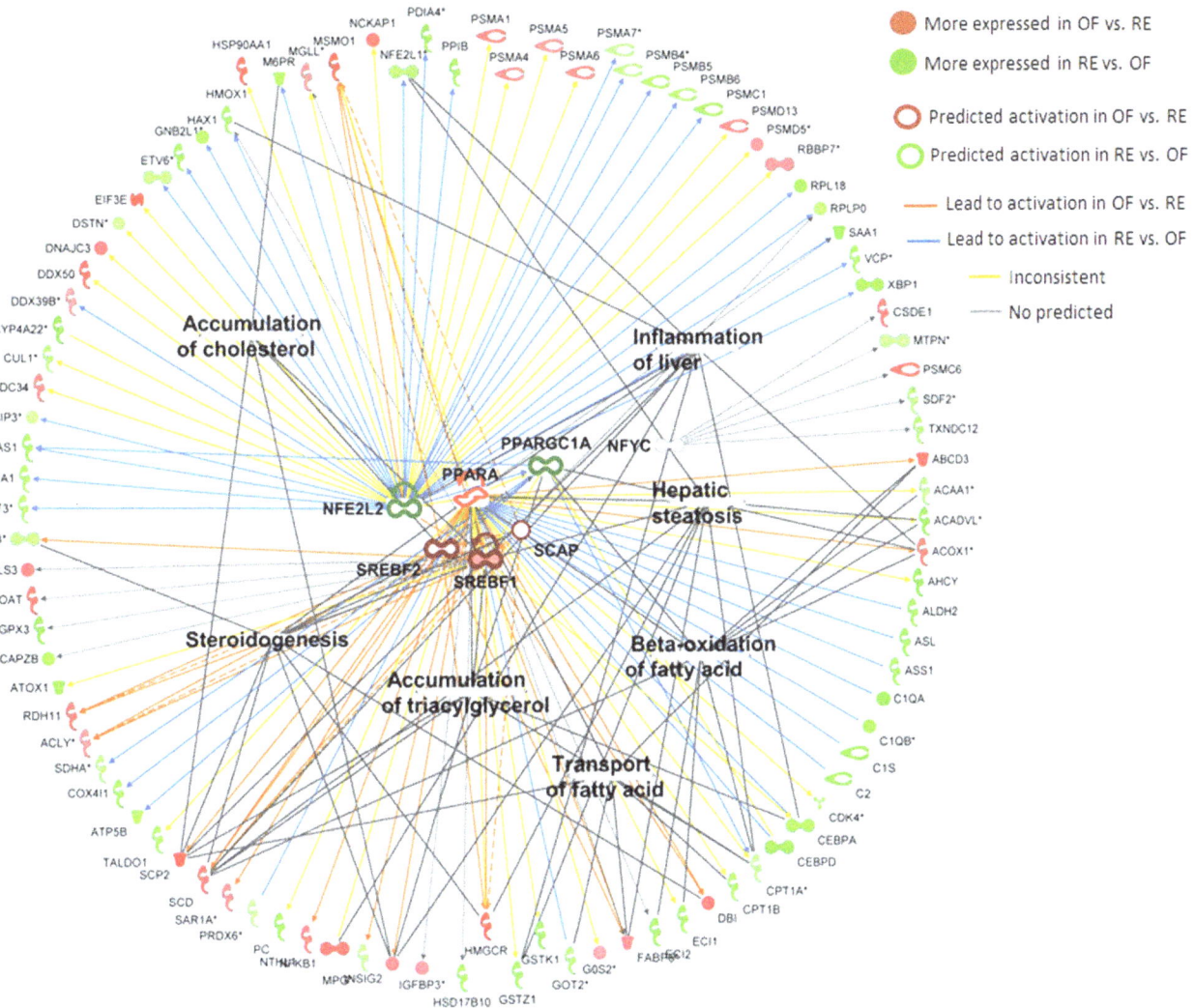

Figure 6. Ingenuity Pathway upstream network analysis of differentially expressed genes (DEG) between OF and RE at −14 d. Upstream regulators are located at the center of the network and down-stream genes are located in the periphery. In the network are also reported the most enriched biological terms among down-stream genes.

signs of greater metabolic flux and utilization of amino acids and fatty acids but also of a more pronounced cellular inflammatory and ER-stress response. Most of those cellular adaptations were confirmed by biomarker analysis specifically during the prepartal period, which strengthened the notion that restricted-energy helped "prime" the liver to cope with the change in physiological state at the onset of lactation.

Clearly, there is a carryover effect of plane of nutrition during late-pregnancy that will result in molecular and physiological adaptations during lactation. Our data support the view of a more robust liver in restricted-fed cows to face the metabolic and inflammatory challenges typical of the early postpartal period. As such, the transcriptomics data provide evidence that plane of dietary energy during late-pregnancy can help prime the liver for the onset of lactation.

Methods

Experimental design and ethic statement

All procedures were conducted under protocols approved by the University of Illinois Institutional Animal Care and Use Commit-

tee. The information about sampling, RNA extraction, and microarray data were published in the original study [10]. Briefly, the data used for this manuscript are from a subset of 8 Holstein dairy cows randomly selected from a larger study [5] fed either a higher-energy diet ad-libitum (>150% of net energy requirements; n = 4; OF) or fed a restricted energy diet (80% of net energy requirements; n = 4; RE) during the entire non-lactating period (~last 65 days prior to parturition). For both groups of cows, the same level of energy and protein in the diet was provided from the day of parturition until the end of the study. The liver biopsies were harvested on day −65, −30, −14, +1, +14, +28, +49 relative to parturition. The microarray data were deposited in the National Center for Biotechnology Information (NCBI) Gene Expression Omnibus (GEO) database (http://www.ncbi.nlm.nih.gov/gds) with accession number GSE3331.

Blood profiling

A large metabolic profiling including 18 parameters was performed in plasma collected every 7 days from −28 to +28 day relative to parturition. Parameters measured were: the indexes of acute phase reaction, such as the positive acute phase protein

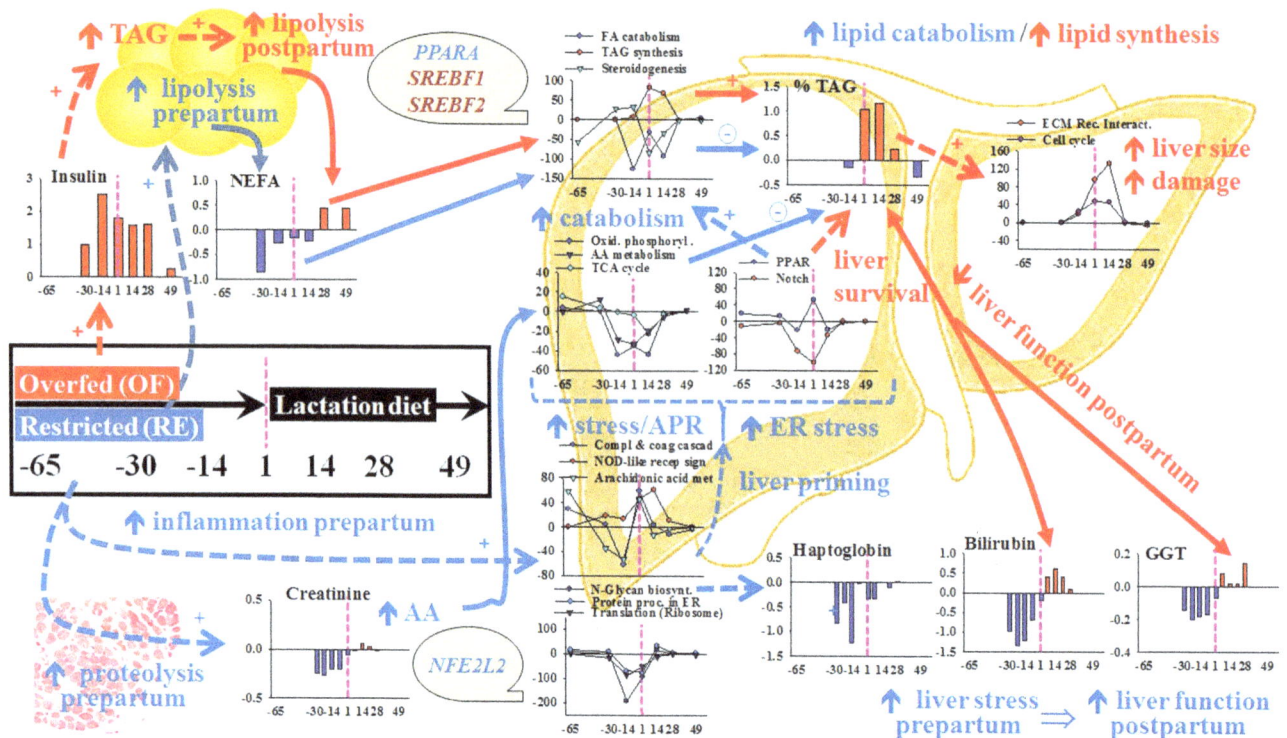

Figure 7. All-encompassing dynamic model proposed based on the main findings from the study. Reported are differential effects of the diet prepartum on adipose tissue, muscle, and liver (depicted by yellow round cells, image of a muscle section, and draw of a stylized liver, respectively). Red (text, lines, symbols and bars in graphs, and arrows) denotes facts that are more pronounced in cows overfed [OF] compared to cows underfed [RE] energy prepartum. Dark blue denotes facts that are more prominent in cows fed energy restricted diet prepartum compared to cows overfed energy prepartum. Data from representative plasma parameters are indicated by bars graphs with fold differences in overfed vs. restricted energy prepartum cows in Y-axis. Data from representative pathways from the Dynamic Impact Approach are shown as lines and scatter plots with the Direction of the Impact in Y-axis. In all cases the X-axis denotes the day relative to parturition. The dotted vertical purple line in all graphs denotes the end of treatment coinciding also with parturition. Solid lines and arrows denote flow of molecules/metabolites. Dotted lines and arrows denote effect. In all cases ⊕ denotes larger activation/amount and ⊖ denote larger inhibition. In text ↑/↓ denote larger or lower in the treatment indicated by the font color (red = overfed and blue = feed restricted) compared to the other treatment. In the light-yellow background round shapes are reported main transcriptional factors potentially involved in controlling the transcriptomics adaptation of the indicated pathways. Summary explanation of the model is reported in the conclusion section of the paper.

haptoglobin and ceruloplasmin; the negative acute phase protein albumin, paraoxonase and the other index related to the negative acute phase protein reaction such as cholesterol, β-carotene, and vitamin A; bilirubin; the liver enzymes aspartate aminotransferase (AST or **GOT**) and γ- glutamyl transpeptidase (**GGT**); minerals (calcium, magnesium, zinc, and phosphorous); creatinine; and vitamin E. The concentration of total proteins was analyzed also previously [10] and the data were re-analyzed in the present study in order to calculate accurately the concentration of globulins which are estimated by the difference between total proteins and albumin. Analysis of all parameters was performed as previously described [14,15].

Statistical analysis

Microarray spots with median intensity ≥3 standard deviation above the median of the background and GenePix 6 flag >100 were applied as filters to ensure high quality data. A total of 106 microarrays were adjusted for dye and array effect (Loess normalization and array centering), duplicated spot intensities were not averaged and were subsequently used for statistical analysis. A mixed model with repeated measures was then fitted to the normalized $\log_2$-transformed adjusted ratios (sample/reference standard) using Proc MIXED (SAS, SAS Inst. Inc., Cary, NC). The model included the fixed effects of time (−65, −30, −14, +1,

+14, +28, +49 d), diet (OF and RE), and interactions of time × diet. Cow was considered as a random effect. The p-values were adjusted for the number of genes tested using Benjamini and Hochberg's false discovery rate (FDR) [65] to account for multiple comparisons. Differences in relative gene expression were considered significant at an FDR-adjusted $p \leq 0.05$ for time × diet. A post-hoc $p \leq 0.05$ was considered significant between diets at each time point.

For metabolic profiling parameters normal distribution was assessed using the procedure UNIVARIATE of SAS. Data not normally distributed were transformed into log scale. Data points with studentised residuals ≥2.5, analyzed by the SAS procedure REGRESSION, were considered outliers and excluded from the subsequent analysis (a total of 6 data points were removed out of ca. 4,000). A MIXED model procedure with a spatial power as a covariance structure was used. The model included diet, time, and diet × time as fixed effects, with cows as random variable. Means between treatments and time point were separated using the PDIFF. Data were deemed to be significant if overall diet × time interaction was $p \leq 0.05$ and tendencies at $p \leq 0.10$. Single point comparisons were determined to be significantly different if $p \leq 0.05$.

Dynamic Impact Approach (DIA) and Enrichment Analysis

The detailed methodology for data analysis using DIA was previously described [12]. Briefly, the whole dataset (File S1) with Entrez gene ID, fold-change and the significance (p-values) of gene expression between the two diets at each time point, plus the overall FDR adjusted p-values were uploaded into DIA. The Kyoto Encyclopedia of Genes and Genomes (**KEGG**) pathways and Gene Ontology (**GO**) biological process category database were used for functional analysis with the DIA. For all the analyses a minimum of 20% genes in the annotated microarray vs. whole genome was used. The GO results from DIA were summarized using REVIGO [66]. For this analysis the GO ID along with the values of the direction of the impact as calculated by DIA were uploaded in REVIGO with analysis performed separately between GO terms induced in OF vs. RE and the one induced in RE vs. OF for −14, +1, and +14 d, the time points with larger number of DEG.

For enrichment analysis, the lists of DEG (overall time × diet FDR<0.05 and p-value) plus the whole annotated microarray as background were uploaded into DAVID bioinformatics resource database (http://david.abcc.ncifcrf.gov/). The Multi-List File function in the functional annotation tool of DAVID was selected to upload separately the list of DEG more expressed in OF vs. RE and the ones more expressed in RE vs. OF for −14, +1, and +14 d only, due to the very low number of DEG in other time points. The results from the default selected databases in DAVID were downloaded using "Functional annotation chart" with an adjusted p-value ≤0.10 (i.e., EASE score, which is defined as a conservative adjustment to the Fisher exact probability see [67]).

KEGG pathway visualization

The most impacted KEGG pathways from DIA and enrichment analysis results were visualized using KegArray tool (http://www.kegg.jp/kegg/download/kcgtools.html). For this purpose, the expression ratios between OF and RE along with Entrez gene ID were used as an input.

Up-stream transcription regulator analysis via Ingenuity Pathway Analysis (IPA)

In order to uncover the main up-stream regulators of the DEG we have taken advantage of the upstream regulator analysis in IPA. The analysis uses an IPA Knowledge base to predict the expected causal effects between up-stream regulators and targets (i.e., DEG). The analysis provides the more plausible prediction of the status of the upstream regulator (i.e., activated or inhibited) by computing an overlap p-value and an activation z-score. For this purpose the whole dataset with Entrez-Gene ID, FDR of the diet × time effect, expression ratio, and p-values between each comparison were uploaded into IPA. All the predicted upstream and their targets were visualized in a network using the network tool. The analysis was performed for DEG at −14, +1, and +14 d due to the number of DEG observed in these comparisons.

Supporting Information

Figure S1 Results of plasma parameters, not significantly affected by prepartum dietary energy (OF = overfed energy and protein and RE = restricted-fed energy and protein prepartum).

Figure S2 Expression of several genes coding for positive and negative acute phase proteins during the whole duration of the study. * indicate significant difference in expression at each time point between cows fed restricted (RE) energy and protein or receiving a higher energy and protein diet (OF) prepartum. Color of * is related to the color of symbol and lines for the gene.

Figure S3 Ingenuity Pathway upstream network analysis of differentially expressed genes (DEG) between liver of cows fed restricted (RE) energy and protein or receiving a higher energy and protein diet (OF) prepartum at +1 d. Up-stream regulators are located at the center of the network and down-stream genes are located in the periphery. Indicated are also the most enriched biological terms in the network.

Figure S4 Ingenuity Pathway upstream network analysis of differentially expressed genes (DEG) between liver of cows fed restricted (RE) energy and protein or receiving a higher energy and protein diet (OF) prepartum at +14 d. Up-stream regulators are located at the center of the network and down-stream genes are located in the periphery.

File S1 Overall dataset with annotation and results of the statistical analysis.

File S2 All results from the Dynamic Impact Approach (DIA) analysis. Reported are the results for the Kyoto Encyclopedia of Genes and Genomes (KEGG) pathways (containing the summary or overall pathways, details of each pathway, and pathways sorted by impact in each time point comparison) and Gene Ontology biological process (GOBP, both details and sorted by impact in each time point comparison).

File S3 KegArray (http://www.genome.jp/kegg/expression/) **results of several of the most affected KEGG pathways by prepartum energy level in the diet for the −14, 1, and 14 d comparison.**

File S4 REVIGO (http://revigo.irb.hr/) **summary of the Dynamic Impact Apporach (DIA) analysis of the Gene Ontology biological processes (GO BP) affected in liver by prepartum dietary energy.** The results are shown as Treemaps separated between terms more activated in OF vs. RE and the ones more activated in RE vs. OF. The dimension of each term is directly proportional to the overal induction. Same color indicate semantic and functional association.

File S5 Complete results from Database for Annotation, Visualization and Integrated Discovery (DAVID) of DEG between liver of cows fed restricted (RE) energy and protein or receiving a higher energy and protein diet (OF) prepaprtum. The analysis was performed separately between genes more expressed in OF vs. RE and genes more expressed in RE vs. OF at −14, 1, and 14 d relative to parturition.

Author Contributions

Conceived and designed the experiments: MB JJL. Performed the experiments: JJL. Analyzed the data: KS MB ET SLR. Contributed reagents/materials/analysis tools: MB ET GB JJL. Wrote the paper: KS MB ET GB JJL.

References

1. Jungermann K, Katz N (1989) Functional specialization of different hepatocyte populations. Physiol Rev 69: 708–764.

2. Drackley JK (1999) ADSA Foundation Scholar Award. Biology of dairy cows during the transition period: the final frontier? J Dairy Sci 82: 2259–2273.

3. Weber C, Hametner C, Tuchscherer A, Losand B, Kanitz E, et al. (2013) Hepatic gene expression involved in glucose and lipid metabolism in transition cows: effects of fat mobilization during early lactation in relation to milk performance and metabolic changes. Journal of Dairy Science 96: 5670–5681.

4. Trevisi E, Amadori M, Cogrossi S, Razzuoli E, Bertoni G (2012) Metabolic stress and inflammatory response in high-yielding, periparturient dairy cows. Research in Veterinary Science 93: 695–704.

5. Dann HM, Litherland NB, Underwood JP, Bionaz M, D'Angelo A, et al. (2006) Diets during far-off and close-up dry periods affect periparturient metabolism and lactation in multiparous cows. Journal of Dairy Science 89: 3563–3577.

6. Sullivan TM, Micke GC, Perry VE (2009) Influences of diet during gestation on potential postpartum reproductive performance and milk production of beef heifers. Theriogenology 72: 1202–1214.

7. Remppis S, Steingass H, Gruber L, Schenkel H (2011) Effects of Energy Intake on Performance, Mobilization and Retention of Body Tissue, and Metabolic Parameters in Dairy Cows with Special Regard to Effects of Pre-partum Nutrition on Lactation - A Review. Asian-Australasian Journal of Animal Sciences 24: 540–572.

8. McCarthy SD, Waters SM, Kenny DA, Diskin MG, Fitzpatrick R, et al. (2010) Negative energy balance and hepatic gene expression patterns in high-yielding dairy cows during the early postpartum period: a global approach. Physiological genomics 42A: 188–199.

9. Loor JJ, Dann HM, Everts RE, Oliveira R, Green CA, et al. (2005) Temporal gene expression profiling of liver from periparturient dairy cows reveals complex adaptive mechanisms in hepatic function. Physiol Genomics 23: 217–226.

10. Loor JJ, Dann HM, Guretzky NA, Everts RE, Oliveira R, et al. (2006) Plane of nutrition prepartum alters hepatic gene expression and function in dairy cows as assessed by longitudinal transcript and metabolic profiling. Physiol Genomics 27: 29–41.

11. Bionaz M, Loor JJ (2012) Ruminant Metabolic Systems Biology: Reconstruction and Integration of Transcriptome Dynamics Underlying Functional Responses of Tissues to Nutrition and Physiological State. Gene Regulation and Systems Biology: 109–125.

12. Bionaz M, Periasamy K, Rodriguez-Zas SL, Hurley WL, Loor JJ (2012) A Novel Dynamic Impact Approach (DIA) for Functional Analysis of Time-Course Omics Studies: Validation Using the Bovine Mammary Transcriptome. PloS One 7: e32455.

13. Huang DW, Sherman BT, Lempicki RA (2009) Systematic and integrative analysis of large gene lists using DAVID bioinformatics resources. Nature Protocols 4: 44–57.

14. Bionaz M, Trevisi E, Calamari L, Librandi F, Ferrari A, et al. (2007) Plasma paraoxonase, health, inflammatory conditions, and liver function in transition dairy cows. J Dairy Sci 90: 1740–1750.

15. Bertoni G, Trevisi E, Han X, Bionaz M (2008) Effects of inflammatory conditions on liver activity in puerperium period and consequences for performance in dairy cows. J Dairy Sci 91: 3300–3310.

16. Ling PR, Smith RJ, Kie S, Boyce P, Bistrian BR (2004) Effects of protein malnutrition on IL-6-mediated signaling in the liver and the systemic acute-phase response in rats. Am J Physiol Regul Integr Comp Physiol 287: R801–808.

17. Madero M, Sarnak MJ (2011) Creatinine-based formulae for estimating glomerular filtration rate: is it time to change to chronic kidney disease epidemiology collaboration equation? Curr Opin Nephrol Hypertens 20: 622–630.

18. Braun JP, Lefebvre HP, Watson AD (2003) Creatinine in the dog: a review. Vet Clin Pathol 32: 162–179.

19. Sahoo A, Pattanaik AK, Goswami TK (2009) Immunobiochemical status of sheep exposed to periods of experimental protein deficit and realimentation. J Anim Sci 87: 2664–2673.

20. Hornick JL, Van Eenaeme C, Diez M, Minet V, Istasse L (1998) Different periods of feed restriction before compensatory growth in Belgian Blue Bulls: II. Plasma metabolites and hormones. Journal of Animal Science 76: 260–271.

21. Delgiudice GD, Mech LD, Kunkel KE, Gese EM, Seal US (1992) Seasonal Patterns of Weight, Hematology, and Serum Characteristics of Free-Ranging Female White-Tailed Deer in Minnesota. Canadian Journal of Zoology-Revue Canadienne De Zoologie 70: 974–983.

22. Rajman M, Jurani M, Lamosova D, Macajova M, Sedlackova M, et al. (2006) The effects of feed restriction on plasma biochemistry in growing meat type chickens (Gallus gallus). Comp Biochem Physiol A Mol Integr Physiol 145: 363–371.

23. Jones SJ, Starkey DL, Calkins CR, Crouse JD (1990) Myofibrillar Protein-Turnover in Feed-Restricted and Realimented Beef-Cattle. Journal of Animal Science 68: 2707–2715.

24. Trevisi E, Amadori M, Bakudila AM, Bertoni G (2009) Metabolic changes in dairy cows induced by oral, low-dose interferon-alpha treatment. J Anim Sci 87: 3020–3029.

25. Baird GD (1981) Metabolic modes indicative of carbohydrate status in the dairy cow. Federation proceedings 40: 2530–2535.

26. Dale H, Vik-Mo L, Fjellheim P (1979) A field survey of fat mobilization and liver function of dairy cows during early lactation. Relationship to energy balance, appetite and ketosis. Nordisk veterinaermedicin 31: 97–105.

27. Grum DE, Drackley JK, Hansen LR, Cremin JD Jr (1996) Production, digestion, and hepatic lipid metabolism of dairy cows fed increased energy from fat or concentrate. Journal of Dairy Science 79: 1836–1849.

28. Petit HV, Palin MF, Doepel L (2007) Hepatic lipid metabolism in transition dairy cows fed flaxseed. Journal of Dairy Science 90: 4780–4792.

29. Overton TR, Drackley JK, Ottemann-Abbamonte CJ, Beaulieu AD, Clark JH (1998) Metabolic adaptation to experimentally increased glucose demand in ruminants. Journal of Animal Science 76: 2938–2946.

30. Donkin SS (2012) The Role of Liver Metabolism During Transition on Postpartum Health and Performance; 2012 January 31 - February 1, 2012; Gainesville, Florida. Institute of Food and Agricultural Sciences pp. 97–107.

31. Noguchi R, Kubota H, Yugi K, Toyoshima Y, Komori Y, et al. (2013) The selective control of glycolysis, gluconeogenesis and glycogenesis by temporal insulin patterns. Mol Syst Biol 9: 664.

32. Kusunoki M, Fukuzawa Y, Sakakibara F, Kato K, Okabayashi N, et al. (2003) Correlation between lipid contents, glycogen and pathological amelioration in liver and insulin resistance in high-fat-fed rats treated with the lipoprotein lipase activator NO-1886. Diabetes 52: A341–A341.

33. Bell AW, Burhans WS, Overton TR (2000) Protein nutrition in late pregnancy, maternal protein reserves and lactation performance in dairy cows. The Proceedings of the Nutrition Society 59: 119–126.

34. Muller MJ, Seitz HJ (1981) Starvation-induced changes of hepatic glucose metabolism in hypo- and hyperthyroid rats in vivo. The Journal of nutrition 111: 1370–1379.

35. Lobley GE (1992) Control of the metabolic fate of amino acids in ruminants: a review. Journal of Animal Science 70: 3264–3275.

36. Tucker WB, Hogue JF, Waterman DF, Swenson TS, Xin Z, et al. (1991) Role of sulfur and chloride in the dietary cation-anion balance equation for lactating dairy cattle. Journal of animal science 69: 1205–1213.

37. Spears JW, Lloyd KE, Fry RS (2011) Tolerance of cattle to increased dietary sulfur and effect of dietary cation-anion balance. J Anim Sci 89: 2502–2509.

38. Hebbring SJ, Adjei AA, Baer JL, Jenkins GD, Zhang J, et al. (2007) Human SULT1A1 gene: copy number differences and functional implications. Human molecular genetics 16: 463–470.

39. Akbar H, Bionaz M, Carlson DB, Rodriguez-Zas SL, Everts RE, et al. (2013) Feed restriction, but not l-carnitine infusion, alters the liver transcriptome by inhibiting sterol synthesis and mitochondrial oxidative phosphorylation and increasing gluconeogenesis in mid-lactation dairy cows. J Dairy Sci 96: 2201–2213.

40. Grala TM, Kay JK, Phyn CV, Bionaz M, Walker CG, et al. (2013) Reducing milking frequency during nutrient restriction has no effect on the hepatic transcriptome of lactating dairy cattle. Physiol Genomics.

41. Prinsen BH, Romijn JA, Bisschop PH, de Barse MMJ, Barrett PHR, et al. (2003) Endogenous cholesterol synthesis is associated with VLDL-2 apoB-100 production in healthy humans. Journal of Lipid Research 44: 1341–1348.

42. Etzler ME, Esko JD (2009) Free Glycans as Signaling Molecules. In: Ajit Varki, Richard D Cummings, Jeffrey D Esko, Hudson H Freeze, Pamela Stanley et al. Essentials of Glycobiology. 2nd ed. New York: Cold Spring Harbor Laboratory Press.

43. Fagioli C, Sitia R (2001) Glycoprotein quality control in the endoplasmic reticulum. Mannose trimming by endoplasmic reticulum mannosidase I times the proteasomal degradation of unassembled immunoglobulin subunits. The Journal of biological chemistry 276: 12885–12892.

44. Lu Y, Xu YY, Fan KY, Shen ZH (2006) 1-Deoxymannojirimycin, the alpha1,2-mannosidase inhibitor, induced cellular endoplasmic reticulum stress in human hepatocarcinoma cell 7721. Biochemical and biophysical research communications 344: 221–225.

45. Loor JJ (2010) Genomics of metabolic adaptations in the peripartal cow. Animal 4: 1110–1139.

46. Mandard S, Patsouris D (2013) Nuclear control of the inflammatory response in mammals by peroxisome proliferator-activated receptors. PPAR Res 2013: 613864.

47. Daley CA, Abbott A, Doyle PS, Nader GA, Larson S (2010) A review of fatty acid profiles and antioxidant content in grass-fed and grain-fed beef. Nutrition journal 9: 10.

48. Metz S, VanRollins M, Strife R, Fujimoto W, Robertson RP (1983) Lipoxygenase pathway in islet endocrine cells. Oxidative metabolism of arachidonic acid promotes insulin release. The Journal of clinical investigation 71: 1191–1205.

49. Oikonomopoulou K, Ricklin D, Ward PA, Lambris JD (2012) Interactions between coagulation and complement-their role in inflammation. Seminars in Immunopathology 34: 151–165.

50. Loor JJ, Moyes KM, Bionaz M (2011) Functional adaptations of the transcriptome to mastitis-causing pathogens: the mammary gland and beyond. J Mammary Gland Biol Neoplasia 16: 305–322.

51. Huls M, Russel FG, Masereeuw R (2009) The role of ATP binding cassette transporters in tissue defense and organ regeneration. The Journal of pharmacology and experimental therapeutics 328: 3–9.

52. Hu Y, Sampson KE, Heyde BR, Mandrell KM, Li N, et al. (2009) Saturation of multidrug-resistant protein 2 (mrp2/abcc2)-mediated hepatobiliary secretion: nonlinear pharmacokinetics of a heterocyclic compound in rats after intravenous bolus administration. Drug Metab Dispos 37: 841–846.

53. Federico A, D'Aiuto E, Borriello F, Barra G, Gravina AG, et al. (2010) Fat: a matter of disturbance for the immune system. World J Gastroenterol 16: 4762–4772.

54. Farney JK, Mamedova LK, Coetzee JF, KuKanich B, Sordillo LM, et al. (2013) Anti-inflammatory salicylate treatment alters the metabolic adaptations to lactation in dairy cattle. Am J Physiol Regul Integr Comp Physiol 305: R110–117.

55. Apte UM, McRee R, Ramaiah SK (2004) Hepatocyte proliferation is the possible mechanism for the transient decrease in liver injury during steatosis stage of alcoholic liver disease. Toxicologic pathology 32: 567–576.

56. Varady KA, Roohk DJ, Bruss M, Hellerstein MK (2009) Alternate-day fasting reduces global cell proliferation rates independently of dietary fat content in mice. Nutrition 25: 486–491.

57. Anderson N, Borlak J (2008) Molecular mechanisms and therapeutic targets in steatosis and steatohepatitis. Pharmacological reviews 60: 311–357.

58. Valenti L, Mendoza RM, Rametta R, Maggioni M, Kitajewski C, et al. (2013) Hepatic Notch Signaling Correlates with Insulin Resistance and Non-Alcoholic Fatty Liver Disease. Diabetes.

59. Flach RJ, Qin H, Zhang L, Bennett AM (2011) Loss of mitogen-activated protein kinase phosphatase-1 protects from hepatic steatosis by repression of cell death-inducing DNA fragmentation factor A (DFFA)-like effector C (CIDEC)/fat-specific protein 27. The Journal of biological chemistry 286: 22195–22202.

60. Bionaz M, Chen S, Khan MJ, Loor JJ (2013) Functional Role of PPARs in Ruminants: Potential Targets for Fine-Tuning Metabolism during Growth and Lactation. PPAR Res 2013: 684159.

61. Kersten S, Seydoux J, Peters JM, Gonzalez FJ, Desvergne B, et al. (1999) Peroxisome proliferator-activated receptor alpha mediates the adaptive response to fasting. J Clin Invest 103: 1489–1498.

62. Kurzawski M, Dziedziejko V, Urasinska E, Post M, Wojcicki M, et al. (2012) Nuclear factor erythroid 2-like 2 (Nrf2) expression in end-stage liver disease. Environ Toxicol Pharmacol 34: 87–95.

63. Viana LD, Affonso RJ, Silva SRM, Denadai MVA, Matos D, et al. (2013) Relationship between the Expression of the Extracellular Matrix Genes SPARC, SPP1, FN1, ITGA5 and ITGAV and Clinicopathological Parameters of Tumor Progression and Colorectal Cancer Dissemination. Oncology 84: 81–91.

64. Anderson DE, Hinds MT (2012) Extracellular matrix production and regulation in micropatterned endothelial cells. Biochemical and biophysical research communications 427: 159–164.

65. Benjamini Y, Hochberg Y (1995) Controlling the False Discovery Rate: A Practical and Powerful Approach to Multiple Testing. Journal of the Royal Statistical Society 57.

66. Supek F, Bosnjak M, Skunca N, Smuc T (2011) REVIGO summarizes and visualizes long lists of gene ontology terms. PLoS One 6: e21800.

67. Hosack DA, Dennis G Jr, Sherman BT, Lane HC, Lempicki RA (2003) Identifying biological themes within lists of genes with EASE. Genome biology 4: R70.

5

The *N*-Reductive System Composed of Mitochondrial Amidoxime Reducing Component (mARC), Cytochrome b5 (CYB5B) and Cytochrome b5 Reductase (CYB5R) Is Regulated by Fasting and High Fat Diet in Mice

Heyka H. Jakobs[1], Michal Mikula[2], Antje Havemeyer[1], Adriana Strzalkowska[2], Monika Borowa-Chmielak[2], Artur Dzwonek[2], Marta Gajewska[2], Ewa E. Hennig[2,3], Jerzy Ostrowski[2,3], Bernd Clement[1]*

1 Department of Pharmaceutical and Medicinal Chemistry, Christian-Albrechts-Universität zu Kiel, Kiel, Germany, 2 Department of Genetics, Maria Sklodowska-Curie Memorial Cancer Center and Institute of Oncology, Warsaw, Poland, 3 Department of Gastroenterology and Hepatology, Medical Center for Postgraduate Education, Warsaw, Poland

Abstract

The mitochondrial amidoxime reducing component mARC is the fourth mammalian molybdenum enzyme. The protein is capable of reducing *N*-oxygenated structures, but requires cytochrome b5 and cytochrome b5 reductase for electron transfer to catalyze such reactions. It is well accepted that the enzyme is involved in *N*-reductive drug metabolism such as the activation of amidoxime prodrugs. However, the endogenous function of the protein is not fully understood. Among other functions, an involvement in lipogenesis is discussed. To study the potential involvement of the protein in energy metabolism, we tested whether the mARC protein and its partners are regulated due to fasting and high fat diet in mice. We used qRT-PCR for expression studies, Western Blot analysis to study protein levels and an *N*-reductive biotransformation assay to gain activity data. Indeed all proteins of the *N*-reductive system are regulated by fasting and its activity decreases. To study the potential impact of these changes on prodrug activation *in vivo*, another mice experiment was conducted. Model compound benzamidoxime was injected to mice that underwent fasting and the resulting metabolite of the *N*-reductive reaction, benzamidine, was determined. Albeit altered *in vitro* activity, no changes in the metabolite concentration *in vivo* were detectable and we can dispel concerns that fasting alters prodrug activation in animal models. With respect to high fat diet, changes in the mARC proteins occur that result in increased *N*-reductive activity. With this study we provide further evidence that the endogenous function of the mARC protein is linked with lipid metabolism.

Editor: Antonio Moschetta, IRCCS Istituto Oncologico Giovanni Paolo II, Italy

Funding: This work was supported by N N401 017436 and N N401 532240 grants from Polish Ministry of Science and Higher Education. The funders had no role in study design, data collection and analysis, decision to publish, or preparation of the manuscript.

Competing Interests: The authors have declared that no competing interests exist.

* Email: bclement@pharmazie.uni-kiel.de

Introduction

The mitochondrial amidoxime reducing component (mARC) is the fourth mammalian molybdenum containing enzyme besides sulfite oxidase, xanthine oxidase and aldehyde oxidase [1], [2]. All so far fully sequenced and annotated mammalian genomes encode two isoforms mARC1 and mARC2, which show a high degree of sequence similarities [3]. Both enzymes are capable of reducing *N*-oxygenated structures, but require at least mitochondrial cytochrome b5 (CYB5B) [4], [5] in cell culture or both CYB5B and NADH-cytochrome b5 reductase (CYB5R) [5–7] for activity reconstitution *in vitro* (Figure 1). The activity of the three component system depends on the molybdenum cofactor bound to mARC and hem in CYB5B [3], [5], [6]. It is well accepted that the enzyme plays a major role in *N*-reductive drug metabolism and is thus a counterpart of the main drug oxidizing enzymes cytochrome P450s (P450) and flavin-containing monooxygenases

(FMO). The *N*-reductive reaction of mARC and its partners is likewise involved in the activation of amidoxime prodrugs, precursor molecules of active drugs with improved pharmacokinetic properties [8–10].

However, little is known about the endogenous functions of the enzyme system and further studies on this topic are necessary. An involvement in detoxification, as common for drug metabolizing enzymes, was proven: The enzyme is capable of detoxifying hydroxylated and mutagenic DNA-bases [7] and mARC2 was shown to be downregulated in colorectal carcinoma tissues [11].

Nevertheless, the involvement in other metabolic pathways is also discussed, including NO metabolism where the enzyme is involved in the reduction of NO precursor substrate N^{ω}-hydroxy-L-arginine [6] and furthermore capable of reducing nitrite to NO under anaerobic conditions [12].

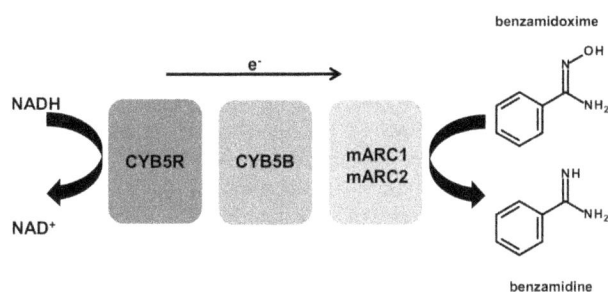

Figure 1. Scheme of the N-reductive reaction with model compound benzamidoxime by mARC, CYB5B and CYB5R. Electrons are transferred from NADH via FADH in CYB5R, hem in CYB5B and Moco in mARC1/2 to benzamidoxime, the latter is reduced to benzamidine.

Moreover, increased mRNA levels for mARC2 have been found in diabetic animal models and in cell culture under high glucose conditions [13]. High abundances of the enzyme after differentiation of 3T3-cells to adipocytes were shown [4], [14] and recently Neve *et al.* provided evidence on its function in lipid synthesis [4].

As the latter findings indicate an involvement in energy metabolism, we studied how changes in energy supply influence the mARC containing enzyme system. Besides cell culture experiments, we studied changes in mice under fasting conditions, high fat diet (HFD) and in leptin receptor-deficient mice. Under the hypothesis that the enzyme system is connected with energy metabolism, changes of the mARC proteins and its partners are assumable in these states. Fasting in mice causes body weight loss, as well as changes in cardiovascular, hormonal and metabolic parameters and hepatic metabolism [15]. In contrast, HFD and leptin-receptor-deficiency results in an obese habitus with most of the symptoms of the metabolic syndrome [16]. In both cell culture and mice experiments, we focused on mRNA and protein levels such as the *in vitro* activity of the enzyme complex. Indeed, changes in the three components of the enzyme system due to glucose in cell culture and due to fasting and HFD in mice were detectable and support the assumption that the enzyme system is connected with energy metabolism. Additionally, we tested whether food deprivation in mice influences N-reductive metabolite concentrations *in vivo*. Potential changes in N-reductive activity due to fasting could imply problems in both preclinical drug testing in rodents and in the application of mARC-activated prodrugs. However, we have been able to dispel these concerns during this study.

Materials and Methods

Materials

Methanol HPLC grade was purchased from J. T. Baker (Deventer, The Netherlands). PBS was from PAA Laboratories GmbH (Pasching, Austria). FBS and DMEM without glucose were from gibco (Life Technologies, Carlsbad, USA). Other chemicals were purchased from Sigma-Aldrich (Munich, Germany), Merck KGaA (Darmstadt, Germany) or Roth (Karlsruhe, Germany) unless otherwise stated.

Cell culture

Human HepG2 and mouse Hepa 1.6 hepatocellular carcinoma cell lines were purchased from ATCC (Manassas, USA). Cells were seeded on 6-well plates in 10% (v/v) FBS DMEM containing

4.5 g/l glucose. Next day medium was switched for 0.5% (v/v) FBS DMEM either with glucose at 1.0/4.5 g/l or glucose deprived. Cells were cultured for another 48 h and then collected for RNA extraction or used in N-reduction assay.

Cellular Protein extraction

Cells were washed with ice cold PBS, scratched and collected in a tube. PBS was removed after centrifugation and cells were resuspended in ice cold lysis buffer (1% (v/v) Nonident P-40, 150 mM NaCl, 50 mM Tris, pH 8.0, protease inhibitor mixture (complete ULTRA; Roche Diagnostics, Laval, Canada)) and shaken for 1 h at 4°C. Lysates were centrifuged and supernatant stored at −80°C for later use in Western Blot.

Animals

Male C57BL/6W mice and B6.V-Lep^{ob}/J mice (ob/ob), homozygous for leptin gene mutation, were housed under temperature (21±2°C) and humidity (55±10%) controlled conditions, with a 12 h light/dark cycle and ad libitum access to food and water.

Ethics statement

Mice were housed in the Department of Genetics and Laboratory Animals Breading, at the Maria Sklodowska-Curie Memorial Cancer Center and Institute of Oncology. Experimental protocol was approved by the 2nd Local Ethical Committee for Animal Research in Warsaw, Poland. For benzamidoxime admistration, each animal was given anesthesia by intraperitoneal administration of pentobarbital at a dose of 40 mg/kg. Animals were euthanized by isoflurane overdose followed by cervical dislocation.

Animal fasting experiment design

All animals were fed regular diet (10% of calories from fat), containing 19.2% protein, 67.3% carbohydrate, and 4.3% fat (D12450B; Research Diets, New Brunswick, USA). In a first experiment 24 animals were used and 12 of them were deprived of food for a period of 18 h before sacrifice followed by liver collection. In a second experiment 28 animals were used and 14 of them were deprived of food for a period of 24 h prior to benzamidoxime administration. Benzamidoxime was reconstituted in water (injection grade). Each animal was given anesthesia by intraperitoneal administration of pentobarbital at a dose of 40 mg/kg. Benzamidoxime was then delivered by intravenous injection at 20 mg/kg dose and 30 minutes later mice were sacrificed followed by immediate collection of liver and blood for plasma. Samples were snap-frozen and stored at −72°C until use.

Animal HFD experiment design

Both one group of C57BL/6W mice (control) and ob/ob-mice were fed regular diet (10% of calories from fat), containing 19.2% protein, 67.3% carbohydrate, and 4.3% fat (D12450B; Research Diets, New Brunswick, USA). A second group of C57BL/6W mice were fed HFD (60% of calories from fat), containing 26.2% protein, 26.3% carbohydrate, and 34.9% fat (D12492; Research Diets, New Brunswick, USA). At 16 weeks of age, mice were sacrificed followed by immediate collection of livers. Samples were snap-frozen and stored at −72°C until use.

Preparation of mice liver homogenates

Mice livers were cut into pieces and homogenized using a Potter S homogenizer (Satorius, Goettingen, Germany) in ice cold buffer containing 0.25 M saccharose, 1 mM EDTA, 10 mM potassium dihydrogene phosphate and 1 mM dithiotreitol with pH 7.4. Samples were frozen and stored at −80°C for later use.

Determination of protein content

For both homogenates and cell culture lysates, protein content was determined using bicinchoninic acid (BCA) protein assay kit (Pierce, Rockford, USA) according to manufacturer's protocol.

RNA extraction and expression studies

Total RNA was isolated from liver and cell culture samples using the RNeasy Plus Mini Kit (Qiagen, Hilden, Germany) or TRIzol Plus RNA Purification Kit (Life Technologies, Carlsbad, USA), respectively, followed by on-column DNAse I digestion. One µg of total RNA and random hexamers were used in cDNA synthesis with Superscript III according to manufacturer's protocol (Life Technologies, Carlsbad, USA). Levels of specific mRNAs were assessed by quantitative real-time PCR (qRT-PCR) using primer pairs at final concentration of 200 nM (Table 1). QRT-PCR was carried out in ABI 7900HT Fast Real-Time PCR System with Sensimix SYBR kit (Bioline, Boston, USA) using standard cycling conditions at 40 cycles consisting of 15 s of denaturation at 95°C and hybridisation for 1 min at 60°C in a 384-well reaction plate. 60S acidic ribosomal protein P0 (RPLP0) mRNA expression was used as reference mRNA for HepG2 samples. Geometric mean of Ct values for Mcoln1(Mucolipin-1) and Hmbs (hydroxymethylbilane synthase) mRNAs was used to normalize gene expression in Hepa 1.6 and liver samples. Mean gene expression was calculated with delta-delta Ct (ddCt) method.

Western Blot analysis

For SDS-PAGE, separation gels containing 12.5% (v/v) of polyacrylamide were used according to the method of Laemmli [17]. Protein samples were diluted to equal protein concentrations and pretreated with β-mercaptoethanol for 10 min at 100°C prior to loading onto the gel. After separation, proteins were blotted on a PVDF transfer membrane (Amersham Hybond P Membrane; GE Healthcare, Buckinghamshire, UK) and blocked for at least one hour in 5% (m/v) milk powder in TBS buffer containing Tween 20 (TBST). For immuno-blot analysis, the following primary antibodies were used: anti-CYB5R3-antibody (HPA001566; Sigma-Aldrich, St. Louis, USA), anti-mARC2-

antibody (HPA015085; Sigma-Aldrich, St. Louis, USA), anti-CYB5B-antibody (HPA007893; Sigma-Aldrich, St. Louis, USA), anti-PCK1-antibody (ab28455; Abcam, Cambridge, UK), anti-MOSC1-antibody (AP9754c; Abgent, San Diego, USA) for of HepG2 lysates or anti-MOSC1-antibody (ABIN503067; antibodies-online GmbH, Aachen, Germany) for murine liver and Hepa 1.6 lysates. Either anti-histone H3-antibody (ab1791; Abcam, Cambridge, UK) or anti-calnexin-antibody (AP03028SU-N; Acris, Herford, Germany) were used as loading control. Incubation with primary antibodies was carried out over night at 4°C; membranes were then washed with TBST. For visualization, a secondary horseradish peroxidase-conjugated anti-rabbit-immunoglobuline-G-antibody (Jackson Immuno Research Laboratories, Suffolk, UK) was used, incubation time was one hour. After washing in TBST, chemiluminescence detection was carried out with Amerscham ECL Plus Western Blotting Detection System (GE Healthcare, Buckinghamshire, UK) according to the manufacturer's instructions.

N-reductive activity assay for mice liver homogenates

N-reductive activity was determined by measuring the reduction of the model compound benzamidoxime to benzamidine. 50 ng of homogenate were incubated in a total volume of 150 µl 100 mM potassium phosphate buffer containing 3 mM benzamidoxime at 37°C. After 3 min of preincubation, the reaction was initiated by addition of 1 mM NADH and stopped after 20 min by addition of 150 µl ice-cold methanol. Samples were centrifuged and the supernatant was analyzed via HPLC.

N-reductive activity assay for cell cultures

For N-reductive activity assay in cell cultures, cell culture medium was removed and cells were gently washed and preincubated for 15 min with benzamidoxime-free incubation buffer (Hanks balanced salt solution and 10 mM HEPES, pH 7.4) at 37°C. The buffer was carefully replaced by 1 ml of incubation buffer containing 5 mM benzamidoxime and incubation was carried out for 120 min. Afterwards supernatant was frozen at −20°C. Later samples were centrifuged to remove cellular contamination and analyzed via HPLC.

Table 1. Primers used for expression studies.

Species	Gene name	Forward (5'→3')	Reverse (5'→3')
Human	RPLP0	GCAATGTTGCCAGTGTCTG	GCCTTGACCTTTTCAGCAA
Human	mARC1	ACTCCAGTGTCTGGGTCCAC	CAGGCCAAATATTGTGGTGA
Human	mARC2	CAGCACAAAATACTGCCCAA	AACCGCTGGAAACACTGAAG
Human	CYB5B	GCTTGTTCCAGCAGAACCTC	GGAACTGTGGCTTGTGATCC
Human	CYB5R3	AGGATGTGCTGGGGTGAC	AGCCCGGACATCAAGTACC
Human	PCK1	GATTGTGTTCTTCTGGATGGT	TGACGCACAAGGTCATTTAAG
Mouse	Hmbs	AAGGGTTTTCCCGTTTGC	TCCCTGAAGGATGTGCCTA
Mouse	Mcoln1	CCACCACGGACATAGGCATAC	GCTGGGTTACTCTGATGGGTC
Mouse	mARC1	CATTTGCCGAGAACTTCTGGG	ATGCAGCTCAGTGGGTCAG
Mouse	mARC2	GAACCTGTCGCGCACTTTG	GGATCTACCCGATCAAGTCCT
Mouse	CYB5B	AGCTTTCAGTTGCATCAGCAC	GAGCCCTCCGTCACCTACTA
Mouse	CYB5R3	CAGGCCGCAACAGGATATCT	AACGACCACACCGTGTGCTA
Mouse	PCK1	GCCTTCCACGAACTTCCTCAC	CTGCATAACGGTCTGGACTTC

Determination of benzamidine in mice plasma and homogenates

Plasma was mixed with two aliquots and homogenates were mixed with twenty aliquots of acetonitrile and shaken for 30 min at 4°C. After centrifugation for 30 min, the supernatant was dried using a Christ Alpha 2–4 freeze dryer (Martin Christ Gefrier-trocknungsanlagen GmbH, Osterode am Harz, Germany). The resulting lyophilisate was resolved in mobile phase for HPLC analysis (10 mM octylsulfonate sodium salt, 17% acetonitrile), shaken for 45 min at 4°C and cleared again by centrifugation for 30 min. The resulting supernatant was analyzed by HPLC.

Quantification of the metabolite benzamidine by HPLC

In case of N-reductive activity assay with cell cultures, benzamidine was quantified using a Phenomenex Gemini (150×4.6 mm) 5 µM C_{18} column with a Phenomenex C_{18} 4×3 mm guard-column (Phenomenex, Aschaffenburg, Germany). The mobile phase consisted of 50 mM ammonium acetate buffer, pH 7.0 and 10% (v/v) methanol. Flow rate was 1 ml/min at room temperature and detection was carried out at 229 nm. Retention time for benzamidine was 5.2±0.1 min and for benzamidoxime 12.0±0.2 min.

In case of in vitro N-reductive activity assay with homogenates and quantification of benzamidine in plasma and homogenates of animals that were treated with benzamidoxime, quantification was carried out as follows: A LiChroCHART 250×4 mm column with LiChrospher 60 RP-select B (5 µm) and a RP-select B 4×4 mm guard column (Merck kGaA, Darmstadt, Germany) was used. The mobile phase consisted of 10 mM sodium octylsulfonate sodium salt and 17% (v/v) acetonitrile. Flow was kept isocratically at 1 ml/min at room temperature and detection was carried out at 229 nm. Retention time for benzamidine was 25.5±0.2 min and for benzamidoxime 6.9±0.1 min.

Statistical analyses

Differences in pair wise comparisons were evaluated using the U-test in GraphPad Prism 5 (GraphPad Software, Inc.; CA, USA). A p-value of less than 0.05 was considered significant (*), (**) = p-value <0.001(**), (***) = p-value <0.0001.

Results

Glucose deprivation downregulates mARC complex components abundance and its activity in cell culture

To test if the mARC-containing enzyme system is influenced by the state of nutrient depletion, we used in vitro human HepG2 and mouse Hepa 1.6 hepatocellular carcinoma cell line models cultured with and without glucose. Glucose withdrawal is regarded as a condition that could mimic physiological fasting in cell culture [18]. Hepatocellular carcinoma cell lines were chosen as the liver is the main organ of energy metabolism. Cells were cultured in cell culture medium without (0.0 g/l) or with either 1.0 g/l or 4.5 g/l glucose and gene and protein expression measurements were performed. Gene expressions for mARC1, mARC2, CYB5R and CYB5B were clearly enhanced in HepG2 cells that were grown either in 1.0 g/l or 4.5 g/l glucose. For Hepa 1.6 cells only 4.5 g/l glucose substantially increased mARC2, CYB5R and CYB5B mRNA abundance without effect on mARC1 transcription level (Figure 2A). Transcripts changes for HepG2 cell line were mirrored in protein abundances where mARC1, CYB5R and CYB5B protein levels measured with Western Blot were higher in cells cultured with glucose. mARC2 was not detectable in any cell lysates. The protein levels examined in Hepa 1.6 cells correspond-

ed with mRNA changes observed for culture conditions with 4.5 g/l glucose. In these cells mARC1 protein could not be detected (Figure 2B). Both hepatoma cell lines were next incubated with model compound benzamidoxime and the reduced metabolite benzamidine was determined in cell culture supernatant to determine N-reductive activity. Cell lines grown in medium containing 1.0 g/l or 4.5 g/l glucose showed significantly higher N-reductive activity compared to cells grown without glucose (Figure 2C).

Fasting in mice down-regulates mARC complex components abundance and its activity

To study the influence of food deprivation on the mARC protein and its electron transfer partner proteins in vivo, mice experiments with 18 h and 24 h food deprivation were carried out. Livers were examined, as the liver is a center of metabolic conversions and energy metabolism and changes due to food deprivation are likely to occur in this organ. Gene expression and protein abundances for CYB5B and CYB5R were clearly decreased in the fasting group after both 18 and 24 h of food deprivation. Marker protein PCK1 was enhanced in the fasting groups of both experiments and thus confirmed the fasting state (Figure 3 & Figures S1A, S2). Gene expression for both mARC1 and mARC2 was enhanced in the fasted groups (Figures 3A & 4A); however no changes in protein levels for mARC2 were detectable after 18 h of fasting (Figure 3B & Figure S1A). In contrast, after 24 h of fasting, decreased abundance of mARC2 was detectable in the fasted group compared to control (Figure 4B). For mARC1, an antibody with reactivity against mARC1 detected a protein with increased abundance upon 18 h of fasting but with decreased protein levels with 24 h of fasting. However, the molecular weight of the detected protein was about 65 kDa and thus not the predicted molecular weight of 37 kDa of mARC1 (Figures 3B & 4B, Figure S1A). The N-reductive activity was determined in an in vitro assay with liver homogenates, where the reduction of the model compound benzamidoxime to benzamidine is analyzed. The N-reductive activity was clearly reduced in mice that underwent 18 h or 24 h of fasting compared to control (Figure 4C & Figure S1B).

Fasting in mice does not influence concentrations of the metabolite of the N-reduction

To study whether fasting influences the concentration of the metabolite of the N-reductive reaction (benzamidine), benzamidoxime was administered to mice that fasted for 24 h or had full access to food (control). First, we performed a pilot study to set a time of sampling after benzamidoxime administration that indicated rapid metabolite decline in a time dependent manner in both plasma and liver tissue, where mother compound was undetectable in plasma after 1 hour (Figure S3). Based on this data we collected liver and plasma samples 30 minutes after benzamidoxime dosing and then level of benzamidine were determined with HPLC. However, no statistically significant differences for the metabolite benzamidine could be found in plasma or in liver homogenates of fasted and control animals (Figure 5).

HFD but not hyperphagia influences mARC abundance and its activity in mice

To test the influence of HFD and hyperphagic behavior on the N-reductive system we used liver tissue samples of HFD-fed and ob/ob-mice described in recent studies by Nesteruk et al.

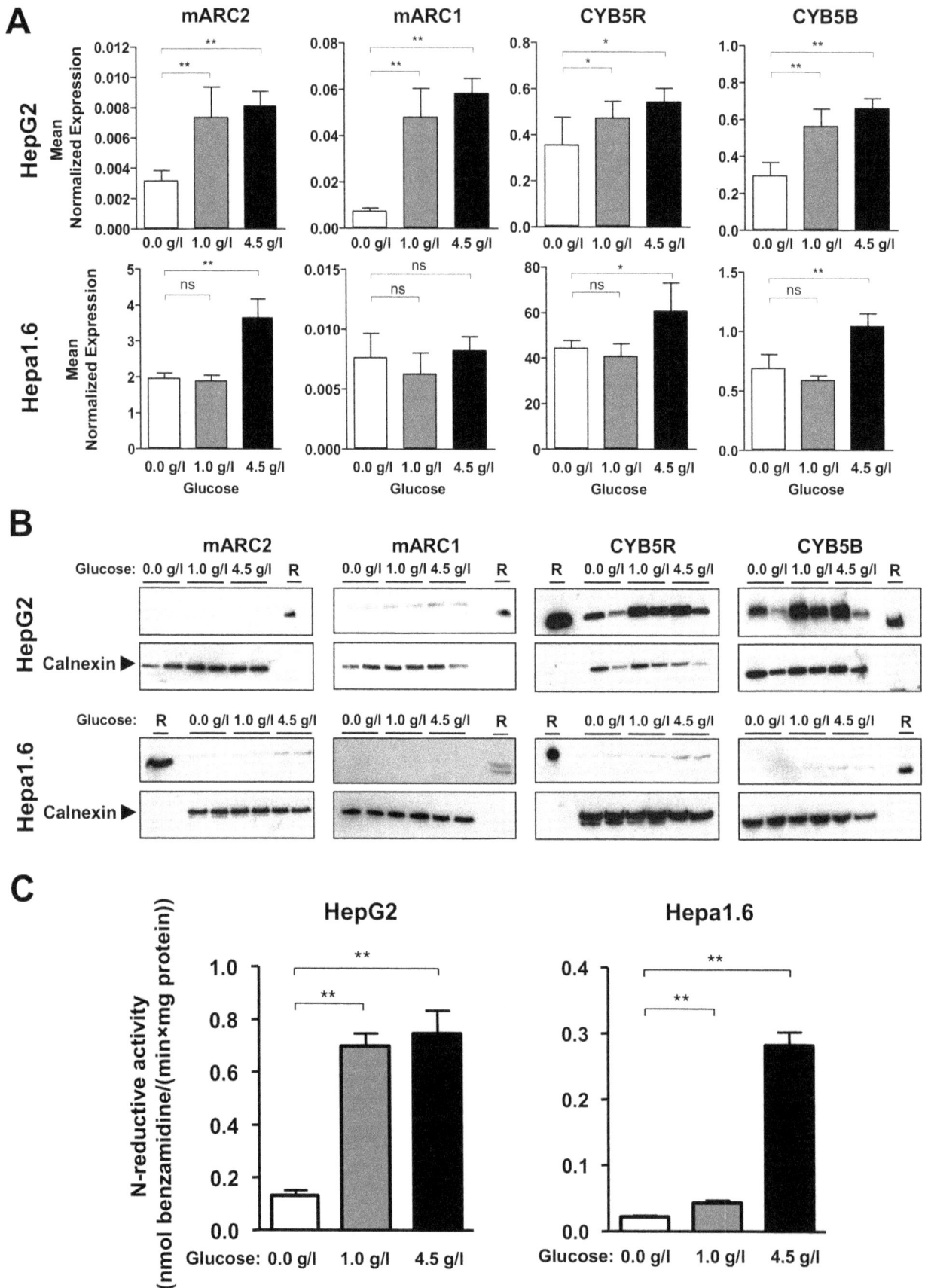

Figure 2. Glucose-dependent changes with expression, abundance and activity of the *N*-reductive system in HepG2 and Hepa 1.6 cell lines. A Expressions of mARC2, mARC1, CYB5R and CYB5B in HepG2 and Hepa 1.6 cell lines cultured without or with 1.0/4.5 g/l glucose

determined by qRT-PCR. Statistical significance was assessed by the U-test. *p*-values <0.05 were considered significant (*); (**) *p*-value <0.001, ns = not significant. **B** Protein levels of mARC2, mARC1, CYB5R and CYB5B in hepatoma cell lines cultured without or with 1.0/4.5 g/l glucose, examined by Western Blot. R = recombinant proteins/control **C** *N*-reductive activity determined in in hepatoma cell lines by determining the reduction of model compound benzamidoxime. The resulting metabolite benzamidine was quantified by HPLC analysis. Determined activities are means ± SD of six biological samples, each measured as duplicates. Statistical significance was assessed by the U-test. *p*-values <0.05 were considered significant (*); (**) *p*-value <0.001.

and Hennig *et al.* [19], [20]. Livers of HFD-fed mice showed decreased mRNA expressions for mARC1, CYB5B and CYB5R, but no changes for mARC2 compared to control (Figure 6A). Contrary, an increased protein abundance for mARC2 was detected, whereas no differences in the abundance of CYB5B and CYB5R could be demonstrated. An antibody with the reactivity against mARC1 detected a protein with increased abundance due to HFD. However the molecular weight of about 65 kDa of the detected protein did not fit the predicted molecular weight of mARC1 with 37 kDa (Figure 6B & Figure S4). Thus, as with mice experiments with fasting, mARC mRNA and protein level do not show same tendencies.

The *N*-reductive activity, determined with the reduction of the model compound benzamidoxime to benzamidine, was also enhanced in livers of HFD-fed mice (Figure 6C). In contrast to the HFD-fed mice, no changes with respect to protein abundance or *N*-reductive activity compared to control were detectable for obese hyperphagic ob/ob-mice (Figure 6 B–C & Figure S4).

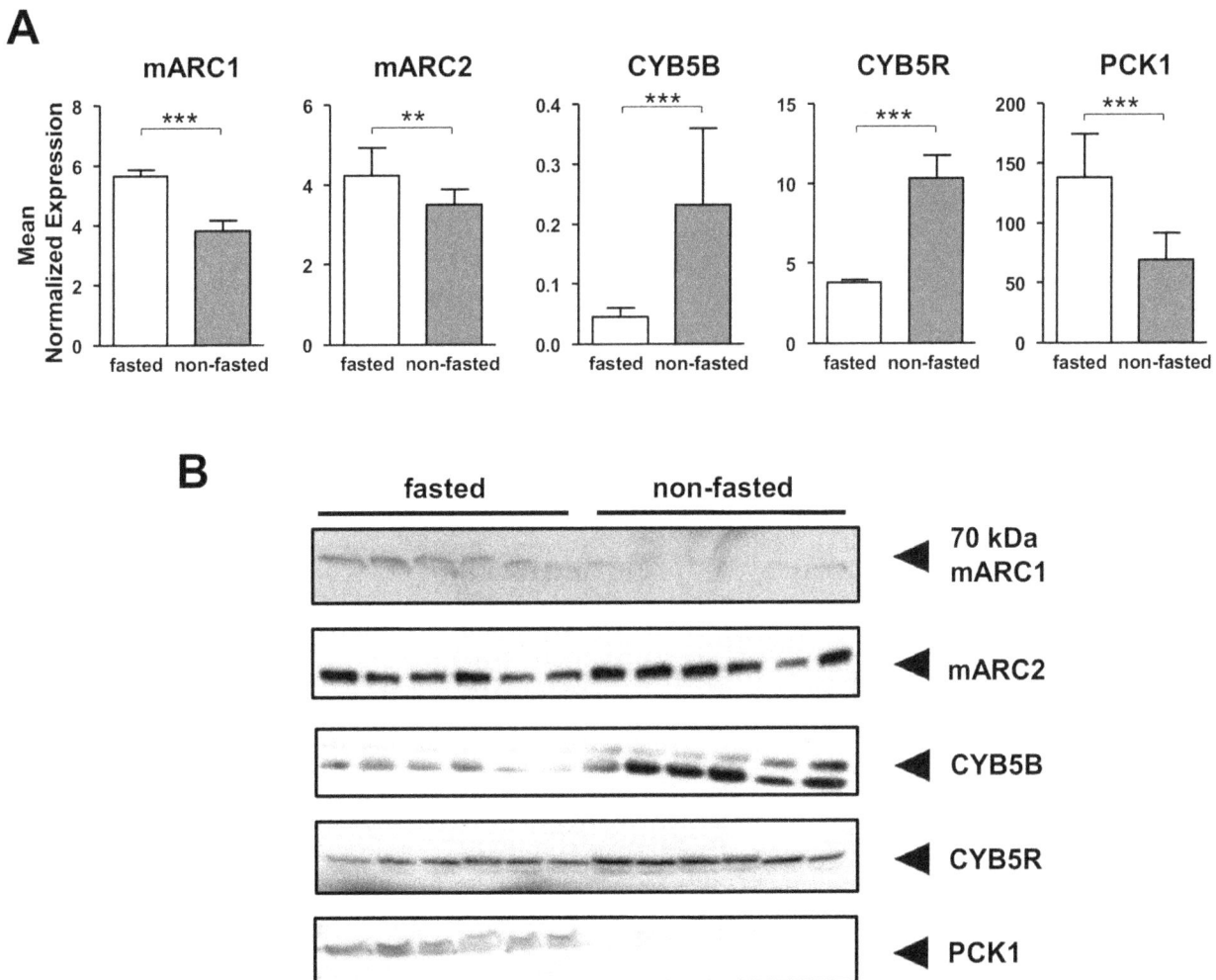

Figure 3. Effects of fasting on expression and protein abundance of the *N*-reductive system in mice. Two groups of 12 C57BL/6W mice were fed with regular diet, one group were food deprived for 18 h (fasted) before sacrifice and liver collection, the second had full access to food and water (non-fasted). **A** Expressions of mARC1, mARC2, CYB5B, CYB5R and PCK1, determined with qRT-PCR, normalized on geometric mean of Ct values for Mcoln1 (Mucolipin-1) and Hmbs (hydroxymethylbilane synthase). Statistical significance was assessed by the U-test. *p*-values <0.05 were considered significant; (**) *p*-value <0.001; (***) *p*-value <0.0001. **B** Protein levels of mARC1, mARC2, CYB5B, CYB5R and PCK1 examined by Western Blot. Each sample consisted of equal protein amount of two individuals.

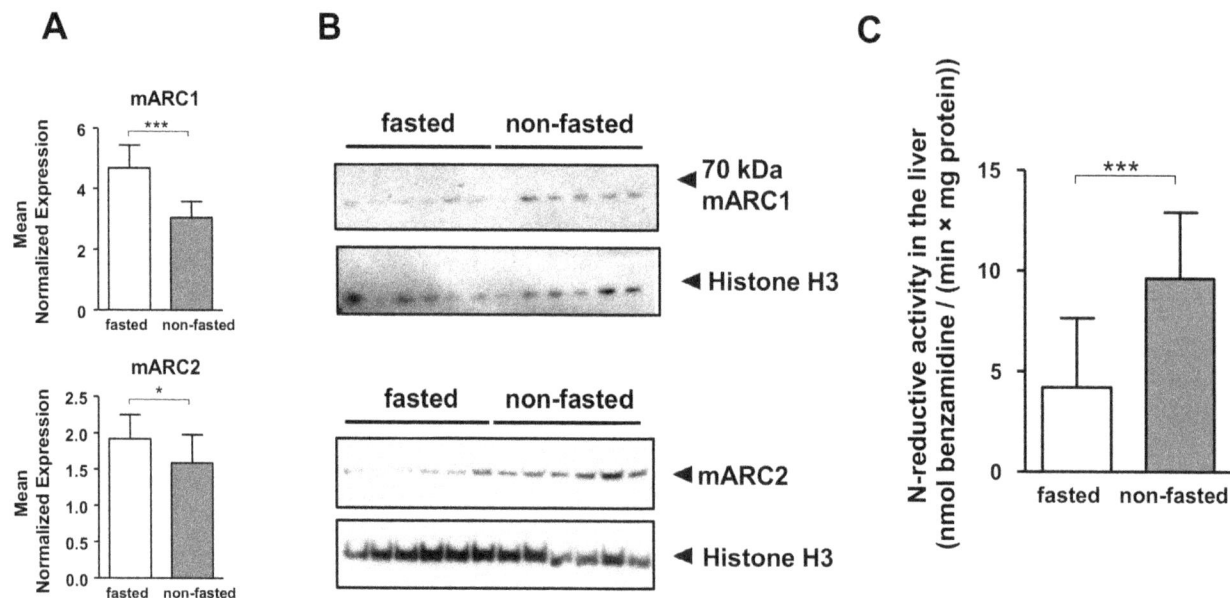

Figure 4. 24 h fasting decreases mARC protein levels and N-reductive activity in mice. Two groups of 14 C57BL/6W mice were fed with regular diet, one group were food deprived for 24 h (fasted) before sacrifice and liver collection, the second had full access to food and water (non-fasted). **A** Expressions of mARC1 and mARC2, determined with qRT-PCR, normalized on expression of Mcoln1 (Mucolipin-1) and Hmbs (hydroxymethylbilane synthase). Statistical significance was assessed by the U-test. p-values <0.05 were considered significant (*); (**) p-value $<$ 0.001; (***) p-value <0.0001. **B** Protein levels of mARC1, mARC2 and histone H3 (loading control) examined by Western Blot. Each sample consisted of equal protein amount of two individuals. **C** N-reductive activity determined by the reduction of model compound benzamidoxime in liver homogenate. The resulting metabolite benzamidine was quantified by HPLC analysis. Determined activities are means $\pm$ SD of 14 biological samples, each measured as duplicates. Statistical significance was assessed by the U-test. p-values <0.05 were considered significant (*); (***) p-value <0.0001.

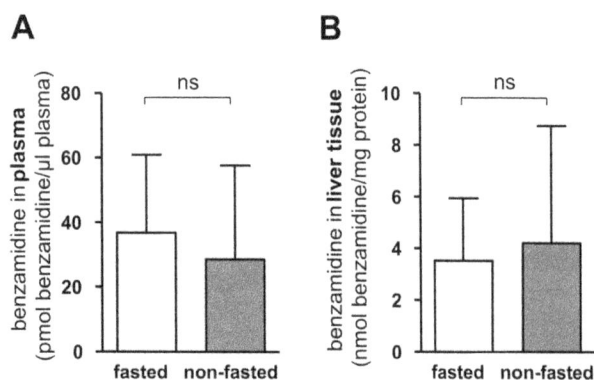

Figure 5. Metabolite concentrations of the N-reductive reaction are not affected by fasting in mice. Two groups of 14 C57BL/6W mice were fed with regular diet, one group were food deprived for 24 h (fasted); the second had full access to food and water (non-fasted). After anesthesia, 20 mg/kg benzamidoxime was injected. 30 min later animals were sacrificed followed by plasma liver collection **A** Metabolite concentrations in plasma 30 min after benzamidoxime injection, determined by HPLC analyses after sample work-up. **B** Metabolite concentrations in liver homogenates 30 min after benzamidoxime injection, determined by HPLC analyses after sample work-up. Determined activities are means $\pm$ SD of 12 biological samples, each measured as duplicates. Statistical significance was assessed by the U-test. p-values <0.05 were considered significant, ns = not significant.

Discussion and Conclusion

Glucose deprivation downregulates mARC complex components abundance and its activity in human and murine hepatoma cell lines

Expressions, protein abundance for all three proteins of the N-reductive system and N-reductive activity increase both in mouse Hepa 1.6 and human HepG2 cell culture fasting models in a glucose-dependent manner, except mARC1 mRNA in Hepa 1.6. However, in mouse Hepa 1.6 mARC1 and in human HepG2 mARC2 were not detectable on protein levels (Figure 2). In both cases, proteins were probably below detection limit. In Hepa 1.6 determined trancript level of mARC1 was about 400 times lower than for mARC2. For HepG2, mARC2 mRNA was about 7 times lower relativly to mARC1 (Figure 2A) and BioGPS database [21] describes as well only faint expression of mARC2 in HepG2 cell line compared to mARC1 (BioGPS gene numbers 64757 and 54996).

Yet, we can demonstrate that all three components of the N-reductive system are regulated in a glucose-dependent manner and support the hypothesis that the enzyme system is linked with energy metabolism. Our data is in consistence with studies showing mARC2 to be regulated glucose-dependent in kidney cell culture [13].

Fasting in mice down-regulates mARC complex components abundance and its activity

Expression for CYB5B and CYB5R was clearly reduced in both experiments under fasting conditions, mirrored by the proteins' abundances (Figure 3 & Figures S1–S2). Expression for mARC2 showed opposite tendencies and increased with fasting (Figure 3A & Figure 4A). Specifically, with 18 h of fasting no changes were

A

B

C

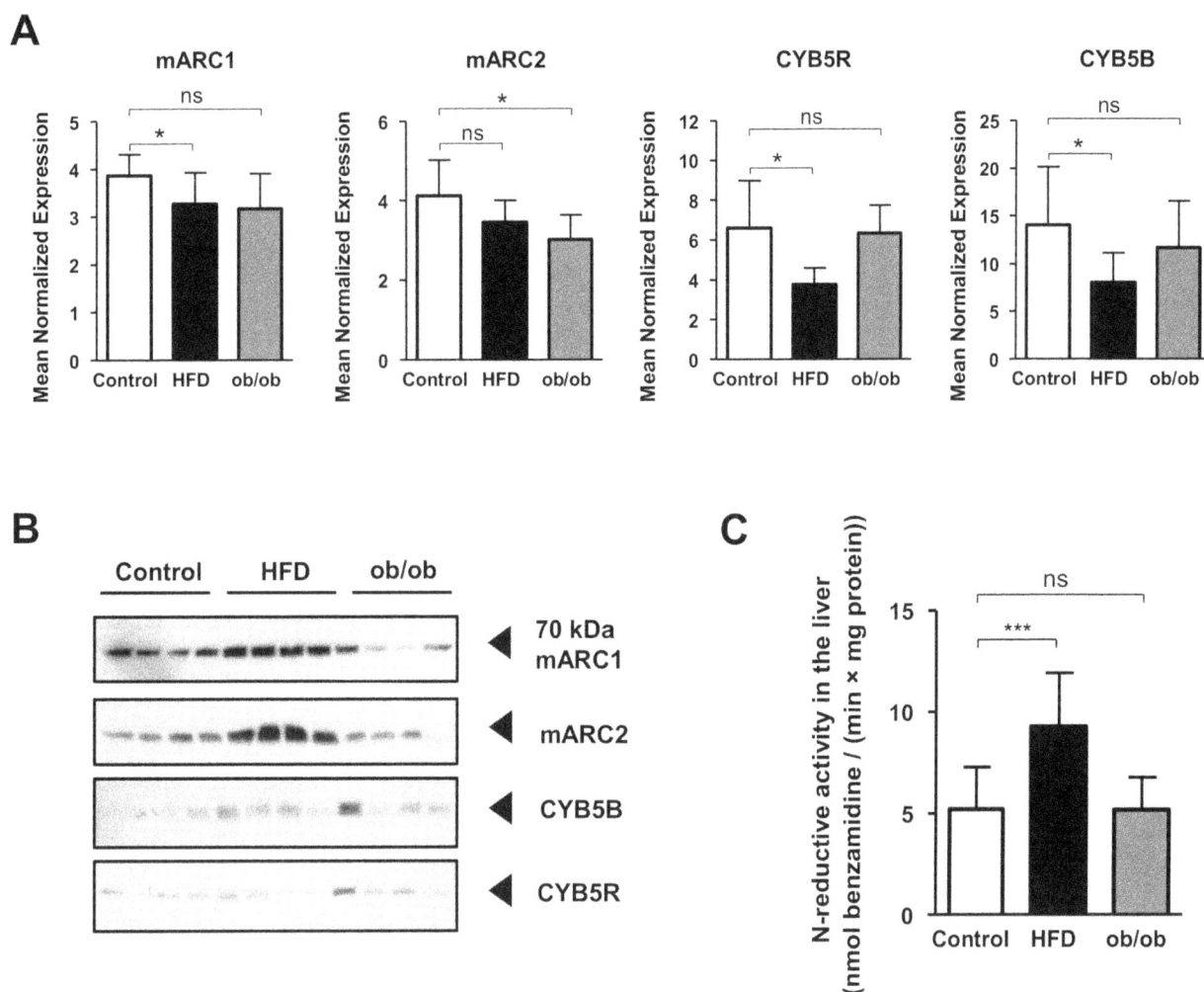

Figure 6. HFD but not hyperphagia increases mARC abundance and N-reductive activity in mice. Both a group of C57BL/6W and ob/ob-mice were fed with regular diet, another group of C57BL/6W mice was fed with HFD. Mice were sacrificed and livers collected. **A** Expressions of mARC1, mARC2, CYB5R and CYB5B determined with qRT-PCR, normalized on expression of Mcoln1 (Mucolipin-1) and Hmbs (hydroxymethylbilane synthase). Expression shown as means ± SD of 6 biological samples. Statistical significance was assessed by the U-test. p-values <0.05 were considered significant (*). **B** Protein levels of mARC1, mARC2, CYB5B and CYB5R in liver homogenates examined by Western Blot. **C** N-reductive activity determined by the reduction of model compound benzamidoxime in liver homogenate. The resulting metabolite benzamidine was quantified by HPLC analysis. Determined activities are means ± SD of four biological samples, each measured as duplicates. Statistical significance was assessed by the U-test. p-values <0.05 were considered significant (*); (***) p-value <0.0001; ns = not significant.

detected while 24 h fasting resulted in decreased protein levels for mARC2 (Figures 3B & 4B, Figure S1). With mARC1, mRNA expression increased with 18 h fasting was mirrored by increased protein levels of a protein assumed to be mARC1. In contrast 24 h of fasting resulted in decreased protein levels although mRNA levels were significantly enhanced. N-reductive activity was clearly reduced by fasting in both experiments (Figure 4C & Figure S1B).

With respect to detection of mARC1 protein levels, a protein was detected at about 65 kDa, which does not fit the predicted molecular weight of 37 kDa for mARC1. Wahl *et al.* already demonstrated the same phenomenon and detected a protein with a mARC1-specific antibody at this molecular weight in murine tissues [3]. Moreover, the detected protein follows the same pattern as mARC2 upon 24 h of fasting and HFD in mice. Thus we assume the protein to be mARC1 with an increased molecular weight as a result of unknown posttranslational modifications as already hypothesized by Wahl *et al.* [3]. However, it is also

possible that the detected signal is not related to mARC1 but stems from an unspecific reaction of the antibody.

The protein levels of the protein assumed to be mARC1 are increased with 18 h of fasting but decreased after 24 h of fasting, indicating a fast breakdown of the protein with prolonged fasting time. Thus protein turnover is different for the three components of the N-reductive system. This phenomenon is as well described for another three component system composed of cytochrome b5 and its reductase together with stearyl-CoA desaturase. The latter undergoes more rapid turnover than the other two members of the system for regulatory reasons [22]. The same mechanism could also be assumable for mARC-containing enzyme system. Reasons or mechanisms for the initial increase of the assumed mARC1 protein due to fasting followed by its decrease with prolonged fasting time are not yet understood, as well as the interplay of mARC1 and mARC2 during fasting.

Surprisingly, the abundances of mARC2 and mARC1 mRNA and protein levels in mice show opposite tendencies due to 24 h of fasting (Figure 4A–B). This may indicate a posttranscriptional or posttranslational regulation that is so far unknown. Discrepancies and even opposite expression patterns for mRNA and protein levels have already been described for other proteins [23–25]. These findings indicate that further studies on mARCs must not solely be based on RNA data but also on protein levels to be reliable. Nevertheless, mARC-proteins abundance and complex activity levels after 24 h of fasting match regarding tendencies. With respect to the 18 h fasting period we assume that both reduced levels of CYB5B and CYB5R are responsible for decreased N-reductive activity, regardless of the stable levels for mARC2 and increased levels for the assumed mARC1. CYB5B has been proven to be essential for N-reductive activity *in vitro* enzyme assay and cell culture [4], [5], as well as CYB5R *in vitro*, but only faint amounts of this protein are necessary [5], [6]. The latter might be the reason why others could not prove an involvement of CYB5R in N-reduction in cell culture [4]. In case of the 24 h fasting period, all three components of the enzyme system show decreased abundance in fasted mice compared to control and are thus likely to cause a decline in N-reductive activity.

Concerning the two isoforms of the mARC protein, mARC1 and mARC2, our group [5] already hypothesized that the isoform mainly involved in the N-reductive activity varies from species to species. For pig and rat, mARC2, but not mARC1 was found in the outer mitochondrial membrane (OMM), the site where the mARC protein is mainly localized [1], [4]. In contrast, human OMM was shown to contain mARC1 [26]. In consistence with this, we demonstrated that changes in mARC2 in murine cell line Hepa 1.6 and changes in mARC1 in human cell line HepG2 result in altered N-reductive activity, whereas the corresponding second isoform was below detection limit in both cell lines. Nevertheless, in murine liver tissues, both isoforms seem to be involved in altered N-reductive activities.

Fasting in mice does not influence concentrations of the N-reduction metabolite

No differences in metabolite concentrations (benzamidine) in liver homogenates and plasma after benzamidoxime administration could be found between fasted and non-fasted mice (control) (Figure 5). Thus, though all three components of the N-reductive system are decreased due to fasting, the metabolic conversion is still fast enough and probably complete to results in equal metabolite levels for both states. With respect to testing of new drug candidates, protocols for animal testing in rodents may differ regarding feeding and fasting procedures. However, based on this data, the results should still be comparable regarding the metabolite concentrations of substances that are reduced by the N-reductive system.

With respect to prodrug activation, changes in activity of the proteins due to nutrition state would have been crucial, as the enzyme system is applied in the activation of amidoxime prodrugs [10]. Yet, based on this experiment, we expect no influence of fasting on the *in vivo* activation of mARC substrates with a metabolic conversion comparable to benzamidoxime. Nevertheless, prodrug candidates with lower or even faint conversion rates should be tested regarding influence of fasting on tissue and plasma metabolite levels.

HFD but not hyperphagia influences mARC abundance and activity in mice

Mice fed with HFD showed increased abundance for mARC2 and a protein assumed to be mARC1 and as a consequence increased N-reductive activity, whereas ob/ob-mice did not show any changes of the N-reductive complex (Figure 6B–C & Fig. S4). With respect to mARC1 protein levels, the detected molecular weight did not fit the predicted molecular weight of the target protein as already discussed with fasting experiments. However, the findings are in consistence with Nesteruk *et al.*, who demonstrated increased mARC1 protein levels in HFD fed mice compared to control with quantitative mass spectrometry technique [19]. This indicates besides the points already discussed with fasting experiments that the protein detected by Western Blot analysis at 65 kDa is supposably a form of mARC1. Again and as already discussed with fasting experiments transcripts and protein levels did not match and showed opposite tendencies for mARC1.

With these findings, we thus demonstrate that changes with mARC2 and probably mARC1 are diet-related and not due to the fact, that the animals develop an obese habitus. Livers of control and HFD fed for 16 weeks old mice did not show any signs of steatosis, whereas ob/ob mice developed mild form [20]. Furthermore, serum glucose was enhanced in HFD mice but not in control and ob/ob-mice [20], suggesting a relation between serum glucose and mARC abundance. Data from other experiments with older mice (48 weeks of age) fed with HFD for 42 weeks could only demonstrate minor differences in abundance and activity of the N-reductive enzyme system (Figure S5). HFD fed mice generally gained weight and became obese until about 14th week upon HFD. Beyond that point until 42nd week on HFD mice underwent only minor changes in body weight (Figure S5B). We assume therefore that the N-reductive system is influenced by increased blood glucose, an active fat metabolism and fat turnover as it occurs with growing and weight gaining mice at early stage of HFD administration.

The mARC proteins and its partners CYB5B and CYB5R are related to lipid metabolism

With this study, we demonstrate for the first time, that the N-reductive complex composed of mARCs, CYB5B and CYB5R is regulated by diet. The N-reductive system undergoes changes with increased glucose amount in cell culture and due to diet in mice. The fact that the mARC proteins and its electron transfer partners decrease with fasting and that mARC2 and the protein assumed to be mARC1 increases with HFD but not with obese ob/ob-mice, reveals that the proteins are connected with food intake and energy supply. Other recent studies support this conclusion: A single nucleotide polymorphism (SNP) with the mARC1 locus was shown to associate with changes in plasma concentration of both total cholesterol and low density lipoprotein cholesterol [27] and others found an influence of this SNP on both baseline low density lipoproteins and the response to fenofibrate [28]. Neve and coworkers [4] proved an involvement of the mARC2 protein in lipid synthesis in 3T3-adipocytes. Moreover, the mARC2$^{-/-}$ knock out mouse model exhibits decreased total body fat amount [29].

Taken together, it is evident that the function of the mARC proteins is related to lipid metabolism. However, the endogenous substrates and detailed regulation mechanisms of the mARC proteins are still not known and require further research. The enzyme system was proven to be changed with physiological disorder like cancer [11] and diabetes [13]. Thus more detailed knowledge of the physiological function and mechanisms may provide not only basic information about the mARC proteins but also ideas in the further research of these diseases.

Supporting Information

Figure S1 Effects of fasting on protein abundance of the N-reductive system in mice. Two groups of 12 C57BL/6W mice were fed with regular diet, one group were food deprived for 18 h (fasted) before sacrifice and liver collection, the second had full access to food and water (non-fasted). **A** Protein levels of mARC1, mARC2, CYB5B, CYB5R and PCK1 examined by Western Blot, histone-H3 was applied as loading control. Each sample consisted of equal protein amount of two individuals. **B** N-reductive activity determined by the reduction of model compound benzamidoxime in liver homogenate. The resulting metabolite benzamidine was quantified by HPLC analysis. Determined activities are means ± SD of 12 biological samples, each measured as duplicates. Statistical significance was assessed by the U-test. p-values <0.05 were considered significant (*).

Figure S2 Effects of fasting on protein abundance of the N-reductive system in mice. Two groups of 14 C57BL/6W mice were fed with regular diet, one group were food deprived for 24 h (fasted) before sacrifice and liver collection; the second had full access to food and water (non-fasted). Protein levels of CYB5B, CYB5R and PCK1 examined by Western Blot, histone-H3 was applied as loading control. Each sample consisted of equal protein amount of two individuals.

Figure S3 Depletion of metabolite and mother compound after benzamidoxime administration in mice. A Metabolite concentrations in plasma and liver homogenate after benzamidoxime administration. Metabolite concentrations were determined by HPLC. Concentrations are means ± SD of two biological samples, each measured as duplicates. **B** Representative HPLC-chromatogram of liver homogenate samples taken 30 min after benzamidoxime administration. **C** Representative HPLC-chromatogramm of liver homogenate samples taken 60 min after benzamidoxime administration.

Figure S4 HFD but not hyperphagia increases mARC2 and mARC1 abundance in mice. Both a group of C57BL/6W and ob/ob-mice were fed with regular diet, another group of C57BL/6W mice was fed with HFD. Mice were sacrificed and livers collected. Protein levels of mARC1, mARC2, CYB5B and CYB5R in liver homogenates examined by Western Blot, histone H3 was used as loading control.

Figure S5 HFD has only minor effects on N-reductive complex abundance and activity in older mice. One group of C57BL/6W mice was fed with regular, another with HFD chow; mice were sacrificed at 48 weeks of age and livers collected. **A** Protein levels of mARC2, CYB5B and CYB5R in liver homogenates examined by Western Blot, histone H3 was used as loading control. **B** Development of bodyweight during aging of mice. **C** N-reductive activity determined by the reduction of model compound benzamidoxime in liver homogenate. The resulting metabolite benzamidine was quantified by HPLC analysis. Determined activities are means ± SD of 5–7 biological samples, each measured as duplicates. Statistical significance was assessed by the U-test. p-values <0.05 were considered significant (**) = p < 0,001.

Acknowledgments

We would like to thank Dr. Florian Bittner and Professor Ralf R. Mendel from the University of Technology, Braunschweig, Germany for providing recombinant proteins for mARC1, mARC2, CYB5B and CYB5R and Petra Köster from the University of Kiel for technical assistance.

Author Contributions

Conceived and designed the experiments: HJ MM AH EH JO BC. Performed the experiments: HJ MM AS MBC AD MG. Analyzed the data: HJ MM. Contributed reagents/materials/analysis tools: HJ MM MG JO BC. Contributed to the writing of the manuscript: HJ MM.

References

1. Havemeyer A, Bittner F, Wollers S, Mendel R, Kunze T, et al. (2006) Identification of the missing component in the mitochondrial benzamidoxime prodrug-converting system as a novel molybdenum enzyme. J Biol Chem 281: 34796–34802.
2. Hille R, Nishino T, Bittner F (2011) Molybdenum enzymes in higher organisms. Coord Chem Rev 255: 1179–1205.
3. Wahl B, Reichmann D, Niks D, Krompholz N, Havemeyer A, et al. (2010) Biochemical and spectroscopic characterization of the human mitochondrial amidoxime reducing components hmARC-1 and hmARC-2 suggests the existence of a new molybdenum enzyme family in eukaryotes. J Biol Chem 285: 37847–37859.
4. Neve EPA, Nordling A, Andersson TB, Hellman U, Diczfalusy U, et al. (2012) Amidoxime reductase system containing cytochrome b5 type B (CYB5B) and MOSC2 is of importance for lipid synthesis in adipocyte mitochondria. J Biol Chem 287: 6307–6317.
5. Plitzko B, Ott G, Reichmann D, Henderson CJ, Wolf CR, et al. (2013) The involvement of mitochondrial amidoxime reducing components 1 and 2 and mitochondrial cytochrome b5 in N-reductive metabolism in human cells. J Biol Chem 288: 20228–20237.
6. Kotthaus J, Wahl B, Havemeyer A, Kotthaus J, Schade D, et al. (2011) Reduction of N(ω)-hydroxy-L-arginine by the mitochondrial amidoxime reducing component (mARC). Biochem J 433: 383–391.
7. Krompholz N, Krischkowski C, Reichmann D, Garbe-Schönberg D, Mendel R, et al. (2012) The mitochondrial Amidoxime Reducing Component (mARC) is involved in detoxification of N-hydroxylated base analogues. Chem Res Toxicol 25: 2443–2450.
8. Froriep D, Clement B, Bittner F, Mendel RR, Reichmann D, et al. (2013) Activation of the anti-cancer agent upamostat by the mARC enzyme system. Xenobiotica 43: 780–784.
9. Gruenewald S, Wahl B, Bittner F, Hungeling H, Kanzow S, et al. (2008) The fourth molybdenum containing enzyme mARC: cloning and involvement in the activation of N-hydroxylated prodrugs. J Med Chem 51: 8173–8177.
10. Havemeyer A, Gruenewald S, Wahl B, Bittner F, Mendel R, et al. (2010) Reduction of N-hydroxy-sulfonamides, including N-hydroxy-valdecoxib, by the molybdenum-containing enzyme mARC. Drug Metab Dispos 38: 1917–1921.
11. Mikula M, Rubel T, Karczmarski J, Goryca K, Dadlez M, et al. (2010) Integrating proteomic and transcriptomic high-throughput surveys for search of new biomarkers of colon tumors. Funct Integr Genomics 11: 215–225.
12. Sparacino-Watkins CE, Tejero J, Sun B, Gauthier MC, Thomas J, et al. (2014) Nitrite Reductase and Nitric-oxide Synthase Activity of the Mitochondrial Molybdopterin Enzymes mARC1 and mARC2. J Biol Chem 289: 10345–10358.
13. Malik AN, Rossios C, Al-Kafaji G, Shah A, Page RA (2007) Glucose regulation of CDK7, a putative thiol related gene, in experimental diabetic nephropathy. Biochem Biophys Res Commun 357: 237–244.
14. Newton BW, Cologna SM, Moya C, Russell DH, Russell WK, et al. (2011) Proteomic analysis of 3T3-L1 adipocyte mitochondria during differentiation and enlargement. J Proteome Res 10: 4692–4702.
15. Jensen T, Kiersgaard M, Sorensen D, Mikkelsen L (2013) Fasting of mice: a review. Laboratory Animals 47: 225–240.
16. Panchal SK, Brown L (2011) Rodent models for metabolic syndrome research. J Biomed Biotechnol 2011: 351982.
17. Laemmli UK (1970) Cleavage of structural proteins during the assembly of the head of bacteriophage T4. Nature 227: 680–685.
18. Malhotra P, Boddy CS, Soni V, Saksena S, Dudeja PK, et al. (2013) D-Glucose modulates intestinal Niemann-Pick C1-like 1 (NPC1L1) gene expression via transcriptional regulation. Am J Physiol Gastrointest Liver Physiol 304: G203–10.

19. Nesteruk M, Hennig EE, Mikula M, Karczmarski J, Dzwonek A, et al. (2013) Mitochondrial-related proteomic changes during obesity and fasting in mice are greater in the liver than skeletal muscles. Funct Integr Genomics 14: 245–259.

20. Hennig EE, Mikula M, Goryca K, Paziewska A, Ledwon J, et al. (2014) Extracellular matrix and cytochrome P450 gene expression can distinguish steatohepatitis from steatosis in mice. J Cell Mol Med DOI 10.1111/jcmm.12328.

21. Wu C, Orozco C, Boyer J, Leglise M, Goodale J, et al. (2009) BioGPS: an extensible and customizable portal for querying and organizing gene annotation resources. Genome Biol 10: R130.

22. Paton CM, Ntambi JM (2009) Biochemical and physiological function of stearyl-CoA desaturase. Endocr Rev 31: 364–395.

23. Tian Q, Stepaniants SB, Mao M, Weng L, Feetham MC, et al. (2004) Integrated genomic and proteomic analyses of gene expression in Mammalian cells. Mol Cell Proteomics 3: 960–969.

24. Stark AM, Pfannenschmidt S, Tscheslog H, Maass N, Rösel F, et al. (2006) Reduced mRNA and protein expression of BCL-2 versus decreased mRNA and increased protein expression of BAX in breast cancer brain metastases: a real-time PCR and immunohistochemical evaluation. Neurological Research 28: 787–793.

25. Shebl FM, Pinto LA, García-Piñeres A, Lempicki R, Williams M, et al. (2010) Comparison of mRNA and protein measures of cytokines following vaccination with human papillomavirus-16 L1 virus-like particles. Cancer Epidemiol Biomarkers Prev 19: 978–981.

26. Klein JM, Busch JD, Potting C, Baker MJ, Langer T, et al. (2012) The mitochondrial amidoxime-reducing component (mARC1) is a novel signal-anchored protein of the outer mitochondrial membrane. J Biol Chem 287: 42795–42803.

27. Teslovich TM, Musunuru K, Smith AV, Edmondson AC, Stylianou IM, et al. (2010) Biological, clinical and population relevance of 95 loci for blood lipids. Nature 466: 707–713.

28. Aslibekyan S, Goodarzi MO, Frazier-Wood AC, Yan X, Irvin MR, et al. (2012) Variants identified in a GWAS meta-analysis for blood lipids are associated with the lipid response to fenofibrate. PLOS ONE 7: e48663.

29. Brown SDM, Moore MW (2012) The International Mouse Phenotyping Consortium: past and future perspectives on mouse phenotyping. Mamm Genome 23: 632–640.

6

DRUM: A New Framework for Metabolic Modeling under Non-Balanced Growth. Application to the Carbon Metabolism of Unicellular Microalgae

Caroline Baroukh[1,2]*, Rafael Muñoz-Tamayo[2], Jean-Philippe Steyer[1], Olivier Bernard[2,3]

1 INRA UR050, Laboratoire des Biotechnologies de l'Environnement, Narbonne, France, **2** INRIA-BIOCORE, Sophia-Antipolis, France, **3** LOV-UPMC-CNRS, UMR 7093, Villefranche-sur-mer, France

Abstract

Metabolic modeling is a powerful tool to understand, predict and optimize bioprocesses, particularly when they imply intracellular molecules of interest. Unfortunately, the use of metabolic models for time varying metabolic fluxes is hampered by the lack of experimental data required to define and calibrate the kinetic reaction rates of the metabolic pathways. For this reason, metabolic models are often used under the balanced growth hypothesis. However, for some processes such as the photoautotrophic metabolism of microalgae, the balanced-growth assumption appears to be unreasonable because of the synchronization of their circadian cycle on the daily light. Yet, understanding microalgae metabolism is necessary to optimize the production yield of bioprocesses based on this microorganism, as for example production of third-generation biofuels. In this paper, we propose DRUM, a new dynamic metabolic modeling framework that handles the non-balanced growth condition and hence accumulation of intracellular metabolites. The first stage of the approach consists in splitting the metabolic network into sub-networks describing reactions which are spatially close, and which are assumed to satisfy balanced growth condition. The left metabolites interconnecting the sub-networks behave dynamically. Then, thanks to Elementary Flux Mode analysis, each sub-network is reduced to macroscopic reactions, for which simple kinetics are assumed. Finally, an Ordinary Differential Equation system is obtained to describe substrate consumption, biomass production, products excretion and accumulation of some internal metabolites. DRUM was applied to the accumulation of lipids and carbohydrates of the microalgae *Tisochrysis lutea* under day/night cycles. The resulting model describes accurately experimental data obtained in day/night conditions. It efficiently predicts the accumulation and consumption of lipids and carbohydrates.

Editor: Akos Vertes, The George Washington University, United States of America

Funding: Caroline Baroukh was supported by a Contrat Jeune Scientifique (CJS) INRA-INRIA fellowship. Rafael Muñoz-Tamayo benefited from the support of the ANR Facteur 4 project. The funders had no role in study design, data collection and analysis, decision to publish, or preparation of the manuscript.

Competing Interests: The authors have declared that no competing interests exist.

* Email: caroline.baroukh@supagro.inra.fr

Introduction

Metabolic modeling is a powerful tool for bioprocesses to understand, predict and optimize the synthesis of intracellular molecules of interest [1]. The main interest of this approach relies on the use of the metabolic network knowledge and its associated stoichiometry. The kinetics modeling of each metabolic reaction is thus needed, especially to represent the transient dynamics of the set of intracellular compounds. However, the experimental difficulty to measure along time the dynamics of intracellular compounds hampers the modeling and calibration of the large set of reaction rates associated to the biochemical reactions of the metabolic network [2].

To overcome these hurdles, a commonly used hypothesis is the balanced-growth hypothesis, also called the Quasi-Steady-State Approximation (QSSA). Internal metabolites are assumed not to accumulate inside the microorganisms, which turns out to be a reasonable hypothesis for most of the microorganisms growing under constant conditions. This implies that every substrate uptake leads to microbial growth and products excretion. Thanks to this hypothesis, intracellular models are simplified and thus depend only on the stoichiometry of the network, the reaction reversibility and the uptake rate of the substrates.

Most of the metabolic modeling and analysis frameworks rely on the balanced-growth hypothesis. These frameworks include Flux Balance Analysis (FBA) [3], Dynamical Flux Balance Analysis (DFBA) [4], Elementary Flux Modes (EFM) [5], Flux Coupling Analysis (FCA) [6], Macroscopic Bioreaction Models (MBM) [7], Hybrid Cybernetic Models (HCM) [8] and Lumped Hybrid Cybernetic Models (L-HCM) [9]. Overall, these models predict well biomass growth and excreted products synthesis [4,8,10,11] as long as the balanced-growth hypothesis is verified [12].

However, the balanced-growth hypothesis is unreasonable for microorganisms undergoing permanent environmental fluctuations. Indeed, in this case, the everlasting dynamics of intracellular accumulation and reuse play a key role in the cell metabolism. This is the case for phototrophic microalgae submitted to day/night cycles, which use photons to fix inorganic carbon during the day using photosynthesis. These promising organisms are seen as good candidates for production of third-generation biofuels thanks to their higher productivity compared to classical biofuels [13]. However, many improvements are necessary to become a cost effective and environmental-friendly bioprocess [14]. For that, a deep understanding of microalgae metabolism is necessary.

Microalgae store energy and carbon during the day so as to support growth and maintenance during the night, because of their autotrophic metabolism and the synchronization of their circadian cycle on the daily light [15]. Therefore, intermediate metabolites such as carbohydrates and lipids accumulate during the day and are remobilized during the night (Figure 1D) [16]. This behavior cannot be described under the balanced-growth assumption. One way to circumvent this issue is to represent these

metabolites as product of the cell during the day and substrate during the night. Therefore applying one of the above-cited QSSA metabolic modeling frameworks could a priori be possible to represent carbon storage and better understand microalgae metabolism submitted to day/night cycles. In literature, only Knoop et al. [17], using the DFBA framework, computed metabolic fluxes for a full day/night cycle. However, determining an optimization function to represent carbon storage during the day and its consumption during the night is not a trivial task. Indeed, the classical optimization function "maximization of biomass production" does not work: when applying it, all the carbon available will go to biomass synthesis, and none to carbon storage. To circumvent this issue, the solution is to either force fluxes to carbon storage or to force the fluxes of biomass synthesis

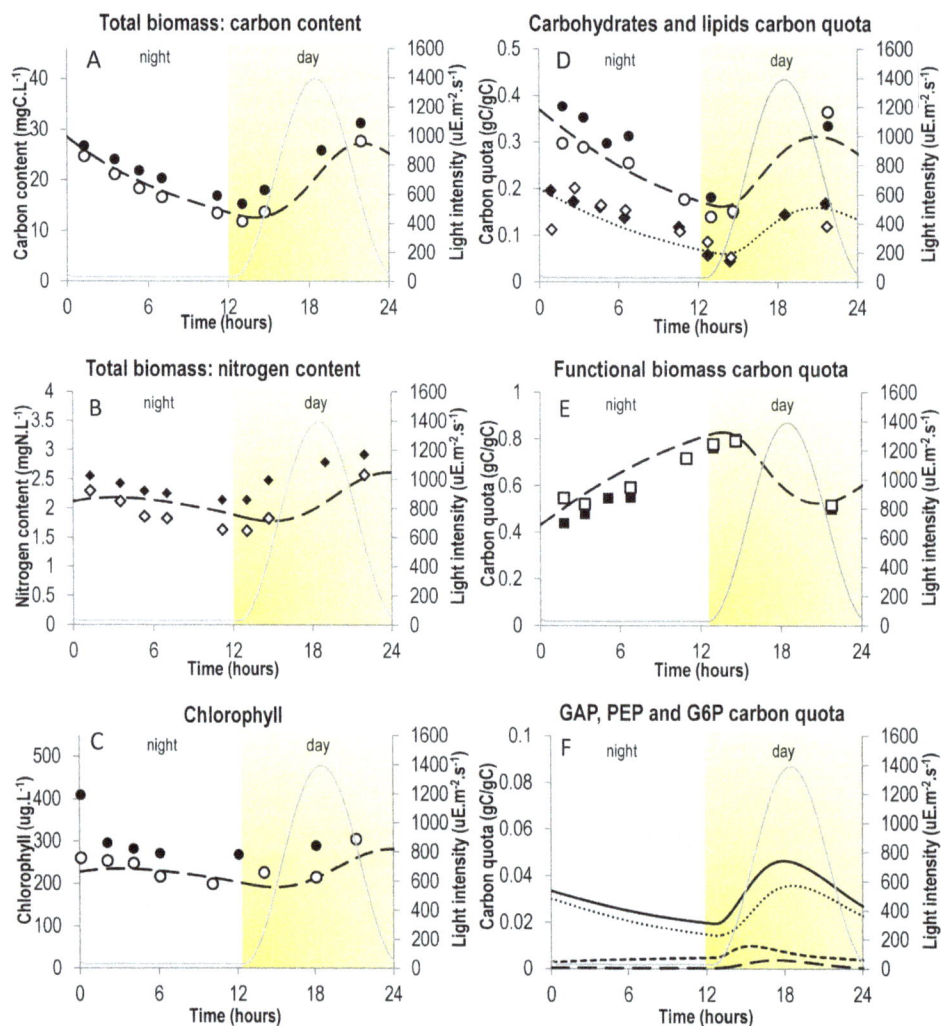

Figure 1. Comparison of simulation results with experimental data. Simulation results were obtained by simulation of system (7) and are represented by dashed or dotted lines. Experimental results were taken from [16] and are represented by dots, diamonds or squares. A. Evolution of total biomass in terms of carbon content. Dashed line: model; Circles: experimental data; Grey line: light intensity. B. Evolution of total biomass in terms of nitrogen content. Dashed line: model; Diamonds: experimental data; Grey line: light intensity. C. Evolution of chlorophyll (computed as a fixed percentage of functional biomass). Dashed line: model; Circles: experimental data; Grey line: light intensity. D. Evolution of "energy and carbon" metabolites. Dashed line and Circles: carbohydrates (CARB); Dotted line and Diamonds: lipids (PA); Grey line: light intensity. Accumulation of carbon and energy metabolites during the day and their consumption during the night for growth and maintenance purpose is well represented. E. Evolution of functional biomass B. Dashed line: model; Squares: experimental data; Grey line: light intensity. F. Evolution of "buffer" metabolites at branching points, as predicted by the model. Dashed line: glyceraldehyde 3-phosphate (GAP); Dotted line: glucose 6-phosphate (G6P); Small-dashed line: phosphoenolpyruvate (PEP); Black line: GAP + PEP + G6P; Grey line: light intensity. Note that their carbon mass quota is relatively small (less than 4%).

and maintenance ($ATP \rightarrow ADP + P_i$) and other futile cycles. In their work, Knoop et al. [17], forced fluxes to carbon storage by changing the biomass composition at each time step. Their method indeed predicted metabolic fluxes dynamically but did not allow predicting the fluxes toward carbon storage and hence the dynamic change of biomass composition. In a context of better understanding and predicting microalgae metabolism for biofuels production, prediction of carbon storage fluxes is essential if one seeks the conditions in which microalgae accumulates more lipids or starch to improve biofuels production yield. Hence, to model such bioprocesses, a metabolic modeling framework that handles non balanced-growth and dynamics behaviors is necessary.

The aim of the present paper is to present DRUM (Dynamic Reduction of Unbalanced Metabolism), a new metabolic modeling framework, which allows to model dynamically intracellular processes where accumulation of metabolites plays a significant role. In a first section, the modeling approach and its mathematical translation are described. Then the approach is applied successfully to the carbon metabolic network of a unicellular microalgae (*Tisochrysis lutea*) in order to illustrate it on a realistic example, where simulation results are compared to experimental data. Finally, assumptions of the present approach and their implications are discussed in a last section along with the perspectives of the present work and the future possible applications.

Method

Let us consider a continuous bioprocess implying microorganisms growing in a perfectly mixed stirred-tank reactor with constant volume, dilution rate D and incoming substrate S_{in}. The microorganisms consume extracellular substrates represented by vector S to synthesize biomass B and produce excreted products represented by the vector P. The metabolic network of the microorganism is represented by the stoichiometric matrix $K \in \mathbb{R}^{n_m \times n_r}$ containing n_m metabolites and n_r reactions.

By applying a mass-balance, the bioprocess can be represented by the Ordinary Differential Equation (ODE) system:

$$\frac{dM}{dt} = \frac{d\begin{pmatrix} S \\ C \\ P \\ B \end{pmatrix}}{dt} = \begin{pmatrix} K_S \\ K_C \\ K_P \\ K_B \end{pmatrix}.v.B - D.\begin{pmatrix} S \\ C \\ P \\ B \end{pmatrix} + D.\begin{pmatrix} S_{in} \\ 0 \\ 0 \\ 0 \end{pmatrix} \quad (1)$$

$$= K.v.B - D.M + D.M_{in}$$

where M represents the metabolites concentration vector composed of biomass B, uptaken substrates S, intracellular metabolites C and excreted products P. Concentrations are expressed in terms of solution concentrations, not concentrations per unit of cell. The kinetics vector $v \in \mathbb{R}^{n_r}$ represents the reactions rates (per biomass unit) of the reactions of the metabolic network. By multiplication to v, biomass B acts as a catalyzer of kinetics v. Due to a lack of experimental data, v is often inferred [2]. The matrices $K_S \in \mathbb{R}^{n_S \times n_r}$, $K_P \in \mathbb{R}^{n_P \times n_r}$, $K_C \in \mathbb{R}^{n_C \times n_r}$ and $K_B \in \mathbb{R}^{1 \times n_r}$ are the stoichiometric matrices of the metabolic network for the substrate, the products, the internal metabolites and the biomass ($n_S + n_C + n_P + 1 = n_m$). They are based on the knowledge of the metabolic network. The stoichiometric coefficients are thus known a priori, they do not need to be determined experimentally. The

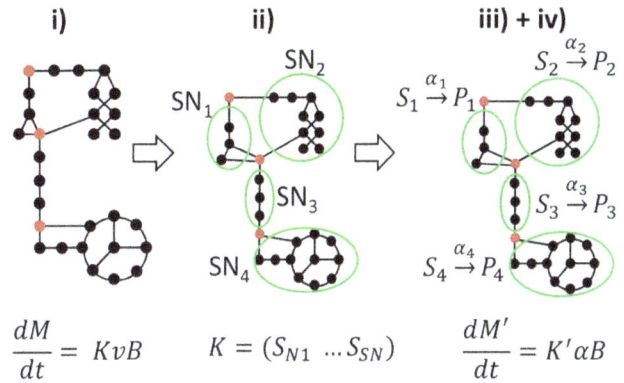

$$\frac{dM}{dt} = KvB \qquad K = (S_{N1} \dots S_{SN}) \qquad \frac{dM'}{dt} = K'\alpha B$$

Figure 2. Modeling approach decomposed into 4 steps. The complete network (step i) is decomposed into sub-networks (SN) assumed at quasi-steady state (step ii). These are reduced to a set of macroscopic reactions ($S \xrightarrow{\alpha} P$) (step iii), for which kinetics are defined (step iv). Linking metabolites interconnecting the SN are allowed to accumulate (red circles) or be reused, which gives the dynamics of the whole network. From step iv, an ordinary differential equation (ODE) system is obtained, representing evolution of the macroscopic scale of the bioprocess as well as intracellular processes and accumulation of metabolites. In the full model described in step i), $K \in \mathbb{R}^{n_m \times n_r}, v \in \mathbb{R}^{n_r}$, while for the resulting model provided by our approach, $K' \in \mathbb{R}^{n_{m'} \times n_E}$ and $\alpha \in \mathbb{R}^{n_E}$, such that $n_{m'} << n_m$ and $n_E << n_r$.

vector M_{in} is the concentration vector of incoming metabolites in the chemostat, composed of incoming substrate S_{in}.

The QSSA implies that internal metabolites do not accumulate ($K_C.v = 0$). In the DRUM approach, instead, we assume that the QSSA is applicable only to groups of metabolic reactions that we call sub-networks (SNs). The remaining metabolites interconnecting the sub-networks, which we name A ($A \subsetneq C$), are not under the quasi-steady-state constraint. They are allowed to accumulate and thus can behave dynamically, which provides the dynamics to the whole network (Figure 2).

The QSSA for sub-networks relies on *i*) the presence of metabolic pathways corresponding to metabolic functions *ii*) the presence of group of reactions regulated together *iii*) the presence of different compartments in a cell (e.g., mitochondrion). Groups of reactions are thus determined taking into account these intracellular mechanisms. It is to be noted that some intracellular reactions can thus belong to several group of reactions. Mathematically, this is represented by redundant columns in the stoichiometric matrix K. The remaining metabolites (A) interconnecting the sub-networks formed using these rules are usually either situated at a branching point between several pathways or are end-products of metabolic pathways (e.g: macromolecules).

The sub-networks correspond mathematically to a partitioning of the stoichiometric matrix K into sub-matrices K_{SNi} formed of grouped reactions:

$$K = \begin{pmatrix} K_{SN_1} & \dots & K_{SN_k} \end{pmatrix} \quad (2)$$

where $K_{SN_i} \in \mathbb{R}^{n_m \times n_{SNi}}$ ($\sum_i n_{SNi} = n_r$) represents the sub-network i composed of i) incoming and outgoing metabolites S_{SNi} and P_{SNi} allowed to accumulate and ii) intermediate metabolites C_{SNi} at quasi-steady state. S_{SNi} and P_{SNi} are either substrates S, products P, biomass B or intracellular metabolites A allowed to accumulate.

Each sub-network is assumed to be in a quasi-steady-state:

$$\forall i = 1..k \quad K_{SN_i}.v_{SN_i} = 0 \tag{3}$$

Under these assumptions and using elementary flux mode analysis [7,12,18], each sub-network can be reduced to a reduced set of macroscopic reactions:

$$\forall i = 1..k \quad v_{SN_i} = E_{SN_i}.\alpha_{SN_i} \quad \alpha_{SN_i} \geq 0$$

$$(K_{S_{SN_i}}.E_{SN_i}).S_{SN_i} \xrightarrow{\alpha_{SN_i}} (K_{P_{SN_i}}.E_{SN_i}).P_{SN_i} \tag{4}$$

where E_{SNi} is the matrix of elementary flux modes of sub-network SN_i and α_{SNi} is the weight vector of the elementary flux modes. α_{SNi} can be interpreted as the kinetics of the macroscopic reactions described by the stoichiometric matrix $K_{SN_i}.E_{SN_i}$ [12].

By grouping all the sub-networks, the following system is obtained:

$$\frac{dM}{dt} = \begin{pmatrix} K_{SN_1} & \cdots & K_{SN_k} \end{pmatrix}.\begin{pmatrix} v_{SN_1} \\ \cdots \\ v_{SN_k} \end{pmatrix}.B - D.M + D.M_{in}$$

$$\cdots = K_{SN_1}.E_{SN_1}.\alpha_{SN_1}.B + \cdots + K_{SN_k}.E_{SN_k}.\alpha_{SN_k}.B - D.M + D.M_{in} \tag{5}$$

$$\cdots = \begin{pmatrix} K_{SN_1}.E_{SN_1} & \cdots & K_{SN_k}.E_{SN_k} \end{pmatrix}.\begin{pmatrix} \alpha_{SN_1} \\ \cdots \\ \alpha_{SN_k} \end{pmatrix}.B - D.M + D.M_{in}$$

$$\cdots = K_E.\alpha.B - D.M + D.M_{in}$$

Only metabolites A are authorized to accumulate. Any other metabolite $C_j \in C \backslash A$ are assumed not to accumulate. Thus:

$$\frac{d(\frac{C_j}{B})}{dt} = 0 \quad \forall C_j \in C \backslash A \tag{6}$$

$C_j \in C \backslash A$ have simple dynamics. Hence a reduced dynamic model is obtained, defined by the metabolites vector $M' \in \mathbb{R}^{n_m}$ and the matrix $K' \in \mathbb{R}^{n_m \times n_E}$, with n_E the number of macroscopic reactions:

$$\frac{dM'}{dt} = \frac{d\begin{pmatrix} S \\ A \\ P \\ B \end{pmatrix}}{dt} = \begin{pmatrix} K'_S \\ K'_A \\ K'_P \\ K'_B \end{pmatrix}.\alpha.B - D.\begin{pmatrix} S \\ P \\ A \\ B \end{pmatrix} + D.\begin{pmatrix} S_{in} \\ 0 \\ 0 \\ 0 \end{pmatrix} \tag{7}$$

$$= K'.\alpha.B - D.M' + D.M'_{in}$$

System (7) is a simplified version of (1) with the same structure but of much lower dimension, where accumulation of some internal metabolites (A) is allowed. Only the kinetics α of the resulting macroscopic reactions need to be determined. Classical

kinetics found in literature are mass-action, power-law, Michaelis-Menten, Hill, cybernetic kinetics [19]. The choice is often arbitrary and the total number of parameters in the kinetics models needs to match the experimental data available so that a model validation is achievable. Once kinetics α are determined, all the metabolic fluxes can be computed using:

$$v = \begin{pmatrix} v_{SN_1} \\ \cdots \\ v_{SN_k} \end{pmatrix} = \begin{pmatrix} E_{SN_1}.\alpha_{SN_1} \\ \cdots \\ E_{SN_k}.\alpha_{SN_k} \end{pmatrix} \tag{8}$$

In the DRUM approach, particular attention has to be drawn to the definition of biomass B, which is no longer the conventional one. Biomass B is usually represented as an average composition of macromolecules present in the cell. With QSSA, any chemical element of substrate S ends up in either biomass B or excreted products P. But in the present approach, accumulation of internal metabolites is allowed. Hence, not all chemical elements from substrate S ends up in biomass B or products P; they can also be present in A. Total biomass (noted X) can then only be determined thanks to a mass-balance on each chemical element:

$$X_Z(t) = \sum_A Z_A.A(t) + Z_B.B(t) \tag{9}$$

where Z correspond to a chemical element ($Z \in \{C; N; O; H; P; S; ...\}$), Z_A and Z_B corresponds to the number of chemical element Z per mole of accumulating metabolites A and biomass B, $A(t)$ and $B(t)$ correspond to the concentrations of A and B at time t, and $X_Z(t)$ correspond to the concentration of chemical element Z in total biomass X at time t.

To sum up, the DRUM approach is based on the following methodology, which is decomposed into a 4-step process (Figure 2):

i)   Find in the literature or build the metabolic network of the microorganism under study.

ii)   Group metabolic reactions into sub-networks assumed to follow the QSSA.

iii)   Reduce each sub-network to a set of macroscopic reaction using elementary modes analysis.

iv)   Define kinetics for macroscopic reactions obtained and deduce an ODE system.

For sake of pedagogy, in the next section, the DRUM approach is illustrated on the carbon metabolism of unicellular microalgae.

Results

1. Metabolic Network

To assess DRUM, experimental data of a continuous culture of *Isochrysis affinis galbana* (clone T-iso, CCAP 927/14) under day/night cycle was used [16]. This microalgae clone, known to accumulate high quantities of lipids was recently renamed *Tisochrysis lutea* [20]. Cultures were grown in duplicates in 5L cylindrical vessels at constant temperature ($22°$) and pH (8.2, maintained by automatic injection of CO_2). The following measurements were performed: nitrates, particulate carbon and nitrogen, chlorophyll, total carbohydrates and neutral lipid concentrations [16].

With regards to the metabolic network, since *Tisochrysis lutea* has not been sequenced yet, no genome-scale metabolic network reconstruction was possible. Using the metabolic network of eukar-

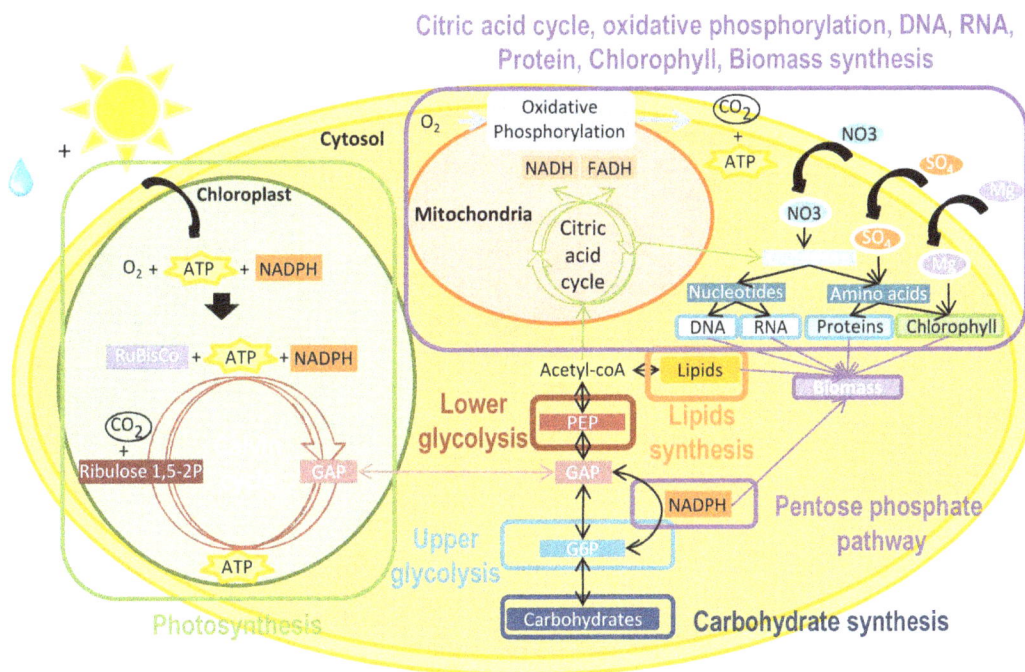

Figure 3. Simplified central carbon metabolic network of a unicellular photoautrotophic microalgae. Central carbon metabolic network is composed of photosynthesis in the chloroplast, transport reaction from the chloroplast to cytosol, glycolysis, carbohydrate synthesis, citric acid cycle, pentose phosphate pathway, lipids synthesis, oxidative phosphorylation, protein, DNA, RNA, chlorophyll and biomass synthesis. Photosynthesis is decomposed into two steps: the light step, which generates energy (ATP and NADPH) and oxygen using light and water and the dark step, which uses the generated energy to incorporate carbon dioxide. The end-product of photosynthesis is a 3 carbon sugar (here glyceraldehyde 3-phoshate written GAP), exported to the cytosol. GAP is situated in the center of glycolysis, and splits it into two parts: upper glycolysis and lower glycolysis. Upper glycolysis generates glucose 6-phosphate (G6P), which is then either invested for carbohydrates synthesis or in the pentose phosphate pathway to generate NADPH. Lower glycolysis generates phosphoenolpyruvate (PEP), which is then invested either in lipids synthesis or in the citric acid cycle, which produces necessary intermediate metabolites for proteins, DNA, RNA, chlorophyll and biomass synthesis. Cofactors (FADH, NADH) generated by citric acid cycle are transformed into energy (ATP) thanks to oxidative phosphorylation.

yotic microalgae available (*Chlorella pyrenoidosa* [21], *Chlamydomonas reinhardtii* [22–27], *Ostreococcus tauri* and *Ostreococcus lucimarinus* [28]), we deduced a core carbon metabolic network common to unicellular photoautotrophic microalgae containing the central metabolic pathways (photosynthesis, glycolysis, pentose phosphate pathway, citric acid cycle, oxidative phosphorylation, chlorophyll, carbohydrates, amino acid and nucleotide synthesis). We did not represent species-specific pathways such as the synthesis of secondary metabolites since we assumed these pathways to have negligible fluxes compare to the main pathways and thus small impact on the other pathways. Indeed, secondary metabolites have very low biomass concentration compared to proteins, lipids, carbohydrates, DNA, RNA and chlorophyll. The reactions of synthesis of the macromolecules (proteins, lipids, DNA, RNA and biomass) were lumped, as classically done, into generic reactions where stoichiometric coefficients of the precursors metabolites were determined for *Tisochrysis* lutea thanks to their measured average quota in those macromolecules [16]. The detailed description of metabolic network reconstruction is available in File S1 section 1.

The resulting metabolic network is composed of the light and dark steps of photosynthesis in the chloroplast, the transport reaction from chloroplast to cytosol, glycolysis, carbohydrate synthesis, citric acid cycle, pentose phosphate pathway, lipids synthesis, oxidative phosphorylation, protein, DNA, RNA, chlorophyll and biomass synthesis (Figure 3). The network is composed of 157 internal metabolites and 162 reactions, including 13 exchange reactions with the environment and 1 internal exchange

reaction (between the chloroplast and the cytosol). List of reactions and metabolites are available in File S1 section 2 and 3.

2. Formation and reduction of sub-networks

Metabolic reactions were grouped by metabolic functions, taking into account cell compartments and metabolic pathways. Six sub-networks were obtained (Figure 4) corresponding to *i*) photosynthesis, *ii*) upper part of glycolysis iii) carbohydrate synthesis *iv*) lower part of glycolysis, *v*) lipids synthesis, *vi*) biomass synthesis. Then, each sub-network was reduced to macroscopic reactions thanks to elementary flux mode analysis [18]. To compute elementary flux modes (EFMs) the software *efmtool* was used [29]. For all six sub-networks, the EFM could be computed easily, and their number was low (less than 30). It should be noted that an EFM analysis of the full network leads to 18776 modes (see File S1 section 4 for more details).

In the following sections, the formation and reduction of each sub-network is developed. The results are summarized in Table 1.

2.1 Photosynthesis. Photosynthesis allows phototrophic organisms to generate cell energy and incorporate carbon autotrophically. The process takes place in the chloroplast and is decomposed into two steps commonly called the light and dark steps. The light step consists in the generation of cell energy (ATP, NADPH) from water and photons, producing oxygen (R1). Thanks to the energy of the light step, the dark step incorporates carbon dioxide through Calvin cycle producing one 3 carbon sugar (3-phosphoglycerate written G3P). Then G3P is transformed

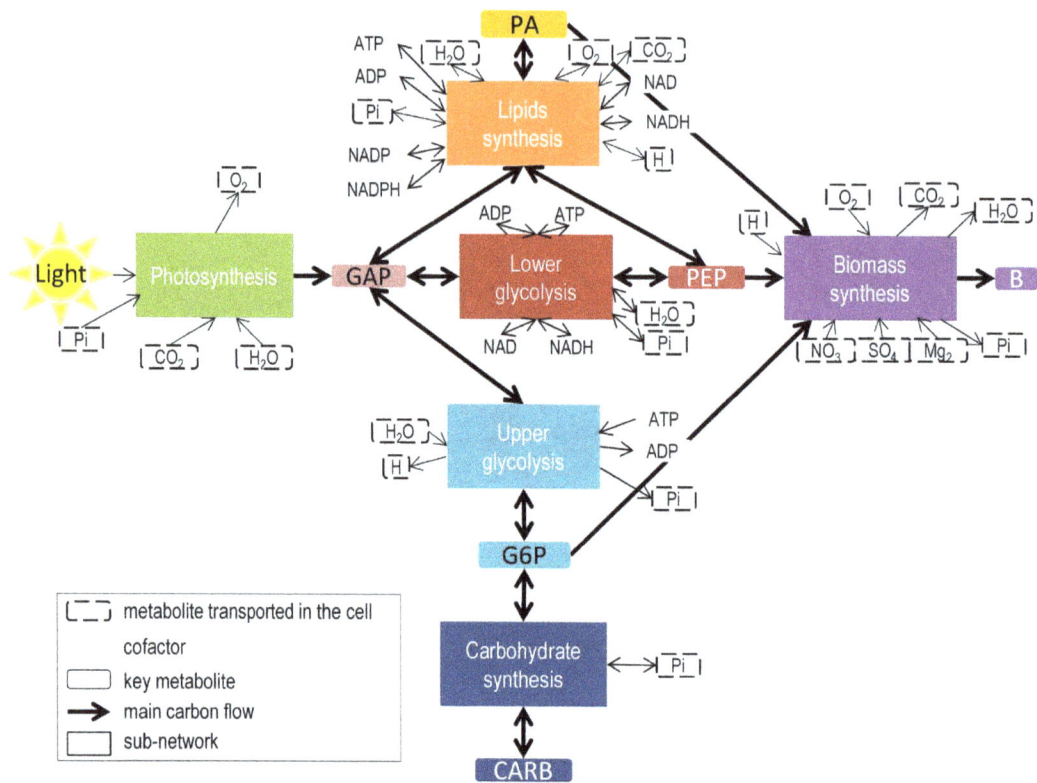

Figure 4. Central carbon metabolic network of a unicellular photoautotrophic microalgae decomposed into 6 sub-networks. The metabolic network was built by deducing a core carbon metabolic network common to unicellular photoautotrophic microalgae containing the central metabolic pathways of the metabolic network of eukaryotic microalgae available (*Chlorella pyrenoidosa* [21], *Chlamydomonas reinhardtii* [22–27], *Ostreococcus tauri* and *Ostreococcus lucimarinus* [28]) and experimental data of [16]. Details of the network reconstruction process and lists of reactions and metabolites are available in File S1 section 1–3. Metabolic reactions were grouped into sub-networks taking into account compartments and metabolic pathways. After reduction, 6 sub-networks were obtained corresponding to i) photosynthesis, ii) upper part of glycolysis iii) carbohydrate synthesis iv) lower part of glycolysis, v) lipids synthesis, vi) biomass synthesis. The resulting metabolites interconnecting the sub-networks and allowed to accumulate are either at branching points of metabolic pathways (glyceraldehyde 3-phosphate (GAP), glucose-6-phosphate (G6P) and phosphoenolpyruvate (PEP)) or end-products of metabolic pathways (lipids (PA), carbohydrates (CARB) and functional biomass (B)) or energy metabolites (ATP, ADP,NADH, NAD, NADPH, NADP) or metabolites transported in the cell (Light, CO_2,O_2,Pi,H_2O,H,NO_3,SO_4,Mg). B corresponds to functional biomass and is composed of proteins, DNA, RNA, chlorophyll and lipids. List of macroscopic reactions for each sub-network is available in Table 1.

Table 1. Definition and reduction of sub-networks formed from metabolic reactions of a unicellular autotrophic microalgae.

Sub-network	Macroscopic reactions	Kinetics
Photosynthesis	30 Light + 3 CO_2 + 2 H_2O + P_i —> GAP + 3 O_2 (MR1)	$v_{MR1} = k_{MR1}*I$
Upper glycolysis	ATP + H_2O —> ADP + P_i + H (MR2)	$v_{MR2} = 0$
	2 GAP + H_2O —> G6P + P_i (MR3)	$v_{MR3} = k_{MR3}*GAP$
	G6P + ATP —> H + ADP + 2 GAP (MR4)	$v_{MR4} = k_{MR4}*G6P$
Lower glycolysis	GAP + ADP + P_i + NAD <—> PEP + ATP + NADH + H_2O + H (MR5)	$v_{MR5} = k_{MR5}*GAP - k'_{MR5}*PEP$
Carbohydrate synthesis	G6P <—> CARB + P_i (MR6)	$v_{MR6} = k_{MR6}*G6P - k'_{MR6}*CARB$
Lipids synthesis	GAP + 16.61 PEP + 2 ADP + 13.46 NAD + 29.3 NADPH + 34.48 H + 2.15 O2 <—> PA + 14.61 Pi + 2 ATP + 13.46 NADH + 29.3 NADP + 4.31 H_2O + 16.61 CO2 (MR7)	$v_{MR7} = k_{MR7}*PEP*GAP - k'_{MR7}*PA$
Biomass synthesis	3.13 PEP + 7.37 O_2 + 4.46 H + 1.31 NO_3 + 1.14 G6P + 0.11 PA + 0.03 SO_4 + 0.0025 Mg —>B + 11.67 CO_2 + 4.23 Pi + 6 H_2O (MR8)	$v_{MR7} = k_{MR8}*PEP*G6P*NO_3$

Each sub-network was decomposed into a set of macroscopic reactions thanks to elementary flux mode analysis. List of reactions, incoming and outgoing metabolites for each sub-network are available in File S1 section 5. I corresponds to light intensity, expressed in $\mu E.m-2.s-1$.

in glyceraldehyde 3-phosphate (GAP) and transported to the cytosol of the cell (R14).

As both the dark and light step of photosynthesis takes place in the chloroplast and they both have the same metabolic function (to incorporate inorganic carbon), the reactions of the two steps were grouped into a sub-network and assumed at quasi-steady state. Elementary flux mode analysis yielded only one Elementary Flux Mode (EFM) (Table 1), giving one macroscopic reaction (MR1). The stoichiometry of the macroscopic reaction obtained is in agreement with literature: a quota of 10 photons are needed per carbon incorporated [24,30].

2.2 Upper glycolysis. As GAP is the end-product of photosynthesis and is situated at the center of glycolysis, glycolysis was split according to GAP into two sub-networks: lower glycolysis and upper glycolysis. In addition, dividing glycolysis into two parts is meaningful since upper glycolysis and lower glycolysis have different metabolic goals. Indeed, upper glycolysis synthesizes glucose 6-phosphate (G6P) to produce reductive power (NADPH) or to produce carbon storage compounds (carbohydrates), whereas lower glycolysis produces phosphenolpyruvate (PEP), which is then invested either in lipids synthesis or in the citric acid cycle to generate precursor metabolites for protein, DNA, RNA, chlorophyll and biomass synthesis.

G6P, instead of glucose, was chosen as the output of upper glycolysis because G6P is at a branching point between two metabolic pathways with different metabolic functions: carbon storage through the synthesis of carbohydrates and synthesis of NADPH reducing power through the pentose phosphate pathway.

Metabolic reactions of upper glycolysis were grouped and assumed at steady-state. Elementary flux mode analysis resulted in

	MR1	MR3	MR4	MR5	MR6	MR7	MR8
Light	-30	-	-	-	-	-	-
CO₂	-3	-	-	-	-	16.61	11.67
O₂	3	-	-	-	-	-2.15	-7.37
Pi	-1	1	-	-1	1	14.61	4.23
SO₄	-	-	-	-	-	-	-0.03
NO₃	-	-	-	-	-	-	-1.31
Mg₂	-	-	-	-	-	-	-0.0025
H₂O	-2	-1	-	1	-	4.31	6
H	-	-	1	1	-	-34.38	-4.46
ATP	-	-	-1	1	-	2	-
ADP	-	-	1	-1	-	-2	-
NADH	-	-	-	1	-	13.46	-
NAD	-	-	-	-1	-	-13.46	-
NADPH	-	-	-	-	-	-29.3	-
NADP	-	-	-	-	-	29.3	-
GAP	1	-2	2	-1	-	-1	-
G6P	-	1	-1	-	-1	-	-1.14
PEP	-	-	-	1	-	-16.61	-3.13
CARB	-	-	-	-	1	-	-
PA	-	-	-	-	-	1	-0.11
B	-	-	-	-	-	-	1

Figure 5. Stoichiometric matrix K′ describing the bioprocess obtained after formation and reduction of metabolic sub-networks. K′ as a much lower dimension (16×8) than the starting metabolic network (157×162). Lines of K′ correspond to kept metabolites whereas columns correspond to macroscopic reactions obtained thanks to elementary flux mode analysis on each sub-networks. K′ can be divided into sub-matrices K$_S$′ (in red), K$_A$′ (in orange) and K$_B$′ (in green), according to the lines corresponding to substrates S, intracellular metabolites allowed to accumulate A and functional biomass R

3 macroscopic reactions (Table 1). Reaction (MR2) corresponds to a futile cycle since energy (ATP) is dissipated without creation of any metabolic product. This occurs when two metabolic pathways run simultaneously in opposite directions and have no overall effect other than to dissipate energy in the form of heat. Reaction (MR3) corresponds to G6P synthesis whereas reaction (MR4) corresponds to its consumption. The two equations cannot be compiled into one reversible reaction because of the irreversibility of the reactions transforming fructose 6-phosphate into fructose 1,6-biphosphate and vice-versa (R17-R18). Stoichiometry agrees with literature, since 1 ATP needs to be invested to transform 6-carbon sugars (G6P) into simpler ones (GAP) before getting 2 ATP back with lower glycolysis [31].

2.3 Lower glycolysis. Lower glycolysis is a cascading set of reactions which generates the key metabolite phosphoenolpyruvate (PEP) and energy cofactors (ATP, NADH) from GAP. Lower glycolysis was cut at PEP instead of acetyl-coa (AcCoA) because of the presence of the anaplerotic reactions (R35, R36), converting oxaloacetate into PEP and vice-versa.

Lower glycolysis was assumed at steady state. One macroscopic reaction (MR5) was obtained with Elementary Flux Mode analysis (Table 1). Stoichiometry is in accordance with literature: after investment of one ATP in the upper part of glycolysis, 2 ATP are returned with one phosphoenolpyruvate [31].

2.4 Carbohydrates synthesis. Carbohydrates (CARB) are complex sugars stored in the cell. They are formed from 6-carbon sugars (here G6P) by reverse glycolysis. All the reactions participating to carbohydrate synthesis were grouped and assumed to be in quasi-steady state. One reversible macroscopic reaction (MR6) was obtained by reduction thanks to elementary flux mode analysis (Table 1).

2.5 Lipids synthesis. Lipids include a broad group of different macromolecules present in a cell. They contain at least one hydrophobic part and are constituted of long carbon chains linked to a sugar by an ether bound. In microalgae, only Triacylglycerols (TAGs) can be transformed into biofuels [32]. Unfortunately, lipid metabolism of microalgae is poorly known and it differs from bacteria and plants [33]. In the present network, lipids are represented by phosphatidic acids (PAs), precursors of many lipids including glycolipids and phospholipids for the membrane and TAGs for carbon storage.

All the reactions participating in lipids synthesis were grouped and assumed at quasi-steady state. One reversible macroscopic reaction (MR7) for the synthesis of PAs was obtained with elementary flux mode analysis (Table 1). Stoichiometric coefficients are non-integers because PAs are composed of two carbon chains with different lengths (C12–C20). To group all PAs under one entity, a generic reaction synthesizing an "average" PA (R123) was used. Its stoichiometric coefficients were determined experimentally using the proportion of the various fatty acids present in the cell (see File S1 section 1.1 for more details).

The macroscopic reaction obtained satisfies balance of the cofactors. For example 2 ADP yield 2 ATP, and 29.3 NADPH yield 29.3 NADP. Interestingly, when lipids are synthesized, some carbon atoms are lost through the production of CO_2 and conversely some carbon atoms are gained when consuming lipids.

2.6 Biomass synthesis. Protein, DNA, RNA and chlorophyll are necessary to synthesize biomass. Hence, all their synthesis reactions were grouped into a sub-network and assumed at quasi-steady state. Reactions for PA synthesis were not included because a dedicated sub-network is already present in the model. Therefore the biomass synthesis sub-network includes citric acid cycle, oxidative phosphorylation, pentose phosphate pathway, N and S assimilation, amino acids synthesis and nucleotide synthesis.

Table 2. Parameters obtained by the calibration of the model.

Parameters	Value
k_{MR1}	$11.07 \times 10^{-3} \ \mu E^{-1}.m^2.s.mM.h^{-1}.mMB^{-1}$
k_{MR3}	$223.53 \ h^{-1}.mM \ B^{-1}$
k_{MR4}	$10.30 \ h^{-1}.mM \ B^{-1}$
k_{MR5}	$436.95 \ h^{-1}.mM \ B^{-1}$
k'_{MR5}	$5.00 \ h^{-1}.mM \ B^{-1}$
k_{MR6}	$70.00 \ h^{-1}.mM \ B^{-1}$
k'_{MR6}	$6.50 \ h^{-1}.mM \ B^{-1}$
k_{MR7}	$4.50 \times 10^3 \ mM^{-1}.h^{-1}.mM \ B^{-1}$
k'_{MR7}	$0.60 \ h^{-1}.mM \ B^{-1}$
k_{MR8}	$2.18 \times 10^4 \ mM^{-2}.h^{-1}.mM \ B^{-1}$

Citric acid cycle takes place in the mitochondrion and transforms PEP into many precursor monomers for nitrogen assimilation, nucleotide and amino acids synthesis. For each run of the cycle, energy cofactors are generated (NADH, FADH2) and can be breathed into ATP thanks to oxidative phosphorylation. ATP is then reinvested into amino acids and nucleotide synthesis, necessary for DNA, RNA, protein and chlorophyll synthesis. Finally, reductive power (NADPH) necessary for nucleotide and amino acids synthesis is synthesized through the pentose phosphate pathway.

The reduction of this sub-network leads to 30 macroscopic reactions, in which 24 yields biomass (File S1 section 6). All macroscopic reactions not synthesizing biomass correspond to futile cycles where carbon is converted to energy, which is then dissipated. In terms of carbon, the 24 macroscopic reactions once normalized by unit of biomass synthesis flux were only different in their consumption of PEP and hence their production of CO_2. A principal component analysis on the EFMs revealed that the difference was mainly due to two metabolic functions (incorporation of nitrogen and alanine synthesis) that could be performed following different pathways, some less energy-efficient than others explaining the difference of CO_2 production (File S1 section 6, Figure S1, Figure S2).

We assumed that the cell was maximizing biomass growth, and hence minimizing carbon loss when synthesizing biomass. Therefore, the elementary flux mode normalized by unit of biomass synthesis flux with the best PEP/CO_2 yield was chosen (Table 1). The resulting macroscopic reaction MR8 consumes PEP and NO_3 for carbon and nitrogen sources, PA for functional

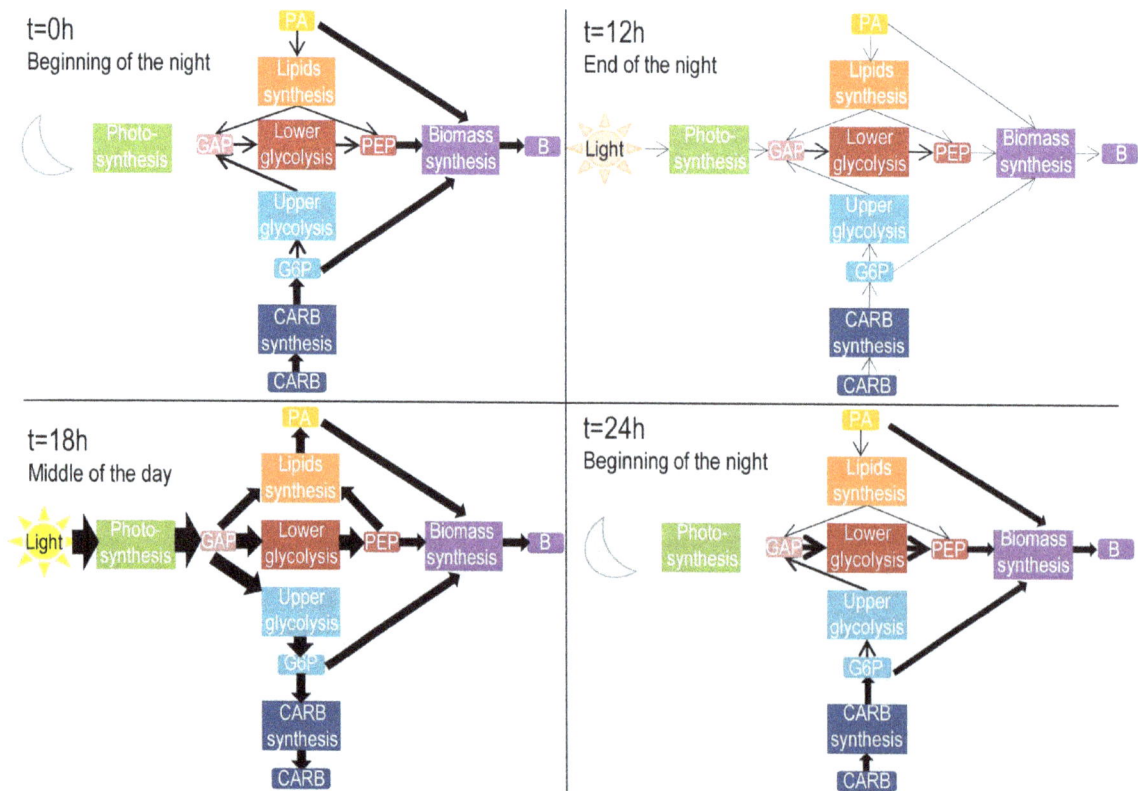

Figure 6. Fluxes between the 6 sub-networks at different time of the day. Fluxes were estimated thanks to model simulations. They were normalized per moles of carbon consumed or produced. Thickness of arrows depends on intensity of the flux. At the beginning of the night (t = 0 h), carbohydrates and lipids are already consumed so as to continue functional biomass growth. Most of carbohydrates and lipids are directly invested for biomass and only few of their carbons are used for PEP synthesis. At the end of the night (t = 12 h), the metabolism is slow, because very few carbons are left for growth and energy. At midday (t = 18 h), when light intensity is at its maximum, slightly less than a third of incoming carbons goes to functional biomass (28,6%). The rest of it is stored into carbohydrates (37,1%) and lipids (34,2%). After one day (t = 24 h), the biological systems has similar fluxes to the beginning (t = 0 h), showing the cyclic behavior of the metabolic network of a unicellular photoautotrophic microalgae submitted to a day/night cycle.

and membrane lipids, G6P for NADPH synthesis through pentose phosphate pathway, SO_4 and Mg for proteins and chlorophyll synthesis and O_2 for ATP synthesis through oxidative phosphorylation. 42.4% of incoming carbon ends up in functional biomass; the rest is breathed through the TCA cycle because of energy demands met thanks to oxidative phosphorylation.

3. Macroscopic reaction kinetics and ODE system

After splitting the network into sub-networks and obtaining the EFMs for each sub-network, a reduced model described by 16 metabolites and 8 macroscopic reactions was obtained. The number of macroscopic reactions is similar to the model of Guest et al [34], where 10 lumped metabolic reactions were obtained. Mathematically, these first two steps of the DRUM approach translated into a reduced stoichiometric matrix K' (Figure 5) of much lower dimension (16×8) than the starting one (157×162). The definition of the reaction kinetics is the final building block of DRUM. For each macroscopic reaction obtained after the reduction step, simple proportional kinetics were assumed (Table 1).

According to section 2, the model is described by the following ODE system:

$$\frac{dM'}{dt} = \frac{d\begin{pmatrix} S \\ A \\ B \end{pmatrix}}{dt} = K'.\alpha.B - D.M' + D.\begin{pmatrix} S_{in} \\ 0 \\ 0 \end{pmatrix} \quad (10)$$

where M' is the vector of kept metabolites (16×1) composed of substrate S, metabolites authorized to accumulate A and functional biomass B; K' is the reduced stoichiometric matrix (16×8) and α is the kinetics vector (8×1) (Figure 5 and Table 1).

As explained in section 2, biomass B corresponds to functional biomass. Total biomass, in terms of particulate carbon and nitrogen, is computed using the following formulae:

$$X_C(t) = \sum_A C_A.A(t) + C_B.B(t)$$
$$X_N(t) = \sum_A N_A.A(t) + N_B.B(t) \quad (11)$$

where $A \in \{CARB; PA; PEP; G6P; GAP\}$, C_A and C_B correspond to the number of carbon atoms per molecule of A and B, N_A and N_B correspond to the number of nitrogen atom per molecule of A and B, $A(t)$ and $B(t)$ correspond to the concentration of A and B at time t, and $X_C(t)$ and $X_N(t)$ correspond to the concentration of carbon and nitrogen in total biomass X. As carbon and nitrogen biomass were measured experimentally, we simulated carbon and nitrogen content of the biomass. However, other chemical elements can be easily computed using the formula above. No additional parameters would be necessary as the above formula only uses chemical element composition and concentrations of A and B. Chemical element composition for A and B is available in section 1.5 of the File S1. In addition, energy cofactors are not taken into account in equation (11), as we assume their contribution negligible in terms of carbon and nitrogen compared to functional biomass and other molecules authorized to accumulate (CARB, PA, PEP, G6P & GAP).

Here, only the core metabolic network of a unicellular autotrophic microalgae was represented. It does not take into account energy necessary for mechanisms not represented by the network, like for instance the turnover of macromolecules and other so-called futile cycles. As it is clearly documented in the

literature [35], energetic cofactors ATP, NADH, NADPH and FADH2 are difficult to balance. Usually, balancing is done through maintenance terms like equation MR2, which are determined so that growth rate and substrate consumption fits experimental data [24,36]. Here, as carbon incorporation was not measured (light absorbed per unit of biomass was not measured, nor was CO_2 dissolved concentration), estimation of maintenance and hence cofactors balance is difficult to perform. We thus decided not to consider the balance of energetic cofactors, and we did not describe their fate (ATP, ADP, NADPH, NADP, NADH, NAD).

The dynamic model has 10 degrees of freedom, each degree represented by a parameter that needs to be calibrated. To estimate parameters, we minimized the squared-error between simulation and experimental measurements (taken as an average of the duplicates) using the following formula:

$$error = \sum_x \sum_t (x_{measured(t)} - x_{simulated}(t))^2$$
$$x \in \{CARB; PA; X_C; X_N\} \quad (12)$$

To minimize the error, the Nelder-Mead algorithm [37] (function *fminsearch* under Scilab (http://www.scilab.org)) was used. To reduce the risk of finding a local minima, several optimizations were performed with random initial parameters set. Then the set fitting the best experimental data was chosen. As very few data were available, all data were used to estimate model parameters. Results of parameter identification are presented in Table 2. The script file of the resulting model in Scilab format and the experimental data are available as File S2 and S3.

4. Simulation

Model simulation reproduces accurately experimental data (see Figure 1). In particular, the model correctly represents lipids and carbohydrates accumulation during the day and their consumption during the night (Figure 1D). The distribution of fluxes during a classical day/night cycle is displayed in Figure 6 and in Video S1.

The model predicts a minimum of carbon storage (lipids and carbohydrates) one hour and a half after sunrise (13h37 and 13h17), when light intensity is sufficient to catch up with carbon loss through respiration. In a similar way, the maximum is reached three hours before sunset (20h50 and 21h02), when light intensity is insufficient to catch up with carbon loss through respiration (Figure 1D). Total carbon biomass follows a similar trend (minimum at 13h19 and maximum at 21h17), suggesting that an adequate harvesting time for biofuels production is three hours before sunset (21 h), when lipids are at their maximum. Interestingly, carbohydrates synthesis begins after and ends before lipids synthesis (respectively 13h31 and 22h08 against 12h58 and 23h26). This is due to the fact that there is a higher carbon demand for functional biomass synthesis from carbohydrates (through G6P) than from lipids: 6.84 carbons from carbohydrates are required per unit of functional biomass against 4.27 carbons from lipids. At midday ($t = 18$ h), when light intensity is at its maximum, carbohydrates and lipids synthesis are also at their maximum. At this time, slightly less than a third of incoming carbons goes to functional biomass (28.6%). The rest goes to carbohydrates (37.1%) and lipids (34.2%) storage (Figure 6).

Contrary to carbon storage, functional biomass carbon quota increases three hours before sunset until two hours after dawn, taking carbon from the lipids and carbohydrates pool (Figure 1D and E, Figure 6). Most of carbohydrates (through G6P) and most of lipids are directly consumed for functional biomass production. Only few of their carbons are used for PEP synthesis (Figure 6). At the end of the night and beginning of the day, the metabolism is really slow, because very few carbons in the storage pools are left for growth (Figure 6). Conversely, functional biomass carbon quota decreases during the day because of its dilution in the total biomass due to carbon storage. These obtained metabolic behavior are in agreement with the description of flux distribution given by Ross and Geider in [38].

Total biomass can be visualized in terms of particulate carbon and nitrogen (Figure 1A and B). Carbon follows a similar trend to carbohydrates and lipids, because carbon is only incorporated through photosynthesis during the day, and is lost during the night because of respiration to meet energy demands for continuing functional biomass growth. The diurnal photosynthetic quotient (moles of oxygen released per mole dioxide fixed) varies between 1.29 and 1.60 (Figure S3), depending on the light intensity, which agrees with the typical range of 1.0–1.8 for algae [22]. During the day, 79% of carbon loss is due to respiration and 21% to lipids synthesis. During the night, 10% of carbon lost by respiration is gained back by lipids consumption.

In the model, nitrogen content has exactly the same trend as functional biomass, since functional biomass is the only intracellular metabolite with nitrogen. It can be observed that there is slight delay in the uptake of nitrogen between the model and experimental data. In experimental data, the minimum is at sunrise and the maximum at sunset, meaning that *Tisochrysis lutea* stops incorporating nitrates as soon as the night starts. This time period corresponds to the period where cells divide [16]. Mocquet et al. in [39] have shown that nitrate uptake is stopped during cell division, which could explain the difference between predicted values and experimental data. However, including such mechanisms at this stage in the model would be debatable. Chlorophyll is also well predicted by the model, validating the hypothesis of a constant ratio with functional biomass.

Finally, it is interesting to look at the evolution of PEP, G6P and GAP concentrations predicted by the model. First, their concentrations are sufficiently low in terms of carbon, showing that carbon storage is mainly done with lipids and carbohydrates. However, their concentrations over time are not constant, and are particularly different between day and night. Indeed, their concentrations are much higher during the day than during the night, giving certain flexibility to the metabolic network when environmental conditions changes rapidly (here light). The ability of metabolic network to face permanent fluctuating environmental conditions consolidates one of the advantages of the DRUM approach. Such flexibility is acquired through certain metabolites, which can accumulate and therefore act as buffers. This could not be achieved with a steady-state assumption.

Discussion

1. Assumptions in the DRUM approach

1.1 QSSA on sub-networks. The main assumption of the DRUM approach is the quasi-steady state assumption on subnetworks of the metabolic network. This assumption is supported by the idea of cell function and cell compartment, often associated to co-regulation and substrate channeling.

Indeed, in a cell, metabolic pathways composed of grouped reactions regulated together are omnipresent. These reactions are often synchronous: intermediate metabolites produced by a reaction are nearly immediately consumed by the next reaction in the cascade. This implies a quasi-steady state for the intermediate metabolites. Many examples of such pathways can be found in literature. One of the most illustrative ones is reactions in cascade where the first reaction of the pathway is submitted to feedback inhibition by the end-product of the last reaction [40].

In addition, spatial and molecule crowding are not negligible phenomena in a cell. When not taken into account, they imply that any intracellular metabolite can be consumed in any reactions of the cell, even if the reaction occurs at a far loci or in a different compartment where the molecule cannot be transported to and needs to be resynthesized. This often leads to erroneous metabolic flux distributions when using flux balance analysis and to a combinatorial explosion of the number elementary flux modes representing the metabolic network. For example, in the case of the metabolic network of *Chlamydomonas reinhardtii* [24], when ATP of the chloroplast is constrained to stay in the chloroplast, the number of EFMs reduces from 4909 to 452. We thought reasonable to assume that reactions inside a same compartment and completing the same metabolic function are synchronous. For example, the light and dark steps of photosynthesis can be assumed synchronized so that all ATP and NADPH produced by the first step are directly consumed in the second step.

An extreme illustration of space phenomena supporting our quasi-state assumption is substrate channeling, where an intermediate metabolite is, instead of being released in the solution, passed from enzyme to enzyme so as to avoid any loss to competing pathways [41]. In this case, the notion of metabolic reaction is difficult to define since the reaction is already a macroscopic reaction composed of synchronous elementary reactions where intermediate metabolites are under QSSA.

Even if regulation, substrate channeling and reactions loci in the cell are not always well-known, we assumed that QSSA is a biologically reasonable assumption for a group of reactions taking place in the same compartment, synthesizing a same pathway endproduct or fulfilling a similar metabolic function. QSSA on subnetwork is a mild way to relax the balanced growth hypothesis, without constraining the full network anymore. In most cases, the main sub-networks will be the same, defined on metabolic functions: upper glycolysis, lower glycolysis, TCA cycle, Calvin cycle (for photoautotrophs), macromolecules synthesis.

It is very important to keep in mind that the DRUM approach does not only split the initial network into sub-networks, but it also duplicates some reactions that take place simultaneously at different part of the cell within different functions. This point is very important in order to keep a sound meaning to the reduced networks derived from the EFM analysis.

1.2 Network splitting into groups of reactions. Network splitting into groups of reactions is performed on the basis of the above-mentioned criteria. However, these intracellular mechanisms are not always well known. Hence, it is difficult to split the network only taking into account experimentally proved report of these phenomena on the microorganism studied. To overcome this hurdle, network splitting was also performed thanks to educated guesses using the topology of the metabolic network, the known metabolic functions of some groups of reactions, the experimentally known accumulating metabolites (e.g., lipids, carbohydrates) and the key topological place of some metabolites. The metabolites A allowed to accumulate are thus end-products of metabolic pathways (e.g., macromolecules) or situated at a branching point between several pathways.

In the case of *Tisochrysis Lutea*, the presence of the chloroplast compartment was used to assume QSSA for photosynthesis. For the rest of the metabolic network, reactions were grouped according to known metabolic functions: carbohydrate synthesis, upper glycolysis, lower glycolysis, lipids synthesis, biomass synthesis. The accumulated metabolites GAP, PEP, G6P were

Table 3. Comparison of existing microalgae models representing carbon storage.

Reference	Modeling type	Macroscopic reactions	Metabolic Fluxes	Metabolites concentrations	Degrees of freedom
[34]	Macroscopic, Dynamic	11	0	7	12
[38]	Macroscopic, Dynamic	5	0	7	18
[53]	Macroscopic, Dynamic	3	0	4	12
[54]	Macroscopic, Dynamic	6	0	7	9
[55]	Macroscopic, Dynamic	4	0	5	15
[56]	Macroscopic, Dynamic	1	0	2	5
[57]	Macroscopic, Dynamic	2	0	3	7
[58]	Macroscopic, Dynamic	11	0	7	8
[59]	Macroscopic, Dynamic	6	0	7	7
[24]	Metabolic, Static	0	160	0	1
[22]	Metabolic, Static	0	484	0	2
[26] & [52]	Metabolic, Static	0	280	7	22
[17]	Metabolic, Static & Dynamic	0	760	9	45
Present approach	Metabolic & Macroscopic, Dynamic	7	162	14	10

To compare the models, our definition of "degrees of freedom" stands for the number of information needed to simulate the models. For macroscopic models, degrees of freedom relate to the kinetic parameters of the model. For FBA models, degrees of freedom relate to the number of constraints needed to determine the flux distribution. Incoming light and biomass composition were not considered as degrees of freedom.
For [56] and [57], no macroscopic reactions are obtained per se, as growth is independent of nutrient uptake. Only population growth is represented ($X \xrightarrow{\mu(X)} 2X$).
For [17], 7 biomass compositions were necessary to perform DFBA. We counted 6 of them as degrees of freedom.

chosen because situated at branching points of several metabolic pathways. Indeed, GAP, the output of photosynthesis, is situated at the middle of glycolysis and is also an output of the pentose phosphate pathway. G6P is situated at the branching point between carbohydrates synthesis and the pentose phosphate pathway. Finally, PEP is situated at the branching point between lipids synthesis, the TCA cycle for precursor metabolites necessary for biomass synthesis and the anaplerotic reactions.

However, the choice of the decomposition is not totally straightforward. The splitting of *Tisochrysis lutea* metabolic network was performed by trial and errors with different possible decompositions. Several possible configurations were tested and the best one was kept. For example, the metabolic network was cut, instead of glyceraldehyde 3-phosphate (GAP) at glycerone-phosphate (DHAP) and instead of phosphoenolpyruvate (PEP) at pyruvate (PYR). To cut at PEP seemed a better choice to fit functional biomass data, but cutting at DHAP did not influence the results since DHAP and GAP are interchangeable metabolites ('DHAP <–> GAP' (EC 5.3.1.1)). Whether the network should be cut at GAP or DHAP could only be answered with additional experimental measurements.

In a general way, only few decompositions work, but some have close performances. Only experimental data will allow favoring one from the other. Still, the presence of these equivalent decompositions is beneficial since it points out the dynamic measurement of metabolites to make so as to discriminate the best model.

The method, in its first developmental stage, is not automatic yet. However, systematic network splitting techniques could be developed. For example, the network could be split according to the metabolites participating in more than a threshold number of metabolic reactions [42]. The network could also be split using flux coupling analysis, where totally coupled reactions could be used as a starting point for sub-networks [43]. Finally, any other network clustering techniques could be used, from metabolic

function annotations to topology [44,45]. In addition, automation of the method will allow discriminating the different possible decompositions. Indeed, the automated decomposition algorithm will yield a finite number of possibilities, which will be explored. For each of them, a finite number of simple kinetics will be tested and their kinetic parameters estimated to fit experimental data. The Akaike Information Criterion could then be used to provide a score for selecting the best candidate model [46], accounting for the tradeoff between fitting and parameter parsimony. However, selecting the best decomposition imposes a computational challenge since global identification procedures, often requires, in practice, expert knowledge to reduce the attraction of local minima.

1.3 Network reduction into macroscopic reactions. Once network splitting into sub-networks was performed, network reduction is straightforward as it consists in computing Elementary Flux Modes (EFMs) for each sub-network and reducing them to macroscopic reactions by keeping only the transport reaction of incoming and outgoing metabolites. This can be performed automatically using softwares like *efmtool* [29] to compute the EFMs and a small script to deduce the macroscopic reactions from the EFMs obtained.

However there is an exponential explosion of the number of Elementary Flux Modes (EFMs) when the number of reactions increases, which implies an exponential explosion of the kinetics parameters to estimate. This could make the approach intractable and annihilate the advantage of DRUM compared to a full kinetics model when using large sub-networks resulting for example from the splitting of a genome-scale metabolic network. To overcome this difficulty, small sub-networks should be favored and there are available methods to reduce the number of EFMs such as the use of experimental data [7], a projection of the EFMs space into the yield space [47] or the clustering of EFMs into phenotypic families [48]. These methods are semi-automatic, well documented and already proved to be efficient to model biological systems [7,11,49]. Flux Balance Analysis (FBA) and by extent Dynamic Flux Balance Analysis (DFBA) can also be seen as methods to reduce the number of EFMs using optimization. Indeed, a solution of FBA corresponds to a positive linear combination of EFMs and the solution for any optimal product/ substrate ratio always coincide with an elementary mode [5]. Thus, when applying DRUM, such above-mentioned methods can be automatically applied if the number of EFMs for some sub-network is too high.

In the case of *Tisochyris lutea*, the biomass synthesis sub-network is composed of 105 reactions. The calculation of the EFMs resulted in 24 macroscopic reactions. Note that the number of macroscopic reactions is already lower than the number of reactions of the original sub-network. For a further reduction, we kept the EFM with best PEP/CO_2 yield when normalized by unit of biomass synthesis flux, which was the same as optimizing biomass growth since we minimized carbon loss through oxidative phosphorylation.

In addition, DRUM drastically reduces the number of EFM compared to a QSSA applied to the whole network thanks to the application of QSSA only on sub-networks. Indeed, as EFMs are only computed on small sub-networks and as the explosion of the number of EFMs is exponential with the number of reactions, the sum of the number of EFMs obtained from each sub-network is smaller than the number of EFMs obtained for a QSSA on the whole network. In the case of *Tisochrysis lutea*, DRUM reduces the number of EFM from 18776 for the whole network down to 11. This implies a low number of degrees of freedom (10 parameters) compared to the other methods (cf Table 3) where degrees of freedom are often hidden in parameters (e.g.: biomass composition) or imposed fluxes (substrate consumption, product formation, biomass growth, maintenance) varying along discrete time instants.

1.4 Macroscopic reactions and their kinetics. Once all macroscopic reactions modes are obtained, their kinetics need to be defined, which is the final step of DRUM. This is a delicate task, and unfortunately there is no unique or systematic way of doing it. The choice is left to the researcher's attention and experience and is also relative to the experimental data available. Classical kinetics found in literature are mass-action, power-law, Michaelis-Menten, Hill, cybernetic kinetics [19], or more complex allosteric regulations kinetics [50]. However, DRUM is an approach looking for a model with a reduced complexity and hence a minimum number of parameters.

In the case of *Tisochrysis Lutea*, since one parameter per reaction turns out to be sufficient to explain the data, we kept this minimum structure to follow a parsimony principle.

In future works, methods such as the one developed by *Curien et al.* [50], based on in vitro reconstitution of the sub network, could provide a way to experimentally determine kinetic models. Alternatively, a multi-level optimization such as in [51] could also be used. It would avoid the need to postulate kinetics and estimate their parameters. Yet, defining the objective function is not a trivial task.

1.5 Total biomass and functional biomass. Biomass B is a variable used to predict the macroscopic biomass production, which is generally measured in dry weight mass or in carbon mass. In metabolic models, biomass B is usually represented as an average composition of macromolecules present in the cell. For example, in the case of *Chlamydomonas reinhardtii*, the biomass is composed of 64.17% of proteins, 27.13% of carbohydrates, 4.53% of lipids, 3.05% of RNA, 1.02% of chlorophyll and 0.11% of DNA in average [24]. An artificial metabolic reaction of biomass synthesis is thus added to the metabolic network, where the

stoichiometric coefficients of the reaction are the measured molar proportions of each macromolecule present in the cell. In system (1), biomass B acts as a growth catalyzer. This reflects the fact that the proteins, nucleic acids and other macromolecules that are part of the biosynthetic apparatus and structural material (e.g., cell walls) catalyze the intracellular reactions and hence growth.

In the DRUM approach, some macromolecules can accumulate and will therefore not appear in biomass B. We assumed that macromolecules catalyzing growth such as proteins do not accumulate and end up in biomass B, which we rename functional biomass B. This relies on the assumption that storage compounds of a cell does not have any other metabolic functions than to store chemical elements (e.g., carbon) so as to supply energy and chemical elements demands to continue growth when these resources are no longer available in the environment. The term αB in (7) is thus still meaningful, since functional biomass B catalyzes growth as the term vB does in (1). An estimation of the total actual biomass can then be obtained by summing up functional biomass B and the storage terms A (cf equation (9)).

2. Comparison to other models

Microalgae models exist for more than 60 years and can be divided into two main categories: dynamical macroscopic models (see [15] for a full review) and static metabolic models [17,22,24,26,52].

To date, there is only 9 macroscopic models representing carbon storage (particularly lipids) in microalgae [34,38,53–59]. However, these models are empirical and do not rely on metabolic knowledge. They describe efficiently some key metabolites, but does not allow to understand the intracellular mechanisms taking place in the cell and stay limited in the number of variables for which accumulation dynamics can be forecasted (Table 3). Only the models of [34] and [58] tried to incorporate some metabolic knowledge. Guest et al [34] used lumped metabolic reactions taken from literature and for which stoichiometric coefficients were determined depending on the environmental conditions. Fleck-Schneider et al [58] used a hybrid modeling technique where ordinary differential equations described the macroscopic scale of the bioprocess whereas flux optimization on a lumped metabolic model was performed at each time-step at the metabolic scale.

For metabolic models, only static flux predictions under constant light were made, where lipids and carbohydrates were at a constant ratio in biomass [17,22,24,26,52]. Even if, sometimes, the influence of light intensity on metabolic fluxes and biomass composition was studied [24,52], only the recent model of Knoop et al [17] tried to simulate, thanks to dynamic flux balance analysis, the evolution of metabolic fluxes during a day/ night cycle. The simulation was performed thanks to a time-dependent biomass reaction based on literature, which allowed forcing the value of the fluxes to the storage compounds. This involves a much higher degree of freedom (45, cf Table 3) than with DRUM (10) since the biomass composition must be postulated at each time instant (or at some key instants and then interpolated). However, a more systematic method for representing carbon accumulation and consumption over time is lacking. Contrary to the work of Knoop and al. [10], DRUM allows predicting at the same time all metabolic fluxes and the change of biomass composition without forcing carbon storage to a given value computed at each time step. This is the real advantage of our method, where we can predict at the same time the macroscopic scale (biomass synthesis, substrate consumption, and products synthesis) and the intracellular scale (metabolic fluxes). To the authors' knowledge, no one managed to predict them dynamically using a metabolic framework managing non-balanced growth.

In relation to the existing microalgae models DRUM, the new framework proposed in this paper, allowed for the first time to predict dynamically at the same time the macroscopic scale of the bioprocess (particulate carbon and nitrogen, Figure 1A and B) and the metabolic scale (lipids, carbohydrates, chlorophyll and all metabolic fluxes, see Figure 1C and D and E, Figure S4) with few parameters to estimate (Table 3). The originality of DRUM lies in the coupling of macroscopic and intracellular modeling approaches as discussed below.

3. Joining the macroscopic and the metabolic scales: a bottom-up approach

Classical modeling approaches of bioprocesses can be sorted into two main categories: modeling at the macroscopic scale, where microorganisms act as catalyzers of macroscopic reactions [60] and modeling at the intracellular scale, which takes into account intracellular mechanisms such as biochemical reactions or genetic regulation.

Macroscopic models have usually a low dimension, allow to account for time varying experimental data and predict well the macroscopic scale of bioprocesses such as substrate consumption and biomass growth [60]. Unfortunately, the number of macroscopic reactions necessary to represent the bioprocess, their expression, their stoichiometric coefficients and their kinetics need to be determined experimentally [61,62]. In addition, macroscopic modeling does not take into account intracellular mechanisms and thus can hardly be used for optimization of intracellular molecules of interest.

On the other hand, intracellular modeling describes accurately mechanisms occurring inside the cell such as reactions between metabolites catalyzed by enzymes, translation and transcription of genes. These models are based on the knowledge of the metabolic, transcriptomic and genomic networks. They allow a better understanding of the cellular mechanisms and seem more appropriate to describe and optimize bioprocesses implying intracellular molecules. However, the use of intracellular models for time varying experiments is hampered by the lack of experimental data required to define and calibrate the kinetic reaction rates of the biochemical reactions [2]. The common assumption found in the literature to overcome this hurdle is the balanced-growth assumption.

While these two modeling approaches bring answers to different objectives, a remaining challenging question is how to couple macroscopic and intracellular models to enlarge the prediction capabilities of the model while keeping a model structure with a low complexity level?

Two strategies can be applied in the attempt to couple the two scales: a top-down approach, where some intracellular mechanisms are included in details in a macroscopic model, or a bottom-up approach where intracellular mechanisms are simplified and linked to the macroscopic scale. The first approach consists in finding and representing in details the preponderant intracellular mechanisms that have an impact at the macroscopic scale. All others intracellular mechanisms are assumed negligible. This approach is thus very microorganism dependent and cannot easily be generalized. Still, even if limited, this approach usually improves the prediction of the macroscopic scale and helps to better understand the bioprocess [38,63].

On the other hand, the reduction of intracellular mechanisms to represent in a simple way the macroscopic scale of a bioprocess is a difficult task, particularly given the lack of knowledge of intracellular mechanisms and the lack of experimental data

available. Still, thanks to the balanced-growth hypothesis, systematic reduction frameworks were already developed for the metabolic scale. Indeed, QSSA allows to link statically [3] or dynamically [7,9] the intracellular scale (metabolic fluxes) to the macroscopic scale (biomass growth). Even if some difficulties still remain (e.g., a high number of elementary flux modes, no accumulation of intracellular metabolites, balance of cofactors), predictions are in good agreement with experimental data and allow insightful understanding and optimization of bioprocesses [7,9,11]. DRUM is the next generation of these existing bottom-up approaches, where dynamics and intracellular accumulation are taken into account, as well as spatial phenomena and regulation to some extent, thanks to the network splitting.

4. Use of DRUM to guide metabolic engineering

Gene deletion studies (GDS) exploit the Gene-Enzyme-Reaction relationship to predict the effect of the deletion of one or several genes on the growth and/or on product synthesis [64–68]. Metabolic engineering can thus be guided thanks to *in silico* models by GDS to find ideal gene targets to improve production yields of molecules of interest. The DRUM approach could extend these approaches at the levels of the metabolic function or of the reaction.

The first level consists in targeting metabolic functions represented by the macroscopic reactions deduced from the EFMs of each sub-networks. Deleting a metabolic function is hence equivalent to delete a macroscopic reaction. In a practical way, as EFMs are minimal metabolic behaviors of the cell [69], targeting an EFM is the same as targeting one of the EFM non-null reactions, since EFMs are non-decomposable vectors by definition [69]. However one needs to be careful that the deletion of one reaction does not affect another EFM using the same reaction.

The second level is the deletion of a reaction in the metabolic network. This could yield the same result as deleting one metabolic function, yet it could also imply accumulation of a previously non-accumulating metabolite hence modifying the decomposition of the sub-networks. It could also imply obtaining different EFMs and hence different macroscopic reactions (e.g.: stoichiometric coefficients). This could require a new decomposition and reduction of the sub-networks, and new kinetics to postulate and parameters to estimate.

For *Tisochrysis* lutea, the goal of our microalgae model was to better apprehend the carbon metabolism of microalgae in day/night cycles. It is clear that such a model has many direct implications for metabolic engineering with microalgae. The fact that cells can store very high amounts of lipids with a daily pattern has clear consequences on the harvesting period (section 3.4). It also indicates the paths and the enzymes to be targeted in order to more efficiently accumulate lipids. For example, we can target the carbohydrates production (MR6) and simulate *de novo* the model to see whether it has an impact on lipids accumulation. The results suggests, as expected, that the carbohydrates storage pool diminished quickly at the expense of the lipids and functional biomass pool (Figure S5, File S1 section 7). In addition, G6P accumulates during the day and is consumed during the night, standing in for the carbohydrates storage pool. The only difference is that at the end of the night, the G6P pool is completely depleted. What is also interesting is that the total carbon biomass X stays the same: only a shift of carbon between the different pools is observed. The day/night cycle growth still occurs and takes place at a similar velocity, which was not straightforward since glucose-6-phosphate concentration could have been too low to allow functional biomass synthesis during the night.

Conclusions

This paper presents DRUM, a new metabolic modeling framework, which allows to predict dynamically the accumulation of intracellular metabolites using metabolic knowledge. The proposed strategy results from a tradeoff between complexity and representativeness. It conciliates intracellular and macroscopic models in a fluctuating environment.

DRUM was applied to the phototrophic unicellular microalgae *Tisochrysis lutea* and led to a model describing well the accumulation of lipids and carbohydrates in the microalgae under day/night cycles.

DRUM helps to better understand intracellular mechanisms at the metabolic level when the biological system undergoes environmental perturbations. In addition, DRUM could be used in dynamic control frameworks to optimize the bioprocess. This was not possible before, as models were static and did not allow accumulation of intracellular metabolites.

Future work will consist in applying the methodology to mixed ecosystems, so as to better understand the interactions taking place between the individual species composing the microbial community. Indeed, even if the scale is different, same philosophical principles can be used to split the metabolic network of a microbial community.

Supporting Information

Figure S1   Projection of elementary flux modes obtained from the biomass synthesis sub-network in the PEP/CO2 yield space. The reduction of the biomass synthesis sub-network leads to 30 macroscopic reactions, in which 24 yields biomass. In terms of carbon, the 24 macroscopic reactions were only different in their consumption of PEP and hence their production of CO_2. A projection in the yield space $PEP = f(CO_2)$ reveals two distinct metabolic behaviors.

Figure S2   Principal component analysis of the elementary flux modes obtained from the biomass synthesis sub-network. The difference in the PEP/CO2 yield is mainly due to two metabolic functions (incorporation of nitrogen (x-axis) and alanine synthesis (y-axis)) that can be performed thanks to different pathways, some less energy-efficient than others explaining the difference in CO_2 production.

Figure S3   Predicted photosynthetic quotient during a day/night cycle. The quotient varies between 1.29 and 1.60, depending on the light intensity, which agrees with the typical range of 1.0–1.8 for algae [22].

Figure S4   Metabolic fluxes of the core network at midday (18 h).

Figure S5   Comparison of the wild type and MR6-deficient *in silico* models. The two models were then simulated for 48 h, one with $k_{carb} = 0$ $h^{-1}.mM$ B^{-1}, the other one with $k_{carb} = 70.00$ $h^{-1}.mM$ B^{-1}. The dilution rate and the incoming substrate concentrations were set at 1 $days^{-1}$ and 4.018 $mgN.L^{-1}$.

File S1   Detailed metabolic network reconstruction process of *Tisochrysis lutea*; list of reactions and metabolites; analysis of the whole metabolic network; list of sub-networks; list of macroscopic reactions obtained for the biomass synthesis sub-network.

File S2   Scilab script of the day/night cycle model of *Tisochrysis Lutea*.

File S3   Experimental data of continuous cultures of *Tisochrysis Lutea*.

Video S1   Predicted metabolic fluxes between sub-networks during a 24 h day/night cycle.

Acknowledgments

C. Baroukh would like to thank Dr Yan Rafrafi for his careful reading and pertinent advices while correcting the paper. Authors would also like to thank Dr Cesar Aceves-Lara for his relevant remarks and useful discussions.

Author Contributions

Conceived and designed the experiments: CB RMT JPS OB. Performed the experiments: CB. Analyzed the data: CB RMT OB. Contributed to the writing of the manuscript: CB RMT OB JPS.

References

1. Stephanopoulos G, Aristidou AA, Nielsen J (1998) Metabolic engineering: principles and methodologies. 1st ed. San Diego: Academic Press - Elsevier USA.
2. Heijnen JJ, Verheijen PJT (2013) Parameter identification of in vivo kinetic models: Limitations and challenges. Biotechnol J 8: 768–775.
3. Orth J, Thiele I, Palsson B (2010) What is flux balance analysis? Nat Biotechnol 28: 245–248.
4. Mahadevan R, Edwards JS, Doyle FJ (2002) Dynamic flux balance analysis of diauxic growth in Escherichia coli. Biophys J 83: 1331–1340.
5. Schuster S, Dandekar T, Fell DA (1999) Detection of elementary flux modes in biochemical networks: a promising tool for pathway analysis and metabolic engineering. Trends Biotechnol 17: 53–60.
6. Burgard AP, Nikolaev V, Schilling CH, Maranas CD (2004) Flux coupling analysis of genome-scale metabolic network reconstructions. Genome Res 14: 301–312.
7. Provost A, Bastin G, Agathos SN, Schneider Y-J (2006) Metabolic design of macroscopic bioreaction models: application to Chinese hamster ovary cells. Bioprocess Biosyst Eng 29: 349–366.
8. Song H-S, Morgan JA, Ramkrishna D (2009) Systematic development of hybrid cybernetic models: application to recombinant yeast co-consuming glucose and xylose. Biotechnol Bioeng 103: 984–1002.
9. Song H-S, Ramkrishna D, Pinchuk GE, Beliaev AS, Konopka AE, et al. (2012) Dynamic modeling of aerobic growth of Shewanella oneidensis. Predicting triauxic growth, flux distributions, and energy requirement for growth. Metab Eng 15: 25–33.
10. Edwards JS, Ibarra RU, Palsson BO (2001) In silico predictions of Escherichia coli metabolic capabilities are consistent with experimental data. Nat Biotechnol 19: 125–130.
11. Zamorano F, Van de Wouwer A, Jungers RM, Bastin G (2013) Dynamic metabolic models of CHO cell cultures through minimal sets of elementary flux modes. J Biotechnol 164: 409–422.
12. Song H-S, Ramkrishna D (2009) When is the Quasi-Steady-State Approximation Admissible in Metabolic Modeling? When Admissible, What Models are Desirable? Ind Eng Chem Res 48: 7976–7985.
13. Wijffels RH, Barbosa MJ (2010) An Outlook on Microalgal Biofuels. Science (80-) 329: 796–799.
14. Lardon L, Helias A, Sialve B, Steyer J, Bernard O (2009) Life-cycle assessment of biodiesel production from microalgae. Environ Sci Technol 43: 6475–6481.
15. Bernard O (2011) Hurdles and challenges for modelling and control of microalgae for CO2 mitigation and biofuel production. J Process Control 21: 1378–1389.
16. Lacour T, Sciandra A, Talec A, Mayzaud P, Bernard O (2012) Diel Variations of Carbohydrates and Neutral Lipids in Nitrogen-Sufficient and Nitrogen-Starved Cyclostat Cultures of Isochrysis Sp. J Phycol 48: 966–975.

17. Knoop H, Gründel M, Zilliges Y, Lehmann R, Hoffmann S, et al. (2013) Flux Balance Analysis of Cyanobacterial Metabolism: The Metabolic Network of Synechocystis sp. PCC 6803. PLoS Comput Biol 9: 1–15.
18. Klamt S, Stelling J (2003) Two approaches for metabolic pathway analysis? Trends Biotechnol 21: 64–69.
19. Young JD, Ramkrishna D (2007) On the matching and proportional laws of cybernetic models. Biotechnol Prog 23: 83–99.
20. Bendif EM, Probert I, Schroeder DC, Vargas C de (2013) On the description of Tisochrysis lutea gen. nov. sp. nov. and Isochrysis nuda sp. nov. in the Isochrysidales, and the transfer of Dicrateria to the Prymnesiales (Haptophyta). J Appl Phycol 25: 1763–1776.
21. Yang C, Hua Q, Shimizu K (2000) Energetics and carbon metabolism during growth of microalgal cells under photoautotrophic, mixotrophic and cyclic light-autotrophic/dark-heterotrophic conditions. Biochem Eng J 6: 87–102.
22. Boyle NR, Morgan JA (2009) Flux balance analysis of primary metabolism in Chlamydomonas reinhardtii. BMC Syst Biol 3: 1–14.
23. Manichaikul A, Ghamsari L, Hom E, Chin C, Murray R, et al. (2009) Metabolic network analysis integrated with transcript verification for sequenced genomes. Nat Methods 6: 589–592.
24. Kliphuis A, Klok AJ, Martens DE, Lamers PP, Janssen M, et al. (2012) Metabolic modeling of Chlamydomonas reinhardtii: energy requirements for photoautotrophic growth and maintenance. J Appl Phycol 24: 253–266.
25. Chang RL, Ghamsari L, Manichaikul A, Hom EFY, Balaji S, et al. (2011) Metabolic network reconstruction of Chlamydomonas offers insight into light-driven algal metabolism. Mol Syst Biol 7: 1–13.
26. Cogne G, Rügen M, Bockmayr A, Titica M, Dussap C-G, et al. (2011) A model-based method for investigating bioenergetic processes in autotrophically growing eukaryotic microalgae: application to the green algae Chlamydomonas reinhardtii. Biotechnol Prog 27: 631–640.
27. Dal'Molin CGDO, Quek L-E, Palfreyman RW, Nielsen LK (2011) AlgaGEM-a genome-scale metabolic reconstruction of algae based on the Chlamydomonas reinhardtii genome. BMC Genomics 12 Suppl 4: 1–10.
28. Krumholz EW, Yang H, Weisenhorn P, Henry CS, Libourel IGL (2012) Genome-wide metabolic network reconstruction of the picoalga Ostreococcus. J Exp Bot 63: 2353–2362.
29. Terzer M, Stelling J (2008) Large-scale computation of elementary flux modes with bit pattern trees. Bioinformatics 24: 2229–2235.
30. Williams PJLB, Laurens LML (2010) Microalgae as biodiesel & biomass feedstocks: Review & analysis of the biochemistry, energetics & economics. Energy Environ Sci: 554–590.
31. Perry JJ, Staley JT, Lory S (2004) Biosynthèse des monomères. Microbiologie, cours et questions de révision. Paris: Dunod. pp. 206–228.
32. Chisti Y (2007) Biodiesel from microalgae. Biotechnol Adv 25: 294–306.
33. Liu B, Benning C (2012) Lipid metabolism in microalgae distinguishes itself. Curr Opin Biotechnol 24: 300–309.
34. Guest JS, van Loosdrecht MCM, Skerlos SJ, Love NG (2013) Lumped Pathway Metabolic Model of Organic Carbon Accumulation and Mobilization by the Alga Chlamydomonas reinhardtii. Environ Sci Technol 47: 3258–3267.
35. Zamorano F, Van de Wouwer A, Bastin G (2010) A detailed metabolic flux analysis of an underdetermined network of CHO cells. J Biotechnol 150: 497–508.
36. Cheung CYM, Williams TCR, Poolman MG, Fell DA, Ratcliffe RG, et al. (2013) A method for accounting for maintenance costs in flux balance analysis improves the prediction of plant cell metabolic phenotypes under stress conditions. Plant J 75: 1050–1061.
37. Nelder J, Mead R (1965) A simplex method for function minimization. Comput J 7: 308–313.
38. Ross O, Geider R (2009) New cell-based model of photosynthesis and photo-acclimation: accumulation and mobilisation of energy reserves in phytoplankton. Mar Ecol Prog Ser 383: 53–71.
39. Mocquet C, Sciandra A, Talec A, Bernard O (2013) Cell cycle implication on nitrogen acquisition and synchronization in Thalassiosira weissflogii (Bacillariophyceae). J Phycol 49: 371–380.
40. Willey J, Sherwood L, Woolverton C (2008) Metabolism: Energy, Enzymes, and Regulation. Prescott, Harley and Klein's Microbiology. Mc Graw Hill higher Education. pp. 167–190.
41. Ovádi J, Saks V (2004) On the origin of intracellular compartmentation and organized metabolic systems. Mol Cell Biochem 256–257: 5–12.
42. Schuster S, Pfeiffer T, Moldenhauer F, Koch I, Dandekar T (2002) Exploring the pathway structure of metabolism: decomposition into subnetworks and application to Mycoplasma pneumoniae. Bioinformatics 18: 351–361.
43. Larhlimi A, David L, Selbig J, Bockmayr A (2012) F2C2: a fast tool for the computation of flux coupling in genome-scale metabolic networks. BMC Bioinformatics 13: 1–9.
44. Verwoerd WS (2011) A new computational method to split large biochemical networks into coherent subnets. BMC Syst Biol 5: 1–25.
45. Barabási A-L, Oltvai ZN (2004) Network biology: understanding the cell's functional organization. Nat Rev Genet 5: 101–113.
46. Akaike H (1974) A new look at the statistical model identification. IEEE Trans Automat Contr 19: 716–723.
47. Song H-S, Ramkrishna D (2009) Reduction of a set of elementary modes using yield analysis. Biotechnol Bioeng 102: 554–568.
48. Song H-S, Ramkrishna D (2010) Prediction of metabolic function from limited data: Lumped hybrid cybernetic modeling (L-HCM). Biotechnol Bioeng 106: 271–284.
49. Ramkrishna D, Song H (2012) Dynamic models of metabolism: Review of the cybernetic approach. AIChE J 58: 986–997.
50. Curien G, Bastien O, Robert-Genthon M, Cornish-Bowden A, Cárdenas ML, et al. (2009) Understanding the regulation of aspartate metabolism using a model based on measured kinetic parameters. Mol Syst Biol 5: 271.
51. Zomorrodi AR, Maranas CD (2012) OptCom: a multi-level optimization framework for the metabolic modeling and analysis of microbial communities. PLoS Comput Biol 8: 1–13.
52. Rügen M, Bockmayr A, Legrand J, Cogne G (2012) Network reduction in metabolic pathway analysis: Elucidation of the key pathways involved in the photoautotrophic growth of the green alga Chlamydomonas reinhardtii. Metab Eng 14: 458–467.
53. Packer A, Li Y, Andersen T, Hu Q, Kuang Y, et al. (2011) Growth and neutral lipid synthesis in green microalgae: a mathematical model. Bioresour Technol 102: 111–117.
54. Mairet F, Bernard O, Lacour T, Sciandra A (2011) Modelling microalgae growth in nitrogen limited photobioreactor for biomass, carbohydrate and neutral lipid productivities. Proc 18th IFAC World Congr 1: 1–6.
55. Quinn J, de Winter L, Bradley T (2011) Microalgae bulk growth model with application to industrial scale systems. Bioresour Technol 102: 5083–5092.
56. Tevatia R, Demirel Y, Blum P (2012) Kinetic Modeling of Photoautotropic Growth and Neutral Lipid Accumulation in terms of Ammonium Concentration in Chlamydomonas reinhardtii. Bioresour Technol 119: 419–424.
57. Yang J, Rasa E, Tantayotai P, Scow KM, Yuan H, et al. (2011) Mathematical model of Chlorella minutissima UTEX2341 growth and lipid production under photoheterotrophic fermentation conditions. Bioresour Technol 102: 3077–3082.
58. Fleck-Schneider P, Lehr F, Posten C (2007) Modelling of growth and product formation of Porphyridium purpureum. J Biotechnol 132: 134–141.
59. Mairet F, Bernard O, Masci P, Lacour T, Sciandra A (2011) Modelling neutral lipid production by the microalga Isochrysis aff. galbana under nitrogen limitation. Bioresour Technol 102: 142–149.
60. Bastin G, Dochain D (1990) On-line estimation and adaptive control of bioreactors. Amsterdam: Elseviers.
61. Bernard O, Bastin G (2005) Identification of reaction networks for bioprocesses: determination of a partially unknown pseudo-stoichiometric matrix. Bioprocess Biosyst Eng 27: 293–301.
62. Bernard O, Bastin G (2005) On the estimation of the pseudo-stoichiometric matrix for macroscopic mass balance modelling of biotechnological processes. Math Biosci 193: 51–77.
63. Koutinas M, Kiparissides A, Silva-Rocha R, Lam M-C, Martins Dos Santos V a P, et al. (2011) Linking genes to microbial growth kinetics: an integrated biochemical systems engineering approach. Metab Eng 13: 401–413.
64. Segre D, Vitkup D, Church G (2002) Analysis of optimality in natural and perturbed metabolic networks. Proc Natl Acad Sci 99: 15112–15117.
65. Burgard AP, Pharkya P, Maranas CD (2003) Optknock: a bilevel programming framework for identifying gene knockout strategies for microbial strain optimization. Biotechnol Bioeng 84: 647–657.
66. Pharkya P, Burgard A, Maranas C (2004) OptStrain: a computational framework for redesign of microbial production systems. Genome Res 14: 2367–2376.
67. Shlomi T, Berkman O, Ruppin E (2005) Regulatory on – off minimization of metabolic flux after genetic perturbations. PNAS 102: 7698–7700.
68. Kim J, Reed JL (2010) OptORF: Optimal metabolic and regulatory perturbations for metabolic engineering of microbial strains. BMC Syst Biol 4: 1–19.
69. Zanghellini J, Ruckerbauer DE, Hanscho M, Jungreuthmayer C (2013) Elementary flux modes in a nutshell: properties, calculation and applications. Biotechnol J 8: 1009–1016.

7

Bacteria Hold Their Breath upon Surface Contact as Shown in a Strain of *Escherichia coli*, Using Dispersed Surfaces and Flow Cytometry Analysis

Jing Geng[1¤], **Christophe Beloin**[2], **Jean-Marc Ghigo**[2], **Nelly Henry**[1]*

1 Laboratoire Jean Perrin (CNRS FRE 3231), UPMC, Paris, France, **2** Institut Pasteur, Unité de Génétique des Biofilms, Département de Microbiologie, Paris, France

Abstract

Bacteria are ubiquitously distributed throughout our planet, mainly in the form of adherent communities in which cells exhibit specific traits. The mechanisms underpinning the physiological shift in surface-attached bacteria are complex, multifactorial and still partially unclear. Here we address the question of the existence of early surface sensing through implementation of a functional response to initial surface contact. For this purpose, we developed a new experimental approach enabling simultaneous monitoring of free-floating, aggregated and adherent cells via the use of dispersed surfaces as adhesive substrates and flow cytometry analysis. With this system, we analyzed, in parallel, the constitutively expressed GFP content of the cells and production of a respiration probe—a fluorescent reduced tetrazolium ion. In an *Escherichia coli* strain constitutively expressing curli, a major *E. coli* adhesin, we found that single cell surface contact induced a decrease in the cell respiration level compared to free-floating single cells present in the same sample. Moreover, we show here that cell surface contact with an artificial surface and with another cell caused reduction in respiration. We confirm the existence of a bacterial cell "sense of touch" ensuring early signalling of surface contact formation through respiration down modulation.

Editor: Vipul Bansal, RMIT University, Australia

Funding: This work was in part supported by a grant from the 'Agence Nationale pour la Recherche,' PIRIbio program Dynabiofilm grant and the Pasteur-Curie PTR/PIC 'Nosocomial Infections.' The funders had no role in study design, data collection and analysis, decision to publish, or preparation of the manuscript.

* Email: nelly.henry@upmc.fr

¤ Current address: Danone Nutricia Research, RD 128, Palaiseau, France

Introduction

Adherent bacteria profoundly differ from planktonic bacteria in physiology and gene expression. From this collective surface-attached life mode, the bacteria gain significant adaptive advantages and exhibit increased resistance to many biocides [1,2,3]. This adhesion-induced physiological shift was suggested very early on by scientists studying bacterial populations in aqueous receptacles [4,5] and has since been confirmed on the basis of molecular biology data. Recently, abundant information on gene expression and metabolic pathway alterations in established biofilms has emerged due to the increasing spread of molecular genetics [6,7,8,9,10,11]. However, the mechanisms of such a transition are not known. The data, obtained on a several hour or day time scale, depict interfering biochemical cascades up- or downregulated in the surface-attached mode of growth compared to the free-floating mode [12]. This reinforces the idea of a surface-attached specific mode of life, but does not enable distinguishing triggering events from further developmental stages that drive biological changes on surfaces. In particular, the respective contributions of the various factors prevailing in biofilms — actual cell surface contact, cell-cell interactions, secreted soluble molecules or extracellular matrix synthesis, together with modi-

fications in the physical and chemical environment due to confinement of cells in a 3D viscoelastic architecture — have not been identified, and their causality remains elusive. In this paper, we focused on the early stage of cell-surface contact formation. Evidence of a direct cell response upon initial adhesion is scarce. Using reporter gene technology and microscope observation in *Pseudomonas aeruginosa* individual cells, Davies and Geesey concluded that attachment of the cell to a glass surface induced *algC* upregulation as early as the first 15 min of contact [13]. In addition, Otto and Silhavy described increased expression of Cpx-regulated genes upon 1 h contact of *Escherichia coli* with artificial surfaces as compared to planktonic cells maintained in suspension; surprisingly, this regulation was observed with stationary phase cells in contact with a hydrophobic surface only [14]. Lately, Li and co-workers showed, in *Caulobacter crescentus*, that formation of physical contact between the bacterium and an artificial surface triggered "just-in-time" adhesin production [15]. These results suggest that bacterial cells possess a 'tactile' machinery which signals formation of surface contact. However, the functional responses put forward in these experiments have also been shown to be upregulated in stationary phase cell populations and in bacteria subjected to various external stresses

— e.g. nutrient deprivation, medium pH or osmolarity changes — raising the question of the direct relationship of these signals with formation of surface contact.

Here we develop an experimental approach aimed at addressing this question in a configuration which enables simultaneous detection of permanent physical contact and relevant biological activity at the single cell level. The principle of the experiments consisted in using dispersed surfaces in the form of micrometric latex particles as an adhesive substrate brought into contact with GFP-expressing bacterial cells in suspension so as to generate a microsystem in which adherent cells co-exist with single planktonic and aggregated cells. The system can then be characterized using flow cytometry, enabling multi-parametric short-time-scale analysis of the mixture [16]. To detect the impact of initial adhesion on cell metabolic activity, we used a fluorescent marker of bacterial respiration, a tetrazolium ion the fluorescence of which can be directly related to cell metabolic activity [17,18]. The experiments were performed in an *E. coli* strain constitutively expressing GFP and curli — a surface multimeric protein structure that fosters surface attachment and self-association [19]. The results indicated that bacterial metabolic activity was affected by formation of a single micrometric contact at the cell surface, either with a synthetic surface or with another cell, as early as the first ten minutes of permanent contact formation, suggesting that bacteria have developed an efficient and fast sense of touch. Interestingly, we observed that both cell-cell and cell-synthetic substrate contact triggered a similar metabolic drop. The implications of these findings on the potential existence and possible nature of a bacterial sense of touch will be discussed below. Clarification of these questions will be useful for a better understanding of the physiological shift induced by bacterial cell development on surfaces, a longstanding concern in microbiology.

Materials and Methods

Bacterial strain and growth conditions

Constitutive *E. coli* curli producers (MG1655*gfp_ompR*234) were in this study obtained by transducing, into *gfp*- tagged MG1655 (MG1655*gfp*), the *ompR*234 mutation that specifies a gain of function allele of *ompR*, a gene encoding an activator of the curli operon [20]. They were grown in lysogeny broth (LB) medium or in M63B1 medium with 0.4% glucose (M63B1Glu) at 37°C in the presence of ampicillin (Amp, 100 μg/ml).

Chemicals and particles

5-cyano-2,3-ditolyl tetrazolium chloride (CTC), Sytox-red and Syto 59 were purchased from Invitrogen; CTC was from the viability kit (B-34956). Monodispersed polystyrene microparticles of 25 μm and 3 μm diameter — (references 07313-5 and 17134-15, respectively) were purchased from Polysciences, GmbH Europe. Particles were used after extensive washing and re-suspension in M63B1 medium.

Flow cytometry

FCM analyses were performed using a Becton-Dickinson flow cytometer (Facscalibur) equipped with a 488 nm argon ion laser and 635 nm laser diode. Under 488 nm excitation, GFP emissions were recorded in fluorescence channel FL-1 (band pass 530/30 nm) and CTC in channel FL-3 (long pass >670 nm). Sytox red was excited using the 635 nm laser diode and emission was collected in channel FL-4 (band pass 661/16 nm). Data were analyzed using CellQuest (BD) multivariate analysis and FlowJo software (Tree Star). Fluorescence values were usually obtained from acquisition of 30,000 events, which optimized precision and

time resolution. Series of successive acquisitions performed on the same sample exhibited standard deviation below 1% of the mean.

Confocal microscopy

Image acquisition was performed with a Nikon A1R confocal laser microscope equipped with an argon and NeHe laser using an oil objective lens (60× by 1.4 numerical aperture). Detectors and filter were set for simultaneous monitoring of GFP, CTC and Sytox red. Images were analyzed in imaging software NIS-Elements (Nikon) and Image J.

Microsystem formation and analysis

We generated a multicellular assemblage microsystem as previously described in detail [16]. Briefly, exponentially growing bacteria in suspension in M63B1Glu of an OD at 600 nm, 1-cm path length about 0.5, were brought into contact with particles at a 1:200 particle to cell ratio in a round-bottom tube and stirred at room temperature on a soft vortex (1000 rpm.min⁻¹). The total volume of this incubator was usually equal to 500 μl. Aliquots of 5 to 20 μl were taken at given incubation times for immediate analysis in FCM. Alternatively, the incubator was used for additional dye staining.

Fluorescent marker incubation procedures

CTC, Sytox Red and Syto 59 were incubated with cells or cell-particle mixtures at 5 mM, 5 nM and 1 μM final concentration, respectively. Fluorescent dye incubation times were usually equal to 30 min for CTC and 10 min for Sytox red and Syto 59.

Results

Combined analysis of CTC reduction and GFP expression provides a single cell respiration index to use in cell assemblages

In order to expose the early bacterial cell response to adhesion, we implemented a strategy consisting of using dispersed surfaces as the adhesive substrate and flow cytometry multiparametric analysis. This approach carried out using GFP-expressing bacteria provides large statistics data sets and time resolution on the order of seconds to determine cell-surface adhesion kinetics. To achieve quantitative monitoring of cell respiration at a single cell level, we first searched for a means of taking advantage of the multiparametric nature of flow cytometry to design a two-color approach to bacterial respiration. For this purpose, we introduced into the experiment a fluorescent marker of bacterial respiration, a tetrazolium ion the fluorescence of which can be directly related to its level of reduction in the electron transfer chain of the bacteria, and thus to the metabolic activity of the cells under study. In its oxidized form, the CTC dye is a non-fluorescent molecule. In contrast, when the compound is reduced via the cell membrane electron transfer chain — in competition with molecular oxygen — it is converted into a water-insoluble product exhibiting characteristic red fluorescence (emission maximum around 602 nm) whose intensity inside the bacterial cell reports the amount of reduced product and thus the cell respiration level, as previously reported [17,18]. In principle, it is possible, using flow cytometry, to measure the individual cell capacity to reduce CTC and to obtain distribution of this property over a perfectly dispersed cell population. Yet, as soon as cells assemble, the single cell information is lost, except if an internal standard enables counting the number of individuals in the assemblage. For this purpose, we analyzed in parallel the flow cytometry profiles of an exponentially growing *E. coli* cell population constitutively expressing GFP and incubated with CTC. Exponentially growing bacteria cultivated at $37°C$ under stirring (hereafter referred to as "fresh cells") were incubated with 5 mM CTC for 30 min just before the flow cytometry test. In parallel, aliquots of the same culture were incubated with Sytox red as a dead cell marker. Incubation time was chosen on the basis of CTC reduction kinetics in cells that indicated that a pseudo-plateau had been reached at that time (see Fig. S1). The same series were prepared with cells taken out of the incubator and left at room temperature for 2 h before the test (bench cells) and with cells previously fixed in 3.7% formaldehyde solution in PBS for 1 h at room temperature (dead cells). FCM dot plots and histograms of GFP, CTC and Sytox red fluorescence intensities collected in their respective emission channels, i.e. FL1, FL3 and FL4, versus forward scattering (FSC), are shown in Fig. 1. Subcellular debris was removed from the analysis by gating the data to an SSC+ region, corresponding to SSC values higher than 10 (a.u. fixed in each channel by the instrument settings chosen at the beginning of the experiment and kept constant for all measurements). Then, we defined for each fluorescence channel a threshold value delimiting positive and negative regions with respect to the considered marker and we determined the corresponding FCM parameters, cell percentage and fluorescence intensity (Table 1). Our results showed that no significant difference in GFP expression or Sytox red labelling was observed between fresh and bench cells. In contrast, both cell distribution between CTC^+ and CTC^- regions and mean fluorescence intensity were noticeably affected (Fig. 1 and Table 1) in bench cells compared with fresh cells, indicating a drop in the reduced CTC production consistent with a cell respiring activity decline caused by a temperature and oxygenation (through sample stirring) decrease. In the extreme situation of the fixed cells, CTC reduction was abolished. Concomitantly, more than 95% of cells exhibited Sytox red labelling, and partial loss of GFP was observed. These results showed that CTC reduction sharply discriminated live from dead cells, as did Sytox red, but also measured graded respiration levels of healthy cells in which GFP content was unchanged on a 2 h time scale. Therefore, GFP intensity could be used as an internal standard to normalize reduced CTC production and define a single-cell respiration index, $f_R = fl_{CTC}/fl_{GFP}$. To check that this normalization introduced no spreading or deformation of the value distribution compared to raw fl_{CTC}, we compared the distributions of f_R and fl_{CTC} values normalized to the mean (Fig. 1D). The plot showed superimposable curves endorsing the two-color parameter f_R as a robust descriptor of respiring activity at a single cell level. In addition, the two-color assessment of bacterial respiration provided a means of deriving a single cell respiration measurement in multiple cell assemblages in which both fl_{GFP} and fl_{CTC} could be determined. In these cases, fl_{GFP} was not only an internal standard, but also served as a cell counter. An assemblage of n cells will have an FL1 fluorescence intensity, Fl_{GFP}^n enabling determination of n like $n = \dfrac{Fl_{GFP}^n}{\overline{fl_{GFP}}}$, with $\overline{fl_{GFP}}$ being the mean fluorescence of the co-existing single cells in the same sample. Based on the fact that only CTC but not GFP content was affected by respiring activity changes, single-cell-reduced-CTC content in the n-cell assemblage was derived using n like $\overline{fl_{CTC}} = \dfrac{Fl_{CTC}}{n}$, where Fl_{CTC} is the FL3 fluorescence of the assemblage, enabling determination of mean single cell respiring activity, and f_R in the assemblage. This will be used hereafter to determine both surface-attached and self-associated cell respiring activity.

Flow cytometry monitoring of cell-particle microsystem-exposed multiple assemblages

To form a cell-particle microsystem in which free-floating, aggregated and adherent cells can be monitored over a short time scale using flow cytometry, we mixed, at a 1:200 ratio, 25 μm diameter polystyrene particles serving as the adhesion substrate (Fig. 2A) with curli expressing $E.\ coli$ bacteria freshly taken from an exponentially growing culture and pipetted (Fig. 2B). At all desired time points, aliquots were removed from the stirred suspension for immediate analysis by flow cytometry. From the time of the mixture ($t = 0$), the cells quickly self-assembled and attached to particle surfaces. The microsystem reached its steady state in about 10 min, exhibiting multiple analyzable subgroups of single, surface-attached and self-aggregated cells at various assemblage degrees, as shown on the FL1 (GFP-fluorescence) versus FSC dot plots (Fig. 2C and D). In the example shown in Fig. 2, 98% of the particles carried between 20 to 200 cells and the self-aggregation number for 90% of the aggregates ranged from 2 to 12 cells at $t = 15$ min.

These results showed that a stable microsystem exhibiting multiple adhesion and self-association degrees could be reproducibly obtained in less than 10 min following mixing of bacteria with particles, with subclasses identifiable and subject to characterization on the basis of their FSC and FL1 parameters in flow cytometry. Unless otherwise stated, the experiments on bacterial respiring activity were performed using this reference microsystem in a steady state.

Differential respiring activity in single free-floating and assembled cells

In order to test the respiring activity of the bacterial cells according to their assemblage status, we introduced CTC into the microsystem previously formed by mixing cells and particles for 15 min. Incubation with CTC was completed 30 min before FCM analysis. Reduction of CTC in healthy respiring cells was visible under microscopy by formation of an intracellular red fluorescent speck, as seen in Fig. 3A. Thus GFP and reduced CTC cell content were analyzed in FCM, enabling derivation of the respiration index, using GFP as a cell counter as explained above, for the three subpopulation types present in the microsystem (Fig 3B, panel a-e). In order to compensate for all set-up fluctuations that might interfere with experiments performed some time ago, we used a respiration index normalized to 1 for free-floating cells ($\overline{f_{R,Norm}} = \dfrac{\overline{f_R}}{f_{R,free}}$). Then, pooling ten experiments, we found $\overline{f_{R,Norm}}$ values equal to 0.78 ± 0.04 and 0.75 ± 0.06 for self-aggregated and surface-attached cells, respectively. These results indicated that both surface-attached and self-aggregated cells exhibited decreased respiring activity compared with free-floating cells. To assess the impact of cell assemblage size n on respiration down-modulation, we delineated n FL1 gates as follows: for all assemblages of n cells present in the sample, either forming an aggregate of n cells or an assemblage of n cells attached to the same particle, an R_n region was defined as including all events having an FL1 value in the interval $\langle Fl_1^n \rangle = \left[(n \times \overline{fl_1}) \pm \dfrac{\overline{fl_1}}{2} \right]$ where $\overline{fl_1}$ is the mean FL1 value of single free-floating bacteria. By doing this, we obtained adjacent gates with no overlap. For the surface-attached cell population, the particle fluorescence background contribution in the FL1 channel was taken into account to define the gate. With the variation in f_R as a function of n, the assemblage size is shown in Fig. 3C for aggregates and surface-attached cells. The respiration decrease

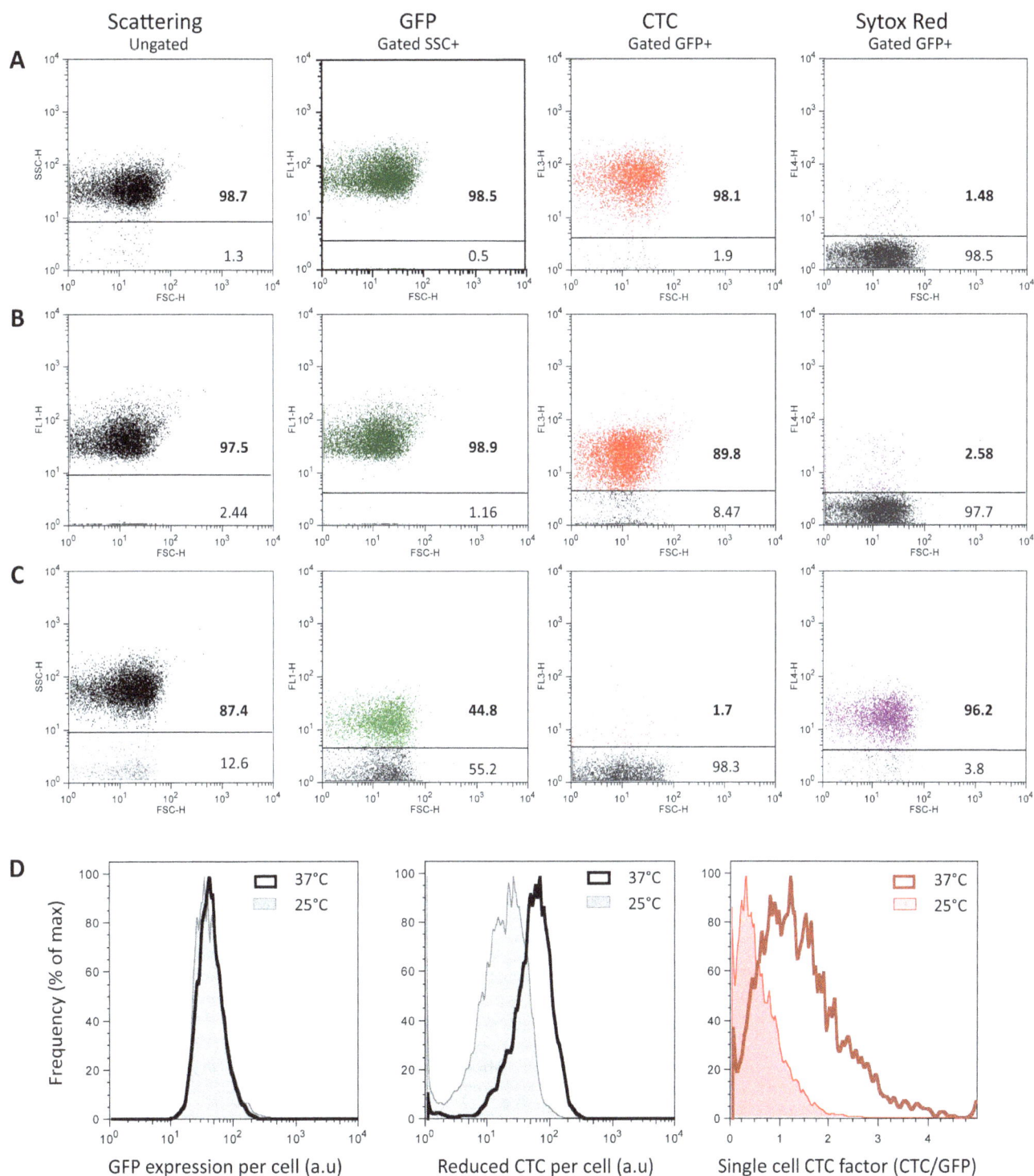

Figure 1. FCM signature of cell respiration down-modulation. Side scattering (SSC, first row) and fluorescence (FL1, second column; FL3, third column; FL4, fourth column) versus forward scattering (FSC) dot plots of bacteria constitutively expressing GFP and labelled with CTC or Sytox red are shown for cells (A) freshly taken from 37°C agitated culture, (B) left to rest at 25°C or (C) fixed in 3.7% formaldehyde. Fluorescence dot plots are from cells exhibiting a positive side scattering signal (>10 a.u.). Cell frequencies below and above a fluorescence intensity threshold taken between 4 and 5 a.u. in each channel are indicated on the dot plots (see also Table 1 for fluorescence intensity data). (D) Histograms of fluorescence intensity distribution for GFP expression (left graph), CTC reduction (middle graph) and single cell respiration index (right graph) for cells freshly taken from 37°C agitated culture and left to rest at 25°C.

was detected in cell aggregates as small as n = 2 and reached its maximum for aggregates of $n = 4$ cells. This suggested that bacteria can sense even single cell-cell contact. For surface-

attached cells, no assemblage size lower than 10 cells per particle with sufficient statistical significance was obtained. We observed that the respiration index decrease experienced by surface-

Table 1. Scattering and fluorescence FCM characteristics of bacterial cells in high (A) and low (B) metabolic state compared with dead cells (C).

Parameter	Gate + boundary (a.u.)	% Cells in gate+ *		Mean intensity in gate+ (a.u.)**
SSC	10.0	A	98.7	37.8±0.8
		B	97.5	36.5±0.9
		C	87.4	58.7±1.6
GFP (FL1)	4.0	A	98.5	47.8±0.7
		B	98.9	45.5±0.9
		C	44.8	15.1±0.5
CTC (FL3)	5.0	A	98.1	64.3±1.2
		B	89.8	25.1±0.9
		C	1.7	-
Sytox Red (FL4)	4.0	A	1.48	-
		B	2.58	-
		C	96.2	18.1±0.5

*A: Alive cells at 37°C; B: Alive cells at 25°C; C: Dead cells.
**Error is standard deviation over at least three measurements.

attached cells was not dependent on the number of bound cells, at least between 10 and 100 cells per particle.

To check the robustness of our approach, we used particles having large-spectrum intrinsic fluorescence, enabling a measurement signal in both FL1 and FL3 channels. We induced these particles — 2.8 µm in diameter — to form a small proportion of stable aggregates of a few individuals by centrifugation (Fig. 3D). We then determined an analogous normalized fluorescence index using the same derivations as those applied to cell assemblages. We observed that particle self-association did not affect the normalized fluorescence index (Fig. 3C, panel a). Although the particle number of aggregation did not exceed n = 6, this result showed that an association in itself does not induce a fluorescence shift.

We also challenged our analysis framework by examining the fluorescence of the different bacterial organizations upon labelling with 1 µM Syto 59, a dye which passively penetrates into cells and stains bacterial DNA after insertion between the base pairs without the requirement for metabolic activity. In this case, free-floating and aggregated cells exhibited the same Syto 59 (FL4 signal) to GFP (FL1 signal) fluorescence ratio (Fig. S2), also supporting the metabolic origin of the CTC fluorescence decrease in assembled cells. These results also indicated that the fluorescent dyes equally labelled free-floating and aggregated cells. The same experimental evidence could not be obtained with surface-attached cells, since Syto 59 itself was bound to the particle surface, inducing too high a fluorescence background that blurred the entire analysis. However, results obtained on cell accessibility to fluorescent dyes in aggregates could reasonably be extrapolated to surface-attached cells.

Thus, our results indicate that bacteria detected the formation of cell surface contact and responded to it by metabolism down-modulation, as confirmed by the decrease in respiring activity.

At this stage, we sought to further analyze the respiration activity decrease observed in particle-attached cells so as to determine whether the observed cell response actually resulted from cell-artificial surface anchorage or from cell-cell associations occurring on the particle surface.

Single cell response to contact

In order to discriminate between cell-cell contact on a surface and cell-surface contact, we needed to form a cell-surface assemblage of lower size. For this purpose, we designed an experimental "single-cell adhesion assay" achieved by using smaller-sized particles 3 µm in diameter and decreasing the cell-to-particle ratio to 4, which provided a majority of particles holding one single cell and a small fraction holding none or two cells. Because of the small particle radius, the cells and particles appeared on the same dot plot (Fig. 4). As previously, we defined gates on the basis of FL1 fluorescence so as to delineate single cell events, either free-floating single cells or surface-attached single cells. Again, we derived the normalized respiration index $f_{R,Norm}$ for the surface-attached cells and found it equal to 0.7±0.1, which indicated that direct cell-synthetic surface contact also induced a decrease in bacterial respiring activity. The respiration index determined on the self-aggregated cells in the single-cell adhesion assay was found equal to 0.79±0.06, in good agreement with previous results obtained in the presence of the 25 µm-radius particles, as expected, since self-aggregation occurred independently from surface attachment.

Short-time-scale single cell response to contact

The results shown above thus indicated that respiring activity decreased upon single cell-cell or cell-substrate contact at the surface of a bacterium after 40 min. This time scale was actually driven by the CTC response time, i.e. reduction rate, and chosen to obtain a sufficiently stable and reduced CTC signal. To evaluate whether a cell response to contact formation occurred earlier than 40 min, we performed kinetic analysis of CTC reduction over the entire period of incubation of the microsystem with CTC. The results displayed in Fig. 5 show that free-floating, aggregated and surface-attached cell respiration indexes diverged from the beginning of incubation. In contrast, the kinetic curves of the reduced CTC labelling frequencies, which report the rate of CTC inflow into the cell independently of the level of reduction, showed no difference between assembled and free-floating cells. Together, these two curve sets suggested that down-modulation of respiring activity had already started after 10 min contact with a

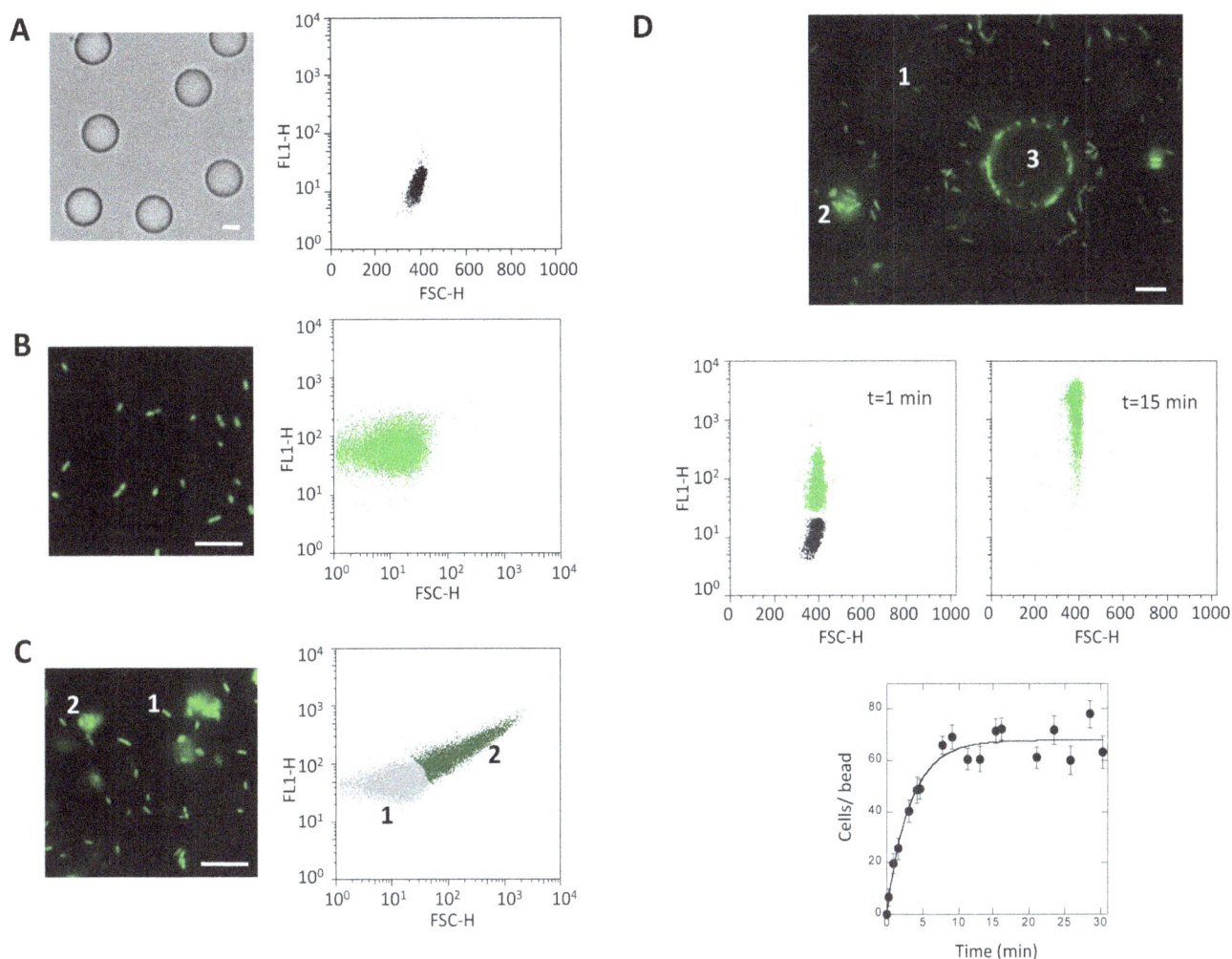

Figure 2. Cell-particle microsystem formation. Microscope image and fluorescence (FL1) versus FSC dot plots of (A) micrometric 25 μm diameter particles alone, (B) single bacteria freshly pipetted to break aggregates, (C) partially aggregated cells as recovered from a usual culture and (D) the cell-particle mixed microsystem comprising single cells (1), aggregated cells (2) and cells adhering to particle surface (3); the two FL1 versus FSC dot plots show the particle signal recorded in the microsystem 1 min and 15 min after bringing cells and particles into contact. Panel D bottom gives cell-particle association kinetics.

facing surface; elsewhere, penetration of the probe did not seem to be affected by contact of the cell either with another cell or with an artificial contact.

Discussion

Bacteria are primarily found in the form of adherent communities, where they exhibit significant remodeling of their properties compared to planktonic cells. Despite the attention that these changes have garnered, the events determining bacterial physiological shift are still not well understood [12,21].

In the work reported here, we sought to detect an early cell biological response to formation of cell-surface adhesive contact. Indeed, the phenotypic alterations observed in adherent communities were mainly examined in systems already established for hours or days during which intricate events occurred, including 3D-extracellular matrix formation [22,23,24,25], morphological changes [25] and quorum sensing communication [26], blurring the initial adhesion step and confusing clarification of the various contributions. To gain access to the initial adhesion phase, we implemented a strategy using dispersed surfaces and flow

cytometry analysis, which has recently been designed to monitor bacterial short-time-scale adhesion [16]. The technique provides large statistical data sets, time resolution on the order of a few tens of seconds and simultaneous analysis of various-sized objects suspended in the same sample — the reason why flow cytometry has gained significant ground in microbial analyses lately [27,28]. In this microsystem, we found that E. coli, engineered to constitutively produce curli fibers on its surface, reached a steady state exhibiting stable fractions of multidimensional aggregates, planktonic and adherent cells.

To quantitatively picture cell metabolic activity within the first hour of surface colonization, we adapted this strategy to simultaneously follow not only adhesion, but also cell respiration on suspended free-floating, adherent and aggregated cells that co-exist in the microsystem formed when micrometric particles are brought into contact with a bacterial cell culture. This was achieved by introduction of a second fluorescent marker — tetrazolium ion CTC [17,18]— in addition to GFP constitutively expressed by the cells. In this way, we were able to use GFP expression as an internal standard and cell counter, which enabled

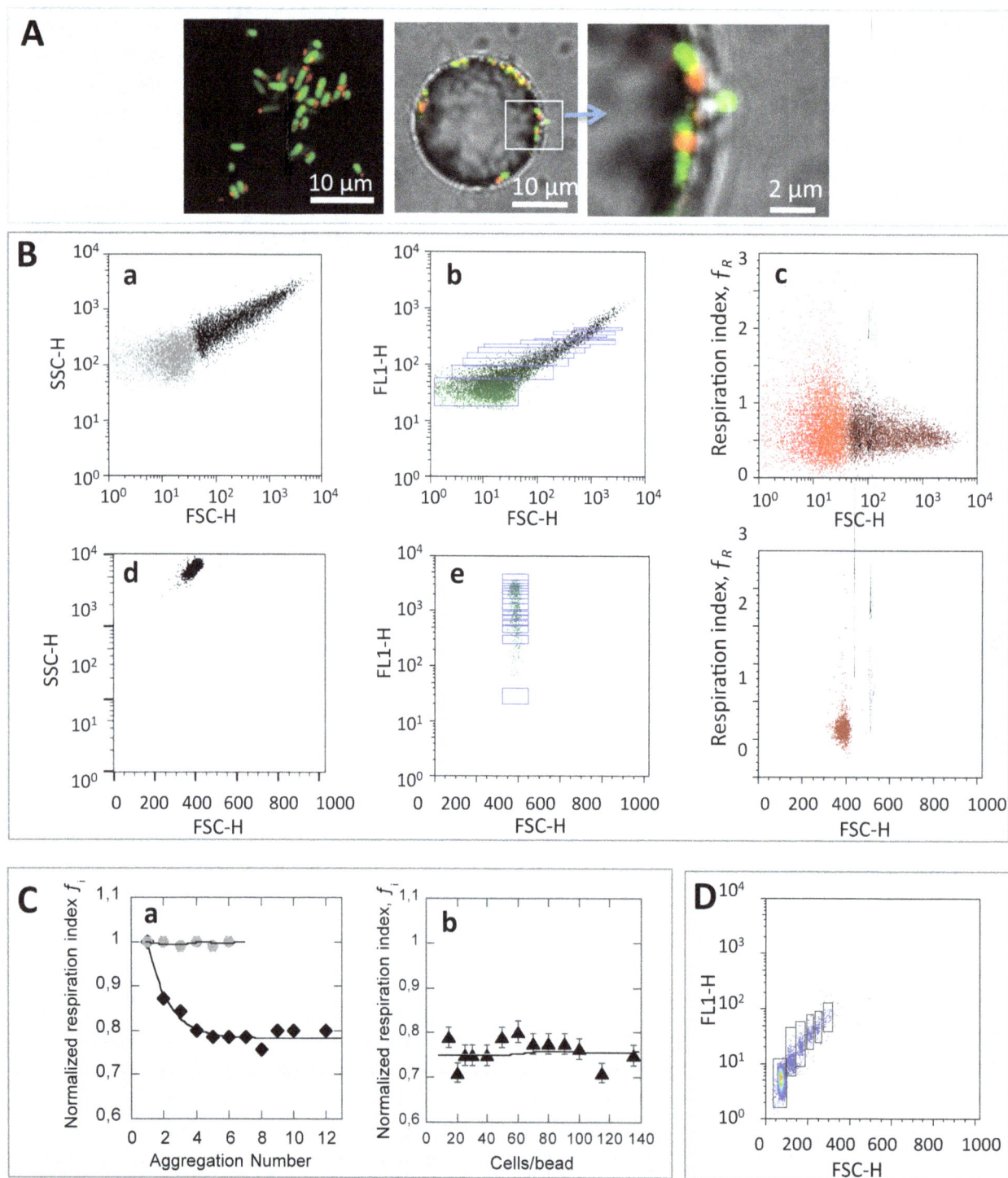

Figure 3. Adherent and aggregated cell respiration down-modulation upon contact. (A) Microscope images of CTC labelled cells in suspension and adhering to the 25 μm diameter particle surface. (B) FCM dot plots: SSC (left panels), FL1 (middle panels) and respiration index f_R (right panels) versus FSC showing cells (upper row, a to c) — single free-floating cells appear as light gray dots, aggregates as dark dots — and cells adhering to particles (lower row, d to f)); FL1 dot plots (middle panels) reporting GFP expression show cell counting gates, from R_2 to R_{12} for aggregates and from R_{10} to R_{135} for particles. Gates were defined as explained in the text based on sample single cell mean FL_1 value. (C) Graphs of normalized respiration index as a function of (a) number of cells ($\blacklozenge$) or 2.8 μm diameter beads ($\bullet$) per aggregate and (b) number of cells per particle. (D) FL1 versus FSC dot plot of 2.8 μm beads used as aggregation controls and population counting gates.

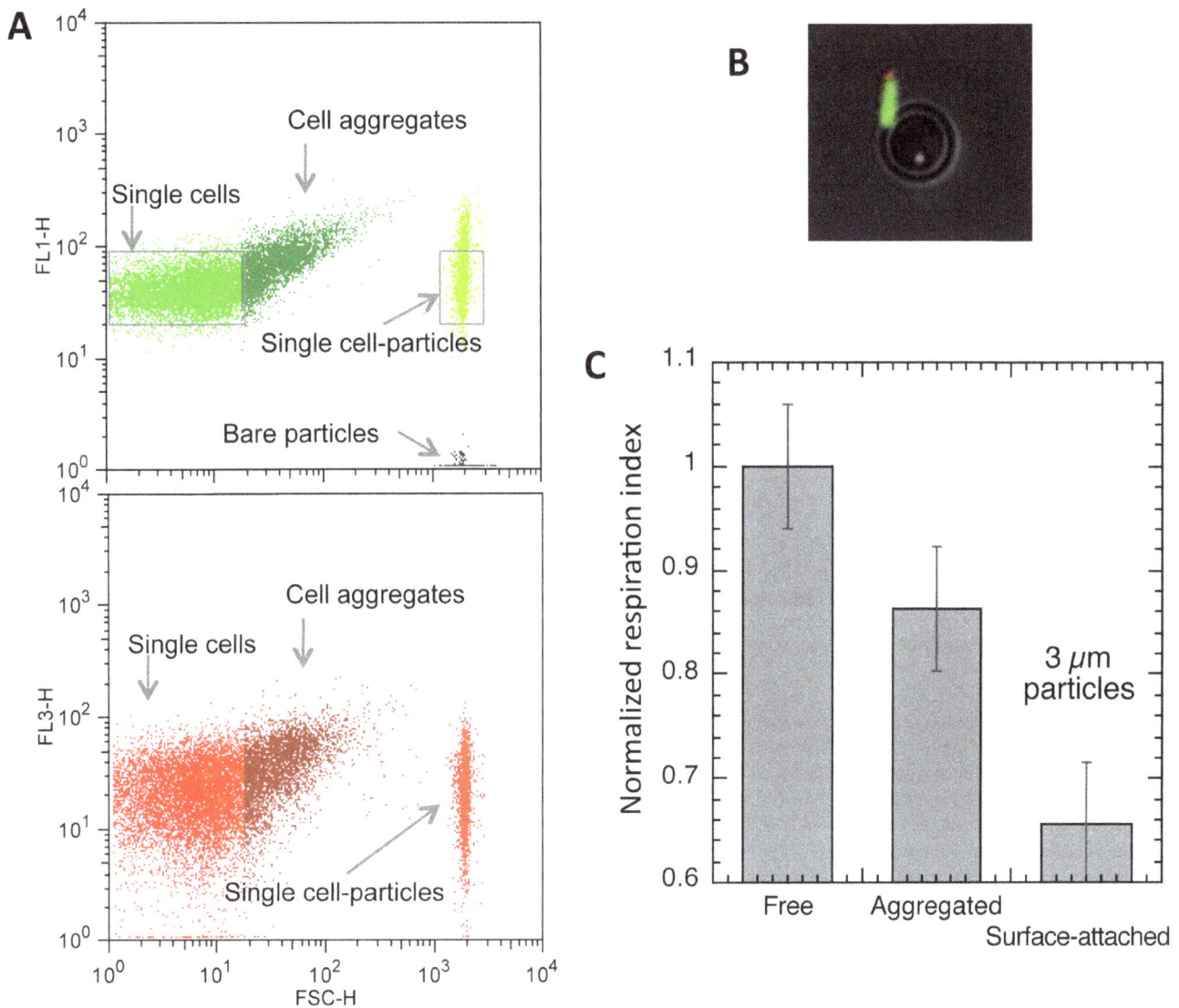

Figure 4. Single cell adhesion assay. (A) FL1 (upper graph) and FL3 versus FSC dot plots of a mixed microsystem of 3 μm particles with single free-floating and aggregated bacteria; microsystem prepared at a cell-to-particle ratio equal to 4 and incubated with 5 mM CTC. (B) Microscope image of a single cell-particle conjugate. (C) Histogram of normalized respiration index of the different cell populations comprised in the same sample, measured 40 min after bringing cells and particles into contact.

unit respiration index determination from CTC and GFP fluorescence intensity measurements.

On the basis of this two-color signature of respiring activity, we demonstrated that CTC reduction monitored at a single cell level in a population of living bacteria could report graded levels of cell respiration within a time scale of 15–20 min.

Implementing this approach in cell-particle mixtures displaying both adherent (particle-attached) and free-floating single cells, we found that the cell-surface association induced a 25 to 30% decrease in the initial cell respiration level.

Interestingly, we also found that cells living in the microsystem in the form of aggregates of various sizes exhibited a similar decrease in respiring activity. Moreover, by examining single-cell/particle associations and low aggregation number (from 2 to 10) cell clusters, we revealed that a single contact was sufficient to initiate this decrease in the respiration level.

These results point to the existence of a bacterial cell "sense of touch", i.e. the capacity to perceive an external object via

formation of a physical contact, implying that bacterial cells have developed molecular pathways to elaborate a biological response to surface physical contact. Such an early single-cell level observation of a substratum-induced response — independently of interfering adhesion secondary events such as confinement, formation of chemical gradients or the onset of extracellular matrix synthesis — had been previously shown by Davies and co-workers in a small number of individuals, based on microscopy monitoring of *algC* gene activation [29]. In comparison, our results rely on data sets of thousands of cells. We show here that surface sensing involves a drop in respiration. Still, we do not know the mechanism behind this process.

Lower and co-workers [30] reported what they called a "tactile" response in *Staphylococcus aureus* after observing — on the basis of atomic force microscopy data — an accumulation of adhesins in the region of cell-substrate contact. Yet, under their experimental conditions, this behavior very likely resulted from thermodynamic equilibration of the free fraction of the binding ligands by

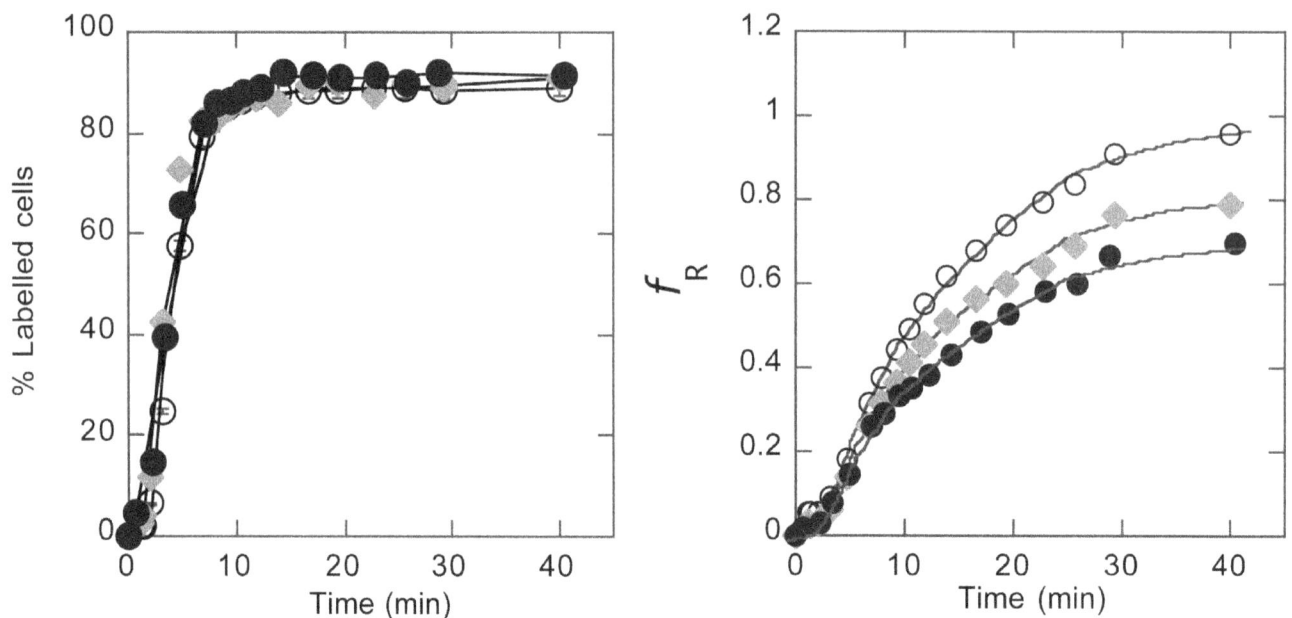

Figure 5. Kinetics of CTC cell influx and reduction. Monitoring (A) CTC positively labelled (FL3>5 a.u.) cell fraction and (B) cell respiration index, f_R as a function of CTC incubation time for free-floating (o), aggregated ($\blacklozenge$) and adherent ($\bullet$) cells. CTC reduction reported by FL3 intensity increase in CTC-labelled cells appears to be the limiting (slowest) step in the process.

diffusion, i.e. a passive process driven by the formation of specific surface receptor/ligand bonds. Our results show a striking parallel between the metabolic down-modulation induced by contact with the synthetic surface of a particle and that induced by cell-cell contact, suggesting that the molecular details of the contact play no determining role in signaling. However, there is no clear-cut conclusion concerning the fundamental difference between cell-artificial surface contact and cell-cell contact in the context of bacterial adhesion. Due to the presence in the cell culture supernatant of a variety of surface active molecules such as proteins and polyelectrolytes, the description of cell/synthetic surface interactions according to a physico-chemical model is hazardous, and the hypothesis of specific interactions between substrate-adsorbed biopolymers and cell surface receptors cannot be ruled out [31]. Again, since experimental data are often the result of biofilms established over several hour or day time scales, at a time when cell-substrate interactions are masked by the predominance of cell-cell interactions, any particular influence of the substrate upon development of the adhesive community is difficult to detect. Here, since we individually and simultaneously monitored each type of contact in the same sample, we were able to demonstrate that formation of cell-cell contact induced a cell response similar to the formation of cell-synthetic surface contact. Recently, based on growth rate, tolerance to antibiotics and extracellular matrix property criteria, Alhede and co-workers [32] pointed out that *Pseudomonas aeruginosa* floating aggregates and surface-attached biofilm shared most phenotypes, suggesting that the substrate per se does not play a determining role in establishment of biofilm properties. These behaviors might, of course, depend on the nature of the adhesive surface, although this point remains difficult to clarify in the context of bacterial adhesion. In our experiments, the polystyrene surface of the particles is expected to be neutral; in fact, it exhibited a slight negative charge (zeta-potential ≈ -10 mV) when suspended in cell culture minimal medium (data not shown). In previous work [16], we showed that the particle surface was quickly converted to being

negatively charged in the presence of cell culture medium independently of the initial physico-chemical properties of the particles, explaining the similar colonization kinetics observed using different particles. This indicates that the substratum initial properties are often masked by a conditioning film in the presence of bacterial cell culture products.

The similarity of the cell response to a single contact with another cell and with a synthetic surface also enables hypothesizing surface sensing mechanisms driven by physical forces. Indeed, the formation of a contact on the cell surface is likely to induce local mechanical stress of cell surface appendages, which could serve to signal contact formation and trigger the biochemical cascade — as shown for flagella [33]. Other issues such as breaking of cell axial symmetry by formation of surface contact or a reduction in the amount of cell surface exposed to external medium and fluid flows could also be taken into consideration when formulating new working hypotheses for elucidating surface sensing mechanisms.

Our results, which associate respiration reduction with surface sensing, also raise the question of the role of that event in surface contact signaling. Recently, a contact-dependent growth inhibition process involving metabolic down-modulation upon cell-cell contact was discovered in *E. coli* [34,35]. This phenomenon — mediated by surface-specific protein recognition in an asymmetric inhibitor/target contact involving the immunity protein and toxin translocation across target cell envelopes [36] — is assuredly different from the one observed here. However, it provides an example of a contact-driven bacterial response pathway in which metabolic parameters are reduced in the absence of nutrient deprivation. Moreover, Aoki and co-workers suggested that a respiration decrease could be a general response to formation of physical contact at the cell surface; they proposed that respiration reduction could enable survival under conditions of oxidative stress. Thus far, to the best of our knowledge, no experimental evidence supporting a molecular mechanism linking respiration reduction and contact formation has been published.

Conversely, other authors detected metabolism stimulation events upon surface adhesion [37]. However, their experiments were performed under 12 h starvation conditions, thus significantly differing from our own experimental conditions.

Here we provide new tools for investigating in more detail this metabolic shift induced by surface contact at the single-cell level. Our approach enables discriminating between sensing mediated by soluble secreted factors and direct physical cell surface contact by monitoring planktonic and attached cells exposed in parallel to the same medium. This is a key issue in the better understanding of the mechanisms driving adherent cell transition towards an altered physiological state. Our results indicate that the reduction in the cell respiration level participates in early signaling triggered upon cell aggregation or settlement on an artificial substrate.

These results raise many questions for further investigation. What is the impact of this reduction on overall cell functioning, and, for instance, on the cell division rate? Is this a way for cells growing in the form of attached communities to achieve greater tolerance to oxidative stress or to various antimicrobials? What are the molecular bases for this contact-dependent behavior? The design of specific short-lifetime genetic reporters, to be monitored in our multiparametric dispersed surface approach, will undoubtedly help to answer these questions.

Supporting Information

Figure S1 CTC reduction kinetics. (A) Fraction of CTC-positively-labelled (FL3>5 a.u.) cells and (B) cell respiration index, f_R as a function of CTC incubation time. Incubation with 5 mM CTC at 37°C under stirring.

Figure S2 A neutral labelling index in free-floating and aggregated cells. Syto 59 (FL4 signal) to GFP (FL1 signal) fluorescence ratio calculated from fluorescence dot plots for free-floating (light grey triangle) and aggregated (black diamond) cells. Incubation with 1 μM Syto59 at 37°C under stirring.

Author Contributions

Conceived and designed the experiments: JG CB JMG NH. Performed the experiments: JG. Analyzed the data: JG NH. Contributed reagents/materials/analysis tools: JG NH. Wrote the paper: NH.

References

1. Stewart PS, Costerton JW (2001) Antibiotic resistance of bacteria in biofilms. Lancet 358: 135–138.
2. Hoiby N, Bjarnsholt T, Givskov M, Molin S, Ciofu O (2010) Antibiotic resistance of bacterial biofilms. Int J Antimicrob Agents 35: 322–332.
3. Anderson GG, O'Toole GA (2008) Innate and induced resistance mechanisms of bacterial biofilms. Curr Top Microbiol Immunol 322: 85–105.
4. van Loosdrecht MC, Lyklema J, Norde W, Zehnder AJ (1990) Influence of interfaces on microbial activity. Microbiol Rev 54: 75–87.
5. Zobell CE (1943) The Effect of Solid Surfaces upon Bacterial Activity. J Bacteriol 46: 39–56.
6. Prigent-Combaret C, Vidal O, Dorel C, Lejeune P (1999) Abiotic surface sensing and biofilm-dependent regulation of gene expression in *Escherichia coli*. J Bacteriol 181: 5993–6002.
7. Kuchma SL, O'Toole GA (2000) Surface-induced and biofilm-induced changes in gene expression. Curr Opin Biotechnol 11: 429–433.
8. Schembri MA, Kjaergaard K, Klemm P (2003) Global gene expression in *Escherichia coli* biofilms. Mol Microbiol 48: 253–267.
9. Beloin C, Valle J, Latour-Lambert P, Faure P, Kzreminski M, et al. (2004) Global impact of mature biofilm lifestyle on *Escherichia coli* K-12 gene expression. Mol Microbiol 51: 659–674.
10. Domka J, Lee J, Bansal T, Wood TK (2007) Temporal gene-expression in *Escherichia coli* K-12 biofilms. Environ Microbiol 9: 332–346.
11. Wood TK (2009) Insights on *Escherichia coli* biofilm formation and inhibition from whole-transcriptome profiling. Environ Microbiol 11: 1–15.
12. Petrova OE, Sauer K (2012) Sticky situations: key components that control bacterial surface attachment. J Bacteriol 194: 2413–2425.
13. Davies DG, Geesey GG (1995) Regulation of the alginate biosynthesis gene *algC* in Pseudomonas aeruginosa during biofilm development in continuous culture. Appl Environ Microbiol 61: 860–867.
14. Otto K, Silhavy TJ (2002) Surface sensing and adhesion of Escherichia coli controlled by the *Cpx*-signaling pathway. Proc Natl Acad Sci U S A 99: 2287–2292.
15. Li G, Brown PJ, Tang JX, Xu J, Quardokus EM, et al. (2012) Surface contact stimulates the just-in-time deployment of bacterial adhesins. Mol Microbiol 83: 41–51.
16. Beloin C, Houry A, Froment M, Ghigo JM, Henry N (2008) A short-time scale colloidal system reveals early bacterial adhesion dynamics. PLoS Biol 6: e167.
17. Kim DH, Yim SK, Kim KH, Ahn T, Yun CH (2009) Continuous spectrofluorometric and spectrophotometric assays for NADPH-cytochrome P450 reductase activity using 5-cyano-2,3-ditolyl tetrazolium chloride. Biotechnol Lett 31: 271–275.
18. Rodriguez GG, Phipps D, Ishiguro K, Ridgway HF (1992) Use of a fluorescent redox probe for direct visualization of actively respiring bacteria. Appl Environ Microbiol 58: 1801–1808.
19. Barnhart MM, Chapman MR (2006) Curli biogenesis and function. Annu Rev Microbiol 60: 131–147.
20. Vidal O, Longin R, Prigent-Combaret C, Dorel C, Hooreman M, et al. (1998) Isolation of an *Escherichia coli* K-12 mutant strain able to form biofilms on inert surfaces: involvement of a new *ompR* allele that increases curli expression. J Bacteriol 180: 2442–2449.
21. Monds RD, O'Toole GA (2009) The developmental model of microbial biofilms: ten years of a paradigm up for review. Trends Microbiol 17: 73–87.
22. Sutherland IW (2001) The biofilm matrix—an immobilized but dynamic microbial environment. Trends Microbiol 9: 222–227.
23. Branda SS, Vik S, Friedman L, Kolter R (2005) Biofilms: the matrix revisited. Trends Microbiol 13: 20–26.
24. Flemming HC, Wingender J (2010) The biofilm matrix. Nat Rev Microbiol 8: 623–633.
25. Serra DO, Richter AM, Klauck G, Mika F, Hengge R (2013) Microanatomy at cellular resolution and spatial order of physiological differentiation in a bacterial biofilm. MBio 4: e00103–00113.
26. Irie Y, Parsek MR (2008) Quorum sensing and microbial biofilms. Curr Top Microbiol Immunol 322: 67–84.
27. Muller S, Nebe-von-Caron G (2010) Functional single-cell analyses: flow cytometry and cell sorting of microbial populations and communities. FEMS Microbiol Rev 34: 554–587.
28. Davey HM, Davey CL (2011) Multivariate data analysis methods for the interpretation of microbial flow cytometric data. Adv Biochem Eng Biotechnol 124: 183–209.
29. Davies DG, Chakrabarty AM, Geesey GG (1993) Exopolysaccharide production in biofilms: substratum activation of alginate gene expression by *Pseudomonas aeruginosa*. Appl Environ Microbiol 59: 1181–1186.
30. Lower SK, Yongsunthon R, Casillas-Ituarte NN, Taylor ES, DiBartola AC, et al. (2010) A tactile response in *Staphylococcus aureus*. Biophys J 99: 2803–2811.
31. Geng J, Henry N (2011) Short Time-Scale Bacterial Adhesion Dynamics Advances in Experimental Medicine and Biology pp. 315–331.
32. Alhede M, Kragh KN, Qvortrup K, Allesen-Holm M, van Gennip M, et al. (2011) Phenotypes of non-attached *Pseudomonas aeruginosa* aggregates resemble surface attached biofilm. PLoS One 6: e27943.
33. Anderson JK, Smith TG, Hoover TR (2010) Sense and sensibility: flagellum-mediated gene regulation. Trends Microbiol 18: 30–37.
34. Aoki SK, Pamma R, Hernday AD, Bickham JE, Braaten BA, et al. (2005) Contact-dependent inhibition of growth in *Escherichia coli*. Science 309: 1245–1248.
35. Aoki SK, Webb JS, Braaten BA, Low DA (2009) Contact-dependent growth inhibition causes reversible metabolic downregulation in *Escherichia coli*. J Bacteriol 191: 1777–1786.
36. Webb JS, Nikolakakis K, Willett JL, Aoki SK, Hayes CS, et al. (2013) Delivery of CdiA nuclease toxins into target cells during contact-dependent growth inhibition. PLoS One 8: e57609.
37. Hong Y, Brown DG (2009) Variation in bacterial ATP level and proton motive force due to adhesion to a solid surface. Appl Environ Microbiol 75: 2346–2353.

8

The Acclimation of *Phaeodactylum tricornutum* to Blue and Red Light Does Not Influence the Photosynthetic Light Reaction but Strongly Disturbs the Carbon Allocation Pattern

Anne Jungandreas[1,2], Benjamin Schellenberger Costa[1], Torsten Jakob[1], Martin von Bergen[3,4,5], Sven Baumann[3,6], Christian Wilhelm[1]*

1 Department of Plant Physiology, Institute of Biology, Faculty of Biosciences, Pharmacy and Psychology, University of Leipzig, Leipzig, Germany, **2** Department of Computational Landscape Ecology, Helmholtz Centre for Environmental Research - UFZ, Leipzig, Germany, **3** Department of Metabolomics, Helmholtz Centre for Environmental Research - UFZ, Leipzig, Germany, **4** Department of Proteomics, Helmholtz Centre for Environmental Research - UFZ, Leipzig, Germany, **5** Department of Biotechnology, Chemistry and Environmental Engineering, University of Aalborg, Aalborg, Denmark, **6** Institute of Pharmacy, Faculty of Biosciences, Pharmacy and Psychology, University of Leipzig, Leipzig, Germany

Abstract

Diatoms are major contributors to the aquatic primary productivity and show an efficient acclimation ability to changing light intensities. Here, we investigated the acclimation of *Phaeodactylum tricornutum* to different light quality with respect to growth rate, photosynthesis rate, macromolecular composition and the metabolic profile by shifting the light quality from red light (RL) to blue light (BL) and vice versa. Our results show that cultures pre-acclimated to BL and RL exhibited similar growth performance, photosynthesis rates and metabolite profiles. However, light shift experiments revealed rapid and severe changes in the metabolite profile within 15 min as the initial reaction of light acclimation. Thus, during the shift from RL to BL, increased concentrations of amino acids and TCA cycle intermediates were observed whereas during the BL to RL shift the levels of amino acids were decreased and intermediates of glycolysis accumulated. Accordingly, on the time scale of hours the RL to BL shift led to a redirection of carbon into the synthesis of proteins, whereas during the BL to RL shift an accumulation of carbohydrates occurred. Thus, a vast metabolic reorganization of the cells was observed as the initial reaction to changes in light quality. The results are discussed with respect to a putative direct regulation of cellular enzymes by light quality and by transcriptional regulation. Interestingly, the short-term changes in the metabolome were accompanied by changes in the degree of reduction of the plastoquinone pool. Surprisingly, the RL to BL shift led to a severe inhibition of growth within the first 48 h which was not observed during the BL to RL shift. Furthermore, during the phase of growth arrest the photosynthetic performance did not change. We propose arguments that the growth arrest could have been caused by the reorganization of intracellular carbon partitioning.

Editor: Rajagopal Subramanyam, University of Hyderabad, India

Funding: The work was supported by the DFG FG 1261 "Specific Light-Driven Reactions in Unicellular Model Algae" Wi 764/19-2. The funders had no role in study design, data collection and analysis, decision to publish, or preparation of the manuscript.

Competing Interests: The authors have declared that no competing interests exist.

* Email: cwilhelm@rz.uni-leipzig.de

Introduction

Diatoms are a highly diverse class of eukaryotic organisms [1] that are widely distributed as phytoplankton species not only in the ocean, but also in freshwater habitats [2]. They have an estimated 40% share of marine primary production [3]. It has been suggested that their ecological success is based on their silica wall, which is formed without assimilated carbon and is therefore less energy demanding than cell walls made from cellulose. A second reason for their ecological success is their capacity to acclimate to dynamic light climates [4] [5]. Diatoms adapt not only to changing light intensities in a very efficient way [6], but also to changes in the light quality. Generally, experiments on the acclimation to light quantity are classified into illumination by photosynthetically non-saturating low light (LL) and saturating high light (HL)

intensities. Acclimation to light quality is denoted as the acclimation to a defined range of wavelengths of photosynthetically absorbed radiation. Accordingly, blue light (BL) can be defined as the spectral range with a center wavelength of approximately 460 nm and red light (RL) with a center wavelength of about 660 nm [7].

In diatoms, the light quality was shown to have an effect on chloroplast migration [8], zygote germination [9] and on the light acclimation reactions of photosynthesis [10–14]. Recently, it was shown that BL is involved in the light acclimation of *Phaeodactylum tricornutum* [15]. They demonstrated that a typical HL phenotype with a reduced Chl *a* content per cell and a high photosynthetic capacity can be generated only in the presence of BL. Additionally, the photoprotective mechanisms, e.g. a high non-photochemical quenching (NPQ) capacity by means of an active xanthophyll cycle together with the presence of Lhcx proteins [16], can be expressed in the presence of BL, but not under RL. It was shown that the HL or LL phenotype is characterized by a specific gene expression profile that reflects the remodeling of the proteome during the light acclimation process [17].

The molecular background of light acclimation in diatoms is still enigmatic [5]. In green algae, it has been shown that the redox state of the plastoquinone (PQ) pool triggers kinases that regulate not only state 1-state 2 transitions [18], but also the activity of plastidal transcription factors and retrograde signaling [19]. Saturating light intensities cause extensive changes in photosynthetic cells that can be used to indirectly sense illumination, including the redox state of thiols, the thylakoid lumen pH and the accumulation of reactive oxygen species or certain metabolites [20]. Recently, it has been shown that artificial manipulation of the redox state of the PQ pool in *P. tricornutum* by DCMU and DBMIB can alter the gene expression profile according to the pattern expected for light acclimation [21]. However, in diatoms, the PQ pool can become largely reduced during the shift from light to darkness [22]. Moreover, this increased redox state was shown to last even during prolonged dark incubation periods [23]. Therefore, it is questionable how the redox state of the PQ pool could be employed as a trigger of the light flux *in vivo*.

Alternatively to the regulation by the redox state of some components of the photosynthetic apparatus, gene expression could also be regulated by the action of photoreceptors. *P. tricornutum* was shown to possess several blue light-sensing cryptochromes and aureochromes as well as red light sensing phytochromes [5] [24–25]. Cryptochromes are a diverse, ubiquitous family of photoreceptors originating from photolyases, which usually have a role in regulating growth and development and serve as circadian clocks [26], while aureochromes are newly described photoreceptors that are restricted to stramenopiles [27] [28–29]. The function of the phytochromes is not clear because, in the water column, blue-green light is predominant and red light is strongly attenuated by the water itself. In particular, far red does not penetrate more than 1 m into the water column and phytochromes would not be able to sense the ratio of red light/far red. In this case the switch of the photoreceptor would always be fixed in the "on" position. Therefore, it is likely that blue light sensors are of major importance for light perception. It was shown that PtCPF1 is located in the nucleus and has 6-4 photolyase activity [30]. The aureochromes PtAUREO1a/b and PtAUREO2 are also nucleus located [31] and PtAUREO1a has been shown to act as a transcription factor of dsCYC2, a protein associated with cell cycle progression [32]. Down regulation of the aureo1a gene also modifies light acclimation [31]. Reduced amounts of aureo1a induce a HL hyper-acclimated state. In contrast to the wild type, transformants with reduced aureo1a concentration were able to acclimate to higher intensities of RL. This was interpreted as a sign that the blue light receptor aureo1a interacts with a RL sensor and might repress HL acclimation in *P. tricornutum* in dependence of the ambient light quality. Putatively, a phytochrome might be such an RL sensor. However, the signaling pathways of these light receptors in diatoms are still largely unknown [5].

Blue light or RL can influence the metabolic network in algae, not only via gene expression and proteome remodeling, but also directly by changing the activity of specific enzymes. BL favors the allocation of carbon into proteins at the expense of carbohydrates in chlorophyta [33–34]. It was also shown that in *Chlamydomonas reinhardtii*, BL directly activates the nitrate reductase [35] and has an influence on cell cycle progression [36]. These direct effects should change the metabolome faster than the action via gene expression and subsequent protein biosynthesis, but they should be slower than direct light effects which alter the redox state of components of the photosynthetic primary reactions (reviewed in [37]). Therefore, it can be hypothesized that changes in the redox state of the PQ pool due to differences in the absorption cross sections of the photosystems can be established in the time range of a few minutes. A direct regulation of enzyme activity by light quality will be mirrored in the metabolome in the time range between several minutes to hours, whereas modified protein biosynthesis takes a minimum of half an hour [37].

In this study, we investigated the influence of BL and RL on the metabolic state of cells of *P. tricornutum* on the times scale of minutes to hours and days. For this purpose we compared changes in the metabolome, the macromolecular structure of the cells and the degree of reduction of the PQ pool in semi-continuously grown BL and RL cultures. The light flux and cell numbers were adjusted to an equal photon capture per chlorophyll and time independent of the light quality. Due to the comparison of metabolic changes on different time scales it is possible to hypothesize on the possible mechanisms of regulation of metabolic activity.

Results

Pre-acclimated cultures of *P. tricornutum*

Before light shift experiments were started, the cultures of *P. tricornutum* were pre-acclimated to RL and BL conditions for at least 7 days, respectively. All cultures were grown at light/dark cycles of 14/10 h. This photoperiod was kept also during the light quality shift experiments. Importantly, the incident light intensity was adjusted to yield the same amount of absorbed photons per cell and time under both light quality conditions. This resulted in similar growth rates, cellular dry weight and Chl *a* content (Table 1) of RL and BL grown cells of *P. tricornutum*. However, differences were observed in the macromolecular composition of the cells. Typically, BL acclimated cells possessed a significantly higher content of proteins but a decreased concentration of carbohydrates in comparison to RL acclimated cells. This led to a significant difference in the C/N ratio of BL and RL acclimated cells. The lipid content was comparable under both light conditions. These results clearly indicate that BL cells show higher N assimilation activity.

Differences between RL and BL acclimated cells were also observed at the level of the primary photosynthetic reaction. The effective quantum yield of PSII (Φ_{PSII}) measured at growth light intensity was slightly but significantly lower in cells under BL conditions compared to cells grown under RL (Table 1). This is a surprising result with respect to the fact that gross oxygen evolution rates measured at growth light intensity did not differ between RL and BL conditions (see below). The maximum values of the non-photochemical quenching of Chl *a* fluorescence and the pool size of XC pigments as indicators of the potential of light protection were significantly higher in BL grown cells in comparison to RL grown cells, although the absorbed photon flux density was the same. No diatoxanthin (Dtx) as the second Ddx/Dtx cycle pigment was detected in any of the samples.

Light shift experiments – long-term acclimation

The light shift experiments of pre-acclimated cultures of *P. tricornutum* from RL to BL and vice versa started 3 h after the beginning of the respective illumination period (t_0) and were followed over a period of 6 days. Day 0 and t_0 depict the starting point of the light switch experiments.

Table 1. Growth rate, photosynthetic parameters and the macromolecular composition of BL and RL light acclimated *P. tricornutum* cultures.

	BL	RL	n_{BL}	n_{RL}	
Growth rate[a]	0.42±0.12	0.42±0.09	15	19	
Chl a[b]	0.58±0.11	0.56±0.08	22	28	
Dry matter[b]	23.9±4.2	21.9±4.5	11	12	
Carbohydrates[c]	30.8±2.5	33.6±3.7	27	28	**
Proteins[c]	44.8±1.7	41.9±2.4	27	28	***
Lipids[c]	14.4±1.8	14.5±1.9	27	28	
C/N[d]	5.9±0.2	6.7±0.1	5	5	***
Φ_{PSII}	0.65±0.002	0.67±0.001	6	6	***
O$_2$ evolution at growth light[e]	64±9	62±7	10	10	
Maximum O$_2$ evolution[e]	238±34	203±31	11	11	*
Maximum non-photochemical quenching	1.05±0.17	0.50±0.12	11	10	***
Xanthophyll cycle pigments[f]	110±3	90±11	4	4	*

Shown are the mean value, standard deviation (±) and the number of replicates (for BL: n_{BL}; for RL: n_{RL}). The asterisks depict the p-values as described in the method part.
[a][d^{-1}];
[b][pg cell^{-1}];
[c][% of dry matter];
[d][g g^{-1}];
[e][µmol O$_2$ mg Chl a^{-1} h^{-1}];
[f][mmol mol Chl a^{-1}].

No differences were observed in the oxygen-based photosynthesis rates during the entire period of observation (Figure 1 A). Hence, during the light shift experiments all *P. tricornutum* cultures showed comparable gross photosynthesis rates of ~60 µmol O$_2$ mg Chl a^{-1} h^{-1} irrespective of the incident light quality. This stands in contrast to the growth rates that were observed to be slightly reduced during the first 48 h after the light shift from BL to RL (~0.3 d^{-1} compared to 0.42 d^{-1} in pre-acclimated cultures), whereas in the shift from RL to BL, the growth was completely arrested during the first 24 h of the light shift (Figure 1 B). Growth rates fully recovered to about 0.4 d^{-1} within the next 48 h until day 3 after the start of the light shift. This result was similar when the growth rates were calculated on the basis of the cell number or on the dry matter yield (data not shown).

Within the six days of light shift experiments, the maximal NPQ values and the xanthophyll cycle pool size (Figure 1 C–D) readjusted to the values observed in the respective pre-acclimated cultures given in Table 1. In other words, maximum NPQ increased after the shift of RL pre-acclimated cultures to BL from 0.46 (±0.09) to 1.02 (±0.09), a typical value for BL pre-acclimated cultures. The Ddx pool increased from 90 (±11) to 111 (±10) mmol Ddx per mol Chl a in the 24 h following the shift and thus reached an amount similar to BL pre-acclimated cultures within just one day. Similarly, the BL to RL shift led to a faster adjustment of the Ddx pool in two days time (from 110±3 to 89±2 mmol Ddx mol Chl a^{-1}) compared to the decrease of the NPQ (from 1.12±0.14 to 0.59±0.11) within three days to levels comparable to RL pre-acclimated cultures.

In conclusion, BL cells shifted to a RL acclimated state within three days, as documented by the growth rates, maximal NPQ values and the xanthophyll cycle pool size (Figure 1 B–D). A similar time frame was observed for the switch from RL acclimated cells to BL (Figure 1 B–D). Overall, the largest changes were observed within the first 24 h. This is in accordance with the

time course of changes of the macromolecular composition of the cells. The FTIR data revealed that 24 h after the shift from RL to BL and vice versa, the cells resembled the macromolecular composition of cells which were pre-acclimated to similar light conditions (data not shown, compare to Table 1). Thus, in further experiments, we aimed to understand the molecular processes of light acclimation during the first day of the time period of the light shift.

Fast fluorescence induction kinetics

Since the photosynthetic redox control is a crucial parameter for light acclimation in higher plants and apparently also in diatoms, changes in the redox state of the PQ pool were characterized. This can be done with the aid of fast fluorescence kinetics of Chl a fluorescence [38]. Here, the normalized J level (F_J') of the fast fluorescence induction was used as a proxy for the relative redox state of the PQ pool. The higher the ratio $F_{2\ ms}'\ F_M'^{-1}$ (F_J'), the more the PQ pool is reduced. This parameter can be measured during the light exposure and therefore reflects the in-situ growth condition. BL acclimated cells showed a slightly lower F_J' (and therefore a more oxidized PQ pool) than RL acclimated ones (Figure 2, 0.68 and 0.73 F_J' respectively; p = 0.09).

The cells shifted from RL to BL conditions showed a lag phase of 5 min. Then the PQ pool became more oxidized as reflected by a significant decrease of F_J', which remained stable after 25 min (Figure 2 A). In contrast, cells that were shifted from BL to RL showed a very fast increase of F_J' (Figure 2 B), indicating that the PQ pool becomes immediately more reduced if RL is driving photosynthesis. This change in F_J' remained stable after about 30 min. One hour after the light shift, the cells were shifted back to their respective pre-acclimated light condition (see below).

The change in the redox state of the PQ pool is not mainly caused by different activities of PSI and PSII, because BL and RL are absorbed mainly by the FCPs (*fucoxanthin/chlorophyll-binding proteins*) that have identical absorption spectra independent of

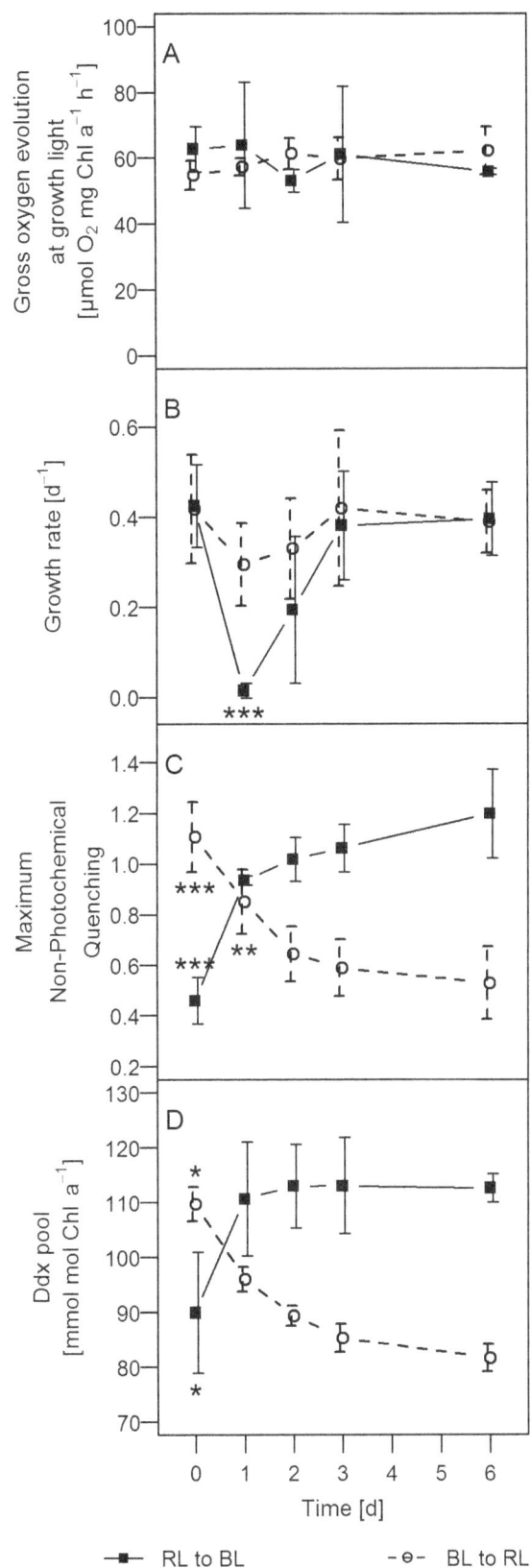

quenching and (D) concentration of the xanthophyll cycle pigment Ddx at growth light conditions. The NPQ values were measured at illumination with saturating light intensities, while all other values are measured at growth light conditions. The asterisks depict the p-values of the respective data point compared to the steady state (BL to RL shift is compared to RL steady state, RL to BL shift is compared to BL steady state). n = 4–5. The exact values are specified in Table S1.

their PS I or PSII attachment. A second argument that the redox state of the PQ pool is not controlled by the light reactions is the observation that the kinetics are different from BL to RL and from RL to BL. A third argument is supported by Figure 2 (right hand side), which shows that the back reaction from BL to RL is not reversible in contrast to the RL to BL. Finally, the time constant for the redox changes is in the range of half an hour, which would be too slow for changes in the energy distribution between both photosystems. Therefore the observed changes are likely to be induced by alternations in metabolic processes.

Light shift experiments – short-term acclimation

After the shift from BL to RL conditions, a 16% increase in the Chl *a* concentration of the algal suspension was observed over a period of 6 h, after which the Chl *a* concentration stagnated until the end of the light period (Figure 3 A). This time course of change of the Chl *a* concentration was also observed in the pre-acclimated cultures under RL and BL, respectively (data not shown). However, after the shift from RL to BL conditions, the increase in Chl *a* concentration was slightly lower during the first 6 h under the new light condition at each sampling point (Figure 3 A). During the next four hours, the Chl *a* concentration of cells shifted to BL decreased again and finally reached the same value as at the starting point of the light shift experiments. This result confirms that growth was arrested in the first 24 h after the RL to BL shift (compare to Figure 1B). In neither experiment did the cell numbers change during the light period (data not shown).

Although cells shifted from RL to BL show impaired Chl *a* biosynthesis and cell division, the effective quantum yield of PSII was only slightly altered and ranged around 0.69 (Figure 3 B), whereas Φ_{PSII} increased slightly from 0.66 ($\pm$0.01) to 0.69 ($\pm$0.01) in cells that were shifted from BL to RL. At the end of the light period, the same values of Φ_{PSII} were observed in the shifted cultures as in the respective pre-acclimated cultures (Table 1), although Chl *a* synthesis was still lower (BL to RL) or completely blocked (RL to BL) during that time (Figure 1A).

Changes in the C-partitioning during the light shift experiments

The analysis of FTIR data allows to follow the changes in the relative pool sizes of carbohydrates, lipids, and proteins. In Figure 4, the relative changes in the macromolecular composition regarding carbohydrates and proteins are presented during the light phase in the shift experiment compared to the controls, which were kept in the same light as before.

In BL pre-acclimated cells the pool sizes of proteins and carbohydrates remained constant, which means that the newly assimilated C is equally directed into the different macromolecular pools during the light phase. Only at the end of the light phase the carbohydrate storage pool increased by 17% at a slight expense of proteins which decreased by 7%. In RL cells, the increase in carbohydrates started earlier in the light phase, but the macromolecular composition remained constant during the first 4 hours. At the end of the light phase, RL and BL pre-acclimated cultures had changed their carbohydrate and protein contents in a comparable manner.

Figure 1. Long-term acclimation to light quality shifts. The acclimation process of *P. tricornutum* cultures was observed for 6 days after the light quality shift from BL to RL and RL to BL. Shown are (A) estimated gross oxygen evolution rates at growth light conditions, (B) growth rate per day, (C) maximum values of non-photochemical

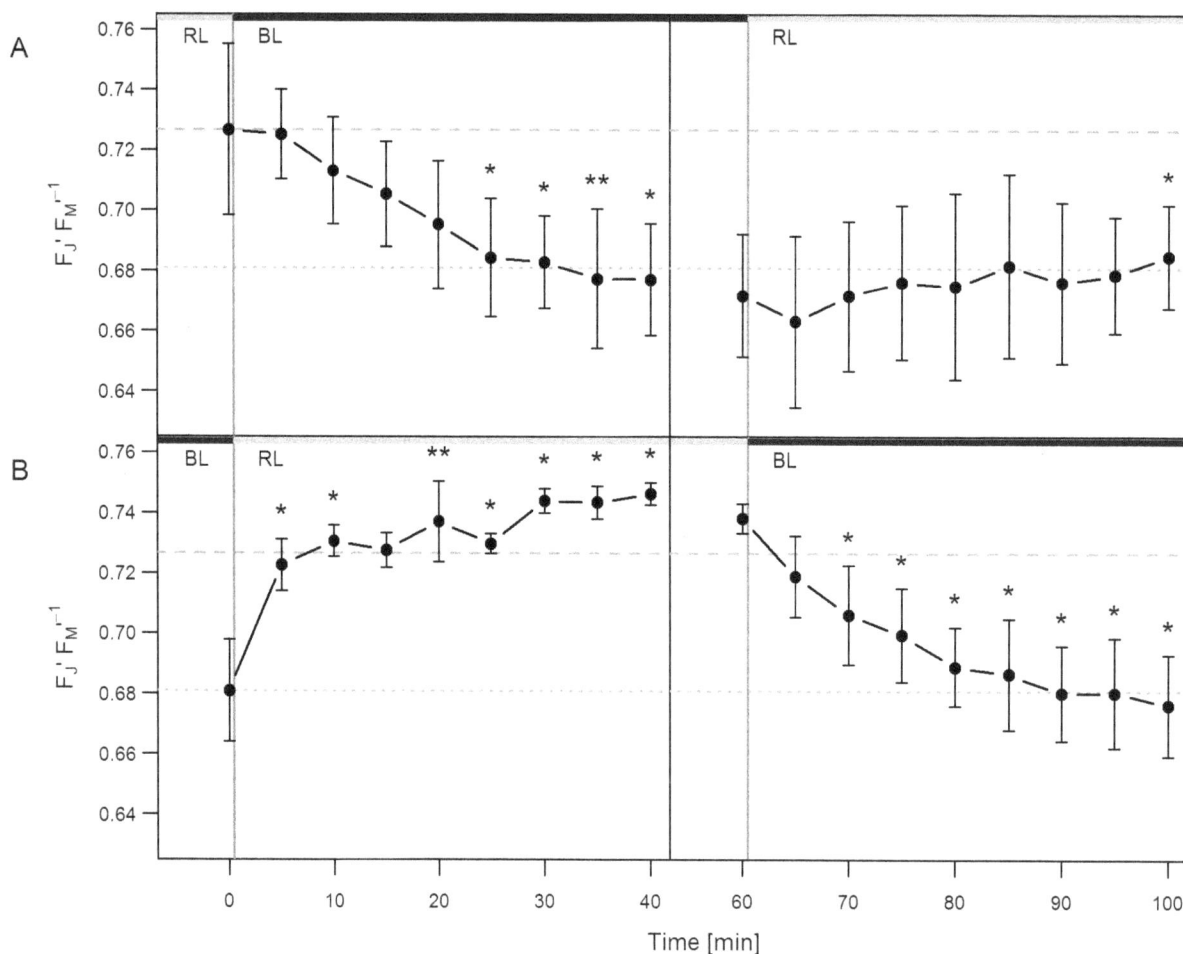

Figure 2. Fast fluorescence induction kinetics. The changes of the J Level of fast fluorescence induction kinetics ($F_J{'} F_M{'}^{-1}$) were recorded during the light quality changes. (A) RL to BL shift of a RL adapted culture (0–40 min) followed by a light shift back to RL (60–100 min) (B) BL to RL shift of a BL adapted culture (0–40 min) followed by a light shift back to BL (60–100 min). The horizontal lines show the J level of BL (dotted) and RL (dashed) adapted cultures. The asterisks depict the p-values of the respective data point compared to $t_{0\ min}$ (0–40 min) or to $t_{60\ min}$ (60–100 min). n = 3. The exact values are specified in Table S2.

In contrast, directly after the BL to RL shift, the cells strongly increased their pool of carbohydrates at the expense of proteins. The steady decrease of proteins (−15% after 10 h) and increase of carbohydrates (+29% after 10 h) led to significant differences from the BL pre-acclimated cultures already 4 h after the shift. The carbohydrate levels after the shift from BL to RL were also higher (significant at 2/4/8/10 h, not shown in Figure 4) and the protein levels lower (significant at 2/8/10 h, not shown in Figure 4) when compared to the RL pre-acclimated cultures.

In RL to BL shifted cultures the opposite effect was observed, albeit to a much lower extent. The protein level significantly increased by about 5% within 4 h following the shift to BL (Figure 4 D) in comparison to RL pre-acclimated cultures (Figure 4 C). It should be emphasized that proteins represent the large fraction of macromolecules in *P. tricornutum*. Thus, the small relative increase shown in Figure 4 D equals to a much larger total increase of proteins (see next paragraph and Figure 5). Moreover, the RL to BL was the only experimental condition where a temporary increase of the protein content was observed. The protein pool proved to be increased when compared to the RL pre-acclimated culture (Figure 4 D) and also the BL pre-

acclimated culture (significant at 6/8/10 h, not shown in Figure 4).

In addition to the relative changes of macromolecular composition, Figure 5 shows the pattern of carbon allocation into the different pools of macromolecules during the first 2 h of the light shift experiments. Within this time period the total changes of macromolecular composition were very small compared to the pool sizes of macromolecules. This caused a relatively high standard deviation of the calculated carbon allocation rates as shown in Figure 5. Nevertheless, the analysis shows that the carbon partitioning was completely different in the comparison of light shift experiments. Thus, the BL to RL shift was characterized by an almost exclusive flow of newly assimilated carbon into carbohydrates whereas almost no carbon is directed into protein synthesis and a net loss of lipids is observed. The shift from RL to BL showed an opposite pattern with a preferred carbon allocation into proteins and lipids at the expense of carbohydrates.

Metabolite distribution

This drastic reorganization of the cell internal carbon flux in dependence on changes in the light quality should also be mirrored in the metabolome. Therefore, a metabolome analysis

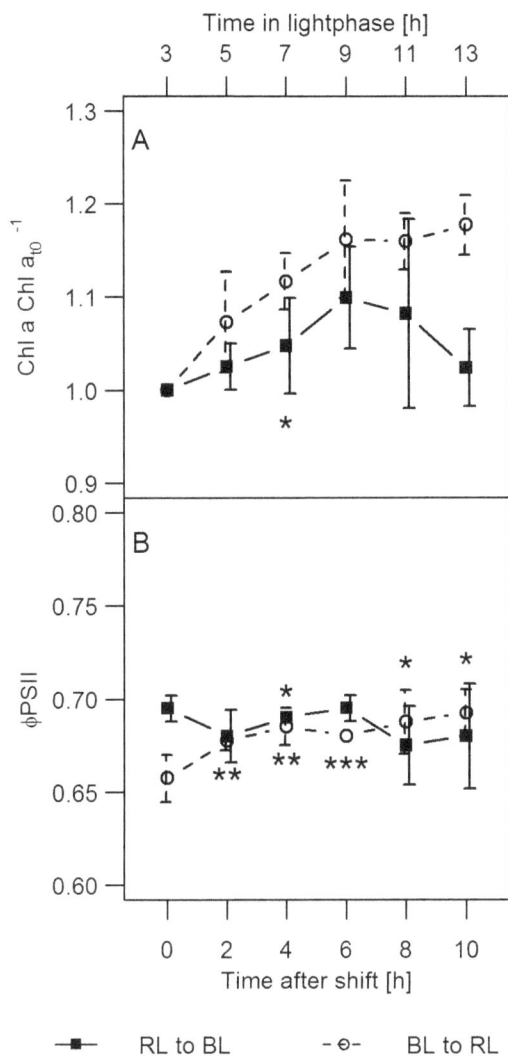

Figure 3. Short-term acclimation to light quality shifts. The changes in (A) the Chl a concentration and (B) Φ_{PSII} of *P. tricornutum* cultures were recorded for 10 h after the light quality shift at t_0 from BL to RL and RL to BL. Additionally, the time in the light phase is given above the plot. The asterisks depict the p-values of the respective data point compared to the steady state the cultures were shifted from at the same time point (Chl a) or the daily mean value when this was constant (Φ_{PSII}). n = 4–8 for the Chl a concentrations and n = 3–4 for $F_V' F_M'^{-1}$. The exact values are specified in Table S3.

was performed including 36 metabolites that were identified and quantified. In addition to a number of amino acids, the metabolites relate in large part to e.g. glycolysis, the TCA cycle, and the pentose phosphate pathway. No metabolites of the photorespiratory pathway were found in any of the samples.

The relative abundance of metabolites in the cultures pre-acclimated to BL and RL conditions were comparable (Tables S6 and S7). The only significant difference was found in a higher histidine and a lower fructose-6-phosphate content in BL acclimated cells compared to RL acclimated cells. To follow the changes in the metabolite profile of *P. tricornutum*, samples were taken 15 min, 30 min, 1 h and 24 h after the respective shift from one to the other light quality condition.

Metabolite profile during the shift from RL to BL

In general, for a number of metabolites, a clear increase in their abundance was observed during the first 30 min after the shift from RL to BL conditions. Moreover, already 15 min after the shift to BL conditions, extensive changes in the metabolite profile were detected (Figure 6 A). The most significant changes were found in the pool of amino acids. With the exception of glutamine and ornithine from the urea cycle, the abundance of most of the other amino acids increased. NH_3 is primarily assimilated into glutamine/glutamate, which represents the precursor for amino acid synthesis [39]. It is therefore likely that the glutamine depletion is due to the increased synthesis of the remaining amino acids. The high levels of most amino acids increased with time (e.g. after 30 min), recovered after 1 h and were found to be lower 24 h after the light shift.

Similarly, a high number of intermediates involved in carbohydrate metabolism were transiently increased after only 15 min BL. This pattern mirrors a transition from an active, carbohydrate-dominated metabolism to a protein biosynthesizing one shortly after the shift. However, the results are not as stringent as for the amino acids. The levels of TCA cycle intermediates clearly increased immediately after the shift to BL. The activated TCA cycle might be explained by an increased need for C skeletons for amino acids. In contrast, apart from an initial increase of glucose-1P and pyruvate, glycolysis intermediate levels were largely unaffected in the first hour. 24 hours after the shift from RL to BL, the relative amount of most metabolites in the central carbon metabolism were increased.

Keeping in mind that in diatoms the final part of the glycolysis is located in the mitochondrion [40], data from the end of these first 24 hours give clear evidence of a strong up-regulation in the reductants in the mitochondrion. This is additionally supported by the high levels of TCA cycle intermediates at the end of this period.

Metabolite profile during the shift from BL to RL

The changes in the metabolic profile after the shift from BL to RL conditions showed an opposite direction in comparison to the changes observed during the shift from RL to BL conditions. Thus, within 15 to 30 min after the start of the light shift, the abundance of all identified amino acids was decreased (Figure 6 B). Subsequently, the amino acid concentrations approached initial values again, although about half of the amino acid levels were still decreased 1 h after the shift to RL.

In contrast to the switch from RL to BL, there was no consistent response from the TCA cycle intermediates. While 2-oxoglutarate and succinate levels immediately increased after the shift to RL, isocitrate and malate levels remained unchanged. Sedoheptulose-7P was clearly increased shortly after the light switch to RL, while erythrose-4P levels dropped. Therefore, the penthose phosphate pathway also showed a strong initial reaction to RL. The levels of the glycolysis intermediates PEP, fructose-6P and glucose-6P increased while pyruvate levels dropped. One day after the light shift from BL to RL, the amount of all amino acids and of the metabolites related to the central carbon metabolism were clearly decreased.

Metabolite ratios

The metabolite ratios were calculated to visualize the dynamics of the metabolite changes during the early shift phase (Figure 7). A high value means that the ratio of abundance between the different amino acids changes drastically between the time points. This indicates that the production and consumption of the different amino acids are not in equilibrium and the dynamic of

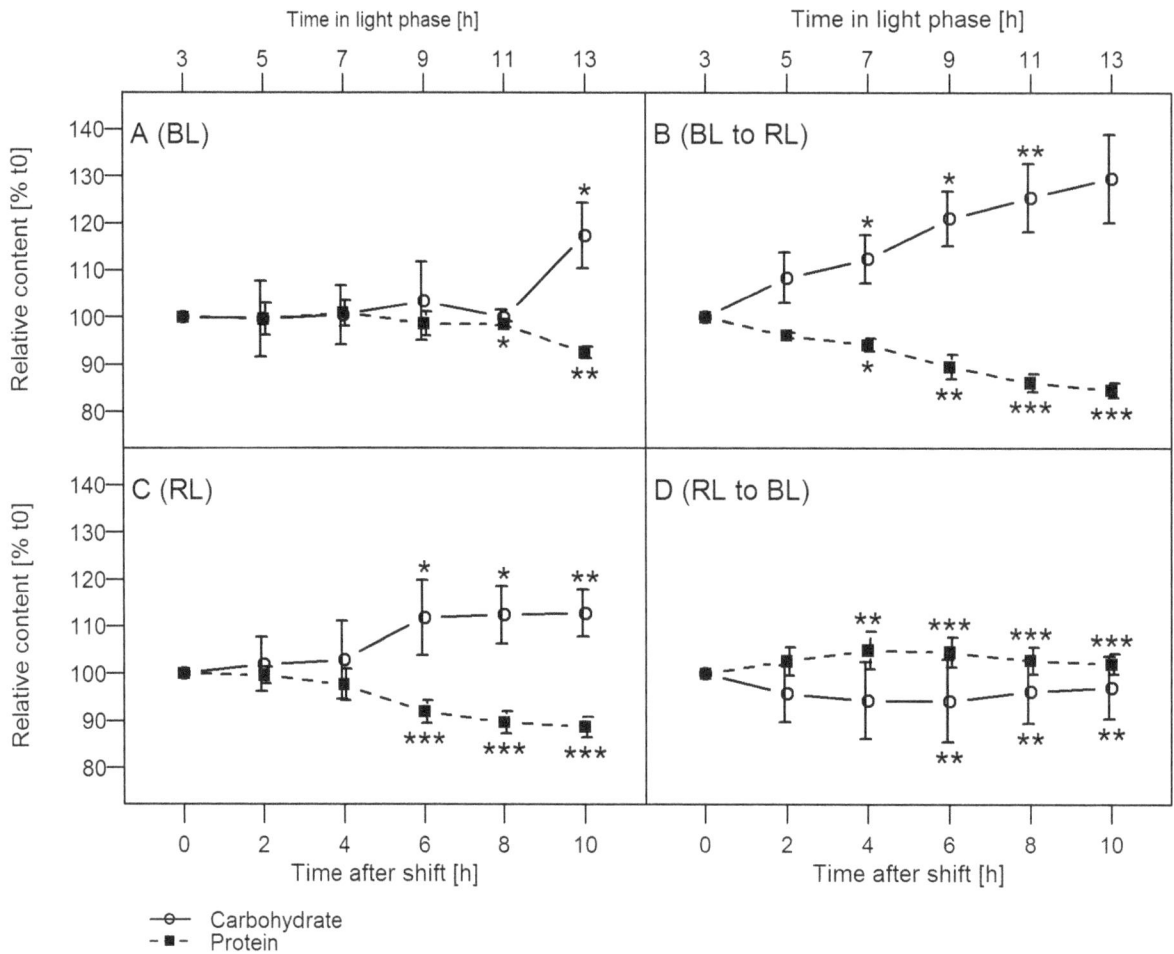

Figure 4. Carbohydrate and protein levels. The carbohydrate and protein levels of *P. tricornutum* cultures during the course of the day were determined for cultures pre-acclimated to (A) BL and (C) RL as well as (B) the changes after a shift from BL to RL and (D) RL to BL. Shown are the relative protein and carbohydrate contents in relation to the values measured at t_0. The asterisks depict the p-values of the respective data point compared to t_0 for pre-acclimated cultures (A, C) or compared to the steady state the cultures were shifted from at the same time point (B, D). n = 3–6. The exact values are specified in Table S4.

metabolome reorganization is high. The value is low under steady state conditions when the flux through the different metabolites is in homeostasis. The same is true for metabolites of the central carbon metabolism.

The metabolite ratios in the pre-acclimated cultures showed very few differences, which confirmed the expectation that an equilibrium in the metabolic pathways is present under both light qualities (left hand side of Figure 7 A, B). During the light shift, the metabolome is strongly disturbed and shows that most significant changes in the metabolome were detected between 15 and 30 min (middle bars in Figure 7 A, B). Moreover, the metabolite ratios confirm that the RL to BL shift induced large changes within the pool of amino acids (Figure 7 A). Interestingly, there was a very fast reorganization in the amino acid pool within the first 15 min and again in the time period from 30 min to 1 h after the start of the light shift. The reorganization between 30 min and 1 h reversed most of the effects that disturbed the metabolism in the first 15 min. The metabolite ratios of the central carbon metabolism were comparable to cells pre-acclimated to BL 24 h after the shift from RL to BL (right hand side of Figure 7 A).

On the other hand, cells shifted from BL to RL conditions showed distinctly less changes in the pool of amino acids (Figure 7

B). Again, the most significant changes occurred in the first 15 min. Although the amount of all amino acids decreased, the overall composition of the amino acid pool stayed quite similar (compare to Figure 6 B). It is noteworthy that 24 h after the shift from BL to RL, cells possessed a significantly different metabolome ratio in comparison to RL pre-acclimated cultures. However, 1 day after the shift to RL no significance was found when compared to the BL pre-acclimated culture, therefore the value is 0 and the respective bar is missing in Figure 7 B.

Discussion

In the pre-acclimated state, the RL and BL acclimated *P. tricornutum* cultures showed similar values for growth rates, cellular dry matter and Chl *a* content (Table 1) in agreement with previously published results [15]. Since the light conditions were carefully adjusted to the same amount of absorbed radiation under RL and BL conditions, the cells used RL and BL with the same quantum efficiency for growth and showed similar photosynthesis rates.

Despite these similarities, the high light protection via NPQ reached much higher levels in BL pre-acclimated cells. This was already demonstrated by [31], who could show that under non-

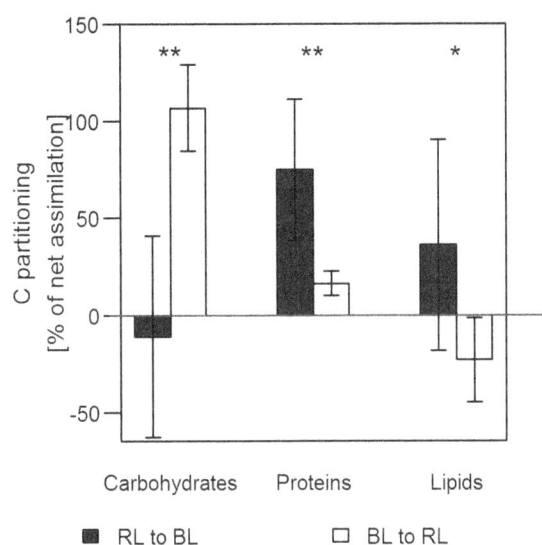

Figure 5. Carbon partitioning. The relative partitioning of C (carbon) into carbohydrates, proteins and lipids was calculated for the 2 h following the light quality changes. Data were calculated on the basis of the FTIR spectra and the net rate of carbon assimilation and, therefore, add up to 100% net carbon assimilation (carbohydrate + protein + lipid bar). Values below zero mean a net loss. Shown are the first 2 h after the light quality change from RL to BL and vice versa. n = 3–6. The exact values are specified in Table S5.

saturating light conditions, the maximum NPQ seems to be largely determined by the light colour and not by light intensity. The present results suggest that the acclimation to the new light quality is completed after about three days (Figure 1). However, the largest changes occurred within the first 24 h.

Although these light quality induced differences did not change the quantum use efficiency of growth in the pre-acclimated cultures, we observed drastic changes of growth performance during light quality shift experiments (Figure 1 B, Figure 3 A). Whereas within the 48 h following the shift from BL to RL only a small decrease of the growth rate of *P. tricornutum* cells was observed, the shift from RL to BL caused a complete arrest of cell division after 24 h. Growth arrest relaxed within the following 48 h which shows that it represents a temporal imbalance of cell growth. To our knowledge this is the first report of a growth arrest induced by light quality changes.

To explain this surprising observation, a direct, BL induced effect (e.g. photoreceptor-mediated arrest of the cell cycle) or an indirect effect induced by a molecular reorganization of the cells could be proposed. In principle, a direct effect of the BL receptor aureochrome 1a was recently demonstrated in *P. tricornutum* [32]. It was shown that the diatom specific cyclin 2, controlling the G1-to-S phase of the cell cycle, is a transcriptional target of aureochrome 1a. Consequently, a BL triggered transcriptional induction of this cyclin with a subsequent onset of cell cycle was observed. In contrast, under RL the expression of cyclin 2 was inhibited and the onset of cell cycle strongly delayed. The results of [32] stand therefore in contrast to the observation of the present study, where inhibition of the cell division was caused by the shift to BL. The results were, however, obtained by two very different experimental setups. The growth inhibition observed after the shift to BL is temporal (~2 days), after which the BL acclimated cells show a similar growth rate as the RL acclimated ones. In contrast, the impaired growth reported by [32] is present in acclimated RL cultures of a higher light intensity than it was used in this study

(not used by us). Indeed, it was reported elsewhere that, while low BL and RL *P. tricornutum* cultures show identical growth rates, higher RL intensities lead to a reduced growth when compared to BL [31]. The photosynthesis rates remained unaffected during the RL to BL shift and no concomitant increase in cellular dry weight was observed. Therefore, a reduced photosynthesis rate can be excluded as the cause of the impaired growth rate after the shift to BL. An alternative explanation of the observed growth arrest comes into consideration which includes (i) a fast reversal of metabolite distribution and (ii) a macromolecular reorganization of the cells which influences the intracellular carbon partitioning under these conditions.

Macromolecular reorganization and carbon partitioning

We observed a reorganization of the macromolecular composition of the cells in response to the light quality shifts (Figure 4). Thus, the shift from BL to RL significantly increased the fraction of carbohydrates at the expense of proteins, whereas the RL to BL shift resulted in an increased protein/carbohydrate ratio. The changes in the macromolecular composition yielded in a corresponding increase of the C/N ratio under RL but in a decrease under BL conditions. In accordance with the metabolite distribution pattern the analysis of the partitioning of the newly assimilated carbon by FTIR revealed a drastic increase of carbohydrate synthesis at the expense of protein and lipid synthesis within two hours after the light shift from BL to RL (Figure 5). In contrast, after the shift from RL to BL the synthesis of carbohydrates stagnated whereas protein synthesis was strongly increased. The induction of protein synthesis would also require the activation of the nitrate reductase by BL. Indeed, this activation was demonstrated [30].

The fast changes in the carbon partitioning could be assumed to also have an influence on the redox poise of the chloroplast and the cytosol. This is deduced from the fact that the synthesis of proteins and lipids requires more electrons per carbon than the synthesis of carbohydrates. Thus, after the shift from BL to RL the electron requirement for the preferred synthesis of carbohydrates instead of proteins and lipids could be decreased. As a consequence, it could be assumed that photosynthetically produced reducing equivalents (e.g. NADPH) accumulate. In contrast, after the RL to BL transition the preferred synthesis of proteins at the expense of carbohydrates should consume more electrons and should drain the pool of reducing equivalents. Support for this hypothesis comes from the changes of the redox state of PQ pool in *P. tricornutum* during light shift experiments (Figure 2). There are a number of studies that show the increase of the degree of reduction of the PQ pool by stromal reductants in diatoms [4] [41] but also in green algae [42]. In accordance with the assumptions stated above, in the present study the observed changes in the J level of fast fluorescence induction correlate with an increased degree of reduction of PQ after the BL to RL shift whereas the PQ pool was less reduced after the RL to BL shift. Since the photosynthesis rates at growth light intensity did not change during the light shift experiments it could be hypothesized that the changes in the degree of reduction of the PQ pool are not primarily under control of the energy balance between the photosystems or the balance between light and dark reactions but can be regarded as a effect of metabolic redox control by the changes in carbon partitioning and of the macromolecular composition of the cells. Such a new model of redox control needs further confirmation by time resolved gene expression studies compared to changes in the proteome and metabolome.

In *P. tricornutum*, the RL to BL shift and the accompanied induction of protein synthesis could be a probable reason for the

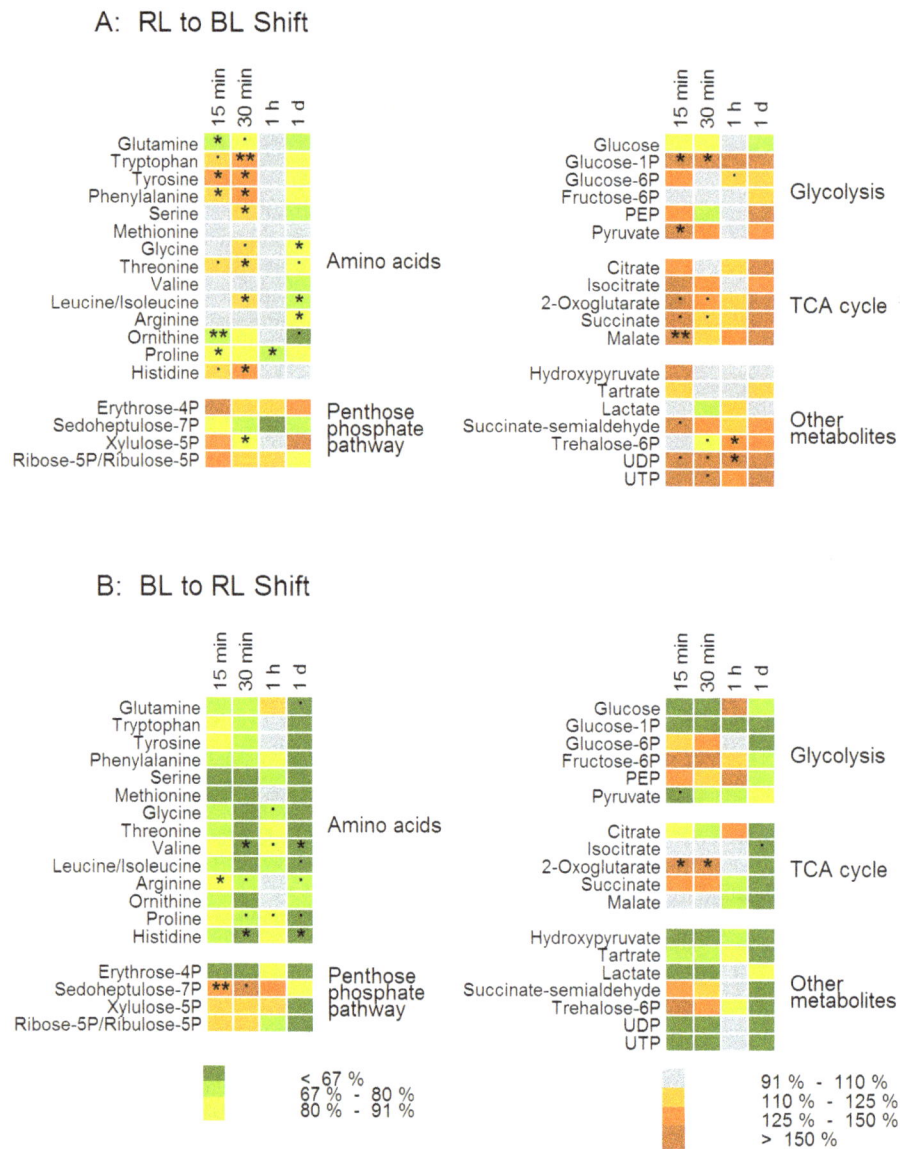

Figure 6. Metabolite profile. The changes in the content of metabolites per cell after a light quality change compared to the pre-acclimated *P. tricornutum* cultures were determined for 4 time points. The asterisks depict the p-values of the respective data point compared to t_0. Additionally '.' marks p<0.1. Significant changes that were lower than 10% (marked grey) were omitted. n = 3.

observed growth arrest as outlined in the following hypothesis. Diatoms are known for their ability to reduce nitrate in excess of their metabolic demand. It was suggested that the subsequent release of reduced nitrogen as ammonium into the surrounding medium functions to maintain the energy balance within the cell [43]. The increased demand of carbon skeletons for protein biosynthesis during the shift from RL to BL might reduce the amount of cellular carbon available for basic metabolism which, in combination with an excretion of reduced nitrogen, could be the reason why the cell growth stopped until this imbalance was readjusted. The inverse shift from BL to RL did not initiate such an enhanced N reduction and the acclimation was performed by a continuous metabolic reorganization. This new hypothesis presented could be tested in a future study by a data set obtained from similar experiments when *P. tricornutum* is grown with ammonium instead of nitrate as the source of nitrogen. Ammonium is known to inactivate the nitrate reductase in diatoms [44]. Thus, if

the activation of the nitrate reductase is responsible for the observed physiological effects during the shift from RL to BL conditions, this acclimation pattern should not be present under cultivation with ammonium.

Fast reversal of metabolite distribution during light quality shifts

The light shift experiments revealed a significant reversal of the distribution of cellular metabolites (Figure 6, Figure 7). Importantly, the initial kinetics of the changes of metabolites was in the range of minutes. To our knowledge, this should be faster than it could be expected due to an altered protein profile as a consequence of gene expression. The time frame given for the effects of transcriptional changes is about 30 min [37]. Although we cannot exclude the interference of a fast control by gene expression, the fast changes of the metabolite levels raise the hypothesis that some of the crucial enzymes are under direct

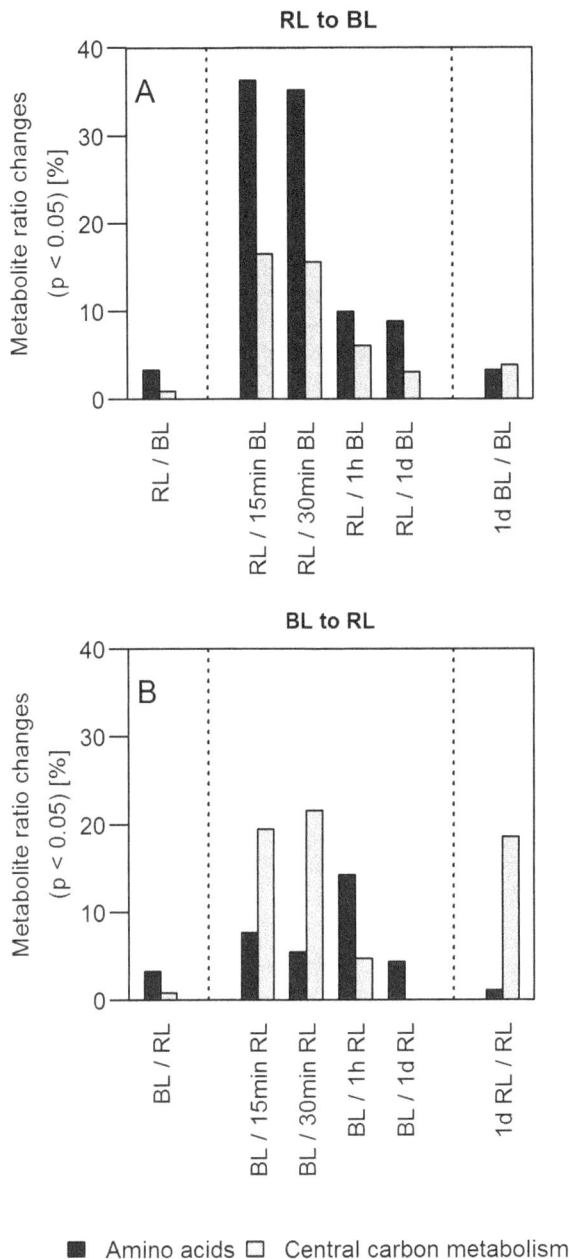

Figure 7. **Changes in the metabolic ratios.** All possible metabolite ratios (e.g. glucose-6P to fructose-6P) of the measured amino acids (black) or of the metabolites of the central carbon metabolism (grey) were calculated for each time point. The figure shows the rate of significant changes of these ratios between the time points given under each bar (e.g. 10% means that 10% of all metabolite ratios between these time points change significantly, p<0.05). This acts as a value for metabolome reorganization. Given are the relevant time points for the (A) RL to BL shift and (B) the BL to RL shift. The left hand bar is similar for (A) and (B) and compares the RL and BL pre-acclimated cultures. The bars in the middle illustrate the changes between shifted cultures to the light quality they were shifted from. The right hand bar compares the shifted cultures at 1 d after the shift to the light quality they were shifted to.

quality dependent changes in the turnover rate of enzymes like e.g. amylase, aldolase, pyruvate kinase, and NAD-dependent isocitrate dehydrogenase [33]. In particular, for pyruvate kinase (PK) large changes in activity were observed. This enzyme is a regulatory site of glycolysis and catalyses the formation of pyruvate from phosphoenolpyruvate (PEP) which is then the precursor of the TCA cycle and thus, also for the synthesis of amino acids. Blue light strongly enhances the turnover rate of PK [33] and should promote the formation of pyruvate. In the present study a corresponding significant increase of the level of pyruvate was detected after the shift from RL to BL. In addition, a clear increase of the metabolites of the TCA cycle suggests an up-regulated activity of this cycle. A high activity of the TCA cycle is required for the synthesis of a number of amino acids and is thus, in accordance with the observed increase in the protein pool.

In contrast, after a BL to RL shift a decrease in the pyruvate/ PEP ratio and the accumulation of glycolytic metabolites upstream of pyruvate indicate a lower activity of PK and thus a down-regulation of the supply of pyruvate through glycolysis. Consistently, the decreased formation of pyruvate could be correlated with the lower level of citrate, the initial metabolite in the TCA cycle. Thus, a down regulation of sugar degradation by glycolysis and a reduced TCA cycle activity should result in an accumulation of carbohydrates and in a reduced synthesis of amino acids and proteins. Indeed, both were observed in the analysis of macromolecular composition of cells during the BL to RL shift experiments.

The RL to BL shift led to a clear decrease of the glutamine concentration. Correspondingly, the levels of about half of the other measured amino acids increased. It is known that glutamate/glutamine acts as the precursor for amino acid synthesis [39]. However, chlorophylls are also derived from glutamic acid. It is therefore possible that the decreased level of this precursor due to the enhanced amino acid synthesis is responsible for the initial suppress of Chl a synthesis.

The exact mode of action of the activity regulation of metabolic enzymes by light quality is not resolved up to now. It should be kept in mind that the proposed regulation of enzymes by light quality was not yet shown for *P. tricornutum*, and should be confirmed in further studies. Importantly, the metabolic changes take place outside of the chloroplast and were observed also in non-green cells [33]. Therefore, the regulation of enzyme activity can be assumed to be independent of the photosynthetic reaction. Instead, it was suggested that a direct regulation of enzyme activity could occur by BL absorbing components, such as flavins like FMN (*flavin mononucleotide*) or FAD (*flavin adenine dinucleotide*). Interestingly, such BL induced activation is known for FAD-containing enzymes, e.g. pyruvate dehydrogenase and nitrate reductase [35]. In contrast, FMN-containing enzymes seem to be inactivated by BL [45]. It is important to note that this mode of enzyme regulation is relevant for the observed changes in the metabolome on the time scale of minutes. On the other hand, it is reasonable to assume that the sustained changes of the macromolecular composition of the cells after light shift experiments were also regulated by *de novo* synthesis of enzymes. Here, in particular BL can be assumed to act on the regulation of transcription and translation of genetic information. This conclusion was deduced from the observation that the effects of a short-term illumination with BL do not disappear immediately after the transfer back to RL illumination [45]. In the present study a similar effect was observed on the changes of the degree of reduction of the PQ pool during shift experiments from RL to BL and back to RL.

regulatory control by light quality. Indeed, in green algae similar changes in the macromolecular composition of the cells were observed in a comparable time frame during light shift experiments with BL and RL [33] [45]. Crude cell extracts showed light

Conclusions

In conclusion we did not find a satisfactory explanation for the observed growth arrest in light shift experiments from RL to BL. However, we can exclude an inhibitory effect by the limitation of the photosynthetic reaction. Moreover, a remarkable result of the present study is the finding that the light acclimation process in *P. tricornutum* is characterized by the disturbance of the cellular carbon allocation revealed by a metabolic reorganization of the cells as a first response to the new light environment. It is hypothesized that this fast initial reaction of the cells in response to light quality changes could be assigned to a direct control of enzyme activity by BL and RL. The second phase of metabolic changes in the range of 30 min to 1 hour would subsequently represent the effects of transcriptional changes [37]. Further studies on gene expression and protein levels need to be done to clear up the effects ascribed to a direct enzyme regulation and transcriptional control via photoreceptors in *P. tricornutum*.

The initial reorganization of the metabolite profile was accompanied by changes in the degree of reduction of the PQ pool. We propose that these changes were not directly due to shifts in light quality. Instead it could be hypothesized that the degree of reduction of the PQ pool is also under metabolic control due to changes in the cellular carbon partitioning.

Finally, we hypothesize that photoreceptors could fulfill both functions: the regulation of the activity of key enzymes of N and C assimilation and the regulation of gene transcription. The proposed participation of photoreceptors on the regulation of cellular carbon allocation will be a subject of future research.

Materials and Methods

Growth conditions

P. tricornutum UTEX 646 was grown as previously described in [15] in modified f/2 medium [46] with 16 g l^{-1} marine salt concentration and no silica. The cells were grown semi-continuously as airlift-cultures in a rectangular culture tank of 3 cm depth under illumination with either blue (465 ± 10 nm) or red (660 ± 10 nm) at light intensities of 24 and 40 µmol photons $\text{m}^{-2} \text{ s}^{-1}$, respectively. This specific intensity of irradiance with BL and RL resulted in a similar amount of photosynthetically absorbed radiation (Q_{Phar}) of 10 µmol photons $\text{m}^{-2} \text{ s}^{-1}$ (calculated as described in [47]). The illumination was provided by Flora LED-panels (CLF Plant Climatics, Wertingen, Germany) or by self-manufactured LED-panels with 14/10 h light/dark cycles. The cultures were diluted daily to achieve a chlorophyll (Chl) *a* concentration of 2 mg l^{-1} at 3 h into the light phase. All experiments were carried out with cultures which were acclimated at least 7 days to the described conditions.

Daily samples for the determination of cell numbers, pigment content and dry matter (DM) as well as FTIR (*fourier-transform infrared*) spectra were always taken 3 h after the start of the light phase. Light shift experiments were also started at this time point.

For the light shift experiments, the illumination of the pre-acclimated cultures was changed from RL to BL or vice versa. Samples for the determination of Chl *a* concentration, effective quantum efficiency (see below), cell numbers and FTIR spectra were taken every 2 h for a period of 10 h after the light quality change to evaluate fast changes during the acclimation process. Additionally, the daily development of these cell characteristics was measured in pre-acclimated RL and BL cultures for the same time points as a reference.

The long-term acclimation process was evaluated by sampling at days 1, 2, 3 and 6 after the light quality change. At these time points, additional measurements were carried out to determine the oxygen-based photosynthetic rates, non-photochemical quenching and diadinoxanthin (Ddx) concentrations.

Growth, pigmentation and photosynthetic parameters

Cell numbers were determined with a particle counter (Z2 Beckman Coulter GmbH, Krefeld, Germany). Dry matter determination was carried out by harvesting 30–50 ml of the culture by centrifugation (3000 *g*, 5 min). The pellets were washed twice with distilled water, frozen in liquid nitrogen, freeze-dried (Labconco FreeZone, ILMVAC GmbH, Ilmenau, Germany) and weighted. To calculate the dry matter per cell, the cell number of the sample was counted prior to the last centrifugation step.

The growth rate was calculated as gain per day based on Chl *a* (growth rate $[\text{d}^{-1}] = \text{Chl } a_{d1} \text{ Chl } a_{d0}^{-1} - 1$). Therefore, a doubling of the Chl *a* content would result in a growth rate of 1, corresponding to an increase of 100%.

For pigment analysis cells were harvested on glass-fiber filters and freeze-dried. Subsequently, the cells were broken with glass beads using a cell homogenizer (Braun, Melsungen, Germany) as explained in [6]. Chl *a* concentration was determined by extraction in 90% acetone according to [48]. Ddx concentration was determined by HPLC (*high pressure liquid chromatography*) using a Waters 600 MS system following the method described in [49].

Oxygen-based photosynthetic rates were measured with a Clark-type electrode (MI730, Microelectrodes Inc., Bedford, NH, USA) at light intensities between 0 and 1500 µmol photons $\text{m}^{-2} \text{ s}^{-1}$ as described in detail in [6]. Simultaneously, fluorescence was detected by a PAM (*pulse amplitude modulation*) fluorometer (PAM 101/103, Walz Effeltrich, Germany). The NPQ was calculated based on the fluorescence as described in [50] at the highest measured light intensity (~ 1500 µmol photons $\text{m}^{-2} \text{ s}^{-1}$).

A Handy PEA (Handy PEA with Liquid Phase Adapter, Hansatech Instruments LTD, UK) was used to record fast fluorescence induction kinetics of algal samples taken directly from the cultures. The so-called OJIP transients depict the course of the fluorescence rise from minimum (F_O) to maximum (F_M) levels when a sample is illuminated with saturating light. Different fluorescence levels become visible that are called J (at 2 ms), I (at 30 ms) and P (at 220 ms) levels by logarithmic scaling [51]. The normalized value of the J level (F_J; $F_J = F_{2 \text{ ms}} F_M^{-1}$) correlates with the redox state of the plastoquinone pool in the thylakoid membrane [38]. To record the OJIP transients of our cultures, light-adapted samples were directly illuminated with 3500 µmol photons $\text{m}^{-2} \text{ s}^{-1}$ for 10 seconds for fluorescence measurement. As the signal was recorded in pre-illuminated cells, all fluorescence designations are marked by a prime (e.g. F_M'). OJIP transients were recorded every 5 min for a period of 40 min after the light quality shift. After 1 h, the cultures were shifted back to their original light quality. Again, the OJIP transients were recorded every 5 min for a period of 40 min. The minimum and maximum values of the measured OJIP transients were also used for the calculation of the effective quantum yield of PSII ($\Phi_{PSII} = (F_M' - F_O') F_M'^{-1}$).

Macromolecular cell composition via FTIR spectroscopy

The macromolecular composition of a dried algae sample was determined via its FTIR spectrum [52]. FTIR spectroscopy is based on the absorption of chemical groups (e.g. $C=O$) at individual wave numbers in the mid-infrared range. As the absorption pattern of given molecules is distinct, this method is able to identify substances contained in a measured sample.

Accordingly it can be used to quantify the amount of carbohydrates, proteins and lipids [52] or other substances [53] in small algae cell samples. Since light driven changes in the macromolecular composition of the cells can be observed already after 30 min, FTIR can follow the distribution of the newly synthesized carbon or cell internal carbon recycling in the time range of hours. For this purpose, the cells were centrifuged twice at 5000 *g for 3 min and re-suspended in distilled water to wash and concentrate them. For each sample, 6 replicates with a volume of 2 µl were placed on a microtiter plate and dried at 40°C. The FTIR spectra were recorded in the range of 4000–700 cm^{-1} with a resolution of 4 cm^{-1} and 32 scans per spot (HTS-XT microtiter plate module and Vector 22 laser unit, Bruker Optics, Germany) [52]. The recorded spectra were processed by using a Blackman-Harris apodization and excluding the CO_2 absorption band, followed by a baseline correction (rubber band method with 64 points).

Spectra of laminarine (carbohydrate reference), glycerol tripalmitate (lipid reference) and bovine serum albumin (protein reference) were used as reference spectra to determine the abundance of carbohydrates, lipids and proteins in the recorded spectra by a PLS (partial least square) regression in the range of 1800–900 cm^{-1}. These three substance classes were assumed to represent 100% of the cellular dry matter.

The carbon (C) and nitrogen (N) content was determined with a CHN elemental analyzer (vario EL, Analytik Jena GmbH) using the dried samples after dry matter measurement.

Calculation of C-partitioning

The rates of net assimilated carbon were calculated based on the measured photosynthetic rates (based on oxygen evolution), dry weight and the cell composition (based on the FTIR spectroscopy results). The amount of C per dry matter and the required electrons per biomass for carbohydrates, proteins and lipids were included as specified in [54]. For a calculation example, see File S1.

Metabolomics

To get an overview of the metabolites of the core metabolic pathways and amino acid concentrations, samples were taken directly prior as well as 15 min, 30 min, 1 h and 24 h after the light quality shift. For each time point, three replicate samples were taken, extracted and analyzed. In addition, the experiments were independently repeated three times.

About 3*10^7 cells (~0.6 mg DM) were filtered on a glass fiber filter, washed with distilled water, frozen in liquid nitrogen, freeze-dried and stored at −80°C. The sampling procedure took 20–25 seconds until freezing. The extraction was carried out in PVC reaction tubes. Glass beads, 1 µg ^{13}C glucose (used as standard for quantification of all measured metabolites) and 2.4 ml methanol were added to the filter. The cells on the filter were homogenized in a Precellys Homogenisator (2×20 sec * 6500 *g). Distilled water (0.6 ml) was added after the extraction. The samples were centrifuged twice at 21000 *g for 1 min to remove the remaining filter pieces, glass beads and bigger cell pieces and frozen again at −80°C until analysis.

Amino acid concentrations from cell lysates were determined using a targeted metabolic approach with the AbsoluteIDQ p150 kit (BIOCRATES Life Sciences AG, Innsbruck, Austria) as described earlier [55]. Briefly, 30 µl of supernatants were prepared according the manufacturer's protocol. FIA-MS/MS analyses were carried out on an Agilent 1100 series binary HPLC system (Agilent Technologies, Waldbronn, Germany) coupled to an 4000 QTrap mass spectrometer (AB Sciex, Concord, Canada) equipped with a TurboIon spray source. Quantification was achieved by

positive and negative multiple reaction monitoring (MRM) detection in combination with the use of stable isotope-labeled and other internal standards. Data evaluation for quantification of metabolite concentrations was performed with the MetIQ software package.

For ion chromatography-tandem mass spectrometry (IC-MS/MS)-based analysis extracts (25 µl) were analyzed on an ICS-5000 (Thermo Fisher Scientific, Dreieich, Germany) coupled to an API 5500 QTrap (AB Sciex). Separation was achieved on an IonPac AS11-HC column (2×250 mm, Thermo Fisher Scientific) with an increasing potassium hydroxide gradient. MS analysis was performed in MRM mode using negative electrospray ionization and included organic acids, carbohydrates and nucleotides involved in central metabolite pathways.

Statistics

The statistical analysis was carried out using R 2.15.2 and RStudio 0.96.331 (R Core Team 2012). The Student's t-test was used for the comparison of two data sets. Data sets were considered paired if the samples were taken from the same culture in a time frame of less than 24 hours. The p-values are marked with * ($p \leq 0.05$), ** ($p \leq 0.01$) and *** ($p \leq 0.001$). The amount of replicates (n) for each experiment is specified in the figure legends.

Supporting Information

Table S1   Long-term acclimation to light quality shifts. The gross oxygen evolution, growth rates, maximum NPQ and Ddx pool were followed for 6 days after the RL to BL and the BL to RL shifts.

Table S2   Fast fluorescence induction kinetics. The changes of the J Level of fast fluorescence induction kinetics ($F_J' F_M'^{-1}$) were recorded during the light quality changes. after 60 min, the shift was reversed.

Table S3   Short-term acclimation to light quality shifts. The changes in the Chl a concentration and Φ_{PSII} of $P.$ $tricornutum$ cultures were recorded for 10 h after the light quality shift.

Table S4   Carbohydrate and protein levels. The changes in the carbohydrate and protein levels of $P.$ $tricornutum$ cultures were recorded for 10 h after the light quality shift and in the respective BL and RL pre-acclimated reference cultures.

Table S5   Carbon partitioning. The relative partitioning of C (carbon) into carbohydrates, proteins and lipids was calculated for the 2 h following the light quality changes.

Table S6   Relative metabolite concentrations in BL and RL pre-acclimated $P.$ $tricurnutum$ cultures (in relative counts * $cell^{-1}$). Significant differences were calculated by a Student's t-test.

Table S7   Amino acid concentrations in BL and RL pre-acclimated $P.$ $tricurnutum$ cultures (in 10^{-3} pmol * $cell^{-1}$). Significant differences were calculated by a Student's t-test.

File S1 The calculation of net C partitioning into biomass for carbohydrates, proteins and lipids. A calculation example is given at the right hand side.

Author Contributions

Conceived and designed the experiments: CW TJ AJ. Performed the experiments: AJ SB BSC. Analyzed the data: AJ BSC SB TJ. Contributed reagents/materials/analysis tools: MvB. Wrote the paper: AJ TJ CW.

References

1. Sims PA, Mann DG, Medlin LK (2006) Evolution of the diatoms: insights from fossil, biological and molecular data. Phycologia 45 (4): 361–402
2. Bowler C, Vardi A, Allen AE (2010) Oceanographic and Biogeochemical Insights from Diatom Genomes. Annual Review of Marine Science 2: 333–365
3. Nelson DM, Tréguer P, Brzezinski MA, Leynaert A, Quéguiner B (1995) Production and dissolution of biogenic silica in the ocean: Revised global estimates, comparison with regional data and relationship to biogenic sedimentation. Global biogeochemical cycle 9 (3): 359–372
4. Grouneva I, Jakob T, Wilhelm C, Goss R (2009) The regulation of xanthophyll cycle activity and of non-photochemical fluorescence quenching by two alternative electron flows in the diatoms *Phaeodactylum tricornutum* and *Cyclotella meneghiniana*. Biochimica et Biophysica Acta (BBA) - Bioenergetics 1787 (7): 929–938
5. Depauw FA, Rogato A, Ribera d'Alcala M, Falciatore A (2012) Exploring the molecular basis of responses to light in marine diatoms. Journal of Experimental Botany 63 (4): 1575–1591
6. Wagner H, Jakob T, Wilhelm C (2006) Balancing the Energy Flow from Captured Light to Biomass Under Fluctuating Light Conditions. New Phytologist 169 (1): 95–108
7. Kirk JTO (1994) Light and photosynthesis in aquatic ecosystems. Cambridge: Cambridge University Press. p. 131
8. Furukawa T, Watanabe M, Shihira-Ishikawa I (1998) Green-and blue-light-mediated chloroplast migration in the centric diatom *Pleurosira laevis*. Protoplasma 203: 214–220
9. Shikata T, Iseki M, Matsunaga S, Higashi S, Kamei Y et al. (2011) Blue and Red Light-Induced Germination of Resting Spores in the Red-Tide Diatom *Leptocylindrus danicus*. Photochemistry and Photobiology 87: 590–597
10. Holdsworth ES (1985) Effect of growth factors and light quality on the growth, pigmentation and photosynthesis of two diatoms, *Thalassiosira gravida* and *Phaeodactylum tricornutum*. Marine Biology 86: 253–262
11. Tremblin G, Cannuel R, Mouget J, Rech M, Robert J (2000) Change in light quality due to a blue-green pigment, marennine, released in oyster-ponds: effect on growth and photosynthesis in two diatoms, *Haslea ostrearia* and *Skeletonema costatum*. Journal of Applied Phycology 12 (6): 557–566
12. Mercado JM, del Pilar Sánchez-Saavedra M, Correa-Reyes G, Lubián L, Montero O et al. (2003) Blue light effect on growth, light absorption characteristics and photosynthesis of five benthic diatom strains. Aquatic Botany 78 (3): 265–277
13. Mouget J, Rosa P, Tremblin G (2004) Acclimation of *Haslea ostrearia* to light of different spectral qualities – confirmation of 'chromatic adaptation' in diatoms. Journal of Photochemistry and Photobiology B: Biology 75: 1–11
14. Wu H, Cockshutt AM, McCarthy A, Campbell DA (2011) Distinctive Photosystem II Photoinactivation and Protein Dynamics in Marine Diatoms. Plant Physiology 156 (4): 2184–2195
15. Schellenberger Costa B, Jungandreas A, Jakob T, Weisheit W, Mittag M, et al. (2013) Blue light is essential for high light acclimation and photoprotection in the diatom *Phaeodactylum tricornutum*. Journal of Experimental Botany 64 (2): 483–493
16. Bailleul B, Rogato A, de Martino A, Coesel S, Cardol P, et al. (2010) An atypical member of the light-harvesting complex stress-related protein family modulates diatom responses to light. Proceedings of the National Academy of Sciences 107 (42): 18214–18219
17. Nymark M, Valle KC, Brembu T, Hancke K, Winge P, et al. (2009) An Integrated Analysis of Molecular Acclimation to High Light in the Marine Diatom *Phaeodactylum tricornutum*. PLoS ONE 4 (11): e7743
18. Bellafiore S, Barneche F, Peltier G, Rochaix J (2005) State Transitions and Light Adaptation Require Chloroplast Thylakoid Protein Kinase STN7. Nature 433 (7028): 892–895
19. Pfannschmidt T, Yang C (2012) The Hidden Function of Photosynthesis: A Sensing System for Environmental Conditions That Regulates Plant Acclimation Responses. Protoplasma 249 (2): 125–136
20. Li Z, Wakao S, Fischer BB, Niyogi KK (2009) Sensing and Responding to Excess Light. Annual Review of Plant Biology 60 (1): 239–260
21. Lepetit B, Sturm S, Rogato A, Gruber A, Sachse M, et al. (2013) High Light Acclimation in the Secondary Plastids Containing Diatom *Phaeodactylum tricornutum* is Triggered by the Redox State of the Plastoquinone Pool. Plant Physiology 161 (2): 853–865
22. Dijkman NA, Kroon BMA (2002) Indications for chlororespiration in relation to light regime in the marine diatom *Thalassiosira weissflogii*. Journal of Photochemistry and Photobiology B: Biology 66 (3): 179–187
23. Jakob T, Goss R, Wilhelm C (1999) Activation of Diadinoxanthin De-Epoxidase Due to a Chlororespiratory Proton Gradient in the Dark in the Diatom *Phaeodactylum tricornutum*. Plant Biology 1 (1): 76–82
24. Armbrust EV, Berges JA, Bowler C, Green BR, Martinez D, et al. (2004) The Genome of the Diatom *Thalassiosira pseudonana*: Ecology, Evolution, and Metabolism. Science 306: 79–86
25. Bowler C, Allen AE, Badger JH, Grimwood J, Jabbari K, et al. (2008) The *Phaeodactylum* Genome Reveals the Evolutionary History of Diatom Genomes. Nature 456: 239–244
26. Chaves I, Pokorny R, Byrdin M, Hoang N, Ritz T, et al. (2011) The Cryptochromes: Blue Light Photoreceptors in Plants and Animals. Annual Review of Plant Biology 62, Nr. (1): 335–364
27. Takahashi F, Yamagata D, Ishikawa M, Fukamatsu Y, Ogura Y, et al. (2007) AUREOCHROME, a photoreceptor required for photomorphogenesis in stramenopiles. Proceedings of the National Academy of Sciences 104 (49): 19625–19630
28. Ishikawa M, Takahashi F, Nozaki H, Nagasato C, Motomura T, et al. (2009) Distribution and phylogeny of the blue light receptors aureochromes in eukaryotes. Planta 230 (3): 543–552
29. Suetsugu N, Wada M (2013) Evolution of Three LOV Blue Light Receptor Families in Green Plants and Photosynthetic Stramenopiles: Phototropin, ZTL/FKF1/LKP2 and Aureochrome. Plant and Cell Physiology 54 (1): 8–23
30. Coesel S, Mangogna M, Ishikawa T, Heijde M, Rogato A, et al. (2009) Diatom PtCPF1 is a new cryptochrome/photolyase family member with DNA repair and transcription regulation activity. EMBO reports 10 (6): 655–661
31. Schellenberger Costa B, Sachse M, Jungandreas A, Rio Bartulos C, Gruber A, et al. (2013) Aureochrome 1a Is Involved in the Photoacclimation of the Diatom *Phaeodactylum tricornutum*. PLoS ONE 8 (9): e74451
32. Huysman MJJ, Fortunato AE, Matthijs M, Schellenberger Costa B, Vanderhaeghen R et al. (2013) AUREOCHROME1a-Mediated Induction of the Diatom-Specific Cyclin dsCYC2 Controls the Onset of Cell Division in Diatoms (*Phaeodactylum tricornutum*). The Plant Cell 25 (1): 215–228
33. Kowallik W (1982) Blue light effects on respiration. Annual Review of Plant Physiology 33 (1): 51–72
34. del Pilar Sánchez-Saavedra M, Voltolina D (1994) The chemical composition of *Chaetoceros* sp. (Bacillariophyceae) under different light conditions. Comparative Biochemistry and Physiology Part B: Comparative Biochemistry 107B (1): 39–44
35. Azuara MP, Aparicio PJ (1983) In Vivo Blue-Light Activation of *Chlamydomonas reinhardii* Nitrate Reductase. Plant Physiology 71 (2): 286–290
36. Münzner P, Voigt J (1992) Blue light regulation of cell division in *Chlamydomonas reinhardtii*. Plant Physiology 99 (4): 1370–1375
37. Eberhard S, Finazzi F, Wollman F (2008) The Dynamics of Photosynthesis. Annual Review of Genetics 42 (1): 463–515
38. Tóth SZ, Schansker G, Strasser RJ (2007) A non-invasive assay of the plastoquinone pool redox state based on the OJIP-transient. Photosynthesis Research 93 (1–3): 193–203
39. Bromke M (2013) Amino Acid Biosynthesis Pathways in Diatoms. Metabolites 3: 294–311
40. Smith SR, Abbriano RM, Hildebrand M (2012) Comparative analysis of diatom genomes reveals substantial differences in the organization of carbon partitioning pathways. Algal Research 1 (1): 2–16
41. Cruz S, Goss R, Wilhelm C, Leegood R, Horton P, et al. (2011) Impact of chlororespiration on non-photochemical quenching of chlorophyll fluorescence and on the regulation of the diadinoxanthin cycle in the diatom *Thalassiosira pseudonana*. Journal of Experimental Botany 62 (2): 509–519
42. Alric J (2010) Cyclic electron flow around photosystem I in unicellular green algae. Photosynthesis Research 106 (1–2): 47–56
43. Lomas MW, Glibert PM (1999) Temperature regulation of nitrate uptake: A novel hypothesis about nitrate uptake and reduction in cool-water diatoms. Limnology and Oceanography 44 (3): 556–572
44. Vergara JJ, Berges JA, Falkowski PG (1998) Diel Periodicity of Nitrate Reductase Activity and Protein Levels in the Marine Diatom *Thalassiosira weissflogii* (Bacillariophyceae). Journal of Phycology 34 (6): 952–961
45. Voskresenkaya NP (1972) Blue Light and Carbon Metabolism. Annual Review of Plant Physiology 23: 219–234
46. Guillard RR, Lorenzen CJ (1972) Yellow-green algae with chlorophyllide c. Journal of Phycology 8 (1), 10–14
47. Gilbert M, Wilhelm C, Richter M (2000) Bio-optical modeling of oxygen evolution using in vivo fluorescence: Comparison of measured and calculated photosynthesis/irradiance (P-I) curves in four representative phytoplankton species. Journal of Plant Physiology 157 (3): 307–314
48. Jeffrey SW, Humphrey GF (1975) New spectrophotometric equations for determining chlorophylls a, b, c1 and c2 in higher plants, algae and natural phytoplankton. Biochemie und Physiologie der Pflanzen 167: 191–194
49. Su W, Jakob T, Wilhelm C (2012) The Impact of Nonphotochemical Quenching of Fluorescence on the Photon Balance in Diatoms Under Dynamic Light Conditions. Journal of Phycology 48 (2): 336–346

50. Bilger W, Björkman O (1990) Role of the Xanthophyll Cycle in Photoprotection Elucidated by Measurements of Light-induced Absorbance Changes, Fluorescence and Photosynthesis in Leaves of *Hedera canariensis*. Photosynthesis Research 25 (3): 173–185

51. Stirbet A, Govindjee (2011) On the relation between the Kautsky effect (chlorophyll a fluorescence induction) and Photosystem II: Basics and applications of the OJIP fluorescence transient. Journal of Photochemistry and Photobiology B: Biology 104 (1–2): 236–257

52. Wagner H, Liu Z, Langner U, Stehfest K, Wilhelm C (2010) The use of FTIR spectroscopy to assess quantitative changes in the biochemical composition of microalgae. Journal of Biophotonics 3 (8–9): 557–566

53. Jungandreas A, Wagner H, Wilhelm C (2012) Simultaneous Measurement of the Silicon Content and Physiological Parameters by FTIR Spectroscopy in Diatoms with Siliceous Cell Walls. Plant and Cell Physiology 53 (12): 2153–2162

54. Kroon BMA, Thoms S (2006) From Electron to Biomass: A Mechanistic Model to Describe Phytoplankton Photosynthesis and Steady-State Growth Rates. Journal of Phycology 42 (3): 593–609

55. Röhmisch-Margl W, Prehn C, Bogumil R, Röhring C, Suhre K, et al. (2011) Procedure for tissue sample preparation and metabolite extraction for high-throughput targeted metabolomics. Metabolomics 8: 133–142

9

Proteomic Analysis Reveals Differences in Tolerance to Acid Rain in Two Broad-Leaf Tree Species, *Liquidambar formosana* and *Schima superba*

Juan Chen[1,9], Wen-Jun Hu[1,9], Chao Wang[2], Ting-Wu Liu[3,1], Annie Chalifour[4], Juan Chen[1], Zhi-Jun Shen[1], Xiang Liu[1], Wen-Hua Wang[1], Hai-Lei Zheng[1]*

1 Key Laboratory of the Coastal and Wetland Ecosystems, Ministry of Education, College of the Environment and Ecology, Xiamen University, Xiamen, Fujian, China, 2 Institute of Urban and Environment, Chinese Academy of Sciences, Xiamen, P.R. China, 3 Department of Biology, Huaiyin Normal University, Huaian, Jiangsu, P.R. China, 4 Department of Biology and Chemistry, City University of Hong Kong, Kowloon, Hong Kong, SAR, China

Abstract

Acid rain (AR) is a serious environmental issue inducing harmful impacts on plant growth and development. It has been reported that *Liquidambar formosana*, considered as an AR-sensitive tree species, was largely injured by AR, compared with *Schima superba*, an AR-tolerant tree species. To clarify the different responses of these two species to AR, a comparative proteomic analysis was conducted in this study. More than 1000 protein spots were reproducibly detected on two-dimensional electrophoresis gels. Among them, 74 protein spots from *L. formosana* gels and 34 protein spots from *S. superba* gels showed significant changes in their abundances under AR stress. In both *L. formosana* and *S. superba*, the majority proteins with more than 2 fold changes were involved in photosynthesis and energy production, followed by material metabolism, stress and defense, transcription, post-translational and modification, and signal transduction. In contrast with *L. formosana*, no hormone response-related protein was found in *S. superba*. Moreover, the changes of proteins involved in photosynthesis, starch synthesis, and translation were distinctly different between *L. formosana* and *S. superba*. Protein expression analysis of three proteins (ribulose-1,5-bisphosphate carboxylase/oxygenase large subunit, ascorbate peroxidase and glutathione-S-transferase) by Western blot was well correlated with the results of proteomics. In conclusion, our study provides new insights into AR stress responses in woody plants and clarifies the differences in strategies to cope with AR between *L. formosana* and *S. superba*.

Editor: Tai Wang, Institute of Botany, Chinese Academy of Sciences, China

Funding: This study was financially supported by the Natural Science Foundation of China (NSFC) (31300505, 30930076, 31260057, 30770192, 30670317), China Postdoctoral Science Foundation (2012M521278), the Foundation of the Chinese Ministry of Education (20070384033, 209084), the Program for New Century Excellent Talents in Xiamen University (NCETXMU X07115), and the Scholarship Award for Excellent Doctoral Student granted by the Chinese Ministry of Education. The funders had no role in study design, data collection and analysis, decision to publish, or preparation of the manuscript.

Competing Interests: The authors have declared that no competing interests exist.

* Email: zhenghl@xmu.edu.cn

9 These authors contributed equally to this work.

Introduction

Acid rain (AR) emerged as a serious environmental issue as a consequence of the increasing industrial activities throughout the world [1]. Forty percent of the territory in China is seriously affected by AR since the late 1970s, especially in southern China [2]. The harmful impacts of AR on plants are observed in a wide array of biological processes. AR decreases seed germination [3], strips the protective wax from leaves [4], induces visible injury symptoms [1], disturbs plant nitrogen metabolism [5]. AR also decreases chlorophyll content and photosynthetic efficiency [3,6], increases reactive oxygen species (ROS) production [7], accelerates the leaching of nutrients from plant foliage [8], which further inhibits tree radial growth, vertical growth and total tree biomass [4,9].

Liquidambar formosana and *Schima superba* are both dominant broad-leaf tree species and are distributed over large surface areas in the forest of southern China [10]. Some field observations and laboratory experiments reported that, when compared with *S. superba*, *L. formosana* was largely injured by AR during the past decade, which had negative impacts on forest ecosystem [9]. Our previous study also found that AR more easily affected some physiological parameters in *L. formosana* seedlings than *S. superba*'s seedlings, e.g., seed germination, seedling growth, photosynthesis, antioxidant system, etc. [3,11]. Thus, *L. formosana* is considered as an AR-sensitive species, while *S. superba* is an AR-tolerant species. Although the differential responses of *L. formosana* and *S. superba* to AR have been analyzed at the morphological and physiological level, a comprehensive elucidation of the molecular mechanisms underlying the different strategies to cope with AR between two tree species is still needed.

Proteomics is a powerful tool for providing new insights into complete proteomes at the organ, tissue and cell levels [12]. A number of proteomic analyses help us understand the molecular mechanisms of plants in responses to various environmental stresses including salinity [13], cold [14], heavy metal [15], etc. In our previous work, a wide array of proteins related to AR-resistance has been identified by comparative proteomic analysis in a model plant, *Arabidopsis thaliana* [5,16]. However, little information is available in proteome analysis for tree species subjected to AR stress, and a comparison between AR-sensitive and AR-resistant broad-leaf tree species has not been fully conducted at the proteome level.

In the present study, we initiated a comparative proteomic study to systematically investigate the changes in protein profile in two broad-leaf tree species, *L. formosana* and *S. superba*, that are different sensitive to AR tolerance, when submitted to simulated AR (pH 3.0) treatment for one month. Based on two-dimensional electrophoresis (2-DE) and mass spectrometry (MS) analysis, a comprehensive inventory of proteins regulated by AR was established in the two tree species. The overall objectives of this study are (1) to provide valuable insights into AR stress responses in woody plants; (2) to clarify the differences in strategies to cope with AR between *L. formosana* and *S. superba*.

Materials and Methods

Plant materials and treatments

Seeds of *L. formosana* and *S. superba* were purchased from Tree Seed Centre of Shuyang County in Jiangsu Province, China. The seeds have been mixed together when they were collected from independent individuals and families. Seeds were surface-sterilized with 0.5% hypochlorite for 30 min, then washed thoroughly with distilled water. Then the seeds were germinated in a soil/vermiculite (1:1) mixture in an environmentally controlled growth chamber. For each species, three weeks old healthy seedlings with similar size were randomly transplanted into individual pots, each with a dimension of 24 cm (open top) × 13 cm (height) × 15 cm (flat bottom), and filled with soil/vermiculite (1:1) mixture. Fifteen seedlings were planted in one pot. All seedlings were cultivated in the same controlled growth chamber with a daily temperature regime of 28/25°C (day/night), relative humidity of 60–70% and a 12-h photoperiod at 210 μmol m^{-2} s^{-1} photosynthetically active radiation (PAR). Three months later, the seedlings were divided into control group (CK) and simulated AR treatment group (AR). Each group had at least three replicates. The control group and simulated AR treatment group were sprayed once per day with the control (pH 5.6) solution and AR (pH 3.0) solution, respectively. The ion compositions of the control solution was adapted from Fan and Wang [4], AR solution was made from control solution and the pH was adjusted by adding a mixture of H_2SO_4 and HNO_3 in the ratio of 5:1. The final concentration of H_2SO_4 and HNO_3 were 0.45 and 0.09 mM, respectively, which represents the average ion composition of rainfall in southern China [4]. After one month of treatment, a portion of fresh leaves was used for measuring some physiological parameters such as necrosis percentage, chlorophyll content, net photosynthetic rate, H_2O_2 content and so on. The remaining leaves were stored at −80°C for proteomic and Western blot analysis.

Necrosis percentage and chlorophyll content measurements

At least 30 fully expanded leaves of each species were randomly selected from control and AR treatment groups and photographed using a digital camera. The necrosis percentage was calculated as described previously [16]. Chlorophyll was extracted from approximately 0.1 g fresh leaf slices directly into 10 ml ice-cold

acetone (80%, v/v). The chlorophyll content (mg g^{-1} fresh weight (FW)) was determined as described previously [17].

Net photosynthetic rate and chlorophyll fluorescence measurements

Three seedlings per species were randomly chosen from different pots and at least two fully emerged leaves per plant were selected for net photosynthetic rate (Pn) and chlorophyll fluorescence measurements. Pn was performed with a portable photosynthesis system (LI-6400, Li-Cor Inc., Lincoln, Nebraska, USA), as described previously [18]. According to the method of Liu et al [16], leaf chlorophyll fluorescence was measured using a pluse-amplitude-modulation fluorometer (PAM-2100, Heinz Walz, Effeltrich, Germany).

Measurement of proline, malondialdehyde (MDA) and ROS production

Proline content was measured according to the method of Jiang et al [19]. The level of lipid peroxidation was measured by estimating MDA content using thiobarbituric acid (TBA) reaction [20]. Superoxide radical ($O_2^{\cdot-}$) and H_2O_2 content was measured following the method of Chen et al [11].

Protein extraction

Protein extraction was performed using phenol-based protocol described by Liu et al [5], with slight modifications. Briefly, frozen leaves (1.0 g) were ground with a mortar and pestle with liquid nitrogen, the ground powder was homogenized in pre-cooled extraction buffer (20 mM Tris-HCl, pH 7.5, 250 mM sucrose, 10 mM ethylene diamine tetraacetic acid (EDTA), 1 mM phenylmethyl-sulfonyl fluoride, 1 mM 1,4-dithiothreitol (DTT) and 1% Triton X-100) on ice. Then an equal volume of ice-cold Tris-HCl (pH 7.5) saturated phenol was added and the mixture was centrifuged (15,000 g, 4°C) for 15 min. The phenol phase was collected and proteins were precipitated with ammonium acetate in methanol for 10 h at −20°C. After centrifugation, the supernatant was discarded and the pellet was washed for three times using cold acetone containing 10 mM DTT. The washed protein pellets were dissolved in a lysis buffer (8 M urea, 2 M thiourea, 4% (w/v) 3-[(3-cholamidopropyl)dimethylammonio]-1-propane sulfonate (CHAPS), 1% (w/v) DTT and 0.5% (w/v) IPG buffer pH 4–7) at room temperature. The total protein concentration of the lysates was determined using a Bio-Rad protein assay reagent (Bio-Rad, Hercules, CA, USA).

Two-dimensional electrophoresis, image and data analysis

Two-dimensional electrophoresis was conducted according to the methods of Bai et al [13] and Hu et al [21]. Isoelectric focusing (IEF) was done using an Ettan IPGphor system (GE Healthcare) PROTEAN electrophoresis system and immobilized IPG dry gel strips with a linear pH range (18 cm long, pH 4–7 linear) (GE Healthcare Amersham Bioscience, Little Chalfont, UK). Protein samples (800 μg) were loaded during the rehydration step at room temperature for 12 h. IEF was performed at 300 voltage (V), 500 V and 1,000 V for 1 h, a gradient to 8,000 V over 4 h, and kept at 8,000 V for a total of 80,000 volt-hours (Vh) at 20°C. Subsequently, focused strips were equilibrated in equilibration buffer as described by Yang et al [12]. For the second dimension, proteins were separated on 15% SDS polyacrylamide gels. Proteins spots were detected by staining the gels with Coomassie Brilliant Blue R-250. The 2-DE gels were scanned with a scanner (Uniscan M3600, China) and the gel images were analyzed with PDQuest software (Version 8.01, Bio-Rad, Hercules, CA), on the basis of their relative volume. Only those protein spots with significant (more than 2-fold change) and reproducible changes in three replicates were selected for next MS analysis.

In-gel digestion, protein identification and classification analysis

The protein spots, which were differentially displayed under AR treatment, were excised from the preparative 2-D gels and digested by trypsin. After digestion, the peptide solution was collected and peptide mass fingerprint (PMF) was acquired using matrix-assisted laser desorption/ionization-time-of-flight mass spectrometry (MALDI-TOF MS) analysis (ReFlexTM III, Bruker, Bremen, Germany) as described previously [16]. The PMF spectra were used in online searches combined with the Mascot program search engine (http://www.matrixscience.com) and National Center for Biotechnology Information (NCBI) protein database (http://www.ncbi.nlm.nih.gov) (NCBInr, 17751536 entries, downloaded on April 17, 2012). PMF search parameters were set up as described previously [22]. Proteins with a MOWSE score >73 were considered as positive identifications. The identified proteins were searched with against the UniProt (http://www.uniprot.org) and/or NCBI protein database (http://www.ncbi.nlm.nih.gov) for updated annotation and homologous proteins identification. Afterwards, the successfully identified proteins were further classified using Functional Catalogue software (http://mips.gsf.de/projects/funcat).

Western blot analysis

Western blot analysis was performed as described previously [11]. Total proteins (40 μg) extracted from L. formosana and S. superba leaves were separated by 12% w/v standard sodium dodecyl sulfate polyacrylamide gel electrophoresis and then electrophoretically blotted to polyvinylidene difluoride membrane for 50 min. The membranes were blocked over-night with Western Blocking Buffer (TIANGEN, China). Protein blots were probed with primary antibodies of Rubisco large subunit (RuBISCO LSU) (AS03037-200, Agrisera, Sweden), ascorbate peroxidase (APX) (AS08368, Agrisera, Sweden) and glutathione-S-transferase (GST) (AS09479, Agrisera, Sweden), at dilution of 1:5000, 1:2000 and 1:1000, respectively, for 4 h at room temperature with agitation. Next, the membranes were washed in phosphate buffered saline with Tween-20 solution (PBST) solution (50 mM Tris-HCl, pH 8.0, 150 mM NaCl, 0.05% Tween 20, v/v) three times and incubated with anti-rabbit IgG horseradish peroxidase (HRP) conjugated to alkaline phosphatase (Abcam, U.K., 1:5000 dilution) for 1 h at room temperature to detect primary antibodies. β-actin (1:5000, Santa Cruz, California, USA) was used as an internal control. After several washes with PBST solution, membranes were incubated in an enhanced chemiluminescence (ECL) substrate detection solution (TIANGEN, China) according to the manufacturer's instructions. Images of protein blots were obtained using a CCD imager (FluorSMax Bio-Rad, Hercules, CA, USA). The optical density values of the protein signals were quantified using the Quantity One software (Bio-Rad, Hercules, CA, USA).

Statistical analysis

For physiological measurements, at lease four independent repetitions were used. Values in figures and tables were expressed as means ± se. The statistical significance of the data was analyzed using a univariate analysis of variance (One-way ANOVA, Duncan's multiple range test, $p < 0.05$) with the SPSS 20.0 package (SPSS, Chicago, Illionis USA). For proteomic experiment, protein samples for 2-DE gel image analysis were extracted from three independent seedlings grown in three different pots in the same growth chamber. Thus, for each species, three independent biological replicates were performed in 2-DE gel image analysis. Statistic analysis for protein spot on 2-DE gels was performed using Student's t-test ($p < 0.05$) provided by PDQuest software as mentioned earlier.

Results

Morphological and physiological responses of L. formosana and S. superba to AR

As shown in Fig. 1A and B, remarkable yellowing symptom and significant necrosis emerge in L. formosana leaves after one-month exposure to AR. There was an important decrease in total chlorophyll content in AR-treated L. formosana leaves, however, no statistically significant change was found in S. superba (Fig. 1C). Pn and Fv/Fm in AR-treated L. formosana seedlings were remarkably inhibited, whereas AR slightly decreased Pn and Fv/Fm in S. superba (Fig. 1D and E). After AR treatment, proline content in L. formosana and S. superba increased by 76.0% and 19.7%, respectively, compared with the control (Fig. 2A). MDA contents in AR-treated L. formosana and S. superba increased by 89.4% and 44.8%, respectively (Fig. 2B). As shown in Fig. 2C and D, the levels of H_2O_2 and $O_2^{\cdot-}$ in both L. formosana and S. superba were significantly stimulated by AR. In particular, compared with the control, H_2O_2 and $O_2^{\cdot-}$ content increased by 83.3% and 67.8% in L. formosana, and by 38.4% and 44.7% in S. superba, respectively (Fig. 2C and D).

Identification of AR-responsive proteins

To investigate the differentially expressed proteins in L. formosana and S. superba exposed to AR treatment, comparative proteomic analysis was performed on Coomassie-stained 2-DE maps shown in Fig. 3. Over 1000 protein spots reproducibly separated and matched between control and AR gels, 74 protein spots in L. formosana, and only 34 protein spots in S. superba had at least a 2-fold greater abundance in either AR or control (Fig. 3). The identified proteins in L. formosana and S. superba by MALDI-TOF MS analysis are presented in Tables 1 and 2, respectively. Because the complete annotated sequences of L. formosana and S. superba genomes are not yet available, all identified proteins were functionally classified by UniProt and NCBI databases according to their homology with other proteins. Functional annotations in databases existed for the majority of the protein spots, while 12 proteins (spots L63–L74) in L. formosana and 6 proteins (spots S29–S34) in S. superba were annotated as predicted or unknown proteins (Tables 1 and 2).

Of the 74 protein spots identified in L. formosana, the abundances of 53 proteins were increased and those of 21 proteins were decreased in response to AR (Table 1). In S. superba, 21 proteins were increased in their abundances and 11 proteins were decreased under AR stress (Table 2). Among these affected proteins, phosphoglycerate kinase (PGK, spot L32, L8, S6) and ATP synthase CF1 alpha subunit (spot L22, S11), were decreased in their abundances in both L. formosana and S. superba (Tables 1 and 2). Remarkably, after AR treatment, abundance of maturase K was increased in L. formosana, but decreased in S. superba (Tables 1 and 2). Further analysis on the results revealed that some proteins were represented by more than one spot. These proteins included phosphoglycerate kinase (PGK, spot L32, L8), ATP synthase beta subunits (spot L22, L26) and maturase K (spot L57–59) in L. formosana and ribulose-1,5-bisphosphate

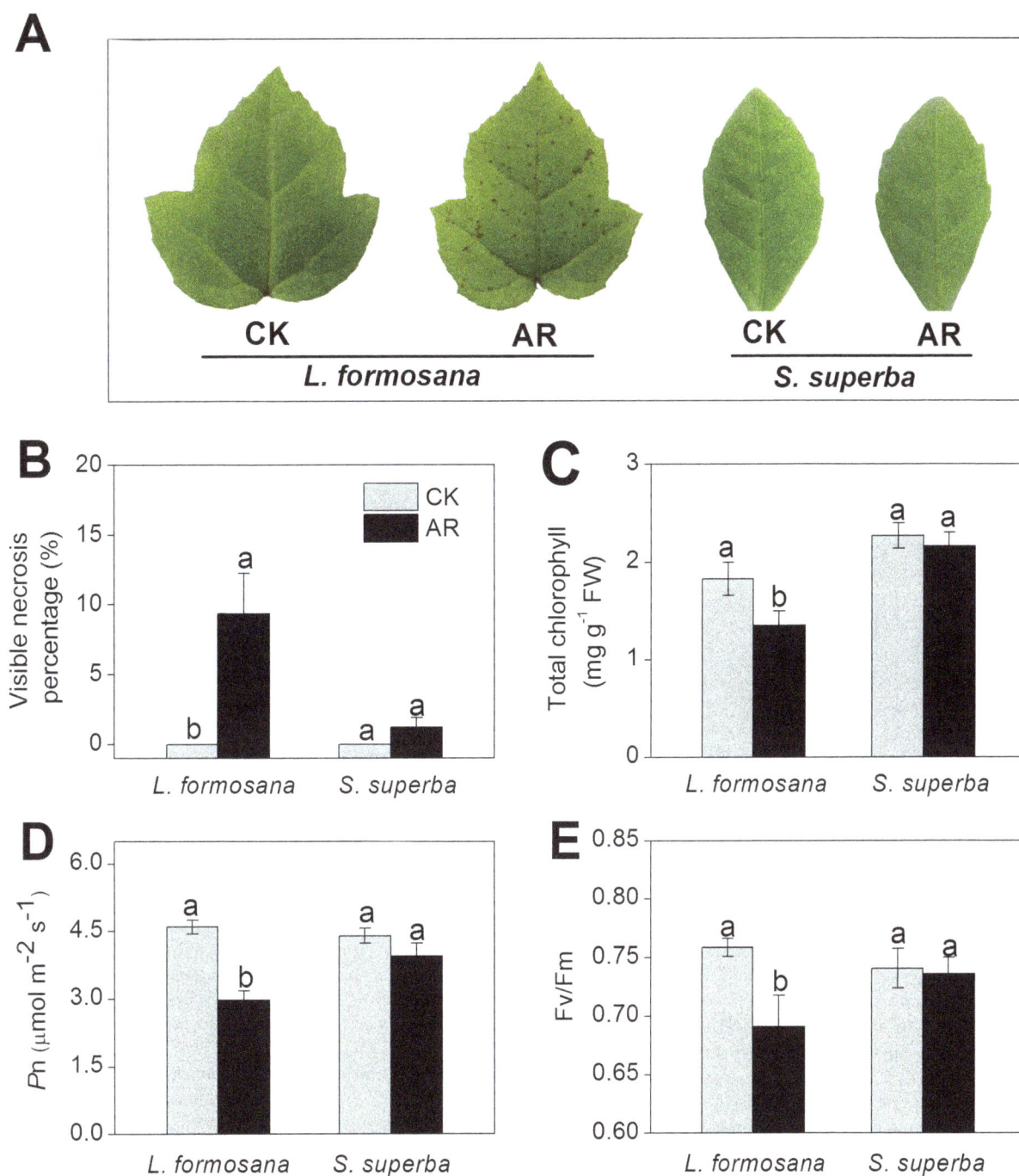

Figure 1. Effects of one-month AR on morphology and photosynthesis of *L. formosana* and *S. superba*. The pH of AR solution was adjusted to 3.0 by adding a mixture of H_2SO_4 and HNO_3 in the ratio of 5:1. The final concentration of H_2SO_4 and HNO_3 were 0.45 and 0.09 mM, respectively. (A) Leaf injury phenotype. (B) Leaf necrosis percentage. (C) Total chlorophyll content. (D) Net photosynthetic rate (Pn). (E) Quantum efficiency of open PSII centers in a dark-adapted state (Fv/Fm). Columns labeled with different letters indicate significant differences at $p < 0.05$.

carboxylase/oxygenase (Rubisco) large subunit (spot S9, S13, S14) in *S. superba*. The multiple spots might represent isoforms or different post-translation modification of individual proteins [13].

Functional classification of AR-responsive proteins

In order to obtain annotation of AR-responsive protein, all identified proteins were further classified according to their biological function and cellular component categories in UniProt (http://www.uniprot.org) and/or NCBI protein database (http://www.ncbi.nlm.nih.gov). AR-responsive proteins were found to be involved in a wide range of biological processes. After AR treatment, with the exception of photosynthesis and energy production related proteins, the abundance of most proteins were decreased in both *L. formosana* and *S. superba* (Fig. 4). As shown in Fig. 5, a higher percentage of proteins were involved in photosynthesis and energy production, which accounted for

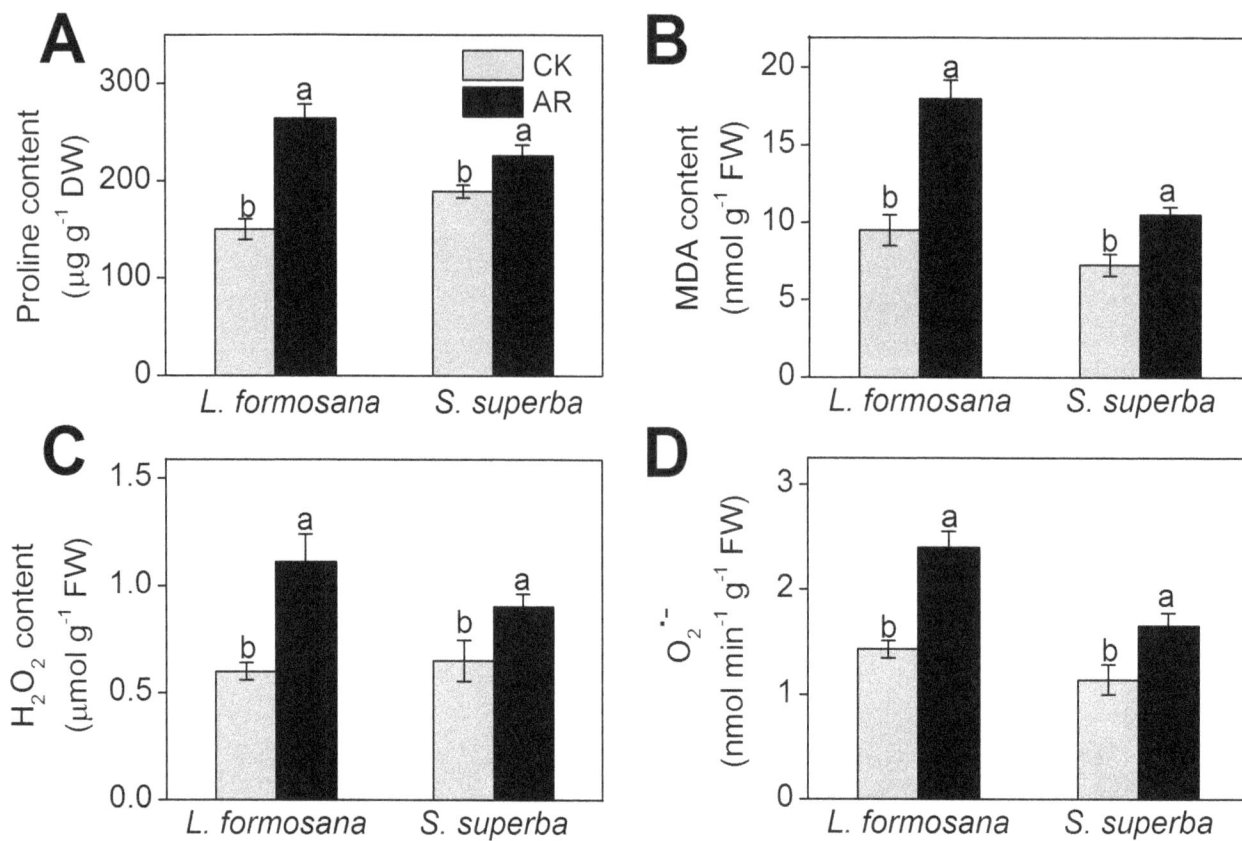

Figure 2. Changes in proline (A), MDA (B), H$_2$O$_2$ (C) and O$_2$$^{-}$ (D) content after AR treatment. Columns labeled with different letters indicate significant differences at $p<0.05$.

24.3% and 29.4% of the total proteins in *L. formosana* and *S. superba*, respectively. The following groups were proteins involved in material metabolism, stress and defense, transcription, post-translational and modification, and signal transduction. In opposition to *L. formosana*, no protein related to hormone response was found in *S. superba*.

Protein expression analysis by Western blot

Our above proteomic results revealed that the abundance of Rubisco was decreased (spot L31, L34), while APX (spot L40) and GST (spot L41) were increased in *L. formosana* under AR treatment (Table 1). As shown in Fig. 6A and B, compared with the control, the protein expression of Rubisco large subunit analyzed by Western blot was significantly decreased in AR-treated *L. formosana*. In contrast, the protein expression level of APX and GST increased 1.3-fold and 1.6-fold, respectively, compared to control, in *L. formosana* (Fig. 6C and D). No significant change in the expression of Rubisco large subunit, APX and GST was observed in *S. superba* (Fig. 6).

Discussion

AR has negative effects on plant growth and development [1,6]. Neves et al [23] found that simulated AR (pH 3.1) caused chlorosis and necrosis in leaves and led to chlorophyll loss and photosynthetic depression in *Eugenia uniflora*. Moreover, chlorophyll content and photosynthesis in *L. formosana* were also remarkably suppressed by AR treatment in this study (Fig. 1). Compared with *S. superba*, reductions on chlorophyll content and

photosynthetic ability by AR were more obvious in *L. formosana* (Fig. 1), which is consistent with the results of previous studies [9,11]. Besides chlorophyll content and photosynthesis, proline content, MDA content and ROS (e.g. H$_2$O$_2$ and O$_2$$^{-}$) production are commonly used as biochemical markers to monitor the damage level in plants under environmental stress [24]. In this study, AR increased proline content, MDA content and ROS production in both *L. formosana* and *S. superba* (Fig. 2), which were consistent with the results obtained by Chen et al [11]. However, the increase in these physiological changes caused by AR was less pronounced in *S. superba* than those in *L. formosana* (Fig. 1 and 2), suggesting that *S. superba*, a tolerant species, had less cell damage than *L. formosana*, a sensitive species.

To further reveal the different strategies to cope with AR between the two species, 74 protein spots in *L. formosana* and 34 protein spots in *S. superba* caused by AR were identified by proteomic analysis in this study. Interestingly, similar results were also reported in previous studies, which reported more changes in protein abundance in sensitive species, *Arabidopsis thaliana*, than in tolerant species, *Thellungiella halophila*, under salt stress [25]. Since *L. formosana* had much higher changes in its protein profile, our results proved that this species is more sensitive to AR than *S. superba*.

Photosynthesis and energy production-related proteins

Photosynthesis is an essential metabolic process of plants and is vulnerable to environmental stress. It is well known that AR can remarkably reduce photosynthesis [3,23]. In this study, two light reaction-related proteins, including chlorophyllide a oxygenase

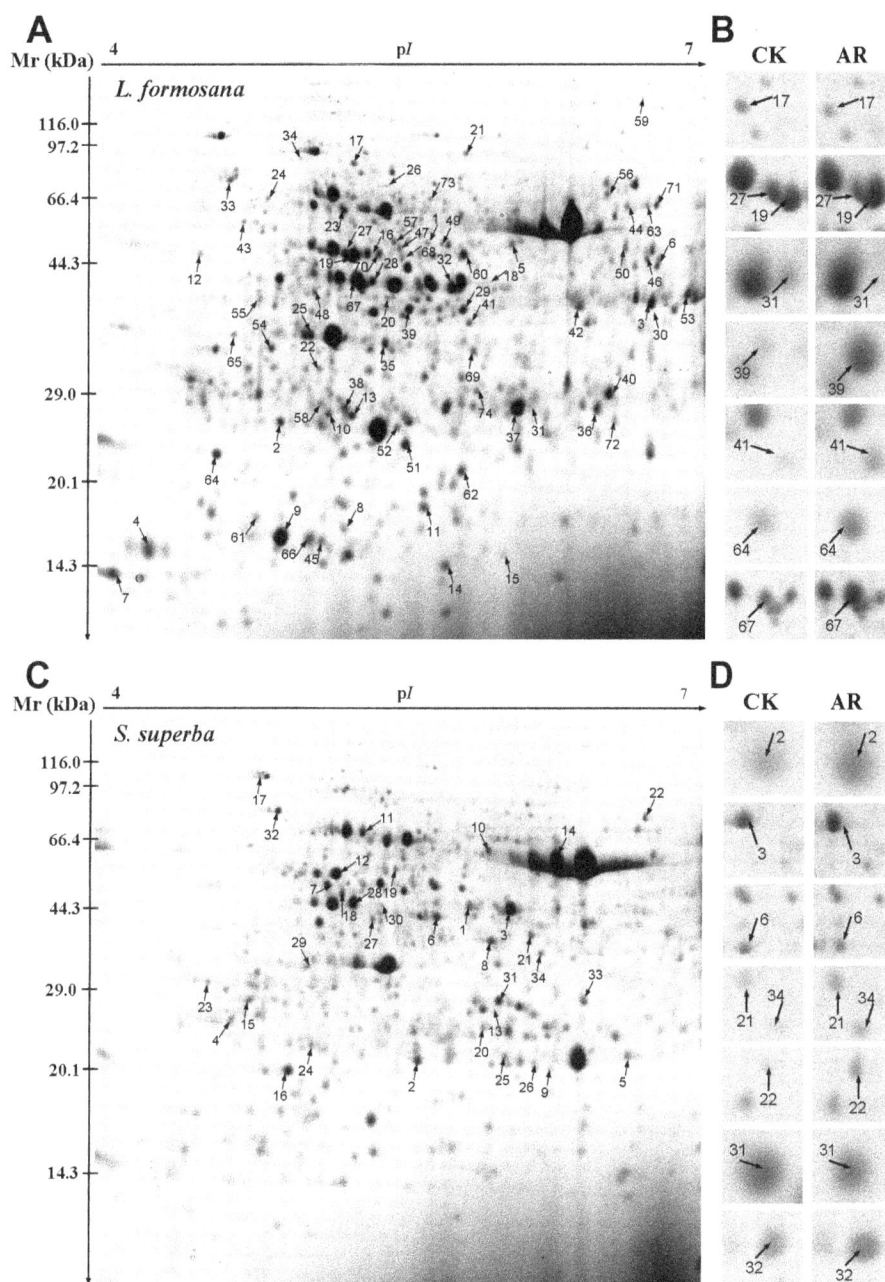

Figure 3. 2D gel analysis of proteins extracted from *L. formosana* and *S. superba* leaves. The numbers assigned to the proteins spots correspond to those listed in Tables 1 and 2. (A) Representative 2-DE gels of *L. formosana* in which 74 spots showing at least 2-fold changes ($p<0.05$) under AR were identified by MALDI-TOF MS. (B) Close-up views of differentially expressed protein spots in *L. formosana* (highlighted by arrows). (C) Representative 2-DE gels of *S. superba* in which 34 spots showing at least 2-fold changes ($p<0.05$) under AR were identified by MALDI-TOF MS. (D) Close-up views of differentially expressed protein spots in *S. superba* (highlighted by arrows).

(CAO, spot L21) and photosystem II (PSII) stability/assembly factor HCF136 (spot L20), were identified in *L. formosana*. CAO, that converts chlorophyll a to chlorophyll b, regulates the stabilization of light-harvesting chlorophyll a/b proteins [26]. Villeth et al [27] reported that pathogen infection could decrease the accumulation of PSII stability/assembly factor HCF136, which is important for the accurate assembly of PSII. In this study, AR remarkably decreased the abundance of CAO and PSII stability/assembly factor HCF136 in *L. formosana*, but no light reaction-related proteins was depressed by AR in *S. superba* (Table 1

and 2). These results suggest that photosynthesis apparatus of *L. formosana* is more sensitive to AR stress than *S. superba*.

It has been reported that the expression of Calvin cycle enzymes were down-regulated in *Arabidopsis* under salinity stress [25]. Our previous study also reported that the reduction in photosynthesis was linked to Calvin cycle enzymes in AR-treated *Arabidopsis* [16]. Consistent with previous results, our proteomic data from this study confirmed that the abundances of Calvin cycle-related proteins including phosphoribulokinase (PPK, spot L28), Rubisco (spot L31, L34) and Rubisco activase (spot L27) were significantly

Table 1. Identification of AR-responsive proteins in *L. formosana*.

Spot[a]	NCBI accession[b]	Protein identity[c]	Thero. Da/pI[d]	Exper. Da/pI[e]	SC[f]	MP/TP[g]	Score[h]	C[i]	Organism matched
Material metabolism									
L1	gi|297835714	glucose-1-phosphate adenylyltransferase (GPAT)	57.66/6.54	46.40/5.48	15%	6/7	81	D	*Arabidopsis lyrata* subsp
L2	gi|226500818	shikimate kinase family protein	30.34/5.87	27.41/4.74	36%	7/8	110	U	*Zea mays*
L3	gi|162458456	zeta-carotene desaturase	63.49/8.62	39.22/6.95	10%	5/5	74	U	*Zea mays*
L4	gi|356504466	haloalkane dehalogenase-like	45.01/8.43	15.10/4.12	21%	8/10	99	U	*Glycine max*
L6	gi|95117792	glutamate dehydrogenase (GDH)	44.82/6.38	44.21/7.00	23%	8/8	112	U	*Vitis vinifera*
L7	gi|15081239	glycine-rich protein 17 (GRP17)	53.32/10.31	14.09/4.02	21%	5/5	87	U	*Arabidopsis thaliana*
L8	gi|15218536	stearoyl-acyl-carrier protein desaturase-like protein (SACPDLP)	44.36/6.10	17.36/5.09	14%	6/6	97	U	*Arabidopsis thaliana*
L9	gi|38426301	6-phosphogluconate dehydrogenase	51.78/6.58	16.60/4.74	12%	5/5	78	U	*Oryza sativa*
L11	gi|255567778	cysteine synthase (CS)	43.38/7.60	20.47/5.49	21%	9/10	114	D	*Ricinus communis*
L12	gi|3341511	cinnamoyl-CoA reductase	40.63/5.73	47.97/4.40	16%	5/5	81	U	*Saccharum officinarum*
L39	gi|309951612	flavanone 3-hydroxylase (F3H)	41.24/5.39	39.35/5.41	28%	7/9	103	U	*Litchi chinensis*
L46	gi|114795072	chalcone synthase (CHS)	42.83/6.05	45.37/6.91	22%	8/10	84	U	*Pyrus communis*
Photosynthesis and energy production									
L18	gi|1022805	phosphoglycerate kinase (PGK)	41.99/4.93	36.02/5.79	24%	8/11	108	D	*Arabidopsis thaliana*
L19	gi|355329944	actin	40.37/5.67	41.03/5.13	44%	15/28	144	U	*Malus domestica*
L20	gi|225423755	photosystem II stability/assembly factor HCF136	44.47/6.92	34.91/5.31	36%	12/27	109	D	*Vitis vinifera*
L21	gi|357438645	chlorophyllide a oxygenase (CAO)	25.99/8.95	90.68/5.69	30%	4/4	76	D	*Medicago truncatula*
L22	gi|183217735	ATP synthase CF1 alpha subunit	55.62/5.20	32.35/4.95	24%	10/11	138	U	*Guizotia abyssinica*
L23	gi|114421	ATP synthase subunit beta	59.93/5.95	59.97/5.09	25%	12/25	88	U	*Nicotiana plumbaginifolia*
L24	gi|225428086	V-type proton ATPase subunit B	54.37/5.04	66.71/4.68	32%	16/40	107	U	*Vitis vinifera*
L25	gi|147945622	oxygen-evolving enhancer protein (OEE)	34.72/6.08	36.12/4.91	24%	7/8	109	U	*Leymus chinensis*
L26	gi|5758863	ATP synthase beta subunit	53.56/5.16	73.58/5.31	42%	17/24	177	U	*Colchicum autumnale*
L27	gi|158726716	ribulose 1,5-bisphosphate carboxylase/oxygenase activase	48.81/6.10	46.81/5.10	26%	10/13	115	D	*Flaveria bidentis*
L28	gi|15222551	phosphoribulokinase (PPK)	44.72/5.71	43.22/5.22	24%	9/18	87	D	*Arabidopsis thaliana*
L29	gi|79322651	fructose-bisphosphate aldolase	41.95/5.94	39.20/5.66	15%	7/7	96	D	*Arabidopsis thaliana*
L30	gi|2108252	P-glycoprotein-2 (PGP2)	135.75/8.97	38.31/6.98	6%	7/7	77	D	*Arabidopsis thaliana*
L31	gi|162946539	ribulose-1,5-bisphosphate carboxylase/oxygenase small subunit	20.80/8.23	28.36/6.01	32%	5/5	97	D	*Solanum tuberosum*
L32	gi|1022805	phosphoglycerate kinase (PGK)	41.99/4.93	42.14/5.61	24%	8/17	87	D	*Arabidopsis thaliana*
L33	gi|356539332	RuBisCO large subunit-binding protein subunit alpha-like isoform 1	61.73/5.23	79.85/4.53	19%	9/11	111	D	*Glycine max*
L34	gi|146188415	ribulose-1,5-biphosphate carboxylase/oxygenase (Rubisco)	46.17/6.44	87.79/4.87	26%	11/21	104	D	*Podalyria canescens*
L61	gi|297816654	metal ion binding protein	26.66/9.55	16.60/4.74	25%	4/4	73	U	*Arabidopsis lyrata* subsp
Stress and defense									
L35	gi|42568255	TIR-NBS-LRR class disease resistance protein	121.85/6.36	35.08/5.30	7%	7/7	81	U	*Arabidopsis thaliana*
L36	gi|241989446	NBS-LRR class disease resistance protein	19.52/5.43	24.37/6.42	29%	4/4	75	U	*Oryza sativa*

Table 1. Cont.

Spot[a]	NCBI accession[b]	Protein identity[c]	Thero. Da/p[d]	Exper. Da/pl[e]	SC[f]	MP/TP[g]	Score[h]	C[i]	Organism matched
L37	gi|224111296	cc-nbs-lrr resistance protein	149.57/5.73	24.48/5.95	13%	12/16	106	U	Populus trichocarpa
L38	gi|289157416	1-hydroxy-2-methyl-2-(E)-butenyl 4-diphosphate reductase	51.64/5.63	28.42/5.08	16%	5/5	77	U	Artemisia annua
-40	gi|14210363	ascorbate peroxidase (APX)	27.50/5.13	29.55/6.55	22%	4/4	73	U	Zantedeschia aethiopica
L41	gi|110289462	glutathione S-transferase (GST)	18.04/10.13	37.47/5.70	36%	4/4	74	U	Oryza sativa
L43	gi|18404004	TSK-associating protein 1	84.20/4.58	56.88/4.55	10%	6/7	74	U	Arabidopsis thaliana
L44	gi|17530547	class III peroxidase ATP32	35.02/6.88	58.60/6.71	30%	6/6	107	U	Arabidopsis thaliana
L45	gi|356559803	stromal 70 kDa heat shock-related protein	73.88/5.20	16.24/4.95	19%	12/23	100	U	Glycine max
L47	gi|357490825	NBS-LRR resistance protein	136.09/6.05	48.01/5.37	8%	9/9	102	U	Medicago truncatula
Signal transduction									
L48	gi|333441302	phytochrome C	42.23/6.08	41.46/4.94	21%	6/7	86	U	Digoniopterys microphylla
L49	gi|371940268	truncate phytochrome A2 protein	110.22/6.65	48.59/5.56	9%	9/9	103	U	Glycine max
L50	gi|18405351	abscisic acid receptor PYL6	24.06/6.17	46.80/6.69	22%	4/4	74	U	Arabidopsis thaliana
L51	gi|359475476	serine carboxypeptidase-like 18	58.18/6.16	25.30/5.39	18%	7/7	108	U	Vitis vinifera
L62	gi|77553062	cyclic nucleotide-gated ion channel 14	72.47/8.15	23.34/5.66	10%	7/7	91	D	Oryza sativa
Transcription									
L52	gi|187369233	topoisomerase I	102.34/9.50	22.01/5.39	13%	11/16	88	U	Catharanthus roseus
L53	gi|308802618	DNA-damage-inducible protein F	48.73/4.50	34.58/6.99	16%	6/8	80	U	Ostreococcus tauri
L54	gi|20196900	putative RNA helicase A	124.80/6.54	34.89/4.69	8%	8/10	74	U	Arabidopsis thaliana
L55	gi|126022792	RNA polymerase beta subunit	155.17/9.36	41.14/4.63	10%	10/12	104	U	Spinacia oleracea
L56	gi|308808201	minichromosome maintenance protein 10 isoform 1-like	64.87/9.42	65.31/6.50	17%	9/11	95	U	Ostreococcus tauri
L57	gi|323690255	maturase K	45.41/9.80	48.01/5.37	16%	5/5	80	U	Pitraea cuneato-ovata
L58	gi|183529139	maturase K	60.34/9.28	28.91/4.96	12%	7/8	82	U	Jacqueshuberia brevipes
L59	gi|21629786	maturase K	46.79/9.62	134.26/6.81	21%	7/8	100	U	Hyalochlamys globifera
L60	gi|255567202	putative transcription elongation factor s-II	38.76/9.57	46.80/5.67	14%	5/5	77	D	Ricinus communis
Post-translational modification									
L13	gi|224089629	f-box family protein	45.92/6.43	27.54/5.11	16%	5/5	82	U	Populus trichocarpa
L14	gi|308802882	ubiquitin-protein ligase/hyperplastic discs protein	92.22/5.72	14.48/5.58	8%	7/7	81	U	Ostreococcus tauri
L15	gi|304322967	translational elongation factor Tu (EF-Tu)	45.64/6.16	14.88/5.89	20%	6/6	94	D	Floydiella terrestris
L16	gi|225429488	eukaryotic initiation factor 4A-11	47.07/5.38	47.10/5.26	17%	10/10	118	D	Vitis vinifera
L17	gi|30684767	cell division protease ftsH-2	74.28/6.00	83.96/5.15	15%	9/9	123	D	Arabidopsis thaliana
Hormone response									
L5	gi|3024127	S-adenosylmethionine synthase (SAM synthase)	43.43/5.51	48.33/5.91	31%	10/29	82	U	Catharanthus roseus
L10	gi|33342178	ABA inducible protein	17.52/5.95	27.92/5.00	24%	5/5	84	U	Triticum aestivum
L42	gi|350535769	ethylene-responsive transcriptional coactivator	16.08/10.03	39.11/6.27	36%	6/6	112	U	Solanum lycopersicum
Others									

Table 1. Cont.

Spot[a]	NCBI accession[b]	Protein identity[c]	Thero. Da/pI[d]	Exper. Da/pI[e]	SC[f]	MP/TP[g]	Score[h]	C[i]	Organism matched
L63	gi|168047657	predicted protein	113.56/6.45	58.41/6.85	12%	10/13	83	U	Physcomitrella patens subsp
L64	gi|302772723	hypothetical protein	84.02/9.40	24.97/4.47	10%	6/6	79	U	Selaginella moellendorffii
L65	gi|302823293	hypothetical protein ELMODRAFT-449095	84.19/9.43	36.83/4.51	15%	8/11	84	U	Selaginella moellendorffii
L66	gi|167998464	predicted protein	35.11/9.51	16.16/4.89	22%	7/8	101	U	Physcomitrella patens subsp
L67	gi|297797753	predicted protein	41.46/6.47	42.91/5.14	11%	5/5	78	U	Arabidopsis lyrata subsp
L68	gi|147834872	hypothetical protein VITISV-040309	46.87/6.06	46.55/5.38	32%	17/33	116	U	Vitis vinifera
L69	gi|168004878	predicted protein	46.87/5.63	34.19/5.72	27%	9/22	92	U	Physcomitrella patens subsp
L70	gi|168000362	predicted protein	120.48/9.19	47.05/5.29	12%	10/13	86	U	Physcomitrella patens subsp
L71	gi|116779860	unknown	23.62/7.77	59.88/6.98	29%	6/7	99	U	Picea sitchensis
L72	gi|218201086	hypothetical protein OsI-29089	83.26/6.03	27.24/6.61	16%	8/10	86	D	Oryza sativa
L73	gi|49389230	hypothetical protein	38.17/6.67	65.34/5.50	21%	7/7	99	D	Oryza sativa
L74	gi|297721931	Os03g0229600	11.76/9.77	29.86/5.71	41%	5/5	107	D	Oryza sativa

The seedlings were treated with AR (pH 3.0) for one month. The pH of AR solution was adjusted with a mixture of H_2SO_4 and HNO_3 in the ratio of 5:1. The final concentration of H_2SO_4 and HNO_3 were 0.45 and 0.09 mM, respectively.

[a] Spot No. is the unique differentially expressed protein spot number. L, protein spot number. L, protein spot in L. formosana gel. [b] Database accession numbers according to NCBInr. [c] Description of the proteins identified by MALDI-TOF MS. [d] Theoretical mass (kDa) and pI of identified proteins. [e] Experimental mass (kDa) and pI of identified proteins. [f] Amino acid sequence coverage for the identified proteins. [g] Number of the matched peptides and the total searched peptides. [h] Mascot searched score against the database NCBInr. Protein score is −10*Log(P), where P is the probability that the observed match is a random event. Protein scores greater than 73 are significant (p<0.05). [i] Spot abundance change. D decreased abundance of proteins, U increased abundance of protein.

Table 2. Identification of AR-responsive proteins in *S. superba*.

Spot[a]	NCBI accession[b]	Protein identity[c]	Thero. Da/pI[d]	Exper. Da/pI[e]	SC[f]	MP/TP[g]	Score[h]	C[i]	Organism matched
Material metabolism									
S1	gi|62321345	glutamate-ammonia ligase	23.45/5.70	44.39/5.55	20%	6/8	93	U	*Arabidopsis thaliana*
S2	gi|205277664	granule-bound starch synthase I (GBSS)	15.08/6.41	20.84/5.29	34%	4/4	78	U	*Thinopyrum intermedium*
S3	gi|60101355	glutamine synthetase (GS)	31.14/5.75	43.58/5.76	23%	6/8	91	U	*Vigna radiata*
S19	gi|303280145	glycosyltransferase family 7 protein	48.84/9.23	53.35/5.17	16%	5/5	79	U	*Micromonas pusilla*
Photosynthesis and energy production									
S6	gi|255544584	phosphoglycerate kinase (PGK)	50.11/8.74	42.11/5.39	36%	13/25	143	D	*Ricinus communis*
S7	gi|37721507	photosystem II subunit H	2.53/11.72	49.18/4.85	62%	3/4	75	D	*Ixiolirion tataricum*
S8	gi|5708095	ATP synthase gamma chain	33.48/6.12	37.32/5.66	35%	9/20	99	D	*Arabidopsis thaliana*
S9	gi|6688696	ribulose-1,5-bisphosphate carboxylase/oxygenase large subunit	52.03/6.22	20.08/5.97	20%	10/18	95	D	*Morkillia mexicana*
S11	gi|290490212	ATP synthase CF1 alpha subunit protein	55.58/5.04	70.29/5.01	32%	17/27	173	U	*Staphylea colchica*
S12	gi|13430334	rubisco activase	37.25/6.70	51.86/4.85	33%	10/18	99	D	*Zantedeschia aethiopica*
S13	gi|308320553	ribulose-1,5-bisphosphate carboxylase/oxygenase large subunit	30.68/6.24	26.45/5.67	22%	5/6	78	D	*Madia sp*
S14	gi|170664996	ribulose-1,5-bisphosphate carboxylase/oxygenase large subunit	52.85/6.00	53.02/6.05	23%	11/16	125	U	*Lycoseris crocata*
S10	gi|81301612	protein Ycf2	268.41/8.58	61.08/5.64	9%	16/20	99	U	*Nicotiana tomentosiformis*
S20	gi|303283276	beta carbonic anhydrase	26.07/6.21	24.62/5.62	27%	5/5	92	D	*Micromonas pusilla*
Stress and defense									
S16	gi|3328221	thioredoxin peroxidase (TPx)	28.40/6.34	19.81/4.56	37%	5/8	86	U	*Secale cereale*
S17	gi|2654208	heat shock 70	76.27/5.19	105.00/4.39	21%	12/22	116	U	*Spinacia oleracea*
S18	gi|116323	endochitinase 3	37.17/8.72	47.33/4.89	30%	5/6	84	U	*Nicotiana tabacum*
S21	gi|384247250	clavaminate synthase-like protein	41.35/5.60	38.59/5.86	17%	5/5	81	U	*Coccomyxa subellipsoidea*
Signal transduction									
S15	gi|350536755	14-3-3 protein 4	29.44/4.66	28.05/4.37	43%	6/9	99	U	*Solanum lycopersicum*
S22	gi|115393868	phytocyanin-like arabinogalactan-protein (PLA)	18.87/9.17	77.80/7.03	34%	4/4	77	U	*Gossypium hirsutum*
S27	gi|110532561	calmodulin (CaM)	17.10/4.06	41.79/5.06	39%	4/4	81	U	*Aegiceras corniculatum*
S28	gi|224131906	calcium dependent protein kinase 6 (CDPK)	62.93/5.37	45.18/4.96	13%	7/8	93	U	*Populus trichocarpa*
Transcription									
S23	gi|108861639	transposase	15.02/9.08	30.68/4.32	47%	5/7	82	U	*Oligostachyum sulcatum*
S24	gi|379041605	maturase K	40.48/10.01	22.60/4.71	18%	8/11	80	D	*Bromus commutatus*
S25	gi|255660958	pentatricopeptide repeat-containing protein	39.59/5.95	21.25/5.73	21%	8/10	98	U	*Verbena macdougalii*
S26	gi|11993344	marpoflo protein	27.93/9.19	20.25/5.88	19%	5/5	74	D	*Marchantia polymorpha*
Post-translational modification									
S4	gi|225441985	proteasome subunit alpha type-5 isoform 1	26.13/4.65	25.55/4.35	31%	6/8	84	U	*Vitis vinifera*
S5	gi|35545337	mitochondrial import receptor subunit TOM6 homolog isoform 1	6.27/9.36	21.22/6.80	62%	4/5	79	U	*Glycine max*
Others									

Table 2. Cont.

Spot[a]	NCBI accession[b]	Protein identity[c]	Thero. Da/pI[d]	Exper. Da/pI[e]	SC[f]	MP/TP[g]	Score[h]	C[i]	Organism matched
S29	gi|224064392	predicted protein	37.25/5.62	33.24/4.68	37%	7/10	102	U	Populus trichocarpa
S30	gi|388496926	unknown	39.81/8.46	44.78/5.11	24%	8/10	120	D	Lotus japonicus
S31	gi|242052501	hypothetical protein SORBIDRAFT_03g010120	48.42/9.53	27.93/5.71	22%	9/12	106	D	Sorghum bicolor
S32	gi|168071263	predicted protein	23.61/9.95	81.80/4.52	26%	4/4	80	U	Physcomitrella patens subsp
S33	gi|77554095	hypothetical protein LOC_Os12g13240	23.16/10.25	27.91/6.30	34%	6/9	92	U	Oryza sativa
S34	gi|145347277	predicted protein	107.30/5.81	35.29/5.92	9%	7/7	78	U	Ostreococcus lucimarinus

The seedlings were treated with AR (pH 3.0) for one month. The pH of AR solution was adjusted with a mixture of H_2SO_4 and HNO_3 in the ratio of 5:1. The final concentration of H_2SO_4 and HNO_3 were 0.45 and 0.09 mM, respectively.

[a]Spot No. is the unique differentially expressed protein spot number. S, protein spot in S. superba gel. [b] Database accession numbers according to NCBInr. [c] Description of the proteins identified by NCBInr. [d] Theoretical mass (kDa) and pI of identified proteins. [e] Experimental mass (kDa) and pI of identified proteins. [f] Amino acid sequence coverage for the identified proteins. [g] Number of the matched peptides and the total searched peptides. [h]Mascot searched score against the database NCBInr. Protein score = −10*Log(P), where P is the probability that the observed match is a random event. Protein scores greater than 73 are significant (p<0.05). [i] Spot abundance change. D decreased abundance of proteins, U increased abundance of proteins.

decreased in *L. formosana* (Table 1). Rubisco, the CO_2 fixing enzyme in Calvin cycle, is the primary limiting factor of net photosynthesis under environmental stress [16]. Rubisco activase promotes and maintains the catalytic activity of Rubisco [28]. The decreased expression of Rubisco and Rubisco activase in AR-treated *L. formosana* may disturb Calvin cycle activity, leading to the reduction in photosynthetic CO_2 assimilation and thus the inhibition in plant growth. Moreover, the results of Western blot analysis showed that the decrease in protein expression of Rubisco large subunit was more obvious in *L. formosana* than in *S. superba* under AR stress (Fig. 6A and B). These results may explain why AR-induced damage to photosynthesis was more serious in *S. superba* than in *L. formosana* (Fig. 1D and E). Compared with *L. formosana*, less damage of AR to photosynthesis-related proteins probably result from higher tolerance to AR in *S. superba*.

Besides photosynthesis-related proteins, AR also affected the abundances of energy production-related proteins in *L. formosana* and *S. superba*. It is well known that increased ATP production is required in response to abiotic stress in plants [29]. For example, the abundance of ATP synthase was considerably increased in salt-stressed rice [30] and osmotic-stressed wheat [31]. Thus, the increased abundances of ATP synthase subunits in *L. formosana* (spot L22, L23, L26) and *S. superba* (spot S11) in our study demonstrate the prime role of ATP synthase in the adaptation of two tree species to AR stress.

Material metabolism-related proteins

In *L. formosana*, AR affected a series of protein abundances involved in several metabolism processes, including nitrogen metabolism (spot L2, L6, L11), starch and sugar metabolism (spot L1, L9), lipid metabolism (spot L7, L8), secondary metabolism (spot L39, L46), vitamin metabolism (spot L3) and lignin biosynthesis (spot L12) (Table 1). However, only two nitrogen metabolism-related proteins (spot S1, S3) and one starch biosynthesis-related protein (spot S2) were remarkably induced by AR in *S. superba* (Table 2). It is clear that AR disturbed more metabolism processes in *L. formosana* than in *S. superba*.

AR stress can change free amino acid levels and disturbs N metabolism in plants [19]. Cysteine synthase (CS, spot L11), which is responsible for the terminal step of cysteine biosynthesis, is a critical enzyme involved in environmental stress response in plants [12]. The abundance of CS was decreased in *L. formosana* (Table 1); indicating that metabolic processes related to cysteine biosynthesis might be strongly depressed by AR. In addition, two glutamine metabolism-related proteins, such as glutamate dehydrogenase (GDH, spot L6) and glutamine synthetase (GS, spot S3), were identified in *L. formosana* and *S. superba*, respectively. Abiotic stresses, such as salinity, drought and metal toxicity, can up-regulate *GDH* gene expression and enhance GDH activity in plants [32]. GS catalyzes ATP-dependent incorporation of ammonium into glutamate and other reduced N compounds [33]. It has been found that GS accumulation was also stimulated by salt and drought, and this helped improve the tolerance of plants to stresses [25]. In agreement with previous findings, our study found that the abundance of GDH in *L. formosana*, as well as that of GS in *S. superba*, were increased (Table 1), suggesting that GDH and GS play critical roles in N metabolic acclimation of plants when exposed to AR.

Abiotic stress can also affect starch synthesis [34]. In this study, two starch biosynthesis-related proteins including glucose-1-phosphate adenylyltransferase (GPAT, spot L1) in *L. formosana* and granule-bound starch synthase (GBSS, spot S2) in *S. superba* were identified (Tables 1 and 2). Interestingly, the responses of GPAT and GBSS to AR were different in two broad-leaf species.

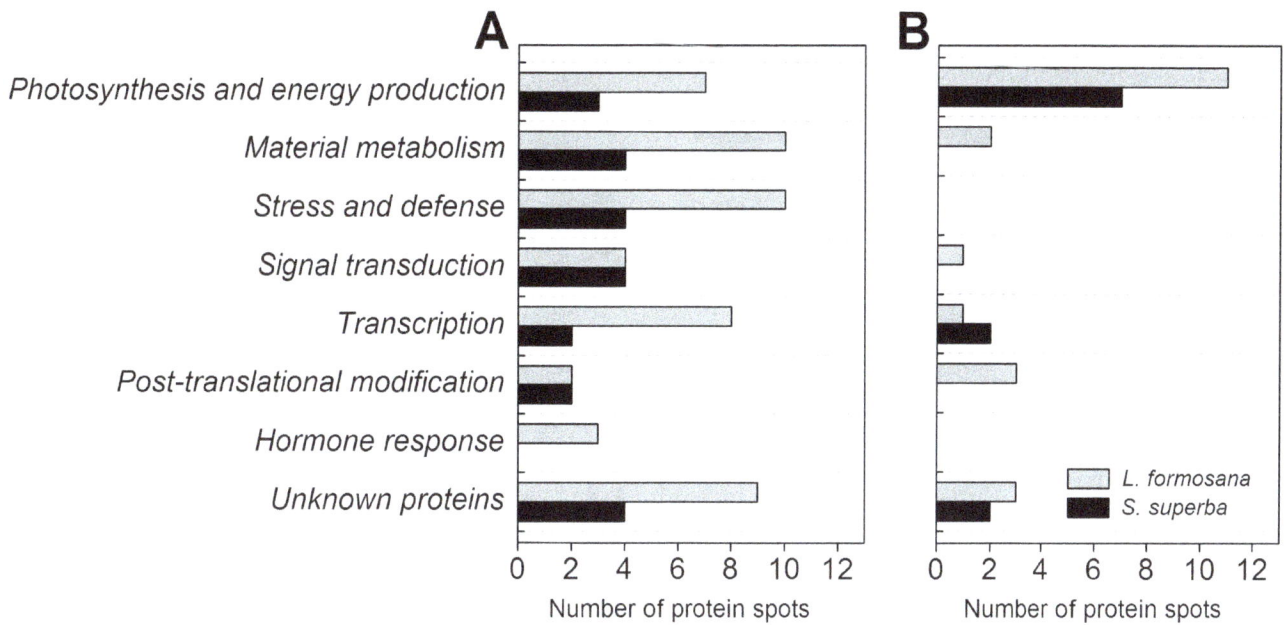

Figure 4. Number of protein spots significantly changed in AR-treated *L. formosana* and *S. superba*. (A) Protein spots increased in their abundances. (B) Protein spots decreased in their abundances.

The abundance of GPAT was decreased in *L. formosana*. However, GBSS, the only enzyme implicated in amylose synthesis [35], was increased in AR-treated *S. superba*. Basically, this result was consistent with the observation where GBSS activity and starch content in rice was found to increase under cold stress [36]. It is well known that starch is required to synthesize sucrose which serves as a carbon and energy source for plant growth and stress response [35]. Thus we believe that the increased abundance of GBSS may contribute to higher AR-tolerance in *S. superba* through enhancing starch synthesis and energy production. It also should be noted that AR stress changed the abundances of lipid metabolism-related proteins including glycine-rich protein 17 (GRP17, spot L7) and stearoyl-acyl-carrier protein desaturase-like protein (SACPDLP, spot L8), as well as secondary metabolism-

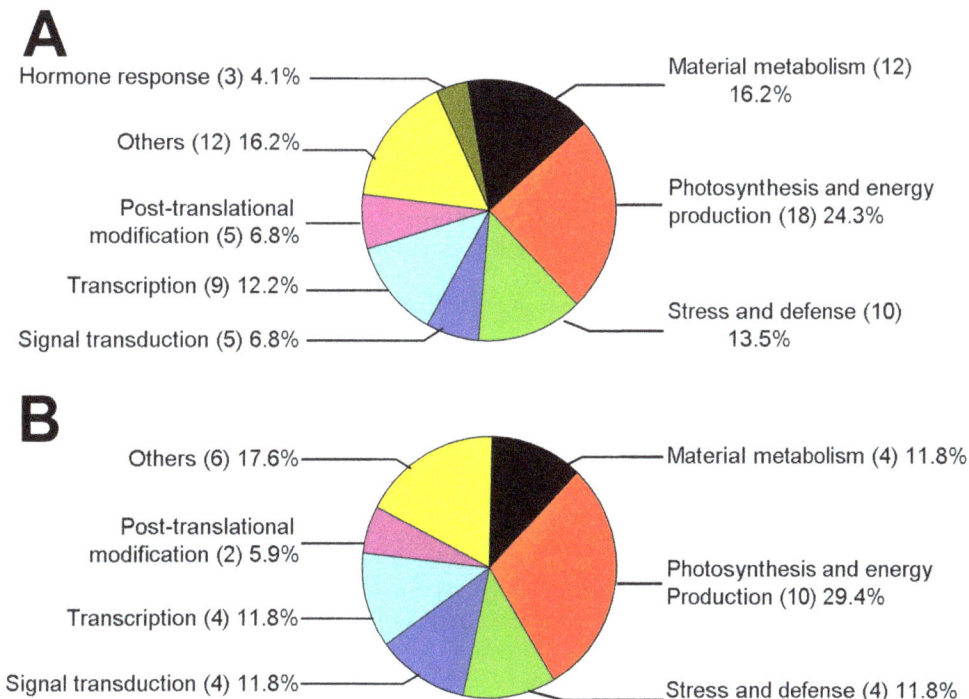

Figure 5. Functional classification of AR-responsive proteins in *L. formosana* (A) and *S. superba* (B). The proportion of identities in each functional group was the sum of this identity accounting for all protein quantities.

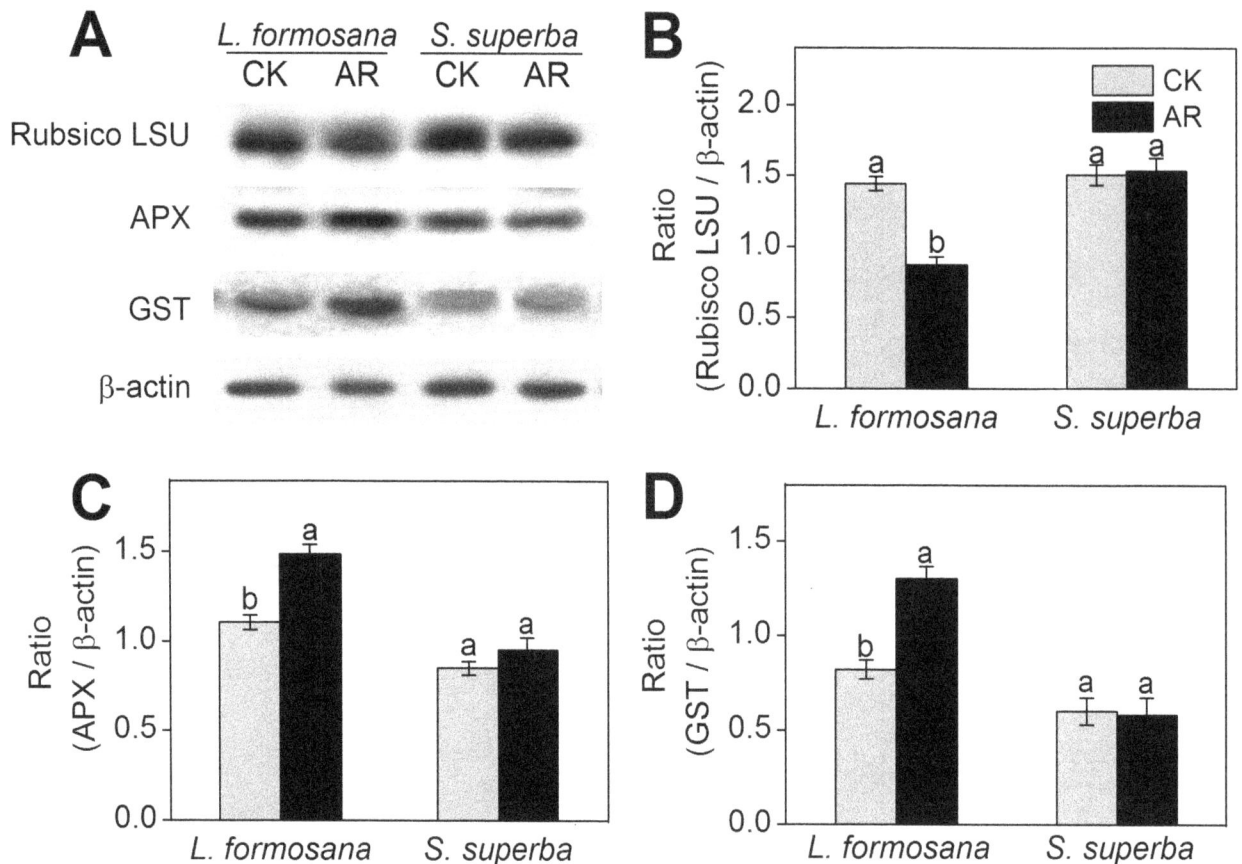

Figure 6. Western blot analysis showing the expression of three protein spots. (A) Expression of rubulose-1,5-bisphoshate carboxylase large subunit (RuBisco LSU), ascorbate peroxidase (APX) and glutathione-S-transferase (GST) in *L. formosana* and *S. superba* seedlings after AR treatment. Relative expression level of RuBisco LSU (B), APX (C) and GST (D) were analyzed with the Quantity One software. β-actin was used as the internal control. Means with different letters indicate significantly difference ($p<0.05$) with regard to AR treatments.

related proteins including flavanone 3-hydroxylase (F3H, spot L39) and chalcone synthase (CHS, L46) in *L. formosana*. The abundance of SACPDLP was increased, which may contribute to lipid synthesis and membrane integrity in AR-treated *L. formosana* [25]. F3H and CHS are two key enzymes that catalyze the biosynthesis of flavonoids and chalcones, both of which play critical roles in enhancing secondary metabolism under environmental stress [37]. The increased abundances of these proteins imply that secondary metabolism pathway may be activated to cope with AR stress in AR-sensitive species, *L. formosana*.

Stress defense-related proteins

The majority of environmental stresses generate a secondary oxidative stress in plants [13]. Oxidative stress occurs when there is a serious imbalance between ROS production and antioxidant defense [24]. Overproduction of ROS induced by a number of adverse environmental factors can attack proteins, lipids, and nucleic acids [38]. Under long-term heavy metal or AR stress, significant accumulation of ROS was observed in previous studies [7,19,39]. For instance, Kovacik et al [38] observed the increased ROS in four *Tillandsia* species under 2 μM Cd^{2+} treatment over 30 days. In this study, H_2O_2 and $O_2^{\cdot-}$ content, and thus oxidative stress, induced by AR was remarkably higher in *L. formosana* than in *S. superba* (Fig. 2C and D).

To avoid oxidative damage, plants developed an antioxidant system consisting of antioxidative enzymes as well as non-enzymatic

antioxidants [38]. In *L. formosana*, three antioxidant-related proteins, including APX (spot L40), GST (spot L41) and class III peroxidase ATP32 (spot L44), were identified by proteomic analysis. APX plays an important role in scavenging H_2O_2 from cells [34]. GST is also an important enzyme that counteracts cellular damage induced by oxidative stress [12]. Enhanced activity of APX and GST has also regularly been detected in plants after exposure to salt, cold, heavy metal, and heat stresses [40]. Our recently published work found that both APX and GST were increased in their abundance in an AR-sensitive conifer tree species, *Piuns massoniana* [41]. Similarity, APX (spot L40) and GST (spot L41) with higher abundances and expression levels (Table 1 and Fig. 6) were also found in AR-treated *L. formosana* in this study, implying that antioxidant defense system was provoked by AR in *L. formosana*. However, the expression levels of APX and GST were not changed in *S. superba* seedlings under AR treatment (Fig. 6). A likely reason is that there may be other pathways that can remove excessive ROS in *S. superba*. Interestingly, thioredoxin peroxidase (TPx, spot S16), which appears to be a key enzyme in H_2O_2 detoxification [42], was found to have an increased abundance in AR-treated *S. superba*. The increased TPx may function in resisting AR-induced oxidative damage by reducing ROS production in *S. superba*. In addition, endochitinase, a glycosyl hydrolase that catalyzes chitin degradation, plays an essential role in forming the fine cell-wall matrix that enhances the physical barrier against abiotic stresses [43]. Tapia et al [44] reported that endochitinase could be induced by heat, drought, and salinity. In this study, the increased abundance of endochitinase (spot S18) in *S. superba* (Table 2) may improve physical interactions at the plasma membrane-cell wall interface to cope with AR stress.

Signal transduction-related proteins

Signal transduction plays a crucial role in triggering a cascade of defense events [12]. Phytochrome is a chromoprotein that regulates the expressions of a large number of light-responsive genes and controls plant growth and development [45]. Recently, phytochrome has been found to modulate both biotic and abiotic stresses, such as salinity, drought, cold or herbivory [46]. Cross-talk between phytochrome-mediated light signals and some stress signaling pathways has been reported in diverse plants [47]. Thus it is possible that phytochrome is involved in the modulation of AR stress. A recent study found that increased abundance of phytochrome was needed to resist cold stress in cucumber [48]. Likewise, AR also increased the abundance of both phytochrome C (spot L48) and truncate phytochrome A2 protein (spot L49) in *L. formosana* but not in *S. superba* in this study (Table 1), which suggest a role of phytochrome signaling in response to AR in this sensitive species. In addition to the phytochrome signaling pathway, antioxidant enzymes have been found to be modulated by phytochromes under stress conditions [46]. In this study, the increased expression of antioxidant enzymes (APX and GST) was observed in *L. formosana* under AR treatment (Fig. 6). We suggest that the enhancement of phytochromes induced by AR modulates the antioxidant system in *L. formosana* seedlings. Further research need to widen our understanding of the role of phytochrome in AR-stressed plants.

Calcium (Ca) also plays a crucial role in regulating plant defense responses to various environmental stimuli [49]. Recently, Kovacik et al [50] found that oxidative stress evoked by hexavalent chromium were evidently suppressed by Ca in *Matricaria chamomilla* using microscopic visualization method. Our previous study also reported that Ca addition dramatically alleviated the negative effects of AR on seed germination, seedling growth and photosynthesis [3]. Free Ca ion within cell is a second messenger for conveying internal and external signals by Ca sensors that subsequently regulate diverse cellular processes in plants [17]. Calmodulin (CaM) and calcium-dependent protein kinase (CDPK), two major types of Ca sensor, play important roles in Ca signaling and further response to diverse stresses in plants [51]. Saijo et al [52] found that over-expression of *OsCDPK* gene enhanced tolerance to cold, salt and drought in transgenic rice. Moreover, 14-3-3 protein can also regulate proteins involved in stress response and activate CDPK signal transduction pathway in plants [13]. In *S. superba*, the abundances of CaM (spot S27), CDPK (spot S28) and 14-3-3 protein (spot S15) were increased after AR treatment, while these proteins were not identified in *L. formosana*. In agreement with the proteomic analysis results, the enhanced gene expression of *CaM* and *CDPK* by real-time quantitative PCR was more obvious in *S. superba* than in *L. formosana* under AR treatment (Fig. S1). These results indicate that AR activated different signaling transduction pathways in two tree species and that Ca sensor-dependent signaling pathway might play a critical role in enhancing AR tolerance in *S. superba*.

AR affected transcription, protein synthesis and modification

Transcription machinery plays an important role in abiotic stress adaptation [25]. Nine proteins (spots L52–L60) and four proteins (spots S23–S26) that involve in gene transcription were identified in response to AR stress in *L. formosana* and *S. superba*, respectively (Tables 1 and 2). Four of these proteins were maturase

K (spots L57–59, S24). In plants, maturase K catalyzes intron RNA binding during reverse transcription and spicing and directly affects gene expression at transcriptional level [53]. The changes in protein expression of maturase K are very complex in plants under abotic stress, depending on plant species and the type of stress [54]. Pandey et al [54] reported that maturase K was induced when plant suffered from high ROS pressure. On the contrary, maturase K protein was down-regulated in salt-treated maize [55]. In this study, the abundance of maturase K (spots L57–59) was increased in *L. formosana*, but was decreased in *S. superba* (Tables 1 and 2). Based on our physiological data, AR induced higher ROS (H_2O_2 and $O_2^{\cdot-}$) production in *L. formosana* than that in *S. superba* (Fig. 2), which might be one reason to explain the increased abundance of maturase K in *L. formosana*. However, the decreased abundance of maturase K in *S. superba* indicates that its gene transcription had been affected by AR, thought no significant visible damage symptom emerged in *S. superba* leaves.

Translational elongation factor Tu (EF-Tu) plays important roles in response to abiotic stresses including high and low temperatures, salinity, and water deficit [56]. Pandey et al [54] reported that the expression of EF-Tu was down-regulated in chickpea with the increased treatment period of water deficit, which was consistent with the decrease in EF-Tu abundance (spot L15) after AR stress in *L. formosana* observed in our study. This result indicates that AR has induced a lower protein synthesis in *L. formosana*. However, EF-Tu was not identified in *S. superba*, instead, proteasome subunit alpha type-5 isoform 1 (spot S4) and mitochondrial import receptor subunit TOM6 homolog isoform 1 (spot S5) were found in AR-treated leaves (Table 2). Proteasome degrades the proteins with translational errors and proteins damaged by stress which can aggregate and become toxic to the cell [57]. Proteasome has been implicated in regulating numerous plant signaling and metabolic pathways under stress condition [57]. Our pervious study found that the expression of 20 S proteasome subunit was up-regulated in *Arabidopsis* after 32 h of AR treatment [16]. In this study, the abundance of proteasome subunit alpha type-5 isoform 1 was also increased in AR-treated *S. superba*, indicating that the control of protein degradation by the proteasome is likely to play an important role in enhancing AR-tolerance in *S. superba*. In addition, mitochondrial import receptor subunit TOM complex mediates the translocation and uptake of nuclear-encoded mitochondrial preproteins from the cytosol [58]. Increased abundance of mitochondrial import receptor subunit TOM6 by AR may accelerate the import of mitochondrial preproteins and enhance cellular metabolism and energy production in mitochondria, finally contributing to improve resistant to AR in *S. superba*.

Hormone response-related proteins

Plant hormones are not only important in plant growth and development, but also closely related to environmental stresses response [59]. It is known that increased ethylene biosynthesis is a general response of plants to stress conditions [60]. S-adenosyl-L-methionine (SAM), a major methyl donor in plants, is used as a substrate for ethylene biosynthesis [60]. SAM synthetase catalyzes SAM biosynthesis from L-methionine and ATP [61]. It has been found that both activity and expression of SAM synthetase were increased in tomato and rice under salt stress [62]. Accordingly, our present results indicate that AR increased the abundance of SAM synthetase (spot L5) in *L. formosana* (Table 1), which may further contribute to ethylene biosynthesis. Another ethylene-related protein, ethylene-responsive transcriptional coactivator (spot L42), which was also increased by AR stress in *L. formosana*, can positively control the expression of ethylene-responsive genes in plants [63]. In addition, ABA inducible protein (spot L10), which is involved in ABA stimulus and abiotic stresses response, was increased in *L. formosana* when exposed to AR. It appeared that ethylene biosynthesis and ABA signaling pathways might be activated in AR-treated *L. formosana*. It is interesting to note that, in *S. superba*, no hormone response-related proteins was identified, partly due to the high resistance to AR stress in *S. superba*.

Conclusion

Using the approach of proteomic analysis, this study investigated the differential responses to AR in two broad-leaf tree species, *L. formosana* and *S. superba*, an AR-sensitive species and an AR-tolerant species, respectively. After AR treatment, more proteins were significantly changed in their abundances in *L. formosana* than in *S. superba*. It should be noted that hormone response-related protein was only found in *L. formosana*. After AR treatment, signaling pathways, energy production and antioxidant system were activated in both *L. formosana* and *S. superba*. Due to higher AR-tolerance in *S. superba*, AR induced less damage to photosynthesis-related proteins in this species. Moreover, the proteins related to starch synthesis and translation were depressed in AR-treated *L. formosana*, but enhanced in *S. superba*, implying that these proteins may greatly contribute to enhance AR-tolerance in *S. superba*. The identification of novel AR-responsive proteins in this study provides not only new insights into AR stress responses,

but also a good starting point for further exploration of the differential AR adaptation strategies between sensitive and tolerant tree species.

Supporting Information

Figure S1   Effects of AR on *CaM* and *CDPK* gene expressions in *L. formosana* and *S. superba*. The relative gene expression level was calculated based on AR/CK.

Table S1   Details of identified proteins and peptides list of each protein in AR-treated *L. formosana*.

Table S2   Details of identified proteins and peptides list of each protein in AR-treated *S. superba*.

Author Contributions

Conceived and designed the experiments: JC (first author) WJH HLZ. Performed the experiments: JC (first author) WJH ZJS TWL. Analyzed the data: CW JC (first author) TWL JC XL. Contributed reagents/materials/analysis tools: JC WJH CW WHW HLZ. Contributed to the writing of the manuscript: JC (first author) WJH AC HLZ. Designed the software used in analysis: CW ZJS XL.

References

1. Larssen T, Lydersen E, Tang D, He Y, Gao J, et al. (2006) Acid rain in China. Environ Sci Technol 40: 418–425.
2. Dai Z, Liu X, Wu J, Xu J (2012) Impacts of simulated acid rain on recalcitrance of two different soils. Environ Sci Pollut Res 20: 4216–4224.
3. Liu TW, Wu FH, Wang WH, Chen J, Li ZJ, et al. (2011) Effects of calcium on seed germination, seedling growth and photosynthesis of six forest tree species under simulated acid rain. Tree Physiol 31: 402–413.
4. Fan HB, Wang YH (2000) Effects of simulated acid rain on germination, foliar damage, chlorophyll contents and seedling growth of five hardwood species growing in China. Forest Ecol Manag 126: 321–329.
5. Liu TW, Jiang XW, Shi WL, Chen J, Pei ZM, et al. (2011) Comparative proteomic analysis of differentially expressed proteins in beta-aminobutyric acid enhanced Arabidopsis thaliana tolerance to simulated acid rain. Proteomics 11: 2079–2094.
6. Sun Z, Wang L, Chen M, Liang C, Zhou Q, et al. (2012) Interactive effects of cadmium and acid rain on photosynthetic light reaction in soybean seedlings. Ecotox Environ Safe 79: 62–68.
7. Kovacik J, Klejdus B, Backor M, Stork F, Hedbavny J (2011) Physiological responses of root-less epiphytic plants to acid rain. Ecotoxicology 20: 348–357.
8. DeHayes DH, Schaberg PG, Hawley GJ, Strimbeck GR (1999) Acid rain impacts on calcium nutrition and forest health. BioScience 49: 789–800.
9. Feng ZW, Miao H, Zhang FZ, Huang YZ (2002) Effects of acid deposition on terrestrial ecosystems and their rehabilitation strategies in China. J Environ Sci-China 14: 227–233.
10. Liu J, Zhou G, Yang C, Ou Z, Peng C (2006) Responses of chlorophyll fluorescence and xanthophyll cycle in leaves of Schima superba Gardn. & Champ. and Pinus massoniana Lamb. to simulated acid rain at Dinghushan Biosphere Reserve, China. Acta Physiol Plant 29: 33–38.
11. Chen J, Wang WH, Liu TW, Wu FH, Zheng HL (2013) Photosynthetic and antioxidant responses of Liquidambar formosana and Schima superba seedlings to sulfuric-rich and nitric-rich simulated acid rain. Plant Physiol Biol 64: 41–51.
12. Yang Q, Wang Y, Zhang J, Shi W, Qian C, et al. (2007) Identification of aluminum-responsive proteins in rice roots by a proteomic approach: cysteine synthase as a key player in Al response. Proteomics 7: 737–749.
13. Bai X, Yang L, Yang Y, Ahmad P, Hu X (2011) Deciphering the protective role of nitric oxide against salt stress at the physiological and proteomic levels in maize. J Proteome Res 10: 4349–4364.
14. Sehrawat A, Gupta R, Deswal R (2013) Nitric oxide-cold stress signalling crosstalk-evolution of a novel regulatory mechanism. Proteomics 13: 1816–1835.
15. Hossain Z, Komatsu S (2012) Contribution of proteomic studies towards understanding plant heavy metal stress response. Front Plant Sci 3: 310.
16. Liu TW, Fu B, Niu L, Chen J, Wang WH, et al. (2011) Comparative proteomic analysis of proteins in response to simulated acid rain in Arabidopsis. J Proteome Res 10: 2579–2589.
17. Huang SS, Chen J, Dong XJ, Patton J, Pei ZM, et al. (2011) Calcium and calcium receptor CAS promote Arabidopsis thaliana de-etiolation. Physiol Plant 144: 73–82.
18. Chen J, Wu FH, Liu TW, Chen L, Xiao Q, et al. (2012) Emissions of nitric oxide from 79 plant species in response to simulated nitrogen deposition. Environ Pollut 160: 192–200.
19. Jiang H, Korpelainen H, Li C (2013) Populus yunnanensis males adopt more efficient protective strategies than females to cope with excess zinc and acid rain. Chemosphere 91: 1213–1220.
20. Montillet JL, Chamnongpol S, Rusterucci C, Dat J, van de Cotte B, et al. (2005) Fatty acid hydroperoxides and H$_2$O$_2$ in the execution of hypersensitive cell death in tobacco leaves. Plant Physiol 138: 1516–1526.
21. Hu WJ, Chen J, Liu TW, Liu X, Chen J, et al. (2013) Comparative proteomic analysis on wild type and nitric oxide-overproducing mutant (nox1) of Arabidopsis thaliana. Nitric Oxide 36C: 19–30.
22. Hu WJ, Chen J, Liu TW, Wu Q, Wang WH, et al. (2014) Proteome and calcium-related gene expression in Pinus massoniana needles in response to acid rain under different calcium levels. Plant Soil doi: 10.1007/s11104-014-2086-9.
23. Neves NR, Oliva MA, da Cruz Centeno D, Costa AC, Ribas RF, et al. (2009) Photosynthesis and oxidative stress in the restinga plant species Eugenia uniflora L. exposed to simulated acid rain and iron ore dust deposition: potential use in environmental risk assessment. Sci Total Environ 407: 3740–3745.
24. Wyrwicka A, Skłodowska M (2006) Influence of repeated acid rain treatment on antioxidative enzyme activities and on lipid peroxidation in cucumber leaves. Environ Exp Bot 56: 198–204.
25. Pang Q, Chen S, Dai S, Chen Y, Wang Y, et al. (2010) Comparative proteomics of salt tolerance in Arabidopsis thaliana and Thellungiella halophila. J Proteome Res 9: 2584–2599.
26. Yamasato A, Nagata N, Tanaka R, Tanaka A (2005) The N-terminal domain of chlorophyllide a oxygenase confers protein instability in response to chlorophyll B accumulation in Arabidopsis. Plant Cell 17: 1585–1597.
27. Villeth GR, Reis FB Jr, Tonietto A, Huergo L, de Souza EM, et al. (2009) Comparative proteome analysis of Xanthomonas campestris pv. campestris in the interaction with the susceptible and the resistant cultivars of Brassica oleracea. FEMS Microbiol Lett 298: 260–266.
28. Portis ARJ (2003) Rubisco activase – Rubisco's catalytic chaperone. Photosynth Res 75: 11–27.
29. Jiang Y, Yang B, Harris NS, Deyholos MK (2007) Comparative proteomic analysis of NaCl stress-responsive proteins in Arabidopsis roots. J Exp Bot 58: 3591–3607.
30. Kim DW, Rakwal R, Agrawal GK, Jung YH, Shibato J, et al. (2005) A hydroponic rice seedling culture model system for investigating proteome of salt stress in rice leaf. Electrophoresis 26: 4521–4539.
31. Flagella Z, Trono D, Pompa M, Di Fonzo N, Pastore D (2006) Seawater stress applied at germination affects mitochondrial function in durum wheat (Triticum durum) early seedlings. Funct Plant Biol 33: 357–366.
32. Beato VM, Teresa Navarro-Gochicoa M, Rexach J, Begona Herrera-Rodriguez M, Camacho-Cristobal JJ, et al. (2011) Expression of root glutamate dehydrogenase genes in tobacco plants subjected to boron deprivation. Plant Physiol Biol 49: 1350–1354.

33. Liu TW, Niu L, Fu B, Chen J, Wu FH, et al. (2013) A transcriptomic study reveals differentially expressed genes and pathways respond to simulated acid rain in Arabidopsis thaliana. Genome 56: 49–60.

34. Wang L, Liang W, Xing J, Tan F, Chen Y, et al. (2013) Dynamics of chloroplast proteome in salt-stressed mangrove Kandelia candel (L.) Druce. J Proteome Res 12: 5124–5136.

35. Cheng J, Khan MA, Qiu WM, Li J, Zhou H, et al. (2012) Diversification of genes encoding granule-bound starch synthase in monocots and dicots is marked by multiple genome-wide duplication events. PLoS ONE 7: e30088.

36. Wang SJ, Liu LF, Chen CK, Chen LW (2006) Regulations of granule-bound starch synthase I gene expression in rice leaves by temperature and drought stress. Biol Plant 50: 537–541.

37. Cheng H, Wang J, Chu S, Yan HL, Yu D (2013) Diversifying selection on flavanone 3-hydroxylase and isoflavone synthase genes in cultivated soybean and its wild progenitors. PLoS ONE 8: e54154.

38. Apel K, Hirt H (2004) Reactive oxygen species: metabolism, oxidative stress, and signal transduction. Annu Rev Plant Biol 55: 373–399.

39. Kovacik J, Babula P, Klejdus B, Hedbavny J (2014) Comparison of oxidative stress in four Tillandsia species exposed to cadmium. Plant Physiol Biochem 80C: 33–40.

40. Kovacik J, Klejdus B, Stork F, Hedbavny J (2012) Physiological responses of Tillandsia albida (Bromeliaceae) to long-term foliar metal application. J Hazard Mater 239–240: 175–182.

41. Hu WJ, Chen J, Liu TW, Simon M, Wang WH, et al. (2014) Comparative proteomic analysis of differential responses of Pinus massoniana and Taxus wallichiana var. mairei to simulated acid rain. Int J Mol Sci 15: 4333–4355.

42. Vieira Dos Santos C, Rey P (2006) Plant thioredoxins are key actors in the oxidative stress response. Trends Plant Sci 11: 329–334.

43. Kwon Y, Kim SH, Jung MS, Kim MS, Oh JE, et al. (2007) Arabidopsis hot2 encodes an endochitinase-like protein that is essential for tolerance to heat, salt and drought stresses. Plant J 49: 184–193.

44. Tapia G, Morales-Quintana L, Inostroza L, Acuna H (2011) Molecular characterisation of Ltchi7, a gene encoding a Class III endochitinase induced by drought stress in Lotus spp. Plant Biol 13: 69–77.

45. Foo E, Ross JJ, Davies NW, Reid JB, Weller JL (2006) A role for ethylene in the phytochrome-mediated control of vegetative development. Plant J 46: 911–921.

46. Carvalho RF, Campos ML, Azevedo RA (2011) The role of phytochrome in stress tolerance. J Integr Plant Biol 53: 920–929.

47. Liu J, Zhang F, Zhou J, Chen F, Wang B, et al. (2012) Phytochrome B control of total leaf area and stomatal density affects drought tolerance in rice. Plant Mol Biol 78: 289–300.

48. Sysoeva MI, Markovskaya EF, Sherudilo EG (2013) Role of phytochrome B in the development of cold tolerance in cucumber plants under light and in darkness. Russ J Plant Physiol 60: 383–387.

49. Tang RH, Han SC, Zheng HL, Cook CW, Choi CS, et al. (2007) Coupling diurnal cytosolic Ca^{2+} oscillations to the CAS-IP_3 pathway in Arabidopsis. Science 315: 1423–1426.

50. Kovacik J, Babula P, Hedbavny J, Klejdus B (2014) Hexavalent chromium damages chamomile plants by alteration of antioxidants and its uptake is prevented by calcium. J Hazard Mater 273C: 110–117.

51. Ho SL, Huang LF, Lu CA, He SL, Wang CC, et al. (2013) Sugar starvation- and GA-inducible calcium-dependent protein kinase 1 feedback regulates GA biosynthesis and activates a 14-3-3 protein to confer drought tolerance in rice seedlings. Plant Mol Biol 81: 347–361.

52. Saijo Y, Hata S, Kyozuka J, Shimamoto K, Izui K (2000) Over-expression of a single Ca^{2+}-dependent protein kinase confers both cold and salt/drought tolerance on rice plants. Plant J 23: 319–327.

53. Ji XL, Gai YP, Zheng CC, Mu ZM (2009) Comparative proteomic analysis provides new insights into mulberry dwarf responses in mulberry (Morus alba L.). Proteomics 9: 5328–5339.

54. Pandey A, Chakraborty S, Datta A, Chakraborty N (2008) Proteomics approach to identify dehydration responsive nuclear proteins from chickpea (Cicer arietinum L.). Mol Cel Proteomics 7: 88–107.

55. Zorb C, Schmitt S, Muhling KH (2010) Proteomic changes in maize roots after short-term adjustment to saline growth conditions. Proteomics 10: 4441–4449.

56. Fu J, Ristic Z (2010) Analysis of transgenic wheat (Triticum aestivum L.) harboring a maize (Zea mays L.) gene for plastid EF-Tu: segregation pattern, expression and effects of the transgene. Plant Mol Biol 73: 339–347.

57. Kurepa J, Smalle JA (2008) Structure, function and regulation of plant proteasomes. Biochimie 90: 324–335.

58. Rapaport D (2002) Biogenesis of the mitochondrial TOM complex. Trends Biochem Sci 27: 191–197.

59. Symons GM, Smith JJ, Nomura T, Davies NW, Yokota T, et al. (2008) The hormonal regulation of de-etiolation. Planta 227: 1115–1125.

60. Wang KL, Li H, Ecker JR (2002) Ethylene biosynthesis and signaling networks. Plant Cell 14 Suppl: S131–151.

61. Ravanel S, Gakiere B, Job D, Douce R (1998) The specific features of methionine biosynthesis and metabolism in plants. Proc Natl Acad Sci USA 95: 7805–7812.

62. Sanchez-Aguayo I, Rodriguez-Galan JM, Garcia R, Torreblanca J, Pardo JM (2004) Salt stress enhances xylem development and expression of S-adenosyl-L-methionine synthase in lignifying tissues of tomato plants. Planta 220: 278–285.

63. Firon N, Pressman E, Meir S, Khoury R, Altahan L (2012) Ethylene is involved in maintaining tomato (Solanum lycopersicum) pollen quality under heat-stress conditions. AoB Plants 2012: pls024.

10

Random Whole Metagenomic Sequencing for Forensic Discrimination of Soils

Anastasia S. Khodakova[1]*, Renee J. Smith[1], Leigh Burgoyne[1], Damien Abarno[1,2], Adrian Linacre[1]

1 School of Biological Sciences, Flinders University, Adelaide, Australia, **2** Forensic Science South Australia, Adelaide, Australia

Abstract

Here we assess the ability of random whole metagenomic sequencing approaches to discriminate between similar soils from two geographically distinct urban sites for application in forensic science. Repeat samples from two parklands in residential areas separated by approximately 3 km were collected and the DNA was extracted. Shotgun, whole genome amplification (WGA) and single arbitrarily primed DNA amplification (AP-PCR) based sequencing techniques were then used to generate soil metagenomic profiles. Full and subsampled metagenomic datasets were then annotated against M5NR/M5RNA (taxonomic classification) and SEED Subsystems (metabolic classification) databases. Further comparative analyses were performed using a number of statistical tools including: hierarchical agglomerative clustering (CLUSTER); similarity profile analysis (SIMPROF); non-metric multidimensional scaling (NMDS); and canonical analysis of principal coordinates (CAP) at all major levels of taxonomic and metabolic classification. Our data showed that shotgun and WGA-based approaches generated highly similar metagenomic profiles for the soil samples such that the soil samples could not be distinguished accurately. An AP-PCR based approach was shown to be successful at obtaining reproducible site-specific metagenomic DNA profiles, which in turn were employed for successful discrimination of visually similar soil samples collected from two different locations.

Editor: Carles Lalueza-Fox, Institut de Biologia Evolutiva - Universitat Pompeu Fabra, Spain

Funding: Funding for this research was provided by the Attorney General's Office of South Australia. The funder had no role in study design, data collection and analysis, decision to publish, or preparation of the manuscript.

Competing Interests: The authors have declared that no competing interests exist.

* Email: anastasia.khodakova@flinders.edu.au

Introduction

Soil can be found on items submitted for forensic analysis, however there is currently no reliable method to compare the DNA content of soils for forensic purposes. Soil, owing to its inherent features, adheres under fingernails, to cars, tools, weapons or items of clothing and can transfer during the commission of a criminal act [1]. Soil can also be useful associative evidence in the investigation of wildlife crimes, such as poaching. The presence of soil is often recorded during the forensic examination of exhibits. Due to the lack of a validated analytical method, or set of techniques for meaningful comparison of soil samples, this evidential type provides only limited value in investigations. There is therefore a need to develop such comparative methodologies.

Traditionally forensic analysis of soils involves comparison of its chemical-physical and biological properties [2]. Over the past decades many studies have been undertaken utilizing the chemical profiles of soil using a wide variety of novel sophisticated and rapid analytical methods such as, FTIR [3], X-ray [4] and elemental analysis [5,6]. These methods are mainly mineralogical techniques and define geological characteristics of soil, which differ across a regional scale. Therefore these techniques may be unable to discriminate soils within a small locality [7]. The potential for discriminating soils at a local scale exists with methods of soil microbial community analysis that have been applied for forensic

purposes [8,9]. Previous attempts at DNA based analysis of soils used DNA fingerprinting techniques which evaluate fragment length variation such as terminal restriction fragment length polymorphism (TRFLP) [7,10], denaturing gradient gel electrophoresis (DGGE) [11], amplified ribosomal DNA restriction analysis (ARDRA) [12] and length heterogeneity-polymerase chain reaction (LH-PCR) [5]. Many fragments in the resultant DNA fingerprint appear identical in length but differ in sequence leading to erroneous conclusions of similarity that would be avoided if the DNA sequences of the fragments were known. These methods are rapid and permit high throughput analysis but have insufficient resolution to discriminate complex soil mixtures [13]. All these methods have potential for use in forensic comparisons, however a lack of reproducibility and the potential for false inclusions has restricted their implementation in a forensic setting.

Development of new platforms for high-throughput DNA sequencing (HTS) has made it more affordable and led to the significant growth of HTS-based studies [14–16]. The application of HTS to soil science has allowed for new insight on the diversity of soil microbial communities inhabiting various biomes [17–19].

Gene-targeted, or locus-specific, sequencing which typically targets the 16S rRNA gene is used for characterization of the taxonomic composition and diversity of microbial communities [20,21]. Shotgun sequencing is primarily a method for studying the functional structure of the communities which aims to examine the entire genetic assemblage and, being amplification-independent, relies on variation and commonality of the collective genomes found in a given environmental sample [22,23]. Shotgun typing allows for a more comprehensive perspective on the whole microbial community but is limited by its propensity to favour identification of the most dominant members over rarer organisms [24]. In order to access the rare species found in such a complex matrix as soil, ultra-deep DNA sequencing is required [25].

Soil samples obtained during forensic investigations, by their nature, put specific requirements on any metagenomic approach. The samples are often small and sufficient amount of the sample should remain after analysis for independent re-testing if required. Soil DNA extraction procedure, as an initial step of metagenomic analysis, should provide high quality DNA with a good yield. Commercially available soil DNA extraction kits is an preferable option offering forensic investigators a means for standardizing soil DNA extraction [26–28]. Gene-targeted sequencing based on PCR amplification technique is able to analyse the minute amounts of template DNA recovered in forensic samples. The need for a relatively large amount of initial DNA template for shotgun sequencing makes this approach less suitable for forensic oriented metagenomic analysis but whole-genome amplification (WGA), using Phi 29 DNA polymerase, represents an effective way of enabling whole-genome shotgun sequencing from small quantities of DNA [29].

The ability to identify DNA from the entire genetic composition of a complex soil mixture is desirable for forensic investigation as the DNA from a wide range of organisms may be present: these include the DNA from bacteria, fungi, nematodes, mammals, plant material, and from insect remains. These can be used to generate a rich DNA profile for comparison and meaningful discrimination between samples. Targeted metagenomics technique that are limited to one particular locus, such as the 16S (small subunit (SSU) of rRNA in prokaryotes [30]), ITS (internal transcribed spacers widely used for fungi [31]), or 18S (nuclear SSU rRNA a widely used phylogenetic marker in eukaryotes [32]), do not detect the variability of the entire soil biota thus providing less information for comparison and differentiation of soil samples.

Metagenomic sequencing approaches have been reported that can reliably differentiate soil microbial communities from different soil types and different land use [17,33]. However from a forensic point of view, discrimination of visually similar soil samples taken from geographically different urban areas (community parks with similar plant cover, residential suburbs, soils of similar land management) is of greater importance [7,34].

We report on the assessment of the ability of random whole metagenomic sequencing approaches to produce reproducible site-specific DNA profiles that can be employed for comparative analysis and discrimination between soils of different locality for application in forensic science. We show an assessment of shotgun and WGA-based sequencing techniques as well as the use of a single arbitrarily primed DNA amplification (AP-PCR) for metagenomic soil DNA analysis [35]. The use of AP-PCR was first reported in the 1990s [36,37] and has been applied to genotyping [38,39] and the study of microbial communities [40].

Materials and Methods

Soil sampling

Soil samples were collected from two different sites in Adelaide in July 2013; Location A (S35 01 43.42 E138 34 16.26) and Location B (S35 00 58.09 E138 32 12.03). These locations are separated by approximately 3 km. For each site, triplicate samples

Table 1. General characteristics of full sequencing data.

Sequencing approach	Average number of reads (range)	Number of Mbp	Average read length, bp ± SD	Failed QC(%)	Number of reads with predicted protein coding regions (range)	Number of reads with predicted rRNA genes (range)	Number of assigned features to M5NR database (%)	Number of assigned features to SEED Subsystems (%)	Number of assigned features to M5RNA database (%)
Shotgun	672 542 (531 108–806 483)	133.6	197±73	20	464 929 (325 410–582 708)	82 151 (62 899–96 886)	35	43	1.3
APPCR	468 187 (74370–1 074 266)	70.7	142±69	24	287 840 (49 902–617 609)	44 896 (5 868–104 247)	26	35	0.0
WGA	911 553 (506 028–2 012 359)	178.5	198±75	20	549 355 (354 930–1 032 625)	96 117 (61 694–187 539)	26	30	0.8

Statistical data represented as mean ± Standard Deviation (SD). Percentage of sequences matching to the M5NR, M5RNA and SEED Subsystems databases was determined with an E-value cut-off of $E<1\times10^{-5}$. QC = quality control.

were taken 1 m apart from the upper 1 cm of the soil layer. The samples collected from Location A and Location B represented a dark loam rich in organic matter and were visually very similar. No specific permits were required for these locations and activities. The field studies did not involve endangered or protected species. The soil samples were placed in individual sterile plastic tubes and stored at $-20°C$ until analysis. DNA extraction was performed within 24 hours of soil sampling.

DNA extraction, amplification and sequencing

Metagenomic DNA was isolated from 50 mg of each soil sample using the ZR Soil Microbe DNA Kit (Zymo Research, USA) following the manufacturer's recommendations. The quality of the DNA extracts was verified by gel electrophoresis in a 1% agarose gel stained with ethidium bromide. DNA concentrations were determined using a Qubit dsDNA HS Assay Kit (Invitrogen, USA) on a Qubit fluorometer (Life technologies, USA).

For each of the six samples, WGA was conducted with 20 ng DNA using Phi29 DNA polymerase (REPLI-g, Qiagen, Germany). The quality of amplification products was determined by 1% agarose gel electrophoresis and by quantification on a Qubit fluorometer (Life technologies, USA) after purification with a QIAquick PCR Kit (Qiagen, Germany).

Amplification of extracted soil DNA was performed with an arbitrarily chosen oligonucleotide primer (sequence 5′-GGAGGTGGTGTTCGAGGG-3′), previously reported for generating soil DNA fingerprints [41]. As a template, 4 ng of metagenomic DNA was used. The 25 µL final reaction volume contained $1×$Hotstar Taq buffer (Qiagen, Germany), 2.5 mM Mg^{2+}, 0.2 mM of each dNTPs, 0.4 µM of the arbitrary chosen primer, and 0.5 U HotstarTaq DNA polymerase (Qiagen, Germany). An initial 15 min denaturation step at 95 °C was followed by 42 cycles of 30 s at 94°C, 30 s at 55°C and 1 min at 72°C. A final extension step of 7 min at 72°C was used to complete the reaction. The quality and concentration of purified PCR products (QIAquick PCR Kit, Qiagen, Germany) were determined as described for WGA procedure.

All the manipulations were performed in dedicated DNA extraction and PCR-mixing hoods using sterile DNA/RNA free water (Ambion, USA) and DNA/RNA free plasticware (Eppendorf, Germany). All the procedures of the extraction and amplification were conducted with the necessary no-template controls, including extraction blank controls.

Library preparation from 100 ng of the corresponding DNA specimen for all three methods under evaluation followed by sequencing was performed at the Australian Genome Research Facility (AGRF, http://www.agrf.org.au/, Adelaide, SA, Australia) using Ion Torrent technology (Ion Torrent PGM Sequencer; Life Technologies, USA) on a separate Ion 318 chip for each of the sequencing approaches.

Processing of sequencing data

Raw sequence datasets were uploaded to the Metagenome Rapid Annotation using Subsystem Technology (MG-RAST) server (http://metagenomics.nmpdr.org/) (Meyer et al., 2008) and filtered from low-quality reads prior to annotation. Metagenomic datasets were annotated to protein genes against the M5NR database and SEED Subsystems database resulting in protein-derived taxonomic and metabolic profiles, respectively. In addition taxonomic profiles were generated by comparison of the metagenomic datasets with the M5RNA ribosomal database also available in MG-RAST. The MG-RAST default annotation parameters such as maximum E-value $<1×10^{-5}$, minimum length of alignment of 15 bp, and minimum sequence identity of

60%, were used to identify the best database matches. Metagenomic profiles were generated at all available MG-RAST taxonomic (phylum to species) and metabolic (level 1 to functions) levels of hierarchy. To adjust the differences in sequencing effort across samples, two common procedure of standardization were taken:

1. In the first approach metagenomic profiles were generated using full datasets of the high-quality reads obtained for each sample. For the metagenomic profiles comparison the relative abundance scores for each taxon and metabolic feature were determined by the percentages of respective reads over the total assigned reads. In the text the relative abundance scores found both for the taxonomic and metabolic features are represented as an average ± SD (standard deviation) across all datasets (if not mentioned otherwise).

2. A second approach was based on comparison of metagenomic profiles generated from randomly subsampled datasets of 49 000 annotated reads per sample.

Metagenomic datasets are freely available on the MG-RAST web-server (http://metagenomics.anl.gov/). The MG-RAST sample IDs are listed in the Table S1.

Statistical analysis of data

The species richness was estimated by rarefaction analysis preformed in MG-RAST. The analysis was performed for total taxa identified with the M5NR protein database in randomly subsampled metagenomic datasets (including Bacteria, Archaea, Eukaryota, Viruses, unclassified and other sequences).

Statistical comparison of metagenomes was conducted on square root transformed data using the statistical package Primer v.6 for Windows (Version 6.1.13, PRIMER-E, Plymouth) [43]. To assess the similarity of the taxonomic and metabolic compositions between soil samples, the Bray-Curtis pair-wise similarity measure was employed. The resulting Bray-Curtis similarity matrices were then used for hierarchical agglomerative clustering (CLUSTER) with the results displayed as group average dendrograms. Similarity profile analysis (SIMPROF) was used to test for multivariate structure in the clusters formed. Non-metric multi-dimensional scaling (NMDS) of Bray-Curtis similarities was performed as an unconstrained ordination method to graphically visualise inter-sample relationships. The program RELATE in the PRIMER package was used to calculate the Spearman rank correlation between Bray-Curtis similarity matrices generated from differently standardised datasets at the same level of taxonomic or metabolic resolution [44].

Metagenome profiles were further analysed using canonical analysis of principal coordinates (CAP) using the PERMANOVA+ version 1.0.3 3 add-on to PRIMER [45] as a constrained ordination method to test for significant differences among the *a priori* groups in multivariate space. All metagenomic profiles were divided into 6 groups according to the sequencing approach applied and origin of the samples. The *a priori* hypothesis of 'no difference' within groups was tested at both taxonomic and metabolic levels using CAP analysis by evaluation of a *P*-value obtained after 9999 permutations. The strength of the association between multivariate data and the hypothesis of group differences was indicated by the value of the squared canonical correlation (δ_1^2). An appropriate number of principal coordinates axes (m) used for the CAP analysis were chosen automatically by the CAP routine to minimize errors of a misclassification. In order to validate the ability of the CAP model to classify correctly the

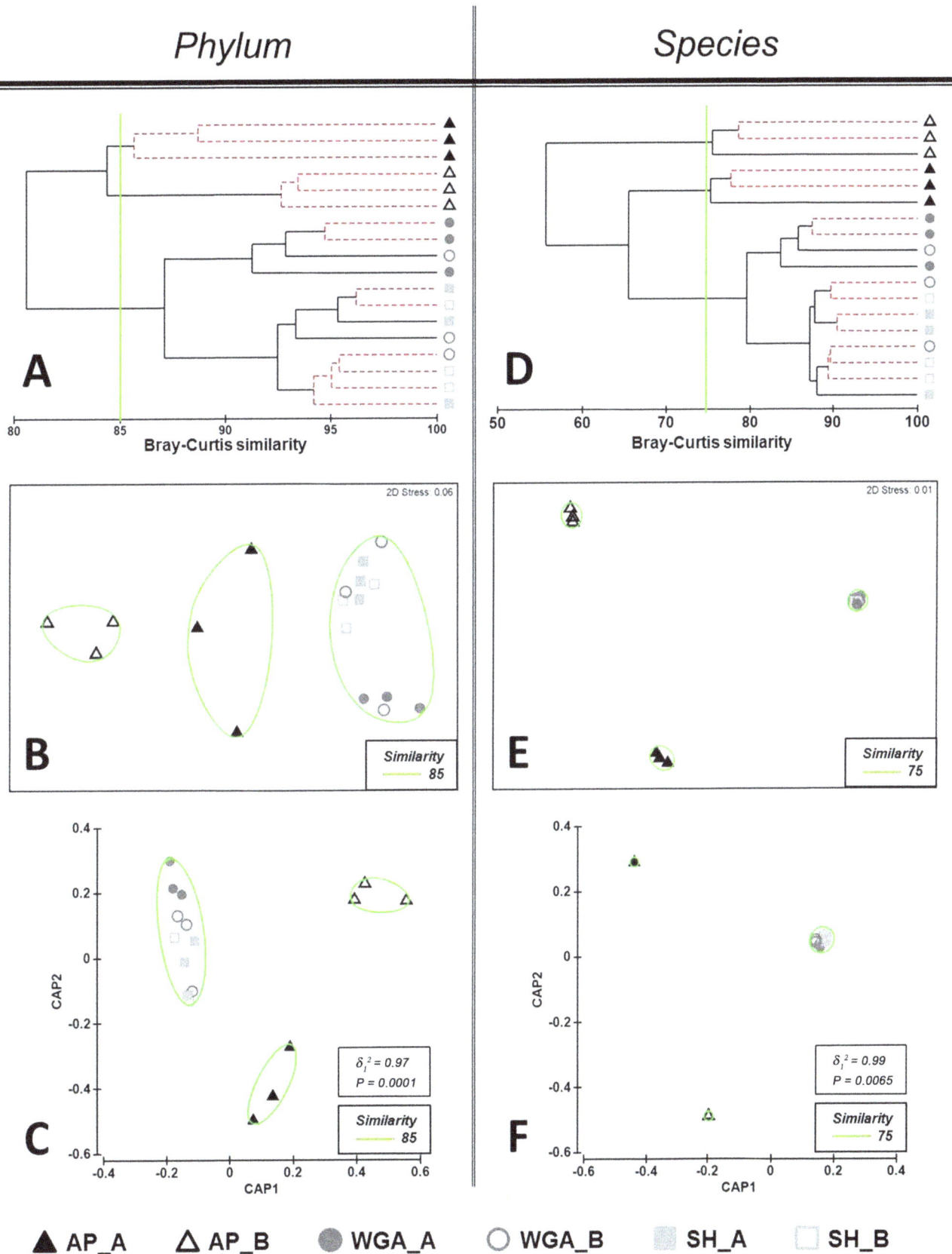

Figure 1. Comparison of the taxonomic soil profiles generated on full datasets at the phylum (A, B, C) and species (D, E, F) resolution levels. Bray-Curtis distance similarity matrix was calculated from the square-root transformed abundance of DNA fragments matching taxa in the M5NR database (E-value $<1 \times 10^{-5}$). The Bray-Curtis matrix was used for generating CLUSTER dendrogram, NMDS and CAP ordination

plots. **CLUSTER analysis (A and D).** Red dotted branches on the CLUSTER dendrogram indicate no significant difference between metagenomic profiles (supported by the SIMPROF analysis, p<0.05). **NMDS unconstrained ordination (B and E).** The NMDS plot displays distances between samples. Data points that are closer to each other represent samples with highly similar metagenomic profiles. **CAP constrained ordination (C and F).** CAP analysis tests for differences among the groups in multivariate space. The significance of group separation along the canonical axis is indicated by the value of the squared canonical correlation (δ_1^2) and P-value. A contour line on the NMDS and CAP ordinations drawn round each of the cluster defines the superimposition of clusters from CLUSTER dendrogram at the selected level of similarity.

samples according to their appropriate groups a cross-validation procedure was performed for the chosen value of m [46].

Results

Notation and general characteristics of sequencing datasets

Obtained datasets were grouped and named according to their sequencing approach and soil sampling sites. Thus samples processed by the AP-PCR approach have a common prefix "AP", shotgun sequenced samples – "SH", and WGA assisted sequencing – "WGA". Each dataset designation identifies the location from where the sample was collected: "_A" – samples collected from location A; and "_B" – samples collected from location B. For example the abbreviation AP_A indicates a sample collected at location A and sequenced by the AP-PCR-based method.

For each soil DNA sample three datasets were generated from the same DNA template using three sequencing approaches. Shotgun metagenome sequencing resulted in an average of 672 542 (531 108–806 843) sequence reads with an average sequence length of 198±73 bases for a total of >133 Mbp of sequence. Sequencing datasets after WGA consisted of an average of 911 554 (506 028–2 012 359) sequences with an average of 198±75 bases in length for a total of >178 Mbp. The AP-based approach gave an average of 468 187 (74 370–1 047 266) reads with an average of 143±69 bases in length for a total of >70.7 Mbp (Table 1). Datasets were annotated using the online MG-RAST server [42]. Approximately 20% of low quality reads were eliminated from each dataset at the filtering step. Only 25–35% of the reads which contained predicted protein coding regions (49 902–689 805 reads per sample), were taxonomically assigned using M5NR protein database. While 30–40% of reads assigned to the SEED Subsystems database were used for generation of metabolic profiles (Table 1). Each of the metagenomic datasets according to the MG-RAST statistics contained approximately 10% of reads with predicted rRNA gene fragments. The subsequent annotation revealed no reads from the AP-based dataset and only 1% of the reads from the SH- and WGA-based datasets matched the M5RNA database.

Taxonomic profiling of metagenomes

The analysis of metagenomic data within MG-RAST occurs both for protein coding genes and ribosomal (rRNA) genes. And therefore analysis of taxonomy can be performed in two ways.

Taxonomic classification of protein gene fragments showed that 85 (±4)% of the annotated reads were assigned to Bacteria, with 4.5 (±2.7)% of reads also matched to Eukaryota and 0.6 (±0.4)% to Archaea. The remaining 10 (±1)% of reads were not assigned. Bacterial taxa *Proteobacteria*, *Actinobacteria* and *Bacteroidetes* dominated in all metagenomic datasets representing close to 70% of protein annotated reads. Additional phyla including *Chloroflexi*, *Planctomycetes*, *Acidobacteria Firmicutes*, *Cyanobacteria*, *Verrucomicrobia* represented less than 5% of reads. Among the eukaryotic taxa, *Ascomicota* was found to be the dominant microorganism 3.0 (±2.6)%. Other eukaryotic taxa such as *Streptophyta*, *Chordata*,

Basidiomycota and *Arthropoda* collectively contributed to the remaining 1% of the annotated reads (Table S2).

Taxonomic classification of the rRNA gene fragments identified only in SH- and WGA-based datasets showed that 78 (±8)% of reads were assigned to bacterial taxa and 14.5 (±6.5)% to eukaryotic taxa (data represented as an average relative abundance of taxa between the samples of SH- and WGA-based datasets). The most abundant bacterial and eukaryotic phyla found were the same as per protein-derived taxonomic classification (described above) namely: *Actinobacteria*, *Proteobacteria*, *Bacteroidetes*, *Ascomycota* and *Streptophyta*. The remaining 7 (±4)% of reads were not assigned (Table S3).

Rarefaction analysis was performed on randomly subsampled metagenomic datasets (49 000 reads per sample) annotated against the M5NR non-redundant protein database. The analysis showed the differences in biodiversity (highest level of taxonomic resolution) of the datasets generated by the three metagenome sequencing approaches (Fig. S1). The SH- and WGA-based datasets demonstrated a similar numbers of identified species from location A and B. A two fold lower number of species were identified in the AP-based dataset.

Metabolic profiling of metagenomes

Metabolic profiles for all datasets were created by matching to the SEED Subsystems database. The most abundant metabolic features found in all datasets, accounting for almost 60% of assigned reads were: clustering-based subsystems; carbohydrates; amino acids and derivatives; protein metabolism; miscellaneous; cofactors; vitamins; prosthetic groups; pigments and DNA metabolism. The relative abundance each of the remaining metabolic features represented less than 5% of reads (Table S4).

Comparison of soil metagenome profiles based on full sequence datasets

Comparison of protein-derived taxonomic profiles. An initial comparison of the taxonomic structures of the metagenomes using lowest (coarsest) resolution profiles derived at the phylum level of taxonomy was performed. CLUSTER analysis with group-average linking based on Bray-Curtis similarity matrices delineated two distinct clusters with similarity of 85% formed by samples from AP-based dataset grouped according to the sites from where the samples were taken (Fig. 1A). These clusters were supported by the SIMPROF analysis that showed statistically significant (p< 0.05) evidence of genuine clustering, as indicated by red dotted branches on the dendrogram (Fig. 1A). Two samples from WGA_A group having 94% profiles similarity also formed such a cluster. Other samples form SH- and WGA-based datasets formed mixed clusters. For example, a sample from the WGA_B group formed a united cluster with a sample from the SH_A group and two samples from the SH_B group (similarity 94%), thus indicating that the samples from two different locations were grouped together incorrectly. One more cluster consisted of two samples from SH_A and SH_B groups with 96% of similarity.

Bray-Curtis distances between metagenomic profiles were then displayed on an NMDS plot (Fig. 1B). NMDS analysis did not reveal a clear visual separation of data. Points denoting samples

Table 2. Results of CAP model cross-validation of soil taxonomic profiles discrimination generated from full sequencing datasets.

Original Group	AP_A	AP_B	WGA_A	WGA_B	SH_A	SH_B
Taxonomy level	*phylum* ($m=6$, $\delta_1{}^2=0.97$, $P=0.0001$)					
% correct	67	100	100	0	67	67
correct/total	2/3	3/3	3/3	0/3	2/3	2/3
Misclassified to group	AP_B	n/a	n/a	SH_B / WGA_A	SH_B	SH_A
Taxonomy level	*class* ($m=5$, $\delta_1{}^2=0.98$, $P=0.0001$)					
% correct	100	100	100	0	67	33
correct/total	3/3	3/3	3/3	0/3	2/3	1/3
Misclassified to group	n/a	n/a	n/a	SH_B	WGA_B	SH_A
Taxonomy level	*order* ($m=3$, $\delta_1{}^2=0.97$, $P=0.0002$)					
% correct	100	100	100	0	67	33
correct/total	3/3	3/3	3/3	0/3	2/3	1/3
Misclassified to group	n/a	n/a	n/a	SH_B	SH_B	SH_A
Taxonomy level	*family* ($m=10$, $\delta_1{}^2=0.99$, $P=0.0034$)					
% correct	100	100	100	0	67	67
correct/total	3/3	3/3	3/3	0/3	2/3	2/3
Misclassified to group	n/a	n/a	n/a	WGA_A / SH_A	SH_B	WGA_B
Taxonomy level	*genus* ($m=11$, $\delta_1{}^2=0.99$, $P=0.01$)					
% correct	100	100	100	0	100	67
correct/total	3/3	3/3	3/3	0/3	3/3	2/3
Misclassified to group	n/a	n/a	n/a	WGA_A / SH_A	n/a	WGA_B
Taxonomy level	*species* ($m=10$, $\delta_1{}^2=0.99$, $P=0.0065$)					
% correct	100	100	100	0	67	67

Table 2. Cont.

Original Group	AP_A	AP_B	WGA_A	WGA_B	SH_A	SH_B
correct/total	3/3	3/3	3/3	0/3	2/3	2/3
Misclassified to group	n/a	n/a	n/a	SH_B WGA_A SH_A	SH_B	WGA_B

from WGA- and SH-based datasets were located much closer together showing a higher similarity of the profiles than points representing AP-based dataset (Fig. 1B). Overlaying clusters on the NMDS plot made visual discrimination of the patterns formed by AP-based dataset easier (Fig. 1B).

It has been noted that the distinct patterns of multi-dimensional datasets could be hidden in the low-dimensional space of NMDS ordination [46]. Consequently for the comparison of our metagenomics datasets, CAP analysis as a constrained ordination method was also performed. CAP analysis tests the hypothesis of whether there is a difference between pre-defined groups. In our research all datasets were divided into 6 groups in accordance with combined factors, including the sequencing approach applied and the origin of soil samples. The results of the CAP ordination at the phylum level demonstrated that the first squared canonical correlation was very large ($\delta_1^2 = 0.97$), indicating the significance of the CAP model. The first canonical axis showed clear separation of the samples within AP-based dataset according to the soil sampling sites. At the same time a close overlapping of the samples from the SH- and WGA-based datasets was observed (Fig. 1C). However, the cross-validation results of the CAP model for the chosen value of $m = 6$ did not confirm the above defined separation of the metagenomic datasets (Table 2). Thus, the most distinct groups, which had a 100% success under cross-validation, were AP_B and WGA_A. One sample from the AP_A group was misclassified to the AP_B group. One sample from each of the SH_A and SH_B groups were misclassified to the SH_B and SH_A groups, respectively. All the samples from the WGA_B group were misclassified to another three different groups (SH_A, SH_B and WGA_A).

It is of note that apart from AP_A and WGA_A groups at the class level of taxonomic resolution the cross-validation of the CAP model showed a 100% correct classification of the samples from AP_B group (Table 2). Additionally one sample from the SH_A group was misclassified to the WGA_B group, whereas two samples from the SH_B group were misclassified to the SH_A and WGA_B groups.

Further CLUSTER analysis, NMDS and CAP ordinations of the metagenomic samples at higher levels of taxonomy demonstrated similar patterns of differentiation as observed at the phylum and class levels (Figs.1, S2,S3). Thus, at the order, family, genus and species levels of resolution two samples from the WGA_A group and two samples from the SH_A group formed separate genuine clusters on the CLUSTER dendrograms (Fig. 1D, S2D, S3A, S3D). Two more genuine mixed clusters were observed consisting of the samples from the SH_B and the WGA_B groups. NMDS and CAP ordinations at all levels of resolution clearly displayed three distinct clusters; two clusters consisting of the samples from the AP_A and the AP_B groups and one mixed cluster of samples from all the other groups (Fig. 1, S2, S3). Cross-validation results of the CAP models at all levels of resolution, starting from the class level, showed an accurate 100% correct classification of samples from the AP-based dataset (Table 2). Despite the visual overlapping of the SH- and WGA-based data points shown on the ordination plots (Fig. 1, S2, S3), the samples from WGA_A group were classified 100% correctly across all levels of taxonomic resolution (Table 2). Of note was that, at the genus level, all samples from the SH_A group were also successfully allocated.

Comparison of taxonomic profiles based on rRNA gene fragment classification. Taxonomic profiles were generated only for the SH- and the WGA-based datasets where the rRNA gene fragments matched to the M5RNA database. The AP-based dataset was excluded from the consecutive comparative analysis

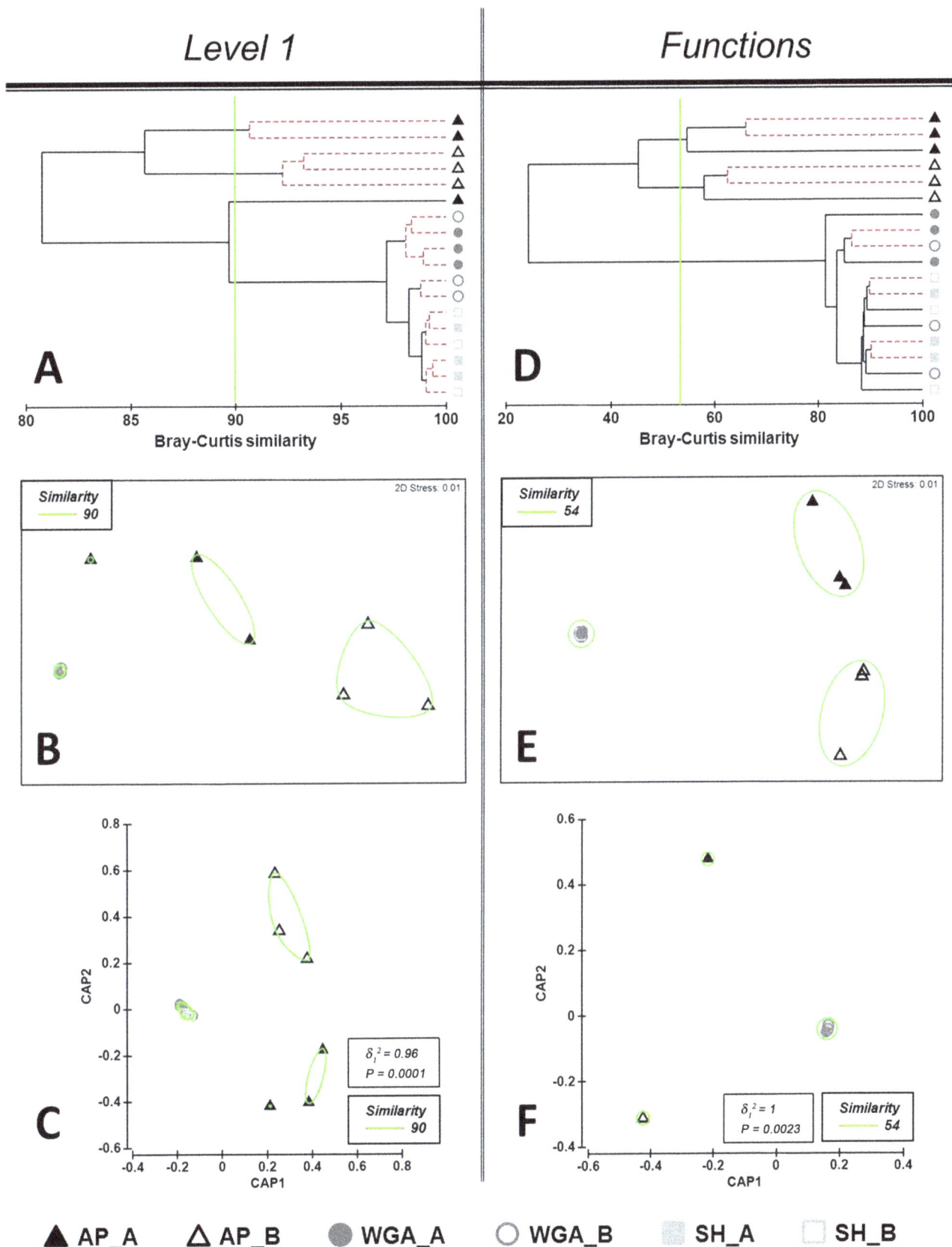

Figure 2. Comparison of the metabolic soil profiles generated on full datasets at the subsystems level 1 (A, B, C) and subsystems function (D, E, F) resolution levels. Bray-Curtis distance similarity matrix was calculated from the square-root transformed abundance of DNA fragments matching taxa in the SEED database (E-value $< 1 \times 10^{-5}$). The Bray-Curtis matrix was used for generating CLUSTER dendrogram, NMDS and

CAP ordination plots. **CLUSTER analysis (A and D).** Red dotted branches on the CLUSTER dendrogram indicate no significant difference between metagenomic profiles (supported by the SIMPROF analysis, p<0.05). **NMDS unconstrained ordination (B and E).** The NMDS plot displays distances between samples. Data points that are closer to each other represent samples with highly similar metagenomic profiles. **CAP constrained ordination (C and F).** CAP analysis tests for differences among the groups in multivariate space. The significance of group separation along the canonical axis is indicated by the value of the squared canonical correlation (δ_1^2) and P-value. A contour line on the NMDS and CAP ordinations drawn round each of the cluster defines the superimposition of clusters from CLUSTER dendrogram at the selected level of similarity.

since no sequence matches to the ribosomal database were found. CLUSTER analysis of rRNA-based taxonomic profiles at the phylum level of resolution demonstrated the formation of four genuine clusters confirmed by SIMPROF analysis (p<0.05) (Fig. S4A). One cluster included three samples from the WGA_A group and one sample from the WGA_B group with similarity of 77%. A second cluster consisted of two samples from the SH_A group and one sample from the SH_B group with similarity of 85%. Two other mixed clusters were formed by the samples from different groups. The pattern formed by the samples from the WGA_A group was also seen on the NMDS and CAP plots with a 100% correct allocation which was confirmed by the results of cross-validation of the CAP model (Fig. S4B, S4C; Table S7). Two separate clusters formed by the samples from the WGA_A and the SH_A groups were observed at the higher levels of taxonomic resolution (genus and species) (Figs. S6). Observed groupings had a 100% correct allocation under cross-validation of the CAP model only at the genus level of classification (Table S7). The latter findings were in full accordance with the allocation of WGA_A and SH_A groups performed using protein-derived taxonomy (Table 2).

Metabolic profiles comparison. CLUSTER analysis of metabolic profiles generated by different sequencing approaches at the lowest level of resolution (level 1) showed that all three samples from the AP_B group formed a separate cluster with a similarity of 92% (Fig. 2A). Two samples from the AP_A group had a similarity of 90%. The third AP_A sample was bundled with the samples from SH- and WGA- based datasets forming a new mixed cluster. Importantly the SH- and WGA-based datasets consisting of 12 metagenomic samples formed one united mixed cluster with a similarity of 97% (Fig. 2A). NMDS and CAP ordinations also showed that all the points associated with the samples from SH- and WGA-based datasets produced a very compact cluster (Fig. 2B and Fig. 2C). However, according to a cross-validation procedure the most distinct groups with 100% allocation success were the AP-based groups and the WGA_A group, whereas misclassification errors were shown for the WGA_B, SH_A and SH_B groups (Table 3). Statistical comparisons of the metabolic profiles at higher resolution levels (level 2, level 3 and function) resulted in similar discriminating success (Fig. 2, S7). CLUSTER analysis showed correct site-specific grouping of the samples from AP-based dataset (Fig. 2D, S7A, S7D). All the profiles produced by SH- and WGA-based methods again formed a single unresolved cluster. NMDS and CAP ordinations demonstrated clear separation of three clusters (Fig. 2, S7), which was also the case for the metagenomic profiles comparison based on protein-derived taxonomy (Fig. 1, S2, S3). In both cases cross-validation results of the CAP model gave 100% correct classification of the samples from the AP_A, AP_B and WGA_A groups and misclassification errors for samples from the SH_A, SH_B and WGA_B groups (Table 2, Table 3).

Comparison of metagenomic profiles based on randomly sub-sampled datasets

Comparison of taxonomic profiles based on rRNA gene fragment classification. CLUSTER analysis and NMDS ordination of rRNA-based taxonomic profiles at the phylum level of taxonomy demonstrated a heterogeneous mixed cluster of the samples from the SH- and WGA-based datasets with an average similarity of 70% (data not shown). Cap analysis showed 100% correct classification of samples from the WGA_A group and misclassification errors for samples from other groups. At the species level of resolution CLUSTER analysis also revealed a single heterogeneous mixed cluster with thea taxonomic profile similarity of approximately 25%. CAP analysis indicated a high degree of misclassification errors.

Comparison of protein-derived taxonomic and metabolic profiles. It has been proposed that in order to enable the comparison of metagenomes based on equal sequencing efforts, the datasets should be randomly sub-sampled to the size of the smallest sample [17,47]. The metagenomic datasets generated by shotgun, WGA-based and AP-PCR-based approaches were re-analysed by MG-RAST at an equivalent sequencing depth of 49 000 annotated reads per sample. Comparison of taxonomic and metabolic profiles generated from sub-sampled datasets at all levels of classification available within MG-RAST was performed by CLUSTER analysis, NMDS and CAP ordination.

Statistical analysis of the sub-sampled metagenomic datasets generated by three metagenome sequencing approaches yielded nearly identical estimates of the overall differences between soil microbial communities from locations A and B as those obtained using full sequence datasets (Figs. S8– S12, Tables S5– S6). This similarity was also confirmed using the RELATE programme which revealed a strong correlation between Bray-Curtis distance matrices (Spearman rank coefficient r>0.9, p<0.0001) generated on both full, and sub-sampled, datasets at all levels of taxonomic and metabolic resolution (Table 4).

Discussion

Numerous ecological studies show that soil microbial communities differ between land uses and vegetation types [17,18,33,48,49]. The discrimination of geographically distinct urban soils with similar land management type and similar plant cover is of great forensic relevance [7,34]. If two soil samples appear very different visually then a simple exclusion can be made but more typically soils appear visually similar and currently no further action is taken. The vast majority of samples submitted for forensic investigation come from urban areas; here we include gardens, parkland and open spaces as well as built-up areas. Thus we focused our study on assessing the ability of three random whole metagenomic sequencing approaches to describe and differentiate the composition of soil microbial communities from two random parklands in 3 km apart within Adelaide residential areas. The vegetation categories of these locations appeared to be very similar, with widespread grass and trees species.

Along with standard metagenomic approaches such as shotgun and WGA, which are widely accepted as the most comprehensive sources of data for studying complex microbial communities, we evaluated AP-PCR as a method for generation of random metagenomic DNA profiles of soils that were then analysed by high throughput DNA sequencing. In this technique an arbitrary chosen oligonucleotide is used as a single primer that targets

Table 3. Results of CAP model cross-validation of soil metabolic profiles discrimination generated from full sequencing datasets.

Original Group	AP_A	AP_B	WGA_A	WGA_B	SH_A	SH_B
Metabolic level *level 1* ($m=2$, $\delta_1^2 = 0.96$, $P = 0.0001$)						
% correct	100	100	100	33	67	33
correct/total	3/3	3/3	3/3	1/3	2/3	1/3
Misclassified to group	n/a	n/a	n/a	SH_A SH_B	SH_B	SH_A WGA_B
Metabolic level *level 2* ($m=11$, $\delta_1^2 = 1$, $P = 0.0002$)						
% correct	100	100	100	33	67	100
correct/total	3/3	3/3	3/3	1/3	2/3	3/3
Misclassified to group	n/a	n/a	n/a	SH_B	SH_B	n/a
Metabolic level *level 3* ($m=12$, $\delta_1^2 = 1$, $P = 0.0009$)						
% correct	100	100	100	33	67	67
correct/total	3/3	3/3	3/3	1/3	2/3	2/3
Misclassified to group	n/a	n/a	n/a	SH_B	SH_B	SH_A
Metabolic level *functions* ($m=10$, $\delta_1^2 = 1$, $P = 0.0023$)						
% correct	100	100	67	0	67	67
correct/total	3/3	3/3	2/3	0/3	2/3	2/3
Misclassified to group	n/a	n/a	WGA_B	SH_B SH_B	SH_B	SH_A

Table 4. RELATE comparison of Bray-Curtis similarity matrices.

Taxonomic level	Spearman rank coefficient	Metabolic level	Spearman rank coefficient
phylum	0.887	level 1	0.652
class	0.944	level 2	0.958
order	0.959	level 3	0.967
family	0.940	functions	0.969
genus	0.965		
species	0.966		

The Bray-Curtis similarity matrices calculated from square root transformed abundance of DNA fragments generated based on full datasets and sub-sampled datasets.

multiple genomic sequences producing a highly primer-sequence-specific profiles. Depending both on the primer chosen and the annealing temperature used there is sequence specific selection of complementary sequences to the primer to be DNA amplified. Based on previous studies the random amplification of polymorphic DNA, normally performed at low stringency conditions (low annealing temperature), becomes more reproducible at high stringency amplification conditions [50]. Amplification with a single arbitrary primer yields an arbitrary product pattern which might possess PCR products from both abundant species and those that are rare, again depending on the affinity of the primer.

The composition of the soil microbial communities was determined from both taxonomic classification of rRNA fragments and the taxonomic assignment of functional gene fragments. Similar taxonomic distribution of dominant microbial phyla was observed across all metagenomic datasets using these two different annotation pipelines. Reads with functional gene fragments were also used for the comparison of metagenomic datasets based on metabolic profiles.

Previous reports have indicated that comparison of metagenomes at low levels of resolution, i.e. analysis based on more broadly defined categories, results in a more conservative estimate of the distances between metagenomic profiles [17]. Low levels of taxonomic or functional classification show less overlap between samples and are therefore also used frequently for metagenomic profile discrimination [51,52]. The results of the metagenomic dataset comparison in the current study are presented at all MG-RAST taxonomic (phylum to species) and metabolic (level 1 to functions) levels of hierarchy. The comparison of metagenomes was performed with a number of unconstrained statistical tools including CLUSTER and NMDS analyses as well as constrained CAP analysis testing a predefined hypothesis that was previously shown to be successful for soil microbial communities discrimination [46,53].

SH- and WGA-based metagemonic sequencing approaches showed incorrect and inconsistent discrimination of soil samples according to sampling sites using both taxonomic (protein and ribosomal) and metabolic classifications. Comparison of the SH- and WGA-based profiles revealed not only misclassification of the samples between the locations but often between repeat analysis of each sequencing approach, with the exception of the WGA_A samples which had a 100% allocation success. The high similarity of the data generated by these methods appears to be driven by the highly similar, or even identical, dominant microorganisms found in the soil samples collected from two distinct sites of similar urban type. This supports the theory that the data generated by shotgun sequencing are commonly shifted towards describing the most abundant taxa leaving the contribution of rare microorganisms undervalued for comparative analysis [54].

A rarefaction analysis was performed to determine microbial species richness of metagenomic datasets produced by three random whole metagenome sequencing approaches for the soil samples from location A and B. The rarefaction curves computed for metagenomic datasets did not reach the plateau phase suggesting that more sequencing effort would be required to achieve species saturation. At the same time the analysis showed that the SH- and WGA-based approaches provided a higher number of species from the same number of sequence reads than the AP-based approach. The AP-PCR utilises primer dependant sequence specific selection of gene fragments and therefore unlikely to amplify all the DNA fragments present in samples. Nevertheless, despite the lower species coverage of soil metagenomes provided by the AP-based approach it allowed for a 100% correct discrimination between soils samples from different locations. This may be as a result of the pre-enrichment mechanisms of AP-PCR that are based on the primer sequence targeting both dominant and rare microorganisms equally. An AP-PCR-based strategy for whole metagenomic profile generation may be compromised by artefacts, including chimeric sequences caused by PCR amplification, which have been reported for gene-targeted (e.g. 16S) sequencing approaches [19]. It is likely that the AP-PCR based approach does not reflect the true picture of the soil microbial community composition. However, we found consistent evidence that an AP-PCR-based whole metagenome sequencing approach was able to discriminate similar soil samples based on differences in both taxonomic and metabolic compositions.

Conclusion

In the research presented here we investigated the ability of whole metagenome analysis techniques to discriminate soil samples of similar land use and vegetation type but collected from different geographical locations. There is currently no agreed evaluation approach leading to an accurate picture of the soil metagenome structure as the true soil microbial community composition [55]. Three methods of whole soil metagenome analysis based on high-throughput DNA sequencing were assessed; shotgun, whole genome amplification and arbitrarily primed PCR. The metagenomic datasets underwent comprehensive statistical analysis using unconstrained and constrained approaches including CLUSTER analysis, NMDS and CAP ordination at all levels of both taxonomic and metabolic classification. The shotgun and WGA-based approaches generated highly similar metagenomic profiles for soil samples such that the soil samples could not be distinguished. An AP-PCR-based approach was shown to be the most powerful technique for obtaining site-specific metagenomic DNA profiles which were able to successfully discriminate between similar soil samples taken from different locations.

The methods presented in this study show a significant step towards possible implementation of forensic soil discrimination using random whole metagenomics for investigation and evidence generation. By increasing the amount of samples analysed from each location and also by increasing the number of distinct geographical locations it will become possible to train algorithms that can then be used for comparison to unknown soil samples obtained as part of criminal investigations. The power of discrimination of these tools is proportional to the amount of samples taken and ultimately the unique metagenomic profile of the different locations. The investigation of temporal microbial variation would further strengthen any tool that is developed. As the sample sizes increase the tool will move from the model developed in this study to one that has sufficient power as a useful investigative tool and ultimately to a method that can be presented in court. The step to being a useful investigative tool for law enforcement can be made from the current study with increased repetition and geographic sampling. For presentation in a court of law the development of a sufficient sample size and distinct geographic profiles will need to be bolstered with a determination of the limitations of the method, including false positive and negative rates. This can be achieved via blind trials, mock case work and a period of casework hardening in order to achieve the levels require for acceptance.

Supporting Information

Figure S1 Rarefaction curves created in MG-RAST. Rarefaction analysis was performed at the species level for each metagenomic protein-derived taxonomic profile based on randomly sub-samples datasets (49 000 reads per sample). The curves for all taxa include Bacteria, Archaea, Eukaryota, Viruses, unclassified and other sequences identified after metagenomic dataset annotation with M5NR database.

Figure S2 Comparison of the soil protein-derived taxonomic profiles generated on full datasets at the class (A, B, C) and order (D, E, F) taxonomic resolution levels. Bray-Curtis distance similarity matrix was calculated from the square-root transformed abundance of DNA fragments matching taxa to the M5NR database (E-value $<1\times10^{-5}$). The Bray-Curtis matrix was used for generating CLUSTER dendrogram, NMDS and CAP ordination plots. **CLUSTER analysis (A and D).** Red dotted branches on the CLUSTER dendrogram indicate no significant difference between metagenomic profiles (supported by the SIMPROF analysis, p<0.05). **NMDS unconstrained ordination (B and E).** The NMDS plot displays distances between samples. Data points that are closer to each other represent samples with highly similar metagenomic profiles. **CAP constrained ordination (C and F).** CAP analysis tests for differences among the groups in multivariate space. The significance of group separation along the canonical axis is indicated by the value of the squared canonical correlation (δ_1^2) and P-value (P<0.05). A contour line on the NMDS and CAP ordinations drawn round each of the cluster defines the superimposition of clusters from CLUSTER dendrogram at the selected level of similarity.

Figure S3 Comparison of the soil rRNA profiles generated on full datasets at the phylum (A, B, C) and class (D, E, F) taxonomic resolution levels. Bray-Curtis distance

similarity matrix was calculated from the square-root transformed abundance of DNA fragments matching taxa in the M5RNA database (E-value $<1\times10^{-5}$). The Bray-Curtis matrix was used for generating CLUSTER dendrogram, NMDS and CAP ordination plots. **CLUSTER analysis (A and D).** Red dotted branches on the CLUSTER dendrogram indicate no significant difference between metagenomic profiles (supported by the SIMPROF analysis, p<0.05). **NMDS unconstrained ordination (B and E).** The NMDS plot displays distances between samples. Data points that are closer to each other represent samples with highly similar metagenomic profiles. **CAP constrained ordination (C and F).** CAP analysis tests for differences among the groups in multivariate space. The significance of group separation along the canonical axis is indicated by the value of the squared canonical correlation (δ_1^2) and P-value (P<0.05). A contour line on the NMDS and CAP ordinations drawn round each of the cluster defines the superimposition of clusters from CLUSTER dendrogram at the selected level of similarity.

Figure S4 Comparison of the soil rRNA profiles generated on full datasets at the phylum (A, B, C) and class (D, E, F) taxonomic resolution levels. Bray-Curtis distance similarity matrix was calculated from the square-root transformed abundance of DNA fragments matching taxa in the M5RNA database (E-value $<1\times10^{-5}$). The Bray-Curtis matrix was used for generating CLUSTER dendrogram, NMDS and CAP ordination plots. **CLUSTER analysis (A and D).** Red dotted branches on the CLUSTER dendrogram indicate no significant difference between metagenomic profiles (supported by the SIMPROF analysis, p<0.05). **NMDS unconstrained ordination (B and E).** The NMDS plot displays distances between samples. Data points that are closer to each other represent samples with highly similar metagenomic profiles. **CAP constrained ordination (C and F).** CAP analysis tests for differences among the groups in multivariate space. The significance of group separation along the canonical axis is indicated by the value of the squared canonical correlation (δ_1^2) and P-value (P<0.05). A contour line on the NMDS and CAP ordinations drawn round each of the cluster defines the superimposition of clusters from CLUSTER dendrogram at the selected level of similarity.

Figure S5 Comparison of the soil rRNA profiles generated on full datasets at the order (A, B, C) and family (D, E, F) taxonomic resolution levels. Bray-Curtis distance similarity matrix was calculated from the square-root transformed abundance of DNA fragments matching taxa in the M5RNA database (E-value $<1\times10^{-5}$). The Bray-Curtis matrix was used for generating CLUSTER dendrogram, NMDS and CAP ordination plots. **CLUSTER analysis (A and D).** Red dotted branches on the CLUSTER dendrogram indicate no significant difference between metagenomic profiles (supported by the SIMPROF analysis, p<0.05). **NMDS unconstrained ordination (B and E).** The NMDS plot displays distances between samples. Data points that are closer to each other represent samples with highly similar metagenomic profiles. **CAP constrained ordination (C and F).** CAP analysis tests for differences among the groups in multivariate space. The significance of group separation along the canonical axis is indicated by the value of the squared canonical correlation (δ_1^2) and P-value (P<0.05). A contour line on the NMDS and CAP ordinations drawn round each of the cluster defines the superimposition of clusters from CLUSTER dendrogram at the selected level of similarity.

Figure S6 Comparison of the soil rRNA profiles generated on full datasets at the genus (A, B, C) and species (D, E, F) taxonomic resolution levels. Bray-Curtis distance similarity matrix was calculated from the square-root transformed abundance of DNA fragments matching taxa in the M5RNA database (E-value $<1 \times 10^{-5}$). The Bray-Curtis matrix was used for generating CLUSTER dendrogram, NMDS and CAP ordination plots. **CLUSTER analysis (A and D).** Red dotted branches on the CLUSTER dendrogram indicate no significant difference between metagenomic profiles (supported by the SIMPROF analysis, p<0.05). **NMDS unconstrained ordination (B and E).** The NMDS plot displays distances between samples. Data points that are closer to each other represent samples with highly similar metagenomic profiles. **CAP constrained ordination (C and F).** CAP analysis tests for differences among the groups in multivariate space. The significance of group separation along the canonical axis is indicated by the value of the squared canonical correlation (δ_1^2) and P-value (P<0.05). A contour line on the NMDS and CAP ordinations drawn round each of the cluster defines the superimposition of clusters from CLUSTER dendrogram at the selected level of similarity.

Figure S7 Comparison of the soil metabolic profiles generated on full datasets at the subsystems level 2 (A, B, C) and level 3 (D, E, F) metabolic resolution levels. Bray-Curtis distance similarity matrix was calculated from the square-root transformed abundance of DNA fragments matching taxa in the SEED database (E-value $<1 \times 10^{-5}$). The Bray-Curtis matrix was used for generating CLUSTER dendrogram, NMDS and CAP ordination plots. **CLUSTER analysis (A and D).** Red dotted branches on the CLUSTER dendrogram indicate no significant difference between metagenomic profiles (supported by the SIMPROF analysis, p<0.05). **NMDS unconstrained ordination (B and E).** The NMDS plot displays distances between samples. Data points that are closer to each other represent samples with highly similar metagenomic profiles. **CAP constrained ordination (C and F).** CAP analysis tests for differences among the groups in multivariate space. The significance of group separation along the canonical axis is indicated by the value of the squared canonical correlation (δ_1^2) and P-value (P<0.05). A contour line on the NMDS and CAP ordinations drawn round each of the cluster defines the superimposition of clusters from CLUSTER dendrogram at the selected level of similarity.

Figure S8 Comparison of the soil protein-derived taxonomic profiles generated on randomly sub-sampled datasets at the phylum (A, B, C) and class (D, E, F) metabolic resolution levels. Bray-Curtis distance similarity matrix was calculated from the square-root transformed abundance of DNA fragments matching taxa in the M5NR database (E-value $<1 \times 10^{-5}$). The Bray-Curtis matrix was used for generating CLUSTER dendrogram, NMDS and CAP ordination plots. **CLUSTER analysis (A and D).** Red dotted branches on the CLUSTER dendrogram indicate no significant difference between metagenomic profiles (supported by the SIMPROF analysis, p<0.05). **NMDS unconstrained ordination (B and E).** The NMDS plot displays distances between samples. Data points that are closer to each other represent samples with

highly similar metagenomic profiles. **CAP constrained ordination (C and F).** CAP analysis tests for differences among the groups in multivariate space. The significance of group separation along the canonical axis is indicated by the value of the squared canonical correlation (δ_1^2) and P-value (P<0.05). A contour line on the NMDS and CAP ordinations drawn round each of the cluster defines the superimposition of clusters from CLUSTER dendrogram at the selected level of similarity.

Figure S9 Comparison of the soil protein-derived taxonomic profiles generated on randomly sub-sampled datasets at the order (A, B, C) and family (D, E, F) taxonomic resolution levels. Bray-Curtis distance similarity matrix was calculated from the square-root transformed abundance of DNA fragments matching taxa in the SEED database (E-value $<1 \times 10^{-5}$). The Bray-Curtis matrix was used for generating CLUSTER dendrogram, NMDS and CAP ordination plots. **CLUSTER analysis (A and D).** Red dotted branches on the CLUSTER dendrogram indicate no significant difference between metagenomic profiles (supported by the SIMPROF analysis, p<0.05). **NMDS unconstrained ordination (B and E).** The NMDS plot displays distances between samples. Data points that are closer to each other represent samples with highly similar metagenomic profiles. **CAP constrained ordination (C and F).** CAP analysis tests for differences among the groups in multivariate space. The significance of group separation along the canonical axis is indicated by the value of the squared canonical correlation (δ_1^2) and P-value (P<0.05). A contour line on the NMDS and CAP ordinations drawn round each of the cluster defines the superimposition of clusters from CLUSTER dendrogram at the selected level of similarity.

Figure S10 Comparison of the soil protein-derived taxonomic profiles generated on randomly sub-sampled datasets at the genus (A, B, C) and species (D, E, F) taxonomic resolution levels. Bray-Curtis distance similarity matrix was calculated from the square-root transformed abundance of DNA fragments matching taxa in the M5NR database (E-value $<1 \times 10^{-5}$). The Bray-Curtis matrix was used for generating CLUSTER dendrogram, NMDS and CAP ordination plots. **CLUSTER analysis (A and D).** Red dotted branches on the CLUSTER dendrogram indicate no significant difference between metagenomic profiles (supported by the SIMPROF analysis, p<0.05). **NMDS unconstrained ordination (B and E).** The NMDS plot displays distances between samples. Data points that are closer to each other represent samples with highly similar metagenomic profiles. **CAP constrained ordination (C and F).** CAP analysis tests for differences among the groups in multivariate space. The significance of group separation along the canonical axis is indicated by the value of the squared canonical correlation (δ_1^2) and P-value (P<0.05). A contour line on the NMDS and CAP ordinations drawn round each of the cluster defines the superimposition of clusters from CLUSTER dendrogram at the selected level of similarity.

Figure S11 Comparison of the soil metabolic profiles generated on randomly sub-sampled datasets at the subsystems level 1 (A, B, C) and subsystems Level 2 (D, E, F) metabolic resolution levels. Bray-Curtis distance similarity matrix was calculated from the square-root transformed abundance of DNA fragments matching taxa in the SEED database (E-value $<1 \times 10^{-5}$). The Bray-Curtis matrix was used for generating CLUSTER dendrogram, NMDS and CAP ordination plots. **CLUSTER analysis (A and D)**. Red dotted branches on the CLUSTER dendrogram indicate no significant difference between metagenomic profiles (supported by the SIMPROF analysis, p<0.05). **NMDS unconstrained ordination (B and E)**. The NMDS plot displays distances between samples. Data points that are closer to each other represent samples with highly similar metagenomic profiles. **CAP constrained ordination (C and F)**. CAP analysis tests for differences among the groups in multivariate space. The significance of group separation along the canonical axis is indicated by the value of the squared canonical correlation (δ_1^2) and P-value (P<0.05)... A contour line on the NMDS and CAP ordinations drawn round each of the cluster defines the superimposition of clusters from CLUSTER dendrogram at the selected level of similarity.

Figure S12   Comparison of the soil metabolic profiles generated on randomly sub-sampled datasets at the subsystems level 3 (A, B, C) and subsystems functions (D, E, F) metabolic resolution levels. Bray-Curtis distance similarity matrix was calculated from the square-root transformed abundance of DNA fragments matching taxa in the SEED database (E-value $<1\times10^{-5}$). The Bray-Curtis matrix was used for generating CLUSTER dendrogram, NMDS and CAP ordination plots. **CLUSTER analysis (A and D)**. Red dotted branches on the CLUSTER dendrogram indicate no significant difference between metagenomic profiles (supported by the SIMPROF analysis, p<0.05). **NMDS unconstrained ordination (B and E)**. The NMDS plot displays distances between samples. Data points that are closer to each other represent samples with highly similar metagenomic profiles. **CAP constrained ordination (C and F)**. CAP analysis tests for differences among the groups in multivariate space. The significance of group separation along the canonical axis is indicated by the value of the squared canonical correlation (δ_1^2) and P-value (P<0.05). A contour line on the NMDS and CAP ordinations drawn round each of the cluster defines the superimposition of clusters from CLUSTER dendrogram at the selected level of similarity.

Table S1   Summary of soil metagenomic samples. All metagenomes are publically available on the MG-RAST server (http://metagenomics.anl.gov/).

Table S2   Protein-derived taxonomic composition of the soil microbial communities. Relative abundances of major taxa (phylum level) derived from taxonomic assignment of protein gene fragments matched to M5NR database.

Table S3   Taxonomic composition of the soil microbial communities based on rRNA gene fragments classification. Relative abundances of major taxa (phylum level) derived from taxonomic assignment of rRNA gene fragments matched to M5RNA database.

Table S4   Metabolic composition of the soil microbial communities. Relative abundances of major metabolic features (level 1) derived from annotation of protein gene fragments to SEED Subsystems database.

Table S5   Results of CAP model cross-validation of soil protein-derived taxonomic profiles discrimination generated from sub-sampled sequencing datasets.

Table S6   Results of CAP model cross-validation of soil metabolic profiles discrimination generated from sub-sampled sequencing datasets.

Table S7   Results of CAP model cross-validation of soil rRNA taxonomic profiles discrimination generated from full sequencing datasets.

Acknowledgments

We would like to thank Dr. Paul Gooding (AGRF, Adelaide) for HTS sequencing and Dr. Sophie Leterme (Flinders University) for helpful assistance in the statistical analysis of the metagenomic data.

Author Contributions

Conceived and designed the experiments: ASK LB AL. Performed the experiments: ASK. Analyzed the data: ASK RJS. Contributed reagents/materials/analysis tools: LB AL. Contributed to the writing of the manuscript: ASK RJS DA AL.

References

1. Fitzpatrick R (2009) Soil: Forensic Analysis. In: Jamieson A, Moenssens A, editors. Wiley Encyclopedia of Forensic Science. The Atrium, Southern Gate, Chichester, West Sussex, United Kingdom: John Wiley&Sons Ltd. 2377–2388.

2. Dawson LA, Hillier S (2010) Measurement of soil characteristics for forensic applications. Surf and Interface Anal 42: 363–377.

3. Cox RJ, Peterson HL, Young J, Cusik C, Espinoza EO (2000) The forensic analysis of soil organic by FTIR. Forensic Sci Int 108: 107–116.

4. Ruffell A, Wiltshire P (2004) Conjunctive use of quantitative and qualitative X-ray diffraction analysis of soils and rocks for forensic analysis. Forensic Sci Int 145: 13–23.

5. Moreno LI, Mills DK, Entry J, Sautter RT, Mathee K (2006) Microbial metagenome profiling using amplicon length heterogeneity-polymerase chain reaction proves more effective than elemental analysis in discriminating soil specimens. J Forensic Sci 51: 1315–1322.

6. Arroyo L, Trejos T, Hosick T, Machemer S, Almirall JR, et al. (2010) Analysis of Soils and Sediments by Laser Ablation Inductively Coupled Plasma Mass Spectrometry (LA-ICP-MS): An Innovative Tool for Environmental Forensics. Environ Forensics 11: 315–327.

7. Macdonald CA, Ang R, Cordiner SJ, Horswell J (2011) Discrimination of soils at regional and local levels using bacterial and fungal T-RFLP profiling. J Forensic Sci 56: 61–69.

8. Horswell J, Cordiner SJ, Maas EW, Martin TM, Sutherland KBW, et al. (2002) Forensic comparison of soils by bacterial community DNA profiling. J Forensic Sci 47: 350–353.

9. Sensabaugh GF (2009) Microbial Community Profiling for the Characterisation of Soil Evidence: Forensic Considerations. In: Ritz K, Dawson L, Miller D, editors. Criminal and Environmental Soil Forensics. Springer. 49–60.

10. Lenz EJ, Foran DR (2010) Bacterial profiling of soil using genus-specific markers and multidimensional scaling. J Forensic Sci 55: 1437–1442.

11. Pasternak Z, Al-Ashhab A, Gatica J (2012) Optimization of molecular methods and statistical procedures for forensic fingerprinting of microbial soil communities. Int Res J Microbiol 3: 363–372.

12. Concheri G, Bertoldi D, Polone E, Otto S, Larcher R, et al. (2011) Chemical elemental distribution and soil DNA fingerprints provide the critical evidence in murder case investigation. PloS One 6: e20222.

13. Rincon-Florez V, Carvalhais L, Schenk P (2013) Culture-Independent Molecular Tools for Soil and Rhizosphere Microbiology. Diversity 5: 581–612.

14. Shokralla S, Spall JL, Gibson JF, Hajibabaei M (2012) Next-generation sequencing technologies for environmental DNA research. Mol Ecol 21: 1794–1805.

15. Loman NJ, Misra RV, Dallman TJ, Constantinidou C, Gharbia SE, et al. (2012) Performance comparison of benchtop high-throughput sequencing platforms. Nat Biotechnol 30: 434–439.

16. Logares R, Haverkamp THA, Kumar S, Lanzén A, Nederbragt AJ, et al. (2012) Environmental microbiology through the lens of high-throughput DNA sequencing: synopsis of current platforms and bioinformatics approaches. J Microbiol Methods 91: 106–113.

17. Fierer N, Leff JW, Adams BJ, Nielsen UN, Bates ST, et al. (2012) Cross-biome metagenomic analyses of soil microbial communities and their functional attributes. Proc Natl Acad Sci USA 109: 21390–21395.

18. Xu Z, Hansen MA, Hansen LH, Jacquiod S, Sørensen SJ (2014) Bioinformatic approaches reveal metagenomic characterization of soil microbial community. PloS One 9: e93445.

19. Wang J, McLenachan P A, Biggs PJ, Winder LH, Schoenfeld BIK, et al. (2013) Environmental bio-monitoring with high-throughput sequencing. Brief Bioinform 14: 575–588.

20. Fierer N, Lauber CL, Ramirez KS, Zaneveld J, Bradford M A, et al. (2012) Comparative metagenomic, phylogenetic and physiological analyses of soil microbial communities across nitrogen gradients. ISME J 6: 1007–1017.

21. Suenaga H (2012) Targeted metagenomics: a high-resolution metagenomics approach for specific gene clusters in complex microbial communities. Environ Microbiol 14: 13–22.

22. Delmont TO, Prestat E, Keegan KP, Faubladier M, Robe P, et al. (2012) Structure, fluctuation and magnitude of a natural grassland soil metagenome. ISME J 6: 1677–1687.

23. Prakash T, Taylor TD (2012) Functional assignment of metagenomic data: challenges and applications. Brief Bioinform 13: 711–727.

24. Fuhrman JA (2012) Metagenomics and its connection to microbial community organization. F1000 Biol Rep 4: 15.

25. Howe AC, Jansson JK, Malfatti SA, Tringe SG, Tiedje JM, et al. (2014) Tackling soil diversity with the assembly of large, complex metagenomes. Proc Natl Acad Sci USA 111: 4904–4909.

26. Young JM, Rawlence NJ, Weyrich LS, Cooper A (2014) Limitations and recommendations for successful DNA extraction from forensic soil samples: A review. Sci Justice 54: 238–244.

27. Sagar K, Singh SP, Goutam KK, Konwar BK (2014) Assessment of five soil DNA extraction methods and a rapid laboratory-developed method for quality soil DNA extraction for 16S rDNA-based amplification and library construction. J Microbiol Methods 97: 68–73.

28. Terrat S, Christen R, Dequiedt S, Lelièvre M, Nowak V, et al. (2012) Molecular biomass and MetaTaxogenomic assessment of soil microbial communities as influenced by soil DNA extraction procedure. Microbiol Biotech 5: 135–141.

29. Binga EK, Lasken RS, Neufeld JD (2008) Something from (almost) nothing: the impact of multiple displacement amplification on microbial ecology. ISME J 2: 233–241.

30. Mao D-P, Zhou Q, Chen C-Y, Quan Z-X (2012) Coverage evaluation of universal bacterial primers using the metagenomic datasets. BMC Microbiol 12: 66.

31. Lewis C, Bilkhu S (2011) Identification of Fungal DNA Barcode Targets and PCR Primers Based on Pfam Protein Families and Taxonomic Hierarchy. Open Applied Informatics: 30–44.

32. Bates ST, Clemente JC, Flores GE, Walters WA, Parfrey LW, et al. (2013) Global biogeography of highly diverse protistan communities in soil. ISME J 7: 652–659.

33. Lauber CL, Ramirez KS, Aanderud Z, Lennon J, Fierer N (2013) Temporal variability in soil microbial communities across land-use types. ISME J 7: 1641–1650.

34. Morrisson A, McColl S, Dawson L (2009) Characterisation and Discrimination of Urban Soils: Preliminary Results from the Soil Forensics University Network. Soil Forensics. In: Ritz K, Dawson L, Miller D, editors. Criminal and Environmental Soil Forensics. Springer. 75–86.

35. Khodakova AS, Burgoyne L, Abarno D, Linacre A (2013) Forensic analysis of soils using single arbitrarily primed amplification and high throughput sequencing. Forensic Sci Int Genet Suppl Ser 4: e39-e40.

36. Welsh J, McClelland M (1990) Fingerprinting genomes using PCR with arbitrary primers. Nucleic Acids Res 18: 7213–7218.

37. Caetano-Anolles G (1993) Amplifying DNA with arbitrary oligonucleotide primers. Genome Res 3: 85–94.

38. Dabrowski W, Czekajlo-Kolodziej U, Medrala D, Giedrys-Kalemba S (2003) Optimisation of AP-PCR fingerprinting discriminatory power for clinical isolates of Pseudomonas aeruginosa. FEMS Microbiol Lett 218: 51–57.

39. Roy S, Biswas D, Vijayachari P, Sugunan A P, Sehgal SC (2004) A 22-mer primer enhances discriminatory power of AP-PCR fingerprinting technique in characterization of leptospires. Trop Med Int Health9: 1203–1209.

40. Franklin RB, Taylor DR, Mills AL (1999) Characterization of microbial communities using randomly amplified polymorphic DNA (RAPD). J Microbiol Methods 35: 225–235.

41. Waters JM, Eariss G, Yeadon PJ, Kirkbride KP, Burgoyne LA, et al. (2012) Arbitrary single primer amplification of trace DNA substrates yields sequence content profiles that are discriminatory and reproducible. Electrophoresis 33: 492–498.

42. Meyer F, Paarmann D, D'Souza M, Olson R, Glass EM, et al. (2008) The metagenomics RAST server - a public resource for the automatic phylogenetic and functional analysis of metagenomes. BMC Bioinform 9: 386.

43. Clarke K, Gorley R (2006) PRIMER V6: User Manual/Tutorial.

44. Clarke K (1993) Non-parametric multivariate analyses of changes in community structure. Aust J Ecol: 117–143.

45. Anderson MJ, Gorley RN CK (2008) PERMANOVA+ for PRIMER: Guide to Software and Statistical Methods.

46. Anderson M, Willis T (2003) Canonical analysis of principal coordinates: a useful method of constrained ordination for ecology. Ecology 84: 511–525.

47. Gilbert JA, Field D, Swift P, Thomas S, Cummings D, et al. (2010) The taxonomic and functional diversity of microbes at a temperate coastal site: a "multi-omic" study of seasonal and diel temporal variation. PloS One 5: e15545.

48. Shange R, Haugabrooks E, Ankumah R, Ibekwe A, Smith R, et al. (2013) Assessing the Diversity and Composition of Bacterial Communities across a Wetland, Transition, Upland Gradient in Macon County Alabama. Diversity 5: 461–478.

49. Uroz S, Ioannidis P, Lengelle J, Cébron A, Morin E, et al. (2013) Functional assays and metagenomic analyses reveals differences between the microbial communities inhabiting the soil horizons of a Norway spruce plantation. PloS One 8: e55929.

50. Atienzar F, Jha A (2006) The random amplified polymorphic DNA (RAPD) assay and related techniques applied to genotoxicity and carcinogenesis studies: a critical review. Mutat Res 613: 76–102.

51. Jeffries TC, Seymour JR, Gilbert JA, Dinsdale E A, Newton K, et al. (2011) Substrate type determines metagenomic profiles from diverse chemical habitats. PloS One 6: e25173.

52. Håvelsrud OE, Haverkamp THA, Kristensen T, Jakobsen KS, Rike AG (2012) Metagenomic and geochemical characterization of pockmarked sediments overlaying the Troll petroleum reservoir in the North Sea. BMC Microbiol 12: 203.

53. Smith RJ, Jeffries TC, Adetutu EM, Fairweather PG, Mitchell JG (2013) Determining the metabolic footprints of hydrocarbon degradation using multivariate analysis. PloS One 8: e81910.

54. Zarraonaindia I, Smith DP, Gilbert JA (2013) Beyond the genome: community-level analysis of the microbial world. Biol Philos 28: 261–282.

55. Delmont TO, Simonet P, Vogel TM (2012) Describing microbial communities and performing global comparisons in the 'omic era. ISME J 6: 1625–1628.

11

Differential Expression of Enzymes Associated with Serine/Glycine Metabolism in Different Breast Cancer Subtypes

Sang Kyum Kim, Woo Hee Jung, Ja Seung Koo*

Department of Pathology, Severance Hospital, Brain Korea 21 PLUS Project for Medical Science, Yonsei University College of Medicine, Seoul, South Korea

Abstract

Purpose: Glycine and serine are well-known, classic metabolites of glycolysis. Here, we profiled the expression of enzymes associated with serine/glycine metabolism in different molecular subtypes of breast cancer and discuss their potential clinical implications.

Methods: We used western blotting and immunohistochemistry to examine five serine-/glycine-metabolism–associated proteins (PHGDH, PSAT, PSPH, SHMT, and GLDC) in six breast cancer cell lines and 709 breast cancer cases using tissue microarray (TMA).

Results: PHGDH and PSPH, associated with serine metabolism, were highly expressed in the TNBC cells. GLDC, associated with glycine metabolism, was highly expressed in HER-2-positive MDA-MB-453 and TNBC-related MDA-MB-435S. TMA showed that the TNBC-type breast cancer tissues highly expressed PHGDH, PSPH, and SHMT1, but not the luminal-A-type tissues ($p < 0.001$). PSPH and SHMT1 expression in the tumor stroma of HER-2-type cancers was the highest, but the luminal-A tissues showed the lowest expression ($p < 0.001$). GLDC was most frequently expressed in cancer cells and stroma of the HER-2-positive cancers and least frequently in TNBC ($p < 0.001$). By Cox multivariate analysis, tumor PSPH positivity (hazard ratio [HR]: 2.068, 95% confidence interval [CI]: 1.049–4.079, $p = 0.036$), stromal PSPH positivity (HR: 2.152, 95% CI: 1.107–4.184, $p = 0.024$), and stromal SHMT1 negativity (HR: 2.142, 95% CI: 1.219–3.764, $p = 0.008$) were associated with short overall survival.

Conclusions: Expression of serine-metabolism–associated proteins was increased in TNBC and decreased in the luminal-A cancers. Expression of glycine-metabolism–associated proteins was high in the tumor and stroma of HER-2-positive cancers.

Editor: Karl X. Chai, University of Central Florida, United States of America

Funding: This research was supported by the Basic Science Research Program through the National Research Foundation of Korea (NRF) funded by the Ministry of Education, Science and Technology (2012R1A1A1002886). This study was supported by a faculty research grant from Yonsei University College of Medicine for 2013 (6-2013-0146). The funders had no role in study design, data collection and analysis, decision to publish, or preparation of the manuscript.

Competing Interests: The authors have declared that no competing interests exist.

* Email: kjs1976@yuhs.ac

Introduction

The "Warburg effect" explains a much higher rate of glycolysis followed by fermentation in the cancer cell mitochondria. According to the Warburg effect, glycolytic intermediates, especially those involved in glycine and serine metabolism, rapidly accumulate in cancer cells [1]. In serine biosynthesis, 3-phosphoglycerate (3PG) produced by glycolysis is oxidized to 3-phosphohydroxypyruvate (pPYR) by phosphoglycerate dehydrogenase (PHGDH), and then pPYR is transaminated to phosphoserine (pSER) by phosphoserine aminotransferase (PSAT). Finally, pSER is dephosphorylated to serine by phosphoserine phosphatase (PSPH). In glycine metabolism, glycine is converted to methylenetetrahydrofolate by glycine decarboxylase (GLDC). On the other hand, serine hydroxymethyltransferase (SHMT) converts serine to glycine reversibly, linking the respective metabolic pathways. According to previous reports, these enzymes are highly expressed in several human tumors: PHGDH in breast cancer and melanoma [2,3] and GLDC in lung cancer [4]. Therefore, they likely play important roles in tumorigenesis [2,3,4,5].

Breast cancer is heterogeneous because it shows disparate clinical, histopathological, and genetic characteristics. Classification of breast cancer is actively underway to identify common characteristics amongst the various subtypes. Breast cancer subtypes include the luminal-A, luminal-B, HER-2, normal-breast-like, and basal-like types defined by gene-profiling analyses [6,7]. Another type, which is negative for estrogen receptor (ER), progesterone receptor (PR), and HER-2—all used as biomarkers in breast cancer treatment—is defined as the triple-negative breast cancer (TNBC) [8].

We believe that metabolic characteristics specific to each subtype could present distinctive hallmarks because each type differs histopathologically, clinically, therapeutically, and prognostically. High levels of glucose transporter 1 (GLUT-1) and

Table 1. Source, clone, and dilution of antibodies.

Antibody	Company	Clone	Dilution
serine/glycine metabolism related proteins			
PHGDH	Abcam, Cambridge, UK	Polyclonal	1:100
PSAT1	Abcam, Cambridge, UK	Polyclonal	1:100
PSPH	Abcam, Cambridge, UK	Polyclonal	1:100
SHMT	Abcam, Cambridge, UK	Polyclonal	1:100
GLDC	Abcam, Cambridge, UK	Polyclonal	1:100
molecular subtype related proteins			
ER	Thermo Scientific, San Siego, CA, USA	SP1	1:100
PR	DAKO, Glostrup, Denmark	PgR	1:50
HER-2	DAKO, Glostrup, Denmark	Polyclonal	1:1500
Ki-67	Abcam, Cambridge, UK	MIB	1:1000

carbonic anhydrase 9 (CA9), glycolysis-associated enzymes, were found in the basal-like type and TNBC subtype [9,10], but expression of glutaminolysis-associated proteins was increased in the HER-2 type [11]. However, few studies have established expression profiling of enzymes associated with serine/glycine metabolism in breast cancer to support a molecular relationship between subtypes and their respective metabolic characteristics. In this study, we profiled the expression of several enzymes associated with serine/glycine metabolism and explored their potential clinical significance. These included PHGDH, PSAT, PSPH, SHMT, and GLDC.

Materials and Methods

Cells and cell culture

Six breast cancer cell lines, MCF-7, MDA-MB-361, MDA-MB-453, MDA-MB-435S, MDA-MB-231, and MDA-MB-468 (all from the American Type Culture Collection), were examined. MCF-7 was maintained in Dulbecco's Modified Eagle's Medium/Nutrient Mixture F12 (DMEM/F12; Gibco) without phenol red, but supplemented with 10 µg/mL insulin (Sigma), 10% fetal bovine serum (Gibco), and 1% penicillin/streptomycin (Gibco). The other cells were maintained in DMEM/F12 containing 10% fetal bovine serum and 1% penicillin/streptomycin. All cells were cultured at 37°C in a humidified atmosphere containing 5% CO_2.

Western blotting

For western blotting, $\sim 8 \times 10^5$ cells were seeded in 60-mm dishes. After 24 h, cells were washed twice with cold phosphate-buffered saline and lysed in the lysis buffer (50 mM Tris-HCL [pH 7.9], 100 mM NaCl, 1 mM EDTA, 2% SDS, 0.1 mM EDTA, and 0.1 mM EGTA) containing a protease and phosphatase inhibitor cocktail (Thermo Scientific). Twenty micrograms of protein was treated with Laemmli sample buffer, heated at 100°C for 5 min, and resolved on 8% or 12% sodium dodecyl sulfate-polyacrylamide gel electrophoresis. Gels were electroblotted onto nitrocellulose membranes (GE Healthcare life-Sciences), which were blocked in 5% non-fat dry milk in TBS-T, and incubated with antibodies for PHGDH, PSAT1, PSPH, SHMT1, GLDC (all at 1:1000, obtained from Abcam), or β-actin (1:2000, Sigma) overnight at 4°C. Membranes were subsequently washed thrice in TBS-T and probed with peroxidase-conjugated goat anti-mouse IgG (1:2000, Santa Cruz) for 1 h at room temperature. Membranes were washed again and developed using a chemilu-

minescent reagent (ECL; GE Healthcare Life Sciences, Inc.). Band densities were measured using TINA imaging software (Raytest, Straubenhardt, Germany).

Patient selection

The institutional review board of Yonsei University Severance Hospital approved this retrospective study. The study population included 709 patients who had been diagnosed histopathologically with invasive ductal carcinoma before tumor excision at Yonsei University Severance Hospital from 2002 to 2006. Patients who had received hormone therapy or chemotherapy before surgery were excluded. This study was approved by the Institutional Review Board (IRB) of Yonsei University Severance Hospital. IRB

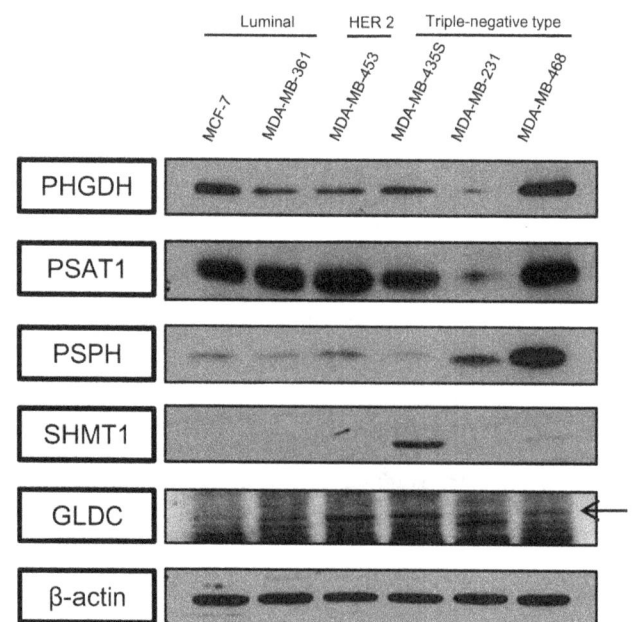

Figure 1. Expression of enzymes associated with serine/glycine metabolism in breast cancer cells detected by western blotting. PHGDH and PSPH levels were increased in the TNBC cell lines and the GLDC levels was increased in MDA-MB-453 of the HER-2 subtype and MDA-MB-435S of the TNBC subtype. SHMT1 was highly expressed in the HER-2 subtype and MDA-MB-435S cells.

waived the informed consent form from patients. Patient records/ information was anonymized and de-identified prior to analysis. Patients' tissue samples were fixed in 10% buffered formalin and embedded in paraffin. Archival tissues stained with hematoxylin and eosin (H&E) were reviewed by three breast pathologists (SK Kim, WH Jung, and JS Koo). Histopathology grading was done according to the Nottingham grading system [12]. Patients' clinicopathological characteristics included age at initial diagnosis, lymph-node metastasis, tumor recurrence, distant metastasis, and survival.

Tissue microarray

A random area was selected on an H&E-stained tissue slide and the corresponding area was marked on the surface of the corresponding paraffin-embedded tissue block. The selected area was punctured using a biopsy needle, and a 3-mm tissue core was extracted and transferred onto a 6×5 recipient block. Two tissue cores were extracted to minimize extraction bias. Each tissue core was assigned a unique tissue microarray location number that was linked to a database recording other clinicopathological data.

Immunohistochemistry

Antibodies used are shown in Table 1. Immunostaining was performed using formalin-fixed, paraffin-embedded (FFPE) tissue sections. Briefly, 5-μm-thick sections were cut using a microtome, transferred onto adhesive slides, and dried at 62°C for 30 min. After incubation with primary antibodies, immunodetection was performed using biotinylated anti-mouse immunoglobulin, followed by streptavidin-conjugated peroxidase from a streptavidin–

biotin kit. 3,3'-Diaminobenzidine was used as the chromogen substrate. Incubation with the primary antibody was omitted for the negative controls. Positive control tissue samples were used as recommended by the manufacturer. Slides were counterstained with Harris hematoxylin.

Interpretation of immunohistochemical staining

Light microscopy was used to visualize the expression of immunohistochemical markers. Pathological status, including ER, PR, and HER-2 positivity were gathered using patients' pathologic reports. A cut-off value of >1% positively stained nuclei was used to define ER and PR positivity [13]. HER-2 staining was analyzed according to guidelines by the American Society of Clinical Oncology (ASCO)/College of American Pathologists (CAP): 0 = no immunostaining; 1+ = weak incomplete membranous staining of <10% of tumor cells; 2+ = complete membranous staining, either uniform or weak, of ≥10% of tumor cells; and 3+ = uniform intense membranous staining of ≥30% of tumor cells [14]. HER-2 immunostaining was considered positive when strong (3+) membranous staining was observed whereas cases with 0 to 1+ were regarded as negative. Cases showing 2+ HER-2 expression were further examined by fluorescent *in situ* hybridization (FISH) for HER-2 amplification.

For measuring immunostaining intensity, we divided breast cancers into four groups as follows: 0 (negative), 1 (weakly positive), 2 (moderately positive), and 3 (strongly positive). For measuring the proportion of stained cells, we divided breast cancers into three groups as follows: 0 (negative), 1 (positive <30%), and 2 (positive >30%). Immunohistochemical values for GLDC, PSAT, PSPH, PHGDH, and SHMT were calculated by multiplying immunostaining intensity to the proportion of stained cells. The final score after multiplication was classified as follows: 0–1 as negative and 2–6 as positive [15]. Ki-67 labeling index (LI) was defined as the percentage of cells with positive nuclei to total number of cancerous cells.

Tumor phenotype classification

Breast cancer phenotypes were classified according to immunohistochemical results for ER, PR, HER-2, and Ki-67 and HER-2 FISH results as follows [16]: *luminal A type*: ER and/or PR positive, HER-2 negative, and Ki-67 LI <14%; *luminal B type*: (HER-2 negative) ER and/or PR positive, HER-2 negative, and Ki-67 LI ≥14%, (HER-2 positive) ER and/or PR positive, HER-2 overexpressed and/or amplified; *HER-2 overexpression type*: ER and PR negative and HER-2 overexpressed and/or amplified; and *TNBC type*: ER, PR, and HER-2 negative.

FFPE tissue microdissection and protein extraction

Hematoxylin-stained, uncovered slides were prepared using FFPE tissue blocks of five breast cancer cases of each molecular subtype. Tumor components or the corresponding stroma was then captured using laser microdissection (LMD 6500, Leica, Wetzlar, Germany). Microdissected FFPE tissues were deparaffinized in xylene and rehydrated in a graded series of alcohol. Total protein was extracted from the captured microdissected FFPE material using the Qproteome FFPE Tissue Kit (Qiagen, Hilden, Germany). Samples were mixed with the FFPE extraction buffer EXB Plus (100 μl per sample), incubated at 100°C for 20 min, at 80°C for 2 h, and finally centrifuged at 14,000 *g* at 4°C for 15 min. The resultant supernatants were measured by the Bradford assay (Bio-Rad, USA) to determine protein concentration.

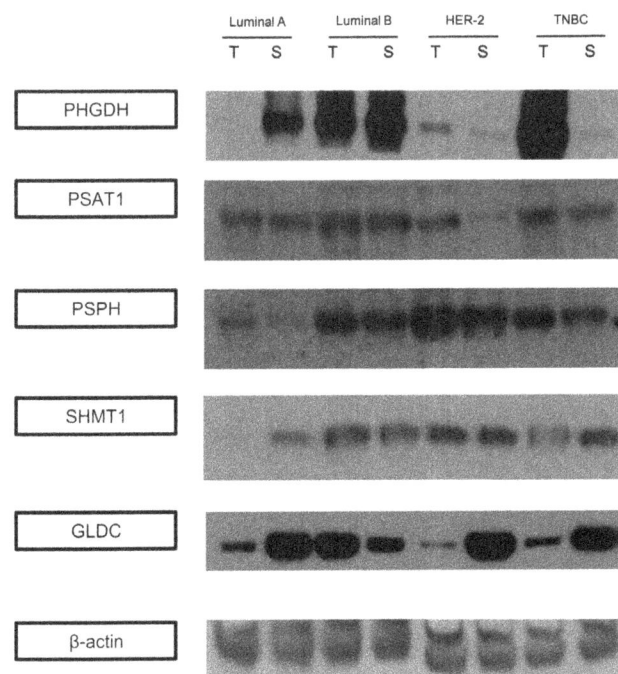

Figure 2. Western blotting detection of enzymes associated with serine/glycine metabolism in breast cancer tissues. PHGDH levels were the highest in the epithelial component of the TNBC and the lowest in the epithelial component of the luminal-A subtype. PSPH and SHMT1 expressions were decreased in the luminal-A subtype. GLDC expression was the highest in the tumor epithelial compartment of the luminal-B subtype and in the stromal compartment of the HER-2 subtype.

Table 2. Patients' clinicopathological characteristics according to breast cancer phenotype.

Parameters	Total (n = 709) (%)	Luminal A (n = 297) (%)	Luminal B (n = 168) (%)	HER-2 (n = 70) (%)	TNBC (n = 174) (%)	P-value
Age (yr, mean ±SD)	49.7±10.9	50.5±10.5	48.4±10.0	52.4±10.0	48.2±12.5	**0.011**
Histologic grade						<0.001
I	119 (16.7)	90 (30.3)	19 (11.3)	1 (1.4)	7 (4.0)	
II	360 (50.6)	179 (60.3)	92 (54.8)	36 (51.4)	53 (30.5)	
III	232 (32.6)	28 (9.4)	57 (33.9)	33 (47.1)	114 (65.5)	
Tumor stage						**0.014**
T1	346 (48.7)	162 (54.5)	86 (51.2)	30 (42.9)	66 (37.9)	
T2	350 (49.2)	127 (42.8)	80 (47.6)	39 (55.7)	104 (59.8)	
T3	15 (2.1)	8 (2.7)	2 (1.2)	1 (1.4)	4 (2.3)	
Nodal stage						0.055
N0	420 (59.1)	170 (57.2)	92 (54.8)	42 (60.0)	115 (66.1)	
N1	188 (26.4)	86 (29.0)	43 (25.6)	13 (18.6)	45 (25.9)	
N2	64 (9.0)	26 (8.8)	18 (10.7)	10 (14.3)	10 (5.7)	
N3	39 (5.5)	15 (5.1)	15 (8.9)	5 (7.1)	4 (2.3)	
Estrogen receptor status						<0.001
Negative	254 (35.7)	5 (1.7)	5 (3.0)	70 (100.0)	174 (100.0)	
Positive	457 (64.3)	292 (98.3)	163 (97.0)	0 (0.0)	0 (0.0)	
Progesterone receptor status						<0.001
Negative	339 (47.7)	48 (16.2)	48 (28.6)	69 (98.6)	174 (100.0)	
Positive	372 (52.3)	249 (83.8)	120 (71.4)	1 (1.4)	0 (0.0)	
HER-2 status						<0.001
0	262 (36.8)	107 (36.0)	24 (14.3)	0 (0.0)	129 (74.1)	
1+	184 (25.9)	119 (40.1)	33 (19.6)	0 (0.0)	32 (18.4)	
2+	141 (19.8)	71 (23.9)	41 (24.4)	16 (22.9)	13 (7.5)	
3+	124 (17.4)	0 (0.0)	70 (41.7)	54 (77.1)	0 (0.0)	
Ki-67 LI (%, mean ±SD)	17.4±18.4	4.7±3.7	19.5±12.4	19.7±12.9	36.1±22.9	<0.001
Tumor recurrence	63 (8.9)	14 (4.7)	13 (7.7)	12 (17.1)	24 (13.8)	**0.001**
Patients' death	59 (8.3)	12 (4.0)	12 (7.1)	12 (17.1)	23 (13.2)	<0.001
Duration of clinical follow-up (months, mean ±SD)	69.9±31.2	72.7±29.6	70.1±30.1	64.9±34.1	67.0±33.5	0.127

TNBC, triple negative breast cancer.

Statistical analyses

Data were analyzed using the SPSS software (Version 12.0; SPSS Inc., Chicago, IL, USA) for Microsoft Windows. Statistical significance was established by the Student's t and Fisher's exact tests for continuous and categorical variables, respectively. A corrected p-value and the Bonferroni method were used for multiple comparisons. Statistical significance was when p<0.05. Kaplan–Meier survival curves and log-rank statistics were used to evaluate time to tumor recurrence and overall survival. Multivariate regression analysis was performed using the Cox proportional-hazards model.

Results

Detection of enzymes associated with serine/glycine metabolism in cell lines

Western blotting of enzymes associated with serine/glycine metabolism in six human breast cancer cell lines is presented in Figure 1. The density of each protein was calculated relative to β-actin and assessed in relation to the molecular subtypes of the tested cell lines: MCF-7 and MDA-MB-361 representing the luminal type; MDA-MB-453, HER-2 type; and MDA-MB-453S, MDA-MB-231, and MDA-MB-468, TNBC type. We confirmed that these proteins were expressed in cell lines of the luminal, HER-2, and TNBC types.

Detection of enzymes associated with serine/glycine metabolism in patients' specimens

Next, expression of the abovementioned proteins was examined in patients' specimens and correlated to the corresponding molecular subtypes. Each tumor tissue was divided into epithelial and the corresponding stromal components by laser microdissection. Protein expression profiles were compared between the epithelial and stromal components in each cancer subtype (Figure 2). All serine-metabolism–associated enzymes were expressed to a greater extent in the epithelial than in the stromal component of the HER-2 and TNBC types. GLDC was expressed to a greater extent in the epithelial rather than in the stromal

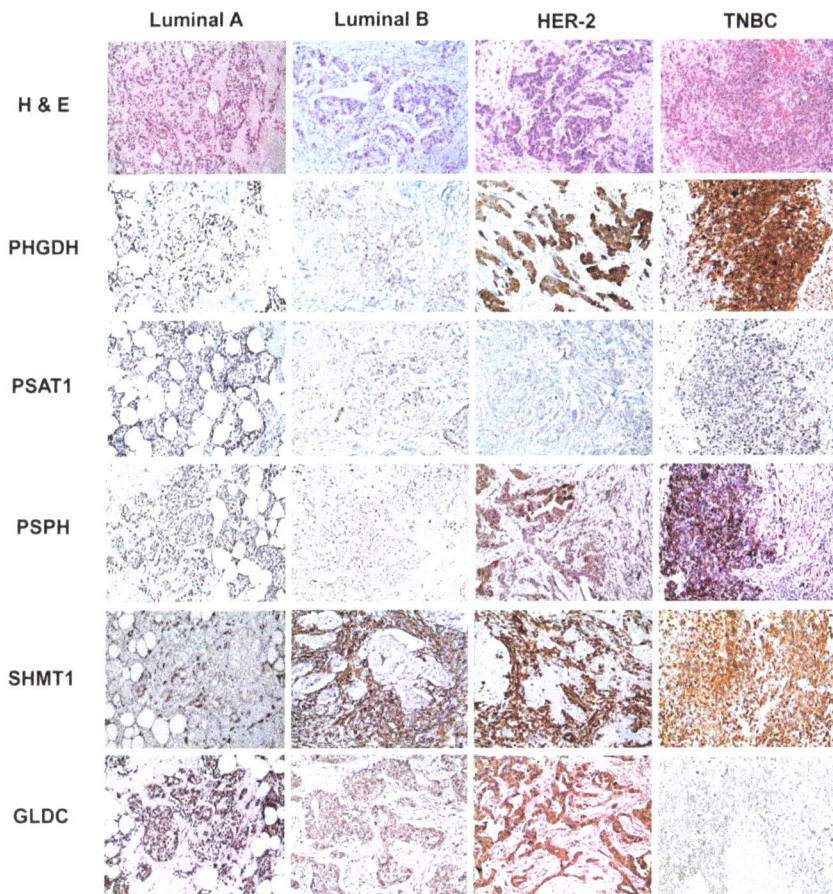

Figure 3. Expression of enzymes associated with serine/glycine metabolism in different molecular breast cancer subtypes. PHGDH levels were increased in the epithelial component of the HER-2 and TNBC subtypes. PSPH was increased in the tumor epithelium of the TNBC subtype and in the HER-2 stroma. SHMT1 expression was high in the epithelial component of the TNBC subtype and in the stromal component of the luminal-B and HER-2 subtypes. GLDC was high in the epithelial and stromal components of the HER-2 subtype.

component of the luminal-B, HER-2, and TNBC types; however, the opposite pattern was observed in the luminal-A cancers.

Patients' clinicopathological characteristics

Breast cancer tissues included in the TMA analysis were classified into luminal A (297 cases, 41.9%), luminal B (168 cases, 23.7%), HER-2 (70 cases, 9.9%), or TNBC (174 cases, 24.5%; Table 2). Notably, the TNBC type demonstrated higher histopathological grading (p<0.001), higher T staging (p = 0.014), and higher Ki-67 LI (p<0.001) than the other subtypes. Whereas, the HER-2 type occurred in older patients

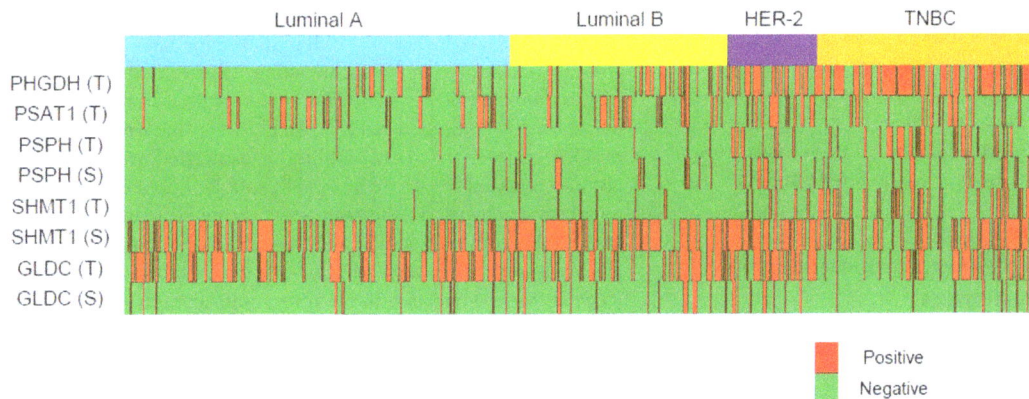

Figure 4. Expression heat map showing levels of enzymes associated with serine/glycine metabolism according to breast cancer molecular subtype.

Table 3. Expression of metabolism-associated enzymes according to breast cancer phenotype.

Parameters	Total (n = 709) (%)	Luminal A (n = 297) (%)	Luminal B (n = 168) (%)	HER-2 (n = 70) (%)	TNBC (n = 174) (%)	P-value
PHGDH in tumor						**<0.001**
Negative	452 (63.8)	253 (85.2)	124 (73.8)	33 (47.1)	42 (24.1)	
Positive	132 (75.9)	44 (14.8)	44 (26.2)	37 (52.9)	132 (75.9)	
PSAT1 in tumor						0.061
Negative	526 (74.2)	229 (77.1)	126 (75.0)	43 (61.4)	128 (73.6)	
Positive	183 (25.8)	68 (22.9)	42 (25.0)	27 (38.6)	46 (26.4)	
PSPH in tumor						**<0.001**
Negative	611 (86.2)	284 (95.6)	157 (93.5)	55 (78.6)	115 (66.1)	
Positive	98 (13.8)	13 (4.4)	11 (6.5)	15 (21.4)	59 (33.9)	
PSPH in stroma						**<0.001**
Negative	625 (88.2)	287 (96.6)	141 (83.9)	54 (77.1)	143 (82.2)	
Positive	84 (11.8)	10 (3.4)	27 (16.1)	16 (22.9)	31 (17.8)	
SHMT1 in tumor						**<0.001**
Negative	598 (84.3)	284 (95.6)	154 (91.7)	59 (84.3)	101 (58.0)	
Positive	111 (15.7)	13 (4.4)	14 (8.3)	11 (15.7)	73 (42.0)	
SHMT1 in stroma						**<0.001**
Negative	310 (43.7)	156 (52.5)	51 (30.4)	19 (27.1)	84 (48.3)	
Positive	399 (56.3)	141 (47.5)	117 (69.6)	51 (72.9)	90 (51.7)	
GLDC in tumor						**<0.001**
Negative	370 (52.2)	138 (46.5)	97 (57.7)	26 (37.1)	109 (62.6)	
Positive	339 (47.8)	159 (53.5)	71 (42.3)	44 (62.9)	65 (37.4)	
GLDC in stroma						**<0.001**
Negative	621 (87.6)	271 (91.2)	135 (80.4)	53 (75.7)	162 (93.1)	
Positive	88 (12.4)	26 (8.8)	33 (19.6)	17 (24.3)	12 (6.9)	

TNBC, triple negative breast cancer.

($p = 0.011$), more frequently recurred, and resulted in higher death rates ($p = 0.001$) than the other types (Table 2).

Expression profiling of metabolism-associated enzymes in relation to cancer phenotypes

Serine-/glycine-metabolism–associated proteins were investigated in cancer tissues by immunohistochemistry (Figure 3, Figure 4 and Table 3). PHGDH, PSPH, and SHMT1 were highly expressed in the epithelial component of the TNBC subtype. However, expression was the lowest in the luminal-A subtype ($p < 0.001$). Similarly, the stromal component expressed the highest PSPH and SHMT1 levels in the HER-2, but the lowest in the luminal-A subtype ($p < 0.001$).

GLDC ranked the highest, in both the epithelial and stromal components, in the HER-2 subtype, but the lowest in TNBC ($p < 0.001$).

Correlations between protein expression and patients' clinicopathological characteristics

After assessing expression profiles of the tested proteins, we examined whether expression profiles were correlated with patients' clinicopathological characteristics (Table 4 and Table 5). High levels of PHGDH, PSPH, SHMT1 in tumor and PSPH in tumor stroma were correlated with high histological grading ($p < 0.001$), ER negativity ($p < 0.001$), PR negativity ($p < 0.001$), and high Ki-67 LI ($p < 0.001$). High levels of SHMT1 and GLDC in

the stromal component were correlated with HER-2 positivity ($p < 0.001$). Expression of stromal PSPH was associated with higher T staging ($p = 0.048$), whereas expression of tumor GLDC was associated with lower T staging ($p = 0.008$) and lower Ki-67 LI ($p = 0.008$).

Correlation between expression profiles of proteins associated with serine or glycine metabolism and glycolysis-associated proteins

Previously, we reported differential expression patterns of glycolysis-associated enzymes, including Glut-1, CA9, and monocarboxylate transporter 4 (MCT4) in tissues derived from different breast cancer subtypes [10,17]. Here, we compared expression profiles of these proteins with those associated with serine or glycine metabolism to identify a possible link between these metabolic pathways in the context of the breast cancer (Table 6 and Table 7). By correlation analyses, we found that expression of glycolysis-associated and serine-/-glycine-metabolism–associated enzymes were correlated in separate tumor components, tumoral and stromal, as follows: tumoral Glut-1 and CA9 and tumoral PHGDH, PSPH, and SHMT1 ($p < 0.05$); tumoral CA9, MCT4, and tumoral PSPH ($p < 0.001$); stromal Glut-1 and tumoral PSPH ($p = 0.018$), stromal PSPH ($p < 0.001$) and stromal GLDC ($p < 0.001$); stromal CA9 and stromal PSPH ($p < 0.001$), SHMT1 ($p < 0.001$), and GLDC ($p < 0.001$); stromal MCT4 and tumoral PHGDH ($p < 0.001$), tumoral PSPH ($p < 0.001$), stromal PSPH

Table 4. Correlations between expression of serine-metabolism–associated enzymes and clinicopathological characteristics.

Parameters	PHGDH in tumor			PSAT1 in tumor			PSPH in tumor			PSPH in stroma		
	Negative n=452 (%)	Positive n=257 (%)	P-value*	Negative n=526 (%)	Positive n=183 (%)	P-value*	Negative n=611 (%)	Positive n=98 (%)	P-value*	Negative n=625 (%)	Positive n=84 (%)	P-value*
Age (yr, mean ±SD)	49.6±10.5	49.7±11.6	7.656	49.6±10.7	49.8±11.6	6.504	49.7±10.8	49.7±11.7	7.896	49.8±10.9	48.9±10.7	4.112
Histologic grade			**<0.001**			1.608			**<0.001**			**0.001**
I/II	353 (78.1)	124 (48.2)		361 (68.6)	116 (63.4)		437 (71.5)	40 (40.8)		434 (69.4)	43 (51.2)	
III	99 (21.9)	133 (51.8)		165 (31.4)	67 (36.6)		174 (28.5)	58 (59.2)		191 (30.6)	41 (48.8)	
ER			**<0.001**			0.392			**<0.001**			**<0.001**
Negative	81 (17.9)	173 (67.3)		177 (33.7)	77 (42.1)		179 (29.3)	75 (76.5)		204 (32.6)	50 (59.5)	
Positive	371 (82.1)	84 (32.7)		349 (66.3)	106 (57.9)		432 (70.7)	23 (23.5)		421 (67.4)	34 (40.5)	
PR			**<0.001**			0.688			**<0.001**			**<0.001**
Negative	150 (33.2)	189 (73.5)		241 (45.8)	98 (53.6)		258 (42.2)	81 (82.7)		283 (45.3)	56 (66.7)	
Positive	302 (66.8)	68 (26.5)		285 (54.2)	85 (46.4)		353 (57.8)	17 (17.3)		342 (54.7)	28 (33.3)	
HER-2			6.784			0.360			5.568			0.120
Negative	359 (79.4)	202 (78.6)		426 (81.0)	135 (73.8)		482 (78.9)	79 (80.6)		503 (80.5)	58 (69.0)	
Positive	93 (20.6)	55 (21.4)		100 (19.0)	48 (26.2)		129 (21.1)	19 (19.4)		122 (19.5)	26 (31.0)	
Tumor stage			0.408			0.160			1.232			**0.048**
T1	232 (51.3)	112 (43.6)		269 (51.1)	75 (41.0)		303 (49.6)	41 (41.8)		315 (50.4)	29 (34.5)	
T2/T3	220 (48.7)	145 (56.4)		257 (48.9)	108 (59.0)		308 (50.4)	57 (58.2)		310 (49.6)	55 (65.5)	
Nodal stage			0.904			3.072			0.936			5.584
N0	257 (56.9)	162 (63.0)		316 (60.1)	103 (56.3)		354 (57.9)	65 (66.3)		371 (59.4)	48 (57.1)	
N1/N2/N3	195 (43.1)	95 (37.0)		210 (39.9)	80 (43.7)		257 (42.1)	33 (33.7)		254 (40.6)	36 (42.9)	
Ki-67 LI (%, mean ±SD)	11.0±11.7	28.6±22.4	**<0.001**	18.0±19.3	15.5±15.6	0.912	15.7±17.5	27.8±20.5	**<0.001**	16.1±17.9	27.3±19.3	**<0.001**
Tumor recurrence			2.184			2.328			0.808			1.192
Absent	416 (92.0)	230 (89.5)		483 (91.8)	163 (89.1)		561 (91.8)	85 (86.7)		573 (91.7)	73 (86.9)	
Present	36 (8.0)	27 (10.5)		43 (8.2)	20 (10.9)		50 (8.2)	13 (13.3)		52 (8.3)	11 (13.1)	
Death			0.536			0.696			0.168			0.088
Survival	421 (93.1)	229 (89.1)		488 (92.8)	162 (88.5)		566 (92.6)	84 (85.7)		579 (92.6)	71 (84.5)	
Death	31 (6.9)	28 (10.9)		38 (7.2)	21 (11.5)		45 (7.4)	14 (14.3)		46 (7.4)	13 (15.5)	

*p-value was corrected by the Bonferroni method.

Table 5. Correlations between expression of enzymes associated with glycine metabolism and clinicopathological characteristics.

Parameters	SHMT1 in tumor			SHMT1 in stroma			GLDC in tumor			GLDC in stroma		
	Negative n=598 (%)	Positive n=111 (%)	P-value*	Negative n=310 (%)	Positive n=399 (%)	P-value*	Negative n=370 (%)	Positive n=339 (%)	P-value*	Negative n=621 (%)	Positive n=88 (%)	P-value*
Age (yr, mean ±SD)	50.2±10.9	46.9±10.7	0.032	49.8±11.7	49.6±10.3	6.448	48.5±11.0	50.9±10.7	0.024	49.6±10.9	50.0±10.8	6.072
Histologic grade			<0.001			1.168			4.600			0.416
I/II	431 (72.1)	46 (41.4)		218 (70.3)	259 (64.9)		245 (66.2)	232 (68.4)		426 (68.6)	51 (58.0)	
III	167 (27.9)	65 (58.6)		92 (29.7)	140 (35.1)		125 (33.8)	107 (31.6)		195 (31.4)	37 (42.0)	
ER			<0.001			6.024			2.176			6.496
Negative	164 (27.4)	90 (81.1)		109 (35.2)	145 (36.3)		140 (37.8)	114 (33.6)		224 (36.1)	30 (34.1)	
Positive	434 (72.6)	21 (18.9)		201 (64.8)	254 (63.7)		230 (62.2)	225 (66.4)		397 (63.6)	58 (65.9)	
PR			<0.001			5.200			0.024			6.568
Negative	253 (42.3)	86 (77.5)		145 (46.8)	194 (48.6)		197 (53.2)	142 (41.9)		298 (48.0)	41 (46.6)	
Positive	345 (57.7)	25 (22.5)		165 (53.2)	205 (51.4)		173 (46.8)	197 (58.1)		323 (52.0)	47 (53.4)	
HER-2			1.024			<0.001			0.128			<0.001
Negative	467 (78.1)	94 (84.7)		269 (86.8)	292 (73.2)		306 (82.7)	255 (75.2)		506 (81.5)	55 (62.5)	
Positive	131 (21.9)	17 (15.3)		41 (13.2)	107 (26.8)		64 (17.3)	84 (24.8)		115 (18.5)	33 (37.5)	
Tumor stage			0.392			2.312			0.008			4.560
T1	300 (50.2)	44 (39.6)		143 (46.1)	201 (50.4)		158 (42.7)	186 (54.9)		304 (49.0)	40 (45.5)	
T2/T3	298 (49.8)	67 (60.4)		167 (53.9)	198 (49.6)		212 (57.3)	153 (45.1)		317 (51.0)	48 (54.5)	
Nodal stage			0.736			6.064			5.176			0.216
N0	345 (57.7)	74 (66.7)		181 (58.4)	238 (59.6)		222 (60.0)	197 (58.1)		377 (60.7)	42 (47.7)	
N1/N2/N3	253 (42.3)	37 (33.3)		129 (41.6)	161 (40.4)		148 (40.0)	142 (41.9)		244 (39.3)	46 (52.3)	
Ki-67 LI (%, mean ± SD)	14.5±16.0	32.9±22.6	<0.001	16.4±19.5	18.1±17.6	1.800	19.6±19.7	14.9±16.7	0.008	17.3±18.8	18.0±15.8	5.808
Tumor recurrence			2.200			0.880			0.136			5.504
Absent	548 (91.6)	98 (88.3)		276 (89.0)	370 (92.7)		328 (88.6)	318 (93.8)		567 (91.3)	79 (89.8)	
Present	50 (8.4)	13 (11.7)		34 (11.0)	29 (7.3)		42 (11.4)	21 (6.2)		54 (8.7)	9 (10.2)	
Death			1.504			0.104			1.392			8.000
Survival	552 (92.3)	98 (88.3)		275 (88.7)	375 (94.0)		334 (90.3)	316 (93.2)		569 (91.6)	81 (92.0)	
Death	46 (7.7)	13 (11.7)		35 (11.3)	24 (6.0)		36 (9.7)	23 (6.8)		52 (8.4)	7 (8.0)	

*p-value was corrected by the Bonferroni method.

Table 6. Correlations between expression of enzymes associated with serine/glycine metabolism and glycolysis.

Parameters	Glut-1 in tumor			CAIX in tumor			MCT4 in tumor		
	Negative n = 499 (%)	Positive n = 210 (%)	P-value*	Negative n = 499 (%)	Positive n = 210 (%)	P-value*	Negative n = 529 (%)	Positive n = 180 (%)	P-value*
PHGDH in tumor			<0.001			<0.001			0.132
Negative	361 (72.3)	91 (43.3)		349 (69.3)	106 (50.5)		350 (66.2)	102 (56.7)	
Positive	138 (27.7)	119 (56.7)		153 (30.7)	104 (49.5)		179 (33.8)	78 (43.3)	
PSAT1 in tumor			2.850			1.968			0.846
Negative	374 (74.9)	152 (72.4)		365 (73.1)	161 (76.7)		385 (72.8)	141 (78.3)	
Positive	125 (25.1)	58 (27.6)		134 (26.9)	49 (23.3)		144 (27.2)	39 (21.7)	
PSPH in tumor			<0.001			<0.001			<0.001
Negative	462 (92.6)	149 (71.0)		447 (89.6)	164 (78.1)		471 (89.0)	140 (77.8)	
Positive	37 (7.4)	61 (29.0)		52 (10.4)	46 (21.9)		58 (11.0)	40 (22.2)	
PSPH in stroma			0.120			2.784			0.240
Negative	449 (90.0)	176 (83.8)		437 (87.6)	188 (89.5)		474 (89.6)	151 (83.9)	
Positive	50 (10.0)	34 (16.2)		62 (12.4)	22 (10.5)		55 (10.4)	29 (16.1)	
SHMT1 in tumor			<0.001			0.018			0.216
Negative	456 (91.4)	142 (67.6)		434 (87.0)	164 (78.1)		455 (86.0)	143 (79.4)	
Positive	43 (8.6)	68 (32.4)		65 (13.0)	46 (21.9)		74 (14.0)	37 (20.6)	
SHMT1 in stroma			0.546			0.768			4.926
Negative	208 (41.7)	102 (48.6)		209 (41.9)	101 (48.1)		230 (43.5)	80 (44.4)	
Positive	291 (58.3)	108 (51.4)		290 (58.1)	109 (51.9)		299 (56.5)	100 (55.6)	
GLDC in tumor			1.332			<0.001			3.576
Negative	253 (50.7)	117 (55.7)		286 (57.3)	84 (40.0)		273 (51.6)	97 (53.9)	
Positive	246 (49.3)	93 (44.3)		213 (42.3)	126 (60.0)		256 (48.4)	83 (46.1)	
GLDC in stroma			0.144			4.746			2.922
Negative	428 (85.8)	193 (91.9)		436 (87.4)	185 (88.1)		466 (88.1)	155 (86.1)	
Positive	71 (14.2)	17 (8.1)		63 (12.6)	25 (11.9)		63 (11.9)	25 (13.9)	

*p-value was corrected by the Bonferroni method.

Table 7. Correlations between expression enzymes associated with serine/glycine metabolism and stromal expression of glycolysis-associated enzymes.

Parameters	Glut-1 in stroma			CAIX in stroma			MCT4 in stroma		
	Negative n=693 (%)	Positive n=16 (%)	P-value*	Negative n=595 (%)	Positive n=114 (%)	P-value*	Negative n=409 (%)	Positive n=300 (%)	P-value*
PHGDH in tumor			1.482			0.618			<0.001
Negative	444 (64.1)	8 (50.0)		387 (65.0)	65 (57.0)		297 (72.6)	155 (51.7)	
Positive	249 (35.9)	8 (50.0)		208 (35.0)	49 (43.0)		112 (27.4)	145 (48.3)	
PSAT1 in tumor			6.000			0.462			4.710
Negative	514 (74.2)	12 (75.0)		449 (75.5)	77 (67.5)		305 (74.6)	221 (73.7)	
Positive	179 (25.8)	4 (25.0)		146 (24.5)	37 (32.5)		104 (25.4)	79 (26.3)	
PSPH in tumor			**0.018**			1.410			**<0.001**
Negative	602 (86.9)	9 (56.3)		517 (86.9)	94 (82.5)		372 (91.0)	239 (79.7)	
Positive	91 (13.1)	7 (43.8)		78 (13.1)	20 (17.5)		37 (9.0)	61 (20.3)	
PSPH in stroma			**<0.001**			**<0.001**			**<0.001**
Negative	619 (89.3)	6 (37.5)		542 (91.1)	83 (72.8)		392 (95.8)	233 (77.7)	
Positive	74 (10.7)	10 (62.5)		53 (8.9)	31 (27.2)		17 (4.2)	67 (22.3)	
SHMT1 in tumor			1.788			4.476			0.558
Negative	586 (84.6)	12 (75.0)		503 (84.5)	95 (83.3)		353 (86.3)	245 (81.7)	
Positive	107 (15.4)	4 (25.0)		92 (15.5)	19 (16.7)		56 (13.7)	55 (18.3)	
SHMT1 in stroma			0.060			**<0.001**			**<0.001**
Negative	308 (44.4)	2 (12.5)		292 (49.1)	18 (15.8)		236 (57.7)	74 (24.7)	
Positive	385 (55.6)	14 (87.5)		303 (50.9)	96 (84.2)		173 (42.3)	226 (75.3)	
GLDC in tumor			0.540			0.114			0.072
Negative	365 (52.7)	5 (31.3)		322 (54.1)	48 (42.1)		230 (56.2)	140 (46.7)	
Positive	328 (47.3)	11 (68.8)		273 (45.9)	66 (57.9)		179 (43.8)	160 (53.3)	
GLDC in stroma			**<0.001**			**<0.001**			**<0.001**
Negative	614 (88.6)	7 (43.8)		552 (92.8)	69 (60.5)		387 (94.6)	234 (78.0)	
Positive	79 (11.4)	9 (56.3)		43 (7.2)	45 (39.5)		22 (5.4)	66 (22.0)	

*p-value was corrected by the Bonferroni method.

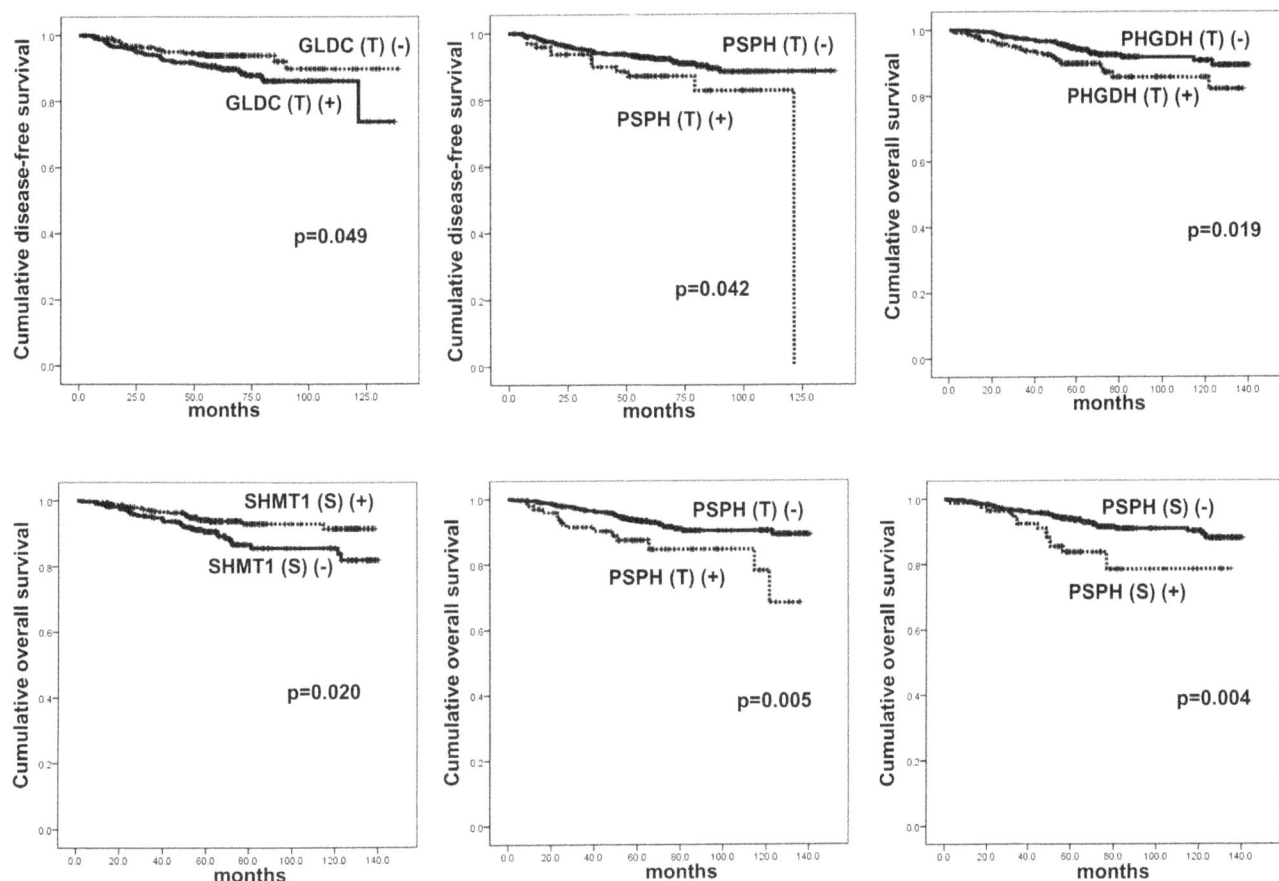

Figure 5. Disease-free and overall survival according to the expression of enzymes associated with serine/glycine metabolism.

(p<0.001), stromal SHMT1 (p<0.001), and stromal GLDC (p< 0.001).

Impact of serine-/glycine-metabolism–associated enzymes on patients' prognosis

Statistics was used to assess the effect of the expression of the metabolism-associated enzymes on patients' survival and prognostic parameters. By univariate analysis, tumoral PSPH positivity (p = 0.042) and tumoral GLDC negativity (p = 0.049) were associated with short disease-free survival (DFS) and tumoral PHGDH positivity (p = 0.019), tumoral PSPH positivity (p = 0.005), stromal PSPH positivity (p = 0.004), and stromal SHMT1 negativity (p = 0.020) were associated with short overall survival (OS) (Table 8 and Figure 5).

By multivariate Cox analysis, high T staging (T1 vs. T2/3, hazard ratio: 2.245, 95% CI: 1.212–4.160, p = 0.01) and lymph-node metastasis (hazard ratio: 2.574, 95% CI: 1.513–4.379, p< 0.001) were associated with short DFS (Table 9). Meanwhile, lymph-node metastasis (hazard ratio: 2.204, 95% CI: 1.284–3.782, p = 0.004), tumoral PSPH positivity (hazard ratio: 2.068, 95% CI: 1.049–4.079, p = 0.036), stromal PSPH positivity (hazard ratio: 2.152, 95% CI: 1.107–4.184, p = 0.024), and stromal SHMT1 negativity (hazard ratio: 2.142, 95% CI: 1.219–3.764, p = 0.008) were associated with short OS.

Next, we examined the correlation between the expression of the enzymes associated with serine/glycine metabolism and survival of breast cancer patients by molecular subtypes. By univariate analysis, patients with each molecular subtype showed

unfavorable OS in accordance with expression profiles of the following proteins (Table S1 in File S1): tumoral PHGDH positivity (p = 0.033), tumoral PSPH positivity (p = 0.020), and stromal SHMT1 negativity (p = 0.026) in the luminal-A; stromal PSPH positivity (p = 0.041) in the HER-2; and stromal SHMT1 negativity (p = 0.040) in the TNBC subtype. In addition, tumoral PSPH positivity (p = 0.042) and stromal SHMT1 negativity (p = 0.009) in the TNBC subtype were associated with unfavorable DFS.

By multivariate Cox analysis (Tabel S2), tumoral PSPH positivity (hazard ratio: 7.067, 95% CI: 1.316–37.95, p = 0.023) and stromal SHMT1 negativity (hazard ratio: 5.300, 95% CI: 1.101–25.50, p = 0.037) were associated with short OS in the luminal-A subtype; lymph-node metastasis (hazard ratio: 17.51, 95% CI: 1.969–155.8, p = 0.010) was correlated with short DFS, and lymph-node metastasis (hazard ratio: 21.49, 95% CI: 2.434–189.7, p = 0.006) and ER negativity (hazard ratio: 14.38, 95% CI: 1.171–176.7, p = 0.037) were correlated with short OS in the luminal-B subtype; lymph-node metastasis (hazard ratio: 6.456, 95% CI: 2.376–17.54, p<0.001), tumoral PHGDH negativity (hazard ratio: 3.358, 95% CI: 1.448–8.866, p = 0.006), tumoral PSPH positivity (hazard ratio: 6.173, 95% CI: 2.323–16.40, p< 0.001), stromal SHMT1 negativity (hazard ratio: 3.312, 95% CI: 1.256–8.736, p = 0.016), and tumoral GLDC negativity (hazard ratio: 4.231, 95% CI: 1.384–12.92, p = 0.011) correlated with short DFS in TNBC subtype. In addition, lymph-node metastasis (hazard ratio: 2.799, 95% CI: 1.119–7.002, p = 0.028), tumoral PHGDH negativity (hazard ratio: 3.624, 95% CI: 1.500–8.757,

Table 8. Impact of expression of enzymes associated with serine/glycine metabolism on disease-free and overall survival tested by log-rank analysis.

Parameters	Number of patients/recurrence/death	Disease-free survival		Overall survival	
		Mean survival (95% CI) months	P-value	Mean survival (95% CI) months	P-value
PHGDH in tumor			0.137		**0.019**
Negative	452/36/31	128 (125–132)		132 (129–135)	
Positive	257/27/28	120 (113–127)		123 (118–128)	
PSAT1 in tumor			0.507		0.225
Negative	526/43/38	126 (123–129)		129 (126–132)	
Positive	183/20/21	124 (118–130)		127 (122–132)	
PSPH in tumor			**0.042**		**0.005**
Negative	611/50/45	128 (125–131)		131 (128–133)	
Positive	98/13/14	107 (99–115)		117 (108–126)	
PSPH in stroma			0.097		**0.004**
Negative	625/52/46	126 (122–130)		131 (128–133)	
Positive	84/11/13	117 (107–127)		116 (106–125)	
SHMT1 in tumor			0.208		0.091
Negative	598/50/46	127 (124–130)		130 (127–133)	
Positive	111/13/13	121 (112–130)		125 (117–132)	
SHMT1 in stroma			0.107		**0.020**
Negative	310/34/35	121 (115–127)		126 (121–130)	
Positive	399/29/24	128 (124–132)		131 (128–134)	
GLDC in tumor			**0.049**		0.377
Negative	370/42/36	122 (116–127)		129 (125–132)	
Positive	339/21/23	130 (126–134)		131 (127–135)	
GLDC in stroma			0.568		0.951
Negative	621/54/52	126 (122–130)		130 (127–132)	
Positive	88/9/7	115 (106–124)		123 (116–130)	

p = 0.004), tumoral PSPH positivity (hazard ratio: 3.880, 95% CI: 1.495–10.07, p = 0.005), and stromal SHMT1 negativity (hazard ratio: 2.605, 95% CI: 1.024–6.626, p = 0.044) correlated with short OS in TNBC subtype.

Discussion

In this study, we examined differential expression patterns of some of the enzymes associated with serine/glycine metabolism in the different molecular subtypes of the breast cancer in separate histopathological tumor compartments (i.e., cancer cell and tumor stroma) and investigated their likely clinical implications. Enzymes associated with serine metabolism, including PHGDH and PSPH were highly expressed in the TNBC cell lines and tissues, but not significantly expressed in the luminal-A subtype. Previous studies have reported increased expression of PHGDH especially in the ER-negative cancers, showing an expression rate about 70% [3]. We also confirmed that PHGDH was highly expressed in the TNBC subtype, representative ER-negative breast cancers. PHGDH was expressed in 68.1% of the ER-negative cancers in comparison with 18.5% in the ER-positive cancers and was associated with high histologic grading, ER negativity, PR negativity, and high Ki-67 LI, which are known as poor prognosticators. Similar to PHGDH, PSPH was frequently expressed in the TNBC subtype and correlated with high

histopathological grading, ER negativity, PR negativity, and high Ki-67 LI. Thus, we hypothesize that a breast cancer subset with high expression of enzymes associated with serine metabolism (PHGDH and PSPH) could present with an aggressive behavior. It is likely that metabolic demand for serine metabolites may increase as the overall metabolic demand increases in breast cancers with aggressive pathological behavior.

Previous reports demonstrated that the glycolysis-associated enzymes, such as Glut-1, CA9, and MCT4, were distinctly expressed in different molecular subtypes: high in the TNBC and basal-like, but low in the luminal-A subtype [9,10]. This study showed that expression of Glut-1 and CA9 were linked to PHGDH and PSPH expression, which could be explained because serine metabolites serve as glycolytic intermediates. Thus, serine metabolism provides α-ketoglutarate, a TCA-cycle intermediate, to cancer cells instead of serine itself, leading to acceleration of mitochondrial metabolism using excess α-ketoglutarate by highly expressed enzymes which drive mitochondrial energy metabolism [3].

Similar to PHGDH and PSPH, SHMT1 was highly expressed in the TNBC subtype and lowest in the luminal-A subtype. As previously reported, glycine metabolism is key in accelerating cancerous cell proliferation; we believe that high SHMT1 levels in the TNBC subtype ensures high proliferative capacity, showing the highest Ki-67 LI among other molecular subtypes [5].

Table 9. Multivariate analysis of breast cancer survival.

Included parameters	Disease-free survival			Overall survival		
	Hazard ratio	95% CI	P-value	Hazard ratio	95% CI	P-value
T stage			**0.010**			0.101
T1 versus T2–3	2.245	1.212–4.160		1.656	0.906–3.024	
N stage			**<0.001**			**0.004**
N0 versus N1–3	2.574	1.513–4.379		2.204	1.284–3.782	
Histologic grade			0.271			0.990
I/II versus III	1.365	0.785–2.374		1.004	0.562–1.791	
ER status			0.129			0.322
Negative versus Positive	1.849	0.836–4.089		1.495	0.675–3.311	
PR status			1.298			0.141
Negative versus Positive	1.298	0.595–2.833		1.839	0.817–4.140	
HER-2 status			0.097			0.112
Negative versus Positive	1.645	0.913–2.962		1.638	0.891–3.012	
PHGDH in tumor			0.367			0.932
Negative versus Positive	0.754	0.409–1.393		0.973	0.520–1.819	
PSPH in tumor			0.069			**0.036**
Negative versus Positive	1.911	0.952–3.836		2.068	1.049–4.079	
PSPH in stroma			0.399			**0.024**
Negative versus Positive	1.345	0.676–2.677		2.152	1.107–4.184	
SHMT1 in stroma			0.141			**0.008**
Negative versus Positive	1.490	0.876–2.532		2.142	1.219–3.764	
GLDC in tumor			1.654			0.634
Negative versus Positive	1.654	0.935–2.925		1.149	0.649–2.033	

GLDC was also expressed at differing levels in different molecular subtypes of breast cancer. The highest GLDC expression was found in the HER-2 and the lowest in the TNBC subtype, in contrast to enzymes involved in the serine metabolism. High GLDC levels have been reported in several human cancers, including non-small-cell lung carcinoma, ovarian cancer, and germ-cell tumors; however, GLDC levels have not been studied in breast cancer [4]. The precise mechanism of increased GLDC in HER-2 subtype could not be postulated, but GLDC levels were previously examined in MCF10A cells after oncogenic transformation by KRAS^{G12D}, PIK3CA^{E545K}, or MYC^{T58A}. All three oncogenes increased GLDC expression by 20-fold, suggesting that oncogene-induced GLDC transcription can commonly be driven by oncogenic Ras, PI3K, and Myc [4]. Therefore, we hypothesize that the TNBC subtype, which is not associated with any well-known driving oncogene, revealed the lowest GLDC level, unlike the HER-2 subtype driven by certain HER-2 oncogenes.

Appreciably, breast cancer stromal components expressed elevated PSPH, SHMT1, and GLDC levels. According to the "reverse Warburg effect" theory, both the stromal and epithelial tumor components are involved in cancer metabolism: oxidative phosphorylation by the functional mitochondria occurs in the epithelial tumor component; glycolysis due to increased autophagic activity leads to dysfunctional mitochondria in the stromal component [18,19,20,21]. Similar to the epithelial tumor component, the stromal component expressed the highest levels of PSPH, SHMT1, and GLDC in the HER-2 and lowest in TNBC subtype. Expression of these enzymes was statistically correlated with expression of glycolysis-associated enzymes such as Glut-1, CA9,

and MCT-4, supporting the possibility of serine/glycine metabolism occurring in the tumor stroma. Although the exact mechanisms underlying high levels of the enzymes associated with serine/glycine metabolism in the HER-2 subtype are unknown, the glutaminolysis-associated enzymes, including glutaminase 1 (GLS1), glutamate dehydrogenase (GDH), and amino acid transporter 2 (ASCT2), have been highly expressed in the HER-2 subtype, suggesting that tumor stromal cells in this subtype are active metabolically [11].

We also examined the potential clinical implications of the expression profiles metabolism-associated enzymes in the different molecular subtypes of breast cancer. Particularly, expression of enzymes associated with serine/glycine metabolism was correlated with patients' prognosis. In the entire patient population irrespective of subtype, tumoral PSPH positivity, stromal PSPH positivity, and stromal SHMT1 negativity were independent factors for poor prognosis. According to the molecular subtype, tumoral PSPH positivity and stromal SHMT1 negativity in the luminal-A subtype and tumoral PHGDH negativity, tumoral PSPH positivity, and stromal SHMT1 negativity in the TNBC subtype were independent factors for poor prognosis. Although correlations between levels of enzymes associated with serine/glycine metabolism and tumor prognosis have remained unknown, several previous studies have shown that high levels of glycolysis-associated enzymes, including Glut-1 and CA9, were correlated with unfavorable prognosis in breast cancer [9,22,23].

Although high SHMT2, but not SHMT1, was correlated with breast cancer patient mortality [5], we found that stromal SHMT1 negativity was associated with poor prognosis. Similarly, tumoral

PHGDH negativity in the TNBC subtype was correlated with poor prognosis. However, further studies are needed to confirm these findings. In conclusion, we identified differential expression of the enzymes associated with serine/glycine metabolism according to breast cancer molecular subtypes. The highest expression of enzymes associated with serine metabolism was found in the TNBC subtype, and the lowest, in the luminal-A subtype. Expression of enzymes associated with glycine metabolism was high in the HER-2 type in both tumor and stromal compartments.

References

1. Warburg O (1956) On the origin of cancer cells. Science 123: 309–314.
2. Mullarky E, Mattaini KR, Vander Heiden MG, Cantley LC, Locasale JW (2011) PHGDH amplification and altered glucose metabolism in human melanoma. Pigment cell & melanoma research 24: 1112–1115.
3. Possemato R, Marks KM, Shaul YD, Pacold ME, Kim D, et al. (2011) Functional genomics reveal that the serine synthesis pathway is essential in breast cancer. Nature 476: 346–350.
4. Zhang WC, Shyh-Chang N, Yang H, Rai A, Umashankar S, et al. (2012) Glycine decarboxylase activity drives non-small cell lung cancer tumor-initiating cells and tumorigenesis. Cell 148: 259–272.
5. Jain M, Nilsson R, Sharma S, Madhusudhan N, Kitami T, et al. (2012) Metabolite profiling identifies a key role for glycine in rapid cancer cell proliferation. Science 336: 1040–1044.
6. Perou CM, Sorlie T, Eisen MB, van de Rijn M, Jeffrey SS, et al. (2000) Molecular portraits of human breast tumours. Nature 406: 747–752.
7. Sorlie T, Perou CM, Tibshirani R, Aas T, Geisler S, et al. (2001) Gene expression patterns of breast carcinomas distinguish tumor subclasses with clinical implications. Proc Natl Acad Sci U S A 98: 10869–10874.
8. Reis-Filho JS, Tutt AN (2008) Triple negative tumours: a critical review. Histopathology 52: 108–118.
9. Pinheiro C, Sousa B, Albergaria A, Paredes J, Dufloth R, et al. (2011) GLUT1 and CAIX expression profiles in breast cancer correlate with adverse prognostic factors and MCT1 overexpression. Histol Histopathol 26: 1279–1286.
10. Choi J, Jung WH, Koo JS (2013) Metabolism-related proteins are differentially expressed according to the molecular subtype of invasive breast cancer defined by surrogate immunohistochemistry. Pathobiology 80: 41–52.
11. Kim S, Kim do H, Jung WH, Koo JS (2013) Expression of glutamine metabolism-related proteins according to molecular subtype of breast cancer. Endocrine-related cancer 20: 339–348.
12. Elston CW, Ellis IO (1991) Pathological prognostic factors in breast cancer. I. The value of histological grade in breast cancer: experience from a large study with long-term follow-up. Histopathology 19: 403–410.
13. Hammond ME, Hayes DF, Dowsett M, Allred DC, Hagerty KL, et al. (2010) American Society of Clinical Oncology/College of American Pathologists guideline recommendations for immunohistochemical testing of estrogen and progesterone receptors in breast cancer. J Clin Oncol 28: 2784–2795.
14. Wolff AC, Hammond ME, Schwartz JN, Hagerty KL, Allred DC, et al. (2007) American Society of Clinical Oncology/College of American Pathologists guideline recommendations for human epidermal growth factor receptor 2 testing in breast cancer. J Clin Oncol 25: 118–145.
15. Won KY, Kim GY, Kim YW, Song JY, Lim SJ (2010) Clinicopathologic correlation of beclin-1 and bcl-2 expression in human breast cancer. Hum Pathol 41: 107–112.
16. Goldhirsch A, Wood WC, Coates AS, Gelber RD, Thurlimann B, et al. (2011) Strategies for subtypes-dealing with the diversity of breast cancer: highlights of the St. Gallen International Expert Consensus on the Primary Therapy of Early Breast Cancer 2011. Ann Oncol 22: 1736–1747.
17. Kim S, Jung WH, Koo JS (2013) The Expression of Glut-1, CAIX, and MCT4 in Mucinous Carcinoma. J Breast Cancer 16: 146–151.
18. Bonuccelli G, Tsirigos A, Whitaker-Menezes D, Pavlides S, Pestell RG, et al. (2010) Ketones and lactate "fuel" tumor growth and metastasis: Evidence that epithelial cancer cells use oxidative mitochondrial metabolism. Cell Cycle 9: 3506–3514.
19. Martinez-Outschoorn UE, Balliet RM, Rivadeneira DB, Chiavarina B, Pavlides S, et al. (2010) Oxidative stress in cancer associated fibroblasts drives tumor-stroma co-evolution: A new paradigm for understanding tumor metabolism, the field effect and genomic instability in cancer cells. Cell Cycle 9: 3256–3276.
20. Pavlides S, Tsirigos A, Vera I, Flomenberg N, Frank PG, et al. (2010) Loss of stromal caveolin-1 leads to oxidative stress, mimics hypoxia and drives inflammation in the tumor microenvironment, conferring the "reverse Warburg effect": a transcriptional informatics analysis with validation. Cell Cycle 9: 2201–2219.
21. Pavlides S, Whitaker-Menezes D, Castello-Cros R, Flomenberg N, Witkiewicz AK, et al. (2009) The reverse Warburg effect: aerobic glycolysis in cancer associated fibroblasts and the tumor stroma. Cell Cycle 8: 3984–4001.
22. Stackhouse BL, Williams H, Berry P, Russell G, Thompson P, et al. (2005) Measurement of glut-1 expression using tissue microarrays to determine a race specific prognostic marker for breast cancer. Breast Cancer Research and Treatment 93: 247–253.
23. Younes M, Brown RW, Mody DR, Fernandez L, Laucirica R (1995) GLUT1 expression in human breast carcinoma: correlation with known prognostic markers. Anticancer Res 15: 2895–2898.

Author Contributions

Conceived and designed the experiments: SKK WHJ JSK. Performed the experiments: SKK WHJ JSK. Analyzed the data: SKK WHJ JSK. Contributed reagents/materials/analysis tools: SKK WHJ JSK. Contributed to the writing of the manuscript: SKK WHJ JSK.

12

Profile of Individuals Who Are Metabolically Healthy Obese Using Different Definition Criteria. A Population-Based Analysis in the Spanish Population

María Teresa Martínez-Larrad[1,2]*, **Arturo Corbatón Anchuelo**[1,2], **Náyade Del Prado**[2], **José María Ibarra Rueda**[2], **Rafael Gabriel**[3], **Manuel Serrano-Ríos**[1,2]

1 Spanish Biomedical Research Centre in Diabetes and Associated Metabolic Disorders (CIBERDEM), Madrid, Spain, **2** Instituto de Investigación Sanitaria del Hospital Clínico San Carlos (IdISSC), Madrid, Spain, **3** Clinical Epidemiology Research Unit, Hospital de La Paz, Madrid, Spain

Abstract

Background: Obesity is associated with numerous metabolic complications such as diabetes mellitus type 2, dyslipidemia, hypertension, cardiovascular diseases and several forms of cancer. Our goal was to compare different criteria to define the metabolically healthy obese (MHO) with metabolically unhealthy obese (MUHO) subjects. We applied Wildman (W), Wildman modified (WM) with insulin resistance (IR) with cut-off point ≥ 3.8 and levels of C- Reactive Protein (CRP) ≥ 3 mg/l; and Consensus Societies (CS) criteria. In these subjects cardiovascular-risk (CV-risk) was estimated by Framingham score and SCORE for MHO and MUHO.

Methods: A cross-sectional study was conducted in Spanish Caucasian adults. A total of 3,844 subjects completed the study, 45% males, aged 35–74 years. Anthropometric/biochemical variables were measured. Obesity was defined as BMI: ≥ 30 Kg/m^2.

Results: The overall prevalence of obesity in our population was 27.5%, (23.7%/males and 30.2%/females). MHO prevalence according to W, WM, and CS definition criteria were: 9.65%, 16.29%, 39.94% respectively in obese participants. MHO has lower waist circumference (WC) measurements than MUHO. The estimated CV-risks by Framingham and SCORE Project charts were lower in MHO than MUHO subjects. WC showed high specificity and sensitivity in detecting high estimated CV risk by Framingham. However, WHR showed high specificity and sensitivity in detecting CV risk according to SCORE Project. MHO subjects as defined by any of the three criteria had higher adiponectin levels after adjustment by sex, age, WC, HOMA IR and Framingham or SCORE risks. This relationship was not found for CRP circulating levels neither leptin levels.

Conclusions: MHO prevalence is highly dependent on the definition criteria used to define those individuals. Results showed that MHO subjects had less WC, and a lower estimated CV-risk than MUHO subjects. Additionally, the high adiponectin circulating levels in MHO may suggest a protective role against developing an unhealthy metabolic state.

Editor: Maria Eugenia Saez, CAEBi, Spain

Funding: This work was supported by grants FEDER 2FD 1997/2309 from the Fondo Europeo para el Desarrollo Regional, Red de Centros RCMN (C03/08), FIS 03/1618, from Instituto de Salud Carlos III-RETIC RD06/0015/0012, Madrid, Spain. The authors also acknowledge CIBER in Diabetes and Associated Metabolic Disorders (ISCIII, Ministerio de Ciencia e Innovación) and Madrid Autonomous Community (MOIR S2010/BMD-2423). Partial support also came from Educational Grants from Eli Lilly Lab, Spain, Bayer Pharmaceutical Co., Spain and Fundación Mutua Madrileña 2008, Spain. The funders had no role in study design, data collection and analysis, decision to publish, or preparation of the manuscript.

Competing Interests: The authors received funding from a commercial source (Eli Lilly Lab and Bayer Pharmaceutical Co.). The authors have declared that no competing interests exist.

* Email: mserrano.hcsc@salud.madrid.org

Introduction

Obesity is a major public health problem in recent decades, because it is a key risk factor of type 2 diabetes, cardiovascular disease, dyslipidemia, hypertension, certain cancers [1,2]. However, a proportion between 20 and 30% of obese individuals may be free of metabolic comorbidities during an unknown variable period of time [3–5]. The existence of a metabolically healthy obese (MHO) phenotype was first proposed by Sims in 2001 [3]. Otherwise, there are several prospective studies aimed at investigating MHO subjects are at lower risk of early mortality of any cause, mostly due to cardiovascular disease [6,7].

Many authors have proposed different diverse definitions of the MHO phenotype according to the presence or absence of specific metabolic abnormalities such as: DM2, dyslipidemia and hypertension in individuals with obesity. Associations and clustering of cardiometabolic risk factors, the clinical phenotype derived from metabolic syndrome (MetS) and the inflammatory biomarkers, have then been widely recently used in categorizing those subjects as metabolically healthy or unhealthy [7,8].

On the basis of proposed MetS criteria, which have a limited value on the diagnosis of a high cardiovascular risk degree in the clinical setting, diverse authors have suggested in the last decade, the use of the definition of MHO and MUHO phenotypes, as a better clinical cardiovascular risk approach.

This issue is controversial and no clear definition criteria are universally accepted [9].

Therefore the purpose of our work was: Firstly, to compare the different accepted MHO definition criteria (table 1): 1) Wildman (W) [4], 2) Wildman modified (WM) using a cut-off point for HOMA-IR ≥3.8 as described in the Spanish population [10], and levels of C-Reactive Protein (CRP) ≥3 mg/l [11], and 3) MetS in accordance with the Consensus Societies (CS) as reported by Alberti KGMM et al. [12]. Secondly, to describe the estimated CV risk associated with each definition.

Design, population

We studied 4,097 subjects from the general Spanish population. Details of recruitment and Study protocols of this population-based survey were previously described [13,14]. In brief, 5,941 men and non-pregnant women aged 35–74 years, from a targeted population of 496,674 subjects from 21 small and middle-sized towns across Spain were invited to participate. All subjects were sent a personalized letter signed by the principal investigator and the authorities of the Regional Public Health Service, explaining the purpose of the study and requesting volunteering for participation. In case of no response, people were again contacted through telephone up to three times.

Two hundred and fifty-three subjects were excluded as they met one or more of the following exclusion criteria: type 1 diabetes, overt heart or hepatic failure; surgery in the previous year, weight changes >5 Kg within the previous 6 months, and hospitalization by whatever reason at the time of participating in our study.

A total of 3,844 subjects completed the study, 1,754 males and 2,090 females. We used standard procedures adapted from the WHO MONICA protocol [15], approved by our Ethics Committee of Clinic San Carlos Hospital. All participants gave written informed consent. Trained interviewers obtained the following data and implemented a medical questionnaire including: age, sex, parity, menopausal status, family history of diabetes, treatment of diabetes, hypertension, and other relevant chronic diseases.

Anthropometric measurements. Included BMI (kg/m^2) and waist circumference (cm) (WC); the cut-off points previously

Table 1. Criteria: Wildman (W), Wildman modified (WM) and Consensus Societies (CS).

WILDMAN	WILDMAN MODIFIED	CONSENSUS SOCIETIES Metabolic syndrome
CARDIOMETABOLIC ABNORMALITIES	**CARDIOMETABOLIC ABNORMALITIES**	**CARDIOMETABOLIC ABNORMALITIES**
1-Elevated blood pressure: Systolic/diastolic blood pressure ≥130/85 mm Hg or antihypertensive medication use	**1-Elevated blood pressure:** Systolic/diastolic blood pressure ≥130/85 mm Hg or antihypertensive medication use	**1-large waist circumference:**≥94 cm in men and ≥ 80 cm in women,
2. Elevated triglyceride level: Fasting triglyceride level ≥150 mg/dL	**2. Elevated triglyceride level:** Fasting triglyceride level ≥150 mg/dL	**2. Elevated triglyceride level:** Fasting triglyceride level (≥1.7 mmol/l),
3. Decreased HDL-C level: HDL-C level <40 mg/dL in men or <50 mg/dl in women or lipid-lowering medication use	**3. Decreased HDL-C level:** HDL-C level <40 mg/dL in men or <50 mg/dl in women or lipid-lowering medication use	**3. Decreased HDL level:** HDL cholesterol level < 1.0 mmol/l in men or <1.3 mmol/l in women,
4. Elevated glucose level: Fasting glucose level ≥100 mg/dL or antidiabetic medication use	**4. Elevated glucose level:** Fasting glucose level ≥100 mg/dL or antidiabetic medication use	**4. Elevated Glucose level:** Fasting glucose levels ≥ 5.6 mmol/l or drug treatment
5. Insulin resistance: HOMA-IR >5.13 (ie, the 90th percentile)	**5. Insulin resistance:** HOMA-IR ≥3.8 (ie, the 90th percentile)	**5. Elevated blood pressure:** systolic ≥130 mmHg and/or diastolic ≥85 mmHg and/or antihypertensive drug treatment or history of hypertension,
6. Systemic inflammation: hsCRP level >0.1 mg/L (ie, the 90th percentile)	**6. Systemic inflammation:** hsCRP level ≥3 mg/L (ie, the 90th percentile)	--------------------
Criteria for body size phenotypes:	**Criteria for body size phenotypes:**	**Criteria for body size phenotypes:**
Normal weight, metabolically healthy: BMI <25.0 Kg/m^2 and <2 cardiometabolic abnormalities	**Normal weight, metabolically healthy:** BMI <25.0 Kg/m^2 and <2 cardiometabolic abnormalities	**Normal weight, metabolically healthy:** BMI < 25.0 Kg/m^2 and <3 cardiometabolic abnormalities
Normal weight, metabolically abnormal: BMI <25.0 Kg/m^2 and ≥2 cardiometabolic abnormalities	**Normal weight, metabolically abnormal:** BMI <25.0 Kg/m^2 and ≥2 cardiometabolic abnormalities	**Normal weight, metabolically abnormal:** BMI < 25.0 Kg/m^2 and ≥3 cardiometabolic abnormalities
Overweight, metabolically healthy: BMI 25.0–29.9 Kg/m^2 and <2 cardiometabolic abnormalities	**Overweight, metabolically healthy:** BMI 25.0–29.9 Kg/m^2 and <2 cardiometabolic abnormalities	**Overweight, metabolically healthy:** BMI 25.0–29.9 Kg/m^2 and <3 cardiometabolic abnormalities
Overweight, metabolically abnormal: BMI 25.0–29.9 Kg/m^2 and ≥2 cardiometabolic abnormalities	**Overweight, metabolically abnormal:** BMI 25.0–29.9 Kg/m^2 and ≥2 cardiometabolic abnormalities	**Overweight, metabolically abnormal:** BMI 25.0–29.9 Kg/m^2 and ≥3 cardiometabolic abnormalities
Obese, metabolically healthy: BMI ≥30.0 Kg/m^2 and <2 cardiometabolic abnormalities	**Obese, metabolically healthy:** BMI ≥30.0 Kg/m^2 and <2 cardiometabolic abnormalities	**Obese, metabolically healthy:** BMI ≥30.0 Kg/m^2 and <3 cardiometabolic abnormalities
Obese, metabolically abnormal: BMI ≥30.0 Kg/m^2 and ≥2 cardiometabolic abnormalities	**Obese, metabolically abnormal:** BMI ≥30.0 Kg/m^2 and ≥2 cardiometabolic abnormalities	**Obese, metabolically abnormal:** BMI ≥30.0 Kg/m^2 and ≥3 cardiometabolic abnormalities

HDL-C: High Density Lipoprotein Cholesterol, HOMA-IR: Homeostasis model assessment of insulin resistance, hsCRP: high sensitivity C Reactive Protein, BMI: Body Mass Index.

reported in Spanish population (94.5/89.5 cm for males/females) [16] were considered to define abdominal obesity. Waist measurements were made with a non stretchable fibre measuring tape while study participants were standing erect in a relaxed position with both feet together on a flat surface. WC was measured as the smallest horizontal girth between the costal margins and the iliac crests at minimal respiration. Hip circumference (HC) was measured at the level of the greater femoral trochanters. These measurements were used to compute WC divided by HC [waist-to-hip ratio (WHR)].

The reliability of the anthropometric measurements was established by comparing values obtained by three different interviewers in a sample (n = 3,844) of individuals.

With regards to alcohol intake, subjects were classified in four groups: 1) no alcohol intake (0 g alcohol/day), 2) 1–14.99 g/day, 3) ≥15–29.99 g/day; 4) ≥30 g/day [17,18]. Smoking habits were recorded as follows: smokers (at least one cigarette per day); non-smokers: never having smoked, and ex-smokers: people who had stopped smoking previous 4 years.

Physical activity was evaluated by asking participants to report their average commitment to various physical activities. We quantified the amount of physical activity by estimating the number of metabolic equivalents (MET) as described (www.cdc.gov). MET estimates were equivalent to the number of hours spent on a particular activity multiplied by a score that was specific for that activity. Subjects were classified in three groups according to their physical activity: low <3 METs; moderate 3.0–6.0 METs; high >6.0 METs.

Procedures and laboratory studies

After an overnight fasting period, 20 ml of blood were obtained from an antecubital vein without compression. Plasma glucose was determined duplicate by a glucose-oxidase method adapted to an Autoanalyzer (Hitachi 704, Boehringer Mannheim, Germany). Total cholesterol, triglycerides and high-density lipoprotein cholesterol (HDL-C) were determined by enzymatic methods using commercial kits (Boehringer, Mannheim, Germany). Low-density lipoprotein cholesterol (LDL-C) was calculated by the Friedewald formula [19]. A 75-g oral glucose tolerance test (OGTT) was performed and interpreted according to the revised 2003 criteria of the American Diabetes Association [20] Diabetes mellitus was diagnosed when fasting plasma glucose was ≥ 7.0 mmol/l or 2-h post glucose ≥11.1 mmol/l. Subjects on anti-diabetic medication were also considered to have diabetes. In non-diabetic subjects, fasting plasma glucose of 5.6–6.9 mmol/l was indicative of impaired fasting glucose (IFG) and 2-h glucose of ≥ 7.8–11.0 mmol/l of impaired glucose tolerance (IGT). Serum insulin concentrations were determined by RIA (Human Insulin Specific RIA kit, Linco Research Inc., St Louis, MO, USA). This assay had a lower detection limit of 2 μU/ml with within and between assay coefficients of variation of <1% and <7.43%, respectively. Cross reactivity with proinsulin was under 0.2%. IR was estimated by homeostasis model assessment of IR (HOMA-IR) using the following formula: fasting insulin (μU/ml) × fasting glucose (mmol/l)/22.5 [21]. In subjects without clinical or biological parameters of IR, the 90th percentile for the HOMA-IR was equal to or greater than 3.8, and this value was considered diagnostic of IR [10].

Leptin and adiponectin serum concentrations were assayed by sensitive/specific RIA as follows: leptin by a highly sensitive RIA (Human Leptin RIA Kit, Linco Research), with a lower detection limit of 0.5 ng/ml to 100 μL, and inter and intra-assays' coefficients of variation were 2%–6% and 3%–7%, respectively. Total adiponectin by a highly specific RIA (Human Adiponectin

Specific RIA kit, Linco Research) with a lower detection of 1 ng/ml. Intra and interassay coefficients of variation were 2% and 2.6%, respectively. The cut-off point were for Leptin 9.23 ng/ml (50th percentile) and for adiponectin 9.7 μg/ml (50th percentile).

CRP was measured by using nephelometry high sensitivity C Reactive Protein (hsCRP) as the latest chemistry enhancement for the Image Inmunochemistry System (Beckman). CRPH reagent provides improved low sensitivity to 0.2 mg/l. The intra-assay and inter-assay coefficients of variation for CRP were 3.5% and 3.3% respectively. The cut-off point was CRP ≥3 mg/l [11].

Study subjects were divided into three categories based on BMI: non obese: BMI <25 Kg/m^2, overweight BMI 25–29.9 Kg/m^2, and obese: BMI ≥30 Kg/m^2.

High CV-risk was estimated as ≥20% with the Framingham risk score [22] and ≥5% with the SCORE project for populations at low CV-risk [23].

For the purposes of this study, we used (Table 1) W criteria [4], WM and CS to define MHO [12] as compared to MUHO subjects.

Statistical analysis

Student t test or ANOVA were used to compare continuous variables expressed as means ± standard deviation (SD). The level of significance was set at 0.05 for all analyses.

Linear regression was used to calculate quantitative variables adjusted for age and sex and their 95% confidence intervals (CI). Otherwise, a logistic regression analysis was performed to evaluate associations of adiponectin, leptin and CRP with MHO. Adjusted Odds Ratios (ORs) and their 95% CI were calculated. The receiver operator characteristic curves (ROC) were conducted to evaluate the performance of the WC, BMI and WHR anthropo-metric parameters in detecting Framingham risk ≥20% and SCORE risk ≥5% for populations at low cardiovascular disease risk. We used the area under the curve (AUC) with 95% confidence intervals (CI). Associations between different defini-tions of MHO and high CV-risk (Framingham ≥20% and SCORE ≥5% risk scores) were studied by estimating crude and adjusted ORs using logistic regression models adjusted by sex and age and stratified by BMI categories. Statistical analysis was done using STATA 11 SE.

Results

The overall prevalence of obesity in our population was 27.5% (n = 1,057) (23.7% in males and 30.2% in females); overweight 45.3% (n = 1,741) (53.1% in males and 38.6% in females) and normal weight 27.2% (n = 1,046) (23.7% in males and 31.2% in females).

The number of obese subjects according to different criteria was as follows: by a) W criteria: 29.11% (n = 1,119), b) WM: 29.37% (n = 1,129), and c) CS criteria: 27.54% (n = 1,059). Among the obese subjects (BMI ≥30 Kg/m^2), a low number was defined as MHO: a) by W criteria: 9.65% (n = 108); b) by WM: 16.29% (n = 184), and c) by CS criteria: 39.94% (n = 423). The prevalence of MHO was 2.81% by W; 4.78% by WM and 11.02% by CS criteria the whole study population (n = 3,844).

In the total population the prevalence of different categories of glucose status was as follows: a) IFG 16.6% (n = 638), b) IGT 8.3% (n = 319) and c) DM2 7.5% (n = 288).

Tables 2, 3 and 4 include the anthropometric parameters of the whole group of participants, in accordance with respective BMI. Overall, MHO subjects had a significantly lower WC and BMI than MUHO. We also observed significant differences in SBP, DBP, HC between groups. Moreover, there is lower abdominal

Table 2. Basic characteristics and anthropometric parameters in individuals with BMI $\geq$30 Kg/m^2 by Wildman criteria, means adjustment by age and sex.

	Metabolically Healthy BMI $\geq$30 Kg/m^2	Metabolically unhealthy BMI $\geq$30 Kg/m^2	p value
	n = 108	n = 1011	
Age (Years)*	52.91 (9.93)	53.59 (9.64)	0.535
Males (%)	40.7	38.5	
Females (%)	59.3	60.5	
	$\bar{X}$ (95% CI)	$\bar{X}$ (95% CI)	
SBP (mmHg)	115.45 (112.06–118.84)	137.28 (136.16–138.41)	<0,001
DBP (mmHg)	74.18 (72.12–76.24)	85.48 (84.80–86.16)	<0,001
BMI (kg/m2)	31.95 (31.50–32.40)	33.46 (33.31–33.61)	<0,001
WC (cm)	96.88 (95.41–98.35)	101.17 (100.60–101.74)	<0,001
WC Males $\geq$94.5 cm or Females $\geq$89.5 cm (%)	82.52 (74.81–90.24)	90.44 (88.40–92.48)	0.041
HC(cm)	107.08 (105.66–108.51)	109.80 (109.33–110.27)	<0.001
WHR	0.91 (0.89–0.92)	0.93 (0.92–0.93)	0.006
	OR	OR	
Framinghan$^\Delta$	1	19.88 (6.76–58.48)	<0.001
SCORE$^\Delta$	1	9.52 (1.84–49.13)	0.007

BMI: Body Mass Index; SBP: Systolic Blood Pressure; DBP: Diastolic Blood Pressure, WC: Waist Circumference; HC: Hip Circumference; WHR: Waist to Hip Ratio;
*Mean $\pm$ (SD). $\bar{X}$: Mean, CI: confidence interval.
$^\Delta$Logistic regression models independent variables: Framinghan and SCORE risk scores adjustment by age and sex. OR: Odd Ratio.

obesity in MHO than MUHO using any of the three definitions. Only 1.81% of MHO individuals by W criteria showed a normal WC in accordance with cut-off points found for our group in Spanish population (less than 94.5/89.5 cm for males/females) [16], by WM: 2.22% and by CS: 3.03%.

Tables 2, 3 and 4 include a logistic regression model with CV-risk (estimated by Framingham and SCORE risks charts) as the independent variable. Framingham and SCORE risks were associated with increased odds of being MUHO to MHO. ORs

Table 3. Basic characteristics and anthropometric parameters in individuals with BMI $\geq$30 Kg/m^2 by Wildman modified criteria, means adjustment by age and sex.

	Metabolically Healthy BMI $\geq$30 Kg/m^2	Metabolically unhealthy BMI $\geq$30 Kg/m^2	p value
	n = 184	n = 945	
Age (Years)*	53.47 (10.04)	53.51 (9.58)	0.965
Males (%)	41.22	39.55	
Females (%)	58.78	60.45	
	$\bar{X}$ (95% CI)	$\bar{X}$ (95% CI)	
SBP (mmHg)	121.61 (118.98–124.24)	137.76 (136.59–138.93)	<0,001
DBP (mmHg)	78.29 (76.68–79.89)	85.57 (84.86–86.29)	<0,001
BMI (kg/m2)	32.14 (31.80–32.48)	33.47 (33.32–33.63)	<0,001
WC (cm)	97.39 (96.27–98.50)	101.82 (101.32–102.32)	<0,001
WC Males $\geq$94.5 cm or Females $\geq$89.5 cm (%)	81.96 (75.95–87.97)	90.72 (88.65–92.79)	0.007
HC(cm)	106.87 (105.78–107.96)	109.97 (109.49–110.46)	<0.001
WHR	0.91 (0.90–0.92)	0.93 (0.92–0.93)	0.083
	OR	OR	
Framinghan$^\Delta$	1	8.54 (4.34–16.79)	<0.001
SCORE$^\Delta$	1	2.74 (1.05–7.14)	0.039

BMI: Body Mass Index; SBP: Systolic Blood Pressure; DBP: Diastolic Blood Pressure, WC: Waist Circumference; HC: Hip Circumference; WHR: Waist to Hip Ratio;
*Mean $\pm$ (SD). $\bar{X}$: Mean, CI: confidence interval.
$^\Delta$Logistic regression models independent variables: Framinghan and SCORE risk scores adjustment by age and sex. OR: Odd Ratio.

Table 4. Basic characteristics and anthropometric parameters in individuals with BMI ≥30 Kg/m² by Consensus Societies criteria, means adjustment by age and sex.

	Metabolically Healthy BMI ≥30 Kg/m²	Metabolically unhealthy BMI ≥30 Kg/m²	p value
	n = 423	**n = 636**	**0.086**
Age (Years)*	52.18 (9.69)	53.82 (9.49)	
Males (%)	38.1	38.2	
Females (%)	61.9	61.8	
	X̄ (95% CI)	X̄ (95% CI)	
SBP (mmHg)	127.67 (126.07–129.27)	138.47 (137.09–139.85)	<0,001
DBP (mmHg)	80.99 (80.00–81.97)	86.20 (85.35–87.05)	<0,001
BMI (kg/m2)	32.60 (32.40–32.80)	33.59 (33.42–33.76)	<0,001
WC (cm)	98.23 (97.58–98.88)	102.31 (101.76–102.87)	<0,001
WC Males ≥94.5 cm or Females ≥89.5 cm (%)	82.09 (78.56–85.63)	92.32 (90.12–94.53)	<0,001
HC(cm)	108.33 (107.68–108.98)	110.26 (109.70–110.83)	<0.001
WHR	0.91 (0.90–0.92)	0.93 (0.92–0.94)	0.086
	OR	OR	
Framinghan^Δ	1	6.71 (4.29–10.48)	<0.001
SCORE^Δ	1	2.19 (1.15–4.16)	<0.001

BMI: Body Mass Index; SBP: Systolic Blood Pressure; DBP: Diastolic Blood Pressure, WC: Waist Circumference; HC: Hip Circumference; WHR: Waist to Hip Ratio;
*Mean ± (SD). X̄: Mean, CI: confidence interval.
^ΔLogistic regression models independent variables: Framinghan and SCORE risk scores adjustment by age and sex. OR: Odd Ratio.

tended to be higher using the W definition as compared to the WM and CS definitions.

Tables 5, 6 and 7 include biochemical characteristics of individuals in accordance with their respective BMI adjusted by age and sex. MUHO subjects had different fasting glucose, 2-h post glucose, HDL-C, triglycerides, fasting insulin, HOMA IR and adiponectin levels when compared with MHO subjects. On the other hand, fasting adiponectin levels were significantly higher in MHO than MUH0 subjects. In addition, CRP serum concentrations were lower in MHO vs MUHO, these differences were only statistically different (p<0.05) when WM criteria were used. On the other hand, there is lower abdominal obesity in MHO than MUHO by three criteria.

Smoking and alcohol intake habits were not significantly different when comparing MHO with MUHO subjects under the three criteria used. Physical activity differed between groups as follows: low grade physical activity (<3 METs) was found for MUHO as compared to MHO subjects no matter which criterion was used. A higher percentage of MHO as compared to MUHO subjects under CS criteria practice moderate (3.0–6.0 METs) and high (>6 METs) physical activity.

The type and prevalence of comorbidities in the MHO subjects are presented in Figure 1.

Abnormalities included in all three criteria are blood pressure, HDL-C, triglycerides and fasting glucose. Arterial hypertension was the most frequently abnormality found in MHO subjects as defined by CS (56%), followed by WM (35%) criteria.

We obtained ROC curves (Figure 2) for BMI, WC and WHR in detecting Framingham risk score (≥20%) and SCORE Project (≥5%). WC showed high specificity and sensitivity in detecting cardiovascular risk according to the Framingham scale. On the other hand, WHR showed high specificity and sensitivity in detecting high cardiovascular risk according to SCORE as shown in Figure 2. In both cases, all anthropometric measurements

correlated with an increased CV risk. Finally, the logistic regression models for W criteria, WM and CS the MHO subjects were associated with elevated levels of adiponectin after adjustment for sex, age, WC, HOMA-IR and CV-risk SCORE project: 1) $OR_{W(adiponectin)}$: 1.04 (95% CI, 1.00–1.07, p = 0.026), 2) $OR_{VM(adiponectin)}$ 1.05 (95% CI, 1.00–1.09, p = 0.015), and 3) $OR_{CS(adiponectin)}$ 1.06 (95% CI 1.00–1.12, p = 0.034).

Adjusted by sex, age, WC, HOMA-IR, and CV Framingham risk score, in all three criteria, MHO subjects were associated with elevated levels of adiponectin: 1) $OR_{W(adiponectin)}$ 1.03 (95% CI, −1.07, p = 0.041), 2) $OR_{VM(adiponectin)}$ 1.04 (95% CI, 1.00–1.08, p = 0.047) and 3) $OR_{CS(adiponectin)}$ 1.08 (95% CI 1.02–1.14, p = 0.002).

Adiponectin is shown as protector of cardiometabolic abnormalities in obese. The risk of developing cardiometabolic abnormalidades is lower in subjects with levels above the median adiponectin. That is the MHO subjects had higher levels than MUHO subjects.

A logistic regression model adjusted by sex, age, WC, HOMA-IR, and CV-risk Framingham risk score or SCORE project for all three definitions criteria used, showed no significant differences in leptin and CRP levels between MHO and MUHO (data not shown).

Discussion

In the current study on a sample of the Spanish population, the prevalence of MHO according to specific definitions criteria among those with obesity was as follows: W: 9.65%; WM: 16.29%, and CS: 39.94%. Some previously published works reported that among obese subjects, the prevalence of MHO ranged from 3.3% to 43% [4,7,24–26]. In a cross-sectional analysis carried out by Pajunen P et al. [27], using the CS definition, the MHO individual prevalence was lower (≈13%) than that found in our study (39.94%).

Table 5. Biochemical characteristics and lifestyle in individuals metabolically healthy and unhealthy with BMI $\geq$30 kg/m^2 by Wildman criteria, adjustment by age and sex.

	Metabolically Healthy BMI $\geq$30 Kg/m^2 $\bar{X}$ (95% CI)	Metabolically unhealthy BMI $\geq$30 Kg/m^2 $\bar{X}$ (95% CI)	p value <0.001
Fasting glucose (mmol/l)	4.76 (4.46–5.05)	5.75 (5.64–5.85)	0.001
Glucose 2 hs (mmol/l)	5.74 (5.23–6.25)	6.63 (6.45–6.82)	<0.001
HDL-C (mmol/l)	1.55 (1.48–1.63)	1.19 (1.17–1.22)	<0.001
Triglycerides (mmol/l)	1.00 (0.81–1.19)	1.69 (1.63–1.76)	<0.001
Fasting insulin (pmol/l)	68.52 (54.55–82.56)	102.55 (97.82–107.28)	<.001
HOMA-IR	2.40 (1.79–3.01)	4.36 (4.15–4.57)	<0.001
CRP (mg/l)	2.57 (1.02–4.12)	3.49 (2.99–3.98)	0.269
Fasting Leptin (ng/ml)	17.96 (15.85–20.08)	19.55 (18.89–20.20)	0.160
Fasting Adiponectin (ug/ml)	12.62 (10.97–14.26)	9.43 (8.67–10.20)	<0.001
Smoking			
Smoker (%)	23.11 (14.36–31.86)	24.57 (21.60–27.55)	0.756
Non-smoker (%)	54.29 (45.38–63.19)	52.92 (49.90–55.94)	0.775
Former smoker (%)	22.43 (13.78–31.09)	21.81 (18.97–24.65)	0.893
Physical Activity			
Low (<3 METs) (%)	33.87 (23.83–43.90)	45.65 (42.17–49.14)	0.029
Moderate (3.0–6.0 METs) (%)	51.23 (40.61–61.85)	43.42 (39.95–46.89)	0.170
High (>6 METs) (%)	15.05 (7.30–22.80)	10.77 (8.53–13.01)	0.297
Alcohol intake			
0 gr (%)	48.08 (38.55–57.61)	43.60 (40.42–46.78)	0.381
0–14.99 gr (%)	27.89 (18.40–37.39)	21.72 (18.77–24.66)	0.222
15–29.99 gr (%)	12.35 (5.31–19.39)	19.03 (16.26–21.79)	0.083
$\geq$30 gr (%)	10.96 (4.30–17.63)	15.36 (12.90–17.82)	0.224

BMI: Body Mass Index; $\bar{X}$: Mean, CI: confidence interval HDL-C: High Density Lipoprotein Cholesterol; HOMA-IR: Homeostasis model assessment of insulin resistance; CRP:C-Reactive Protein, METs: Metabolic Equivalents.

In a 20 year follow up US cohort population, Wildman RP et al. [4] found a 31.7% MHO prevalence using a newly proposed criteria including cut-off point >5.3 (90th percentile). A lower prevalence of about 24% of all obese individuals was found by Hamer M et al. [11] in a representation of the general population study (Health for England and Scottish Health Surveys). In these studies used an adaptation of previous criteria [4,28].

It is noticeable that among southern European countries very little data on MHO prevalence is available. Most recently in Spain, a well designed epidemiological cross-sectional study including 11,520 individuals by Lopez-Garcia E et al. [5] concluded that MHO subjects represent a 6.5% overall and correspond to 28.9% of obese individuals in the Spanish population aged $\geq$18 years. Also in this study, 1.8% of subjects MHO showed a normal WC such as it did in our study. Nevertheless a direct comparison between our study and that of Lopez-Garcia et al. is difficult since their population was younger than ours; and the cut-off point they used for HOMA IR and CRP were different. In a population-based prospective follow-up study with 1,051 subjects, by Soriguer F. et al. in Spain [29], the MHO prevalence ranged from 3.0% to 16.9% depending on the set of chosen criteria. In the Cremona Italian prospective study, during a 15 year follow up of 2,011 subjects, Calori G et al. [30] found 11% of MHO individuals among the obese population and 2% within their total study population. The authors used an IR cut-off point for HOMA-IR $\geq$2.5 as criteria of MHO phenotype, allowing

them to compare these results with those from the NHANES III in US [26], where a 6% MHO total prevalence was found. In comparison another Italian population-based study [31] reported a higher (27.5%) MHO prevalence in a cohort of 681 obese individuals living in Rome and surrounding areas. These discrepancies could be related to the anthropometric characteristics of the population, or to regional lifestyle habits which would show a different impact of the typical variety of components (BMI, WC, IR) used in some MHO definitions [6,32]. It seems, therefore, that the prevalence of the MHO phenotype is highly dependent on the MHO definition and to a certain extent may justify the disparity of results. However, our results show that MUHO individuals tended to have higher WC and HC than MHO counterparts; however, WHR is similar in both phenotypes. This may suggest that MHO and MUHO individuals differ only in terms of the amount and not in the type of adiposity (central vs. peripheral). Therefore, our results may by used to hypothesize that MHO individuals could develop over the years an unhealthy phenotype with further weight gain.

Finally, Van Vliet-Ostaptchouk JV et al. recently compared the defining characteristics of the metabolically healthy obese phenotype across ten population based cohort studies and concluded that there is a "considerable variation in the occurrence of MHO across the different European populations even when unified criteria or definitions were used to classify this phenotype. Further

Table 6. Biochemical characteristics and lifestyle in individuals metabolically healthy and unhealthy with BMI ≥30 kg/m^2 by Wildman modified criteria, adjustment by age and sex.

	Metabolically Healthy BMI ≥30 Kg/m^2 X̄ (95% CI)	Metabolically unhealthy BMI ≥30 Kg/m^2 X̄ (95% CI)	p value
Fasting glucose (mmol/l)	4.78 (4.56–5.00)	5.80 (5.71–5.91)	<0.001
Glucose 2 hs (mmol/l)	6.00 (5.62–6.37)	6.61 (6.42–6.80)	0.003
HDL-C (mmol/l)	1.47 (1.42–1.53)	1.18 (1.16–1.22)	<0.001
Triglycerides (mmol/l)	1.03 (0.88–1.16)	1.73 (1.66–1.79)	<0.001
Fasting insulin (pmol/l)	68.32 (57.81–78.82)	105.61 (100.83–110.38)	<0.001
HOMA-IR	2.40 (1.94–2.85)	4.53 (4.31–4.74)	<0.001
CRP (mg/l)	1.94 (0.95–2.94)	3.83 (3.30–4.35)	0.001
Fasting Leptin (ng/ml)	18.07 (16.64–19.51)	19.60 (18.91–20.29)	0.058
Fasting Adiponectin (ug/ml)	11.14 (10.04–12.25)	9.22 (8.47–9.97)	0.040
Smoking			
Smoker (%)	22.95 (16.26–29.65)	24.15 (21.11–27.20)	0.748
Non-smoker (%)	55.78 (48.97–62.60)	55.83 (49.73–55–93)	0.831
Former smoker (%)	20.94 (14.49–25.22)	22.28 (19.34–25.22)	0.709
Physical Activity			
Low (<3 METs) (%)	35.63 (27.91–45.35)	45.73 (42.15–49.32)	0.019
Moderate (3.0–6.0 METs) (%)	50.42 (42.33–58.51)	43.51 (39.95–47.08)	0.124
High (>6 METs) (%)	14.04 (8.25–19.81)	10.57 (8.2–12.85)	0.273
Alcohol intake			
0 gr (%)	42.53 (35.37–47.31)	46.67 (43.09–50.24)	0.704
0–14.99 gr (%)	27.11 (19.90–34.31)	21.18 (18.18–24.19)	0.136
15–29.99 gr (%)	19.78 (13.39–26.18)	18.37 (15.56–21.19)	0.691
≥30 gr (%)	4.78 (4.56–5.00)	16.15 (13.58–18.72)	0.032

BMI: Body Mass Index; X̄: Mean, CI: confidence interval HDL-C: High Density Lipoprotein Cholesterol; HOMA-IR: Homeostasis model assessment of insulin resistance; CRP:C-Reactive Protein, METs: Metabolic Equivalents.

studies are needed to identify the underlying factors for these differences" [33].

Lifestyle habits. In our study a lower degree of physical activity was observed in MUHO as compared to MHO individuals depending on the criteria used. However this finding is not consistently reproduced in other published reports [25,34,35]. Similarly data on alcohol intake and smoking habits vary widely between different studies [4,25] including ours.

Inflamation. On the other hand, the use of proinflammatory biomarkers as CRP, in the MHO definition criterion, is rarely reported in literature [25,29]. Marquez-Vidal et al. [36] in a population-based study of 881 obese individuals, found that MHO individuals had lower CRP levels, as did our results in MHO subjects defined by WM criteria.

Biomarkers. The potential role of adiponectin must be stressed: as it is one of the major active cytokine molecules produced by white adipose tissue [37] inversely correlated with IR. Otherwise, this cytokine has antiatherogenic and anti-inflammatory properties [38–40]. In this context Aguilar-Salinas et al. [24] reported that high adiponectin levels were associated with MHO phenotype. Hence, these Mexican authors proposed the inclusion of adiponectin concentrations in the criteria to define the MHO phenotype. Our own current results support this proposal.

Cardiovascular Risk. We found that MUHO individuals defined by three criteria presented higher CV-risk using either the Framingham or SCORE risk than MHO subjects.

Several epidemiological studies has addressed the impact of MHO definition criteria on potential CV-risk [11,26], with variable results ranging from none to high CV-risk in MHO individuals. On the other hand, Hamer et al. [11] in a large nationally representative sample initially free of CVD, reported that MHO participants did not have increased risk of CVD compared with the metabolically healthy no obese reference group. Our study showed lower CV-risk in MHO than in MUHO subjects. Also, some of the published results in the literature [30,41] but not all of them [42,43] are consistent with our finding.

Strengths and limitations. A major strength of our study is the high number of carefully phenotyped participants, as well as the availability of new biomarkers such as adiponectin to characterized MHO subjects regardless their definition criteria.

However, there are too some limitations in our work: 1) The cross-sectional design does not allow the establishment of cause-effect relationships. 2) The Framingham risk chart assessment probably overestimates CV- risk in low risk populations such as the Spanish one. Nevertheless, we have tried to attenuate this limitation by using the SCORE project chart, which is widely recommended to estimate CV-risk in low-risk population.

Conclusions

a) Overall, the prevalence of MHO observed in our population is concordant with some of the previous reported data, in literature. b) MHO and MUHO individuals differ only in terms

Table 7. Biochemical characteristics and lifestyle in individuals metabolically healthy and unhealthy with BMI $\geq$30 kg/m^2 by Consensus Societies criteria, adjustment by age and sex.

	Metabolically Healthy BMI $\geq$30 Kg/m^2	Metabolically unhealthy BMI $\geq$30 Kg/m^2	p value
	$\bar{X}$ (95% CI)	$\bar{X}$ (95% CI)	
Fasting glucose (mmol/l)	4.89 (4.77–5.01)	6.00 (5.90–6.11)	<0.001
Glucose 2 hs (mmol/l)	5.66 (5.45–5.88)	6.88 (6.67–7.09)	<0.001
HDL-C (mmol/l)	1.41 (1.38–1.44)	1.12 (1.09–1.15)	<0.001
Triglycerides (mmol/l)	1.11 (1.04–1.19)	1.87 (1.80–1.94)	<0.001
Fasting insulin (pmol/l)	79.94 (73.84–86.04)	105.56 (100.32–110.80)	<0.001
HOMA-IR	2.89 (2.63–3.16)	4.70 (4.48–4.92)	<0.001
CRP (mg/l)	2.87 (2.10–3.63)	3.79 (3.17–4.41)	0.065
Fasting Leptin (ng/ml)	18.99 (18.00–19.98)	19.42(18.64–20.19)	0.505
Fasting Adiponectin (ug/ml)	10.93 (9.89–11.92)	8.99 (8.03–9.96)	0.011
Smoking			
Smoker (%)	21.55 (17.59–25.52)	24.77 (21.14–28.40)	0.239
Non-smoker (%)	55.22 (51.09–59.34)	52.88 (49.25–56.52)	0.403
Former smoker (%)	22.59 (18.60–26.58)	21.33 (17.92–24.74)	0.637
Physical Activity			
Low (<3 METs) (%)	36.48 (31.81–41.14)	47.32 (41.14–51.50)	<0.001
Moderate (3.0–6.0 METs) (%)	48.85 (43.99–53.72)	42.15 (38.02–46.28)	0.038
High (>6 METs) (%)	14.69 (11.17–18.22)	10.14 (7.51–12.77)	0.041
Alcohol intake			
0 gr (%)	43.89 (39.50–48.29)	45.36 (41.50–49.21)	0.621
0–14.99 gr (%)	23.74 (19.59–27.90)	20.25 (16.81–23.69)	0.202
15–29.99 gr (%)	17.97 (14.24–21.70)	18.46 (15.16–21.75)	0.848
$\geq$30 gr (%)	14.10 (10.78–17.41)	15.53 (12.48–18.57)	0.532

BMI: Body Mass Index; $\bar{X}$: Mean, CI: confidence interval HDL-C: High Density Lipoprotein Cholesterol; HOMA-IR: Homeostasis model assessment of insulin resistance; CRP:C-Reactive Protein, METs: Metabolic Equivalents.

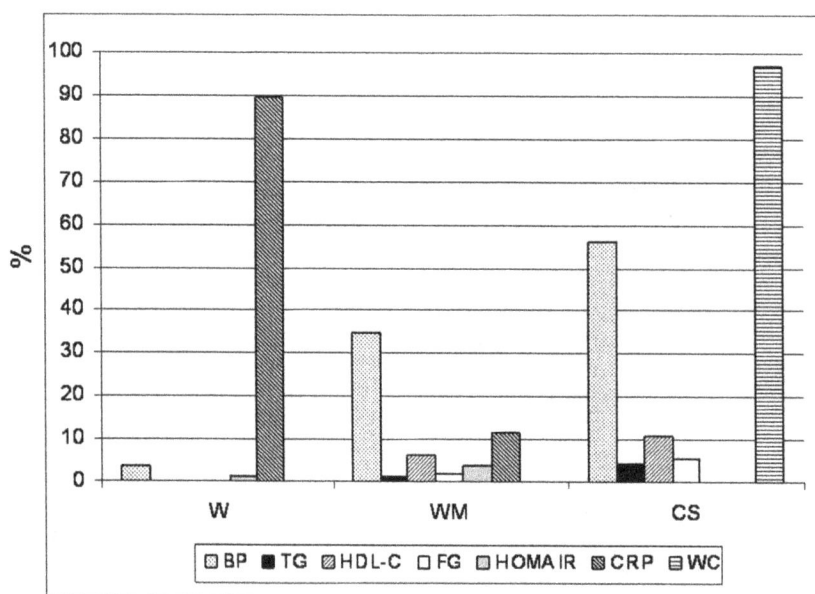

Figure 1. Comorbidities in the MHO subjects by W (Wildman), WM (Wildman modified) and CS (Consensus Societies) criteria according to the data shown in Table 1. BP: Elevated Blood Pressure, TG: Elevated Triglycerides, HDL-C: Decreased HDL C; FG: Elevated Fasting Glucose, HOMA IR: Elevated Homeostasis Model Assessment Insulin Resistance, CRP: Elevated C Reactive Protein, WC: Large Waist Circumference.

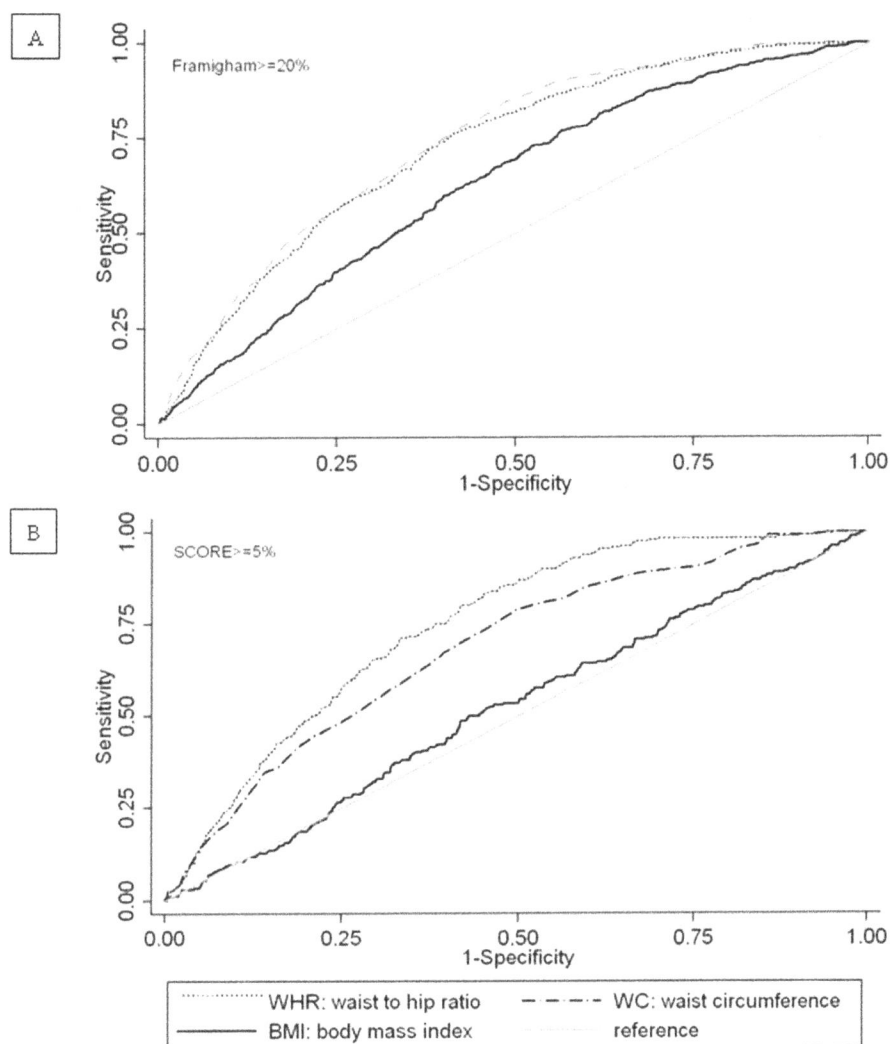

Figure 2

Figure 2. BMI, WC and WHR in detecting Framingham risk score (≥20%) and SCORE Project (≥5%). A.-The AUC was 0.739 (95%CI 0.719–0.759) for WC; 0.724 (95%CI 0.704–0.745) for WHR and 0.631 (95%CI 0.608–0.653) for BMI (p<0.001). B. - The AUC was 0.688 (95%CI 0.655–0.721) for WC; 0.746 (95%CI 0.719–0.773 for WHR, 0.523 (95%CI 0.486–0.561) for BMI (p<0.001).

of amount of adiposity, but not in the type of adiposity (central vs. peripheral). c) Our data show that MHO subjects have a lower estimated CV-risk than MUHO. Likewise, we estimate that amongst the current criteria used to define MHO individuals, WM seems to be the most clinically appropriate one as well as physiopathologically more understandable, since it includes an indirect measure of IR and circulating CRP levels. Insulin resistance linked to obesity is a major risk factor for type 2 diabetes and cardiovascular disease and CRP is an inflammatory marker associated with CV disease in obesity. d) Interestingly enough, we found that high adiponectin circulating levels seem to play a protective role against the risk of developing an unhealthy metabolic state.

Acknowledgments

To Members of the Segovia Insulin Resistance Study Group. We acknowledge Milagros Perez Barba for dedicated and careful technical assistance. The authors also wish to thank Carlos Lorenzo, and María Peiró-Camaró Adán who helped revise the manuscript.

Author Contributions

Wrote the paper: MTML MSR. Study concept and design: MSR. Acquisition of data: ACA JMIR RG. Analysis and interpretation of data: NDP MTML MSR.

References

1. Bray GA, (2003) Risks of obesity. Endocrinol Metab Clin North Am 32 (4): 787–804.

2. Bray GA. (2004) Medical consequences of obesity. J Clin Endocrinol Metab 89 (6): 2583–2589.

3. Sims EA. (2001) Are there persons who are obese, but metabolically healthy? Metabolism 50: 1499–1504.

4. Wildman RP, Mauntner P, Reynols K, McGinn AP, Rajpathak S, et al. (2008) The obese without cardiometabolic risk factor clustering and the normal weight

with cardiometabolic risk factor clustering: prevalence and correlates of 2 phenotypes among the US population (NHANES 1999–2004). Arch Intern Med 168 (15): 1617–1624.

5. Lopez-Garcia E, Guallar-Castillon P, Leon-Muñoz L, Rodriguez-Artalejo F. (2013) Prevalence and determinants of metabolically healthy obesity in Spain. Atherosclerosis 231: 152–157.

6. Primeau V, Coderre L, Karelis AD, M . Brochu, M-E . Lavoie, et al. (2011) Characterizing the profile of obese patients who are metabolically healthy. Int J Obes (Lond) 35 (7): 971–981.

7. Blüher M. (2010) The distinction of metabolically "healthy" from "unhealthy" obese individuals. Curr Opin Lipidol 21: 38–43.

8. Pataky Z, Bobbioni-Harsch E, Golay A. (2010) Open questions about metabolically normal obesity. Int J Obes (Lond) 34: S18–S23.

9. Blüher M. (2012) Are there still healthy obese patients? Curr Opin Endocrinol Diabetes Obes 19: 341–346.

10. Ascaso JF, Romero P, Real TJ, Priego A, Valdecabres C, et al. (2001) Insulin resistance quantification by fasting insulin plasma values and HOMA index in non diabetic population. Med Clin (Barc) 117: 530–533.

11. Hamer M, Stamatakis E. (2012) Metabolically healthy obesity and risk of all cause and cardiovascular disease mortality. J Clin Endocrinol Metab 97 (7): 2482–2488.

12. Alberti KGMM, Eckel R H, Grundy SM, Zimmet P, Cleeman JI, et al. (2009) Harmonizing the Metabolic Syndrome: A Joint Interim Statement of the International Diabetes Federation Task Force on Epidemiology and Prevention; National Heart, Lung, and Blood Institute; American Heart Association; World Heart Federation; International Atherosclerosis Society; and International Association for the Study of Obesity. Circulation 120: 1640–1645.

13. Lorenzo C, Serrano-Rios M, Martinez-Larrad MT, Gabriel R, Williams K, et al. (2002) Prevalence of hypertension in Hispanic and non-Hispanic white populations. Hypertension 39: 203–208.

14. Martínez-Larrad MT, Fernández Pérez C, González Sánchez JL, Lopez A, Fernández-Álvarez J et al. (2005) Prevalence of the metabolic syndrome (ATP-III criteria) population-based study of rural and urban areas in the Spanish province of Segovia. Med Clin (Barc) 125: 481–486.

15. World Health Organization: WHO MONICA project. Part III: Population survey. Section 1: Population survey data component. In: MONICA manual. 1990 Geneva: World Health Organization.

16. Martínez Larrad MT, Fernández-Pérez C, Corbatón-Anchuelo A, Gabriel R, Lorenzo C, et al. (2011) Revised waist circumference cut-off points for the criteria of abdominal obesity in the Spanish population: Multicenter nationwide Spanish population based study. Av Diabetolologia 27(5):168–174.

17. Buja A, Scafato E, Sergi G, Maggi S, Suhad MA, et al. (2010) Alcohol consumption and Metabolic Syndrome in the elderly: Results from the Italian longitudinal study on aging. Eur J Clin Nutr 64: 297–307.

18. Yoon YS, Oh SW, Baik HW, Park HS, Kim WY. (2004) Alcohol consumption and the metabolic syndrome in Korean adults: The 1998 Korean National Health and Nutrition Examination Survey. Am J Clin Nutr 80: 217–224.

19. Fridewald WT, Levy RI, Fredrickson D. (1972) Estimation of the concentration of low-density lipoprotein cholesterol in plasma without use of the preparative ultracentrifugue. Clin Chem 18, 499–502.

20. Diagnosis and classification of diabetes mellitus. American Diabetes Association. (2004) Diabetes Care 27: S5–10.

21. Matthews DR, Hosker JP, Rudenski AS, Naylor BA, Treacher DF, et al. (1985) Homeostasis model assessment: insulin resistance and beta cell function from fasting plasma glucose and insulin concentration in man. Diabetologia 28: 412–419.

22. D'Agostino RB Sr, Vasan RS, Pencina MJ, Wolf PA, Cobain M, et al. (2008) General cardiovascular risk profile for use in primary care: the Framingham Heart Study. Circulation 18: 499–502.

23. Conroy RM, Pyörälä K, Fitzgerald AP, Sans S, Menotti A, De Backer G, et al. (2003) Estimation of ten-year risk of fatal cardiovascular disease in Europe: the Score project. Eur Heart J 24: 987–1003.

24. Aguilar-Salinas CA, Garcia EG, Robles L, Riaño D, Ruiz-Gomez DG, et al. (2008) High adiponectin concentrations are associated with the metabolically healthy obese phenotype. J Clin Endocrinol Metab 93: 4075–4079.

25. Velho S, Paccaud F, Waeber G, Vollenweider P, Marques-Vidal P. (2010) Metabolically healthy obesity: different prevalence using different criteria. Eur J Clin Nutr 64: 1043–1051.

26. Kuk JL, Ardern CI. (2009) Are metabolically normal but obese individuals at lower risk for all-cause mortality? Diabetes Care 32: 2297–2299.

27. Pajunen P, Kotronen A, Korpi-Hyövälti E, Keinänen-Kiukaanniemi S, Oksa H, et al. (2011) Metabolically healthy and unhealthy obesity phenotypes in the general population: the FIN-D2D Survey. BMC Public Health 11: 754.

28. Grundy SM, Cleeman JI, Merz CN, Brewer Jr HB, Clark LT, et al. (2004) Implications of recent clinical trials for the National Cholesterol Education Program Adult Treatment Panel III guidelines. Circulation 110: 227–239.

29. Soriguer F, Gutiérrez-Repiso C, Rubio-Martín E, García-Fuentes E, Almaraz MC, et al. (2013) Metabolically healthy but obese, a matter of time? Findings from the prospective Pizarra study. J Clin Endocrin Metab 98 (6): 2318–2325.

30. Calori G, Lattuada G, Piemonti L, Garancini MP, Ragogna F, et al. (2011) Prevalence, metabolic features and prognosis of metabolically healthy obese Italian individuals: the Cremona Study. Diabetes Care 34: 210–215.

31. Iacobellis G, Ribaudo MC, Zappaterreno A, Iannucci CV, Leonetti F. (2005) Prevalence of uncomplicated obesity in an Italian obese population. Obes Res 13: 1116–1122.

32. Shin MJ, Hyun YJ, Kim OY, Kim JY, Jang Y, et al. (2006) Weight loss effect on inflammation and LDL oxidation in metabolically healthy but obese (MHO) individuals: low inflammation and LDL oxidation in MHO women. Int J Obes (Lond) 30: 1529–1534.

33. van Vliet-Ostaptchouk JV, Nuotio ML, Slagter SN, Doiron D, Fischer K, et al. (2014) The prevalence of metabolic syndrome and metabolically healthy obesity in Europe: a collaborative analysis of ten large cohort studies. BMC Endocrine Disorders 14: 9.

34. Jennings C L, Lambert EV, Collins M, Joffe Y, Levitt NS, et al. (2008) Determinants of insulin resistant phenotypes in normal weight and obese black African women. Obesity 16 (7): 1602–1609.

35. Lee K. (2009) Metabolically obese but normal weight (MONW) and metabolically healthy but obese (MHO) phenotypes in Koreans: characteristics and health behaviours. Asia Pac J Clin Nutr 18 (2): 280–284.

36. Marques-Vidal P, Velho S, Waterworth D, Waeber G, von Känel R, et al. (2012) The association between inflammatory biomarkers and metabolically healthy obesity depends on the definition used. Eur J Clin Nutr 66: 426–435.

37. Basati G, Pourfarzam M, Movahedian A, Samsamshariat SZ, Sarrafzadegan N.(2011) Reduced plasma adiponectin levels relative to oxidized low density lipoprotein and nitric oxide in coronary artery disease patients. Clinics (Sao Paulo) 66: 1129–1135.

38. Ukkola O, Santaniemi M. (2002) Adiponectin: a link between excess adiposity and associated comorbidities. J Mol Med (Berl) 80 (11): 696–702.

39. Juge-Aubry CE, Henrichot E, Meier CA. (2005) Adipose tissue: a regulator of inflammation. Best Pract Res Clin Endocrinol Metab 19 (4): 547–566.

40. Koenig W, Khuseyinova N, Baumert J, Meisinger C, Löwel H. (2006) Serum concentrations of adiponectin and risk of type 2 diabetes mellitus and coronary heart disease in apparently middle aged men. J Am Coll Cardiol 48: 1369–1377.

41. Ogorodnikova AD, Kim M, McGinn A, Muntner P, Khan UI, et al. (2012) Incident Cardiovascular Disease Events in Metabolically Benign Obese Individuals. Obesity 20: 651–629.

42. Arnlöv J, Ingelsson E, Sundström J, Lind L. (2010) Impact of BMI and the Metabolic Syndrome on the Risk of Diabetes in Middle-Aged Men. Circulation 121: 230–236.

43. Flint AJ, Hu FB, Glynn RJ, Caspard H, Manson JE, et al. (2010) Excess weight and the risk of incident coronary heart disease among men and women. Obesity 18: 377–383.

13

A Mur Regulator Protein in the Extremophilic Bacterium *Deinococcus radiodurans*

Amir Miraj Ul Hussain Shah[1❸], Ye Zhao[1❸], Yunfei Wang[1], Guoquan Yan[3], Qikun Zhang[3], Liangyan Wang[1], Bing Tian[1], Huan Chen[2,3]*, Yuejin Hua[1]*

1 Key Laboratory of Chinese Ministry of Agriculture for Nuclear-Agricultural Sciences, Institute of Nuclear-Agricultural Sciences, Zhejiang University, Hangzhou, China, **2** Key Laboratory of Laboratory Medicine, Ministry of Education, Zhejiang Provincial Key Laboratory of Medical Genetics, School of Laboratory Medicine & Life Science, Wenzhou Medical College, Wenzhou, China, **3** Laboratory of Microbiology and Genomics, Zhejiang Institute of Microbiology, Hangzhou, China

Abstract

Ferric uptake regulator (Fur) is a transcriptional regulator that controls the expression of genes involved in the uptake of iron and manganese, as well as vital nutrients, and is essential for intracellular redox cycling. We identified a unique Fur homolog (DR0865) from *Deinococcus radiodurans*, which is known for its extreme resistance to radiation and oxidants. A *dr0865* mutant (Mt-0865) showed a higher sensitivity to manganese stress, hydrogen peroxide, gamma irradiation and ultraviolet (UV) irradiation than the wild-type R1 strain. Cellular manganese (Mn) ion (Mn^{2+}) analysis showed that Mn^{2+}, copper (Cu^{2+}), and ferric (Fe^{3+}) ions accumulated significantly in the mutant, which suggests that the *dr0865* gene is not only involved in the regulation of Mn^{2+} homeostasis, but also affects the uptake of other ions. In addition, transcriptome profiles under $MnCl_2$ stress showed that the expression of many genes involved in Mn metabolism was significantly different in the wild-type R1 and DR0865 mutant (Mt-0865). Furthermore, we found that the *dr0865* gene serves as a positive regulator of the manganese efflux pump gene *mntE* (*dr1236*), and as a negative regulator of Mn ABC transporter genes, such as *dr2283*, *dr2284* and *dr2523*. Therefore, it plays an important role in maintaining the homoeostasis of intracellular Mn (II), and also other Mn^{2+}, zinc (Zn^{2+}) and Cu^{2+} ions. Based on its role in manganese homeostasis, DR0865 likely belongs to the Mur sub-family of Fur homolog.

Editor: Christophe Herman, Baylor College of Medicine, United States of America

Funding: This work was supported by grants from National Natural Science Foundation of China (31100058, 31210103904, 31370102), and a major project for genetically modified organisms breeding from the Ministry of Agriculture of China (2014ZX08009-003-002), a grant from Special Fund for Agro-scientific Research in the Public Interest from the Ministry of Agriculture of China (201103007), a Program for New Century Excellent Talents in University (NCET-10-0739), the Natural Science Foundation and Educational Commission of Zhejiang Province (LY13C010001, Y201329892), a project of the Science and Technology Department of Zhejiang Province of China (2013F20011, 2014F50012, 2014F30033), the Fundamental Research Funds for the Central Universities from Zhejiang University (2013QNA6015), and Key Innovation Team Program of Zhejiang Province (2010R50033). The funders had no role in study design, data collection and analysis, decision to publish, or preparation of the manuscript.

Competing Interests: The authors have declared that no competing interests exist.

* Email: yjhua@zju.edu.cn (YH); chenhuan7809@gmail.com (HC)

❸ These authors contributed equally to this work.

Introduction

Metal ions, such as manganese (Mn^{2+}) and iron (Fe^{2+}), are essential micronutrients for many microorganisms and act as enzyme cofactors for a wide range of proteins in processes such as DNA synthesis, DNA repair, reactive oxygen species (ROS) scavenging and electron transport [1]. However, when in excess, they are toxic to cells. Excess iron induces the over-production of harmful ROS, such as super-oxide anion radicals (O_2^-) and hydrogen peroxide (H_2O_2) [1]. High levels of ROS may target DNA, RNA, proteins and lipids through the hydroxyl radicals (HO•) that are generated from H_2O_2 in the Fenton reaction, which uses divalent ions [2]. Inhibition of RNA and protein synthesis occur when high intracellular levels of manganese are reached [3]. Therefore, microorganisms have evolved efficient mechanisms to maintain metal ion homeostasis [4].

The uptake of metal ions is controlled by the ferric uptake regulator (Fur) or the *Diphtheria* toxin repressor (DtxR) family of proteins [5]. The Fur superfamily comprises different proteins with distinct regulatory roles [6]. Fur and Zur (zinc uptake regulator) [7], which respond to iron (Fe^{2+}) or zinc (Zn^{2+}), respectively, repress, the expression of genes involved in Fe^{2+} or Zn^{2+} uptake. The PerR protein, which has been found in Gram-positive bacteria such as *Bacillus* and *Staphylococcus*, regulates several genes that are involved in the oxidative stress response [8,9]. Another Fur homolog, named Irr, which can repress the heme biosynthesis pathway, was first found in *Bradyrhizobium japonicum* [10,11]. In 2004, Johnston's group identified a new *fur*-like protein named Mur (manganese uptake regulator) in *Rhizobium leguminosarum*. This protein represses the transcription of the *sitABCD* genes in response to Mn^{2+} [12].

Deinococcus radiodurans is a well-known bacterium that has extraordinary resistance to ionizing radiation (IR), ultraviolet radiation (UV), various DNA-damaging agents, oxidative stress and desiccation [13]. Ionizing radiation can directly damage biomacromolecules and also produces ROS, which can attack both proteins and DNA [14]. Recently, it has been shown that *D. radiodurans* has a special Mn/Fe regulatory system, which accumulates exceptionally high levels of intracellular Mn^{2+} and low levels of Fe^{2+}. Mn^{2+} may act as an antioxidant to strengthen or support the antioxidant enzyme system, which protects the bacteria from oxidative stress [15,16]. It was shown that there are three types of Mn^{2+}-dependent transport genes in D. radiodurans: dr1236 (Mn^{2+} efflux genes) [17], dr1709 (Nramp family transporters) and three ATP-dependent transporters (dr2283, dr2284 and dr2523). The genes that are involved in Fe^{2+}-dependent transport encode an ABC-type hemin transporter (drb0016), an ABC-type Fe(III)-siderophore transporter (drb0017), two Fe(II) transporters (dr1219, dr1220) and two DNA protection proteins (Dps) (dr2263, drb0092) [1]. Furthermore, D. radiodurans also has three oxidation-related regulators: OxyR (DR0615), DtxR (DR2539), and a Fur homolog (DR0865) [1]. The DR0615 protein is both a transcriptional activator of the katE and drb0125 genes and a transcriptional repressor of the dps and mntH genes [18]. The DR2539 protein acts as a negative regulator of a Mn^{2+} transporter gene (dr2283) and as a positive regulator of Fe^{2+}-dependent transporter genes (dr1219 and drb0125) [19]. However, the function of the Fur homolog (DR0865) is still unknown.

In this study, we aimed to elucidate the function of the Fur homolog (DR0865) and demonstrate its role in maintaining the homoeostasis of intracellular Mn. The results showed that DR0865 is not only a Mur protein, but is also vital for the homoeostasis of intracellular Mn^{2+}, Zn^{2+} and Cu^{2+} ions.

Results

D. radiodurans gene encodes a putative Fur family protein

There is a potential fur homolog (dr0865), which encodes a protein that contains 132 amino acids, in D. radiodurans genome. A BLASTP analysis showed that DR0865 exhibits 24% identity to Helicobacter pylori Fur (Hpy-Fur) and 26% identity to the E. coli Fur protein. Further comparison with the Hpy-Fur sequence showed that DR0865 has three similar metal-binding domains. Domain I consists of amino acid residues C82, C85, C121 and C124, domain II comprises the residues E70, H77 and H79 and domain III is formed from residues H76, H92, T97, H113 and H78 (Figure S1 in File S1). The predicted structure of DR0865 is based on the crystal structure of Hpy-Fur (Figure 1). Previous data showed that ZnS_4 binding by domain I stabilizes β3-β4-β5 structures. Domain II is a metal sensing site, which can regulate DNA binding ability in response to changes in metal concentrations. Domain III is not necessary for DNA binding, however, mutation of this domain reduces the DNA binding ability [20].

The absence of dr0865 inhibits cell growth

To confirm the specific roles of DR0865 in D. radiodurans, the null mutant of dr0865 (Mt-0865) and the complemented strain (C-0865) were constructed. The coding region of the dr0865 gene was replaced with a kanamycin resistance cassette under a constitutively expressed D. radiodurans groEL promoter (Figure S2 in File S1). As shown in Figure 2, the cell growth of Mt-0865 was approximately two-fold lower than that of the wild type strain at 30°C, whereas the growth rate of the complemented strain C-0865 was similar to that of the wild-type strain. This result indicates that the dr0865 gene is necessary for cell growth and other metabolic activities.

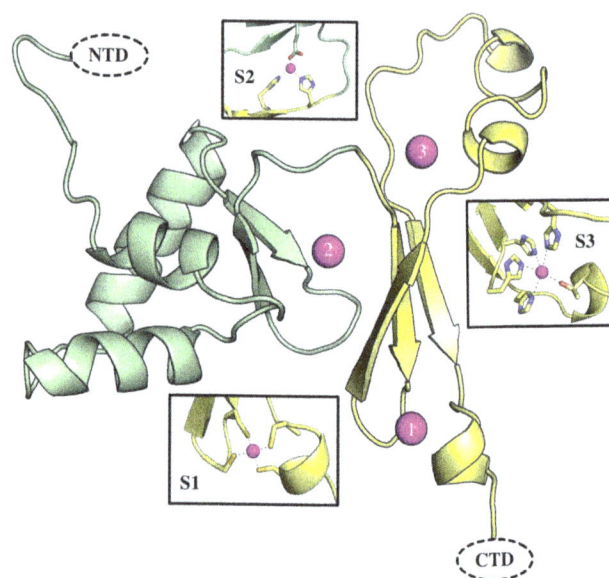

Figure 1. The predicted structure of DR0865 was obtained from homology modeling (Swiss-model) using Hpy-Fur PDB: 2XIG as starting model.

Loss of dr0865 causes Mn (II) ion sensitivity in D. radiodurans

To test whether the growth inhibition was caused by a disruption of ion homeostasis, a metal ion sensitivity assay was carried out as described previously [17]. As shown in Figure 3 and Figure S3 in File S1, the growth of Mt-0865 was strongly inhibited by Mn^{2+} but not by the presence of other metal ions. The C-0865 showed the same growth phenotype as the wild-type R1 strain, which indicates that mutation of the dr0865 gene disrupts Mn^{2+} homeostasis.

To further confirm the Mn sensitivity of Mt-0865, we measured the effect of various concentrations of Mn^{2+} on the growth of Mt-0865 (Figure 4). In comparison with the wild-type R1 strain, the

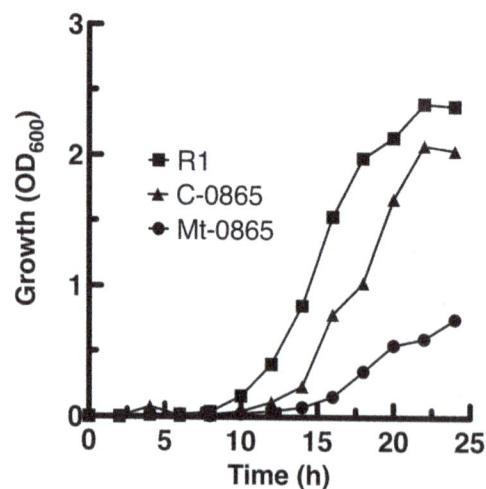

Figure 2. Growth curves of the wild-type R1 strain (black square), complement C-0865 strain (black triangle) and Mt-0865 strain (black circle). Data represent the mean ± standard deviation of three independent experiments.

Metal ions sensitivity assay (10 µl)

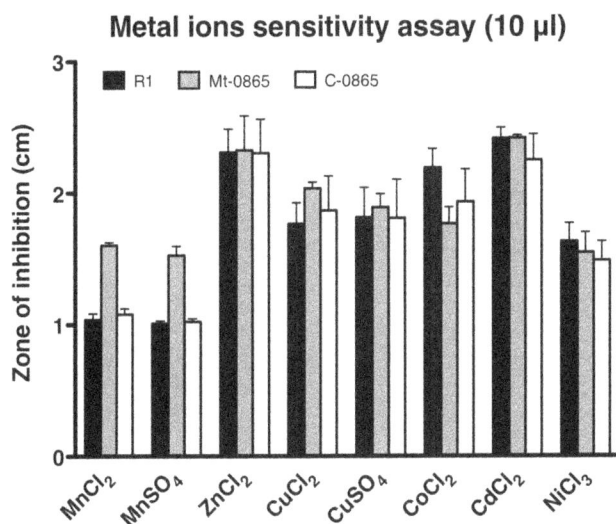

Figure 3. The zone of inhibition of the wild-type R1 (black bar), Mt-0865 (gray bar) and C-0865 strains (white bar) respectively, under various cations stress. Strains were cultured in TGY plates, overlaid with filter discs saturated with 1 M solution of various cations. The zone of inhibition was measured from the edge of the disc after 3 days. Data represent the means ± standard deviation of three independent experiments.

growth of Mt-0865 was inhibited in the presence of low concentrations of Mn^{2+} in TGY medium. When the Mn^{2+} concentration was increased, the growth defect phenotype became more pronounced. An analogous was observed in previous studies, in which the growth of a *Streptococcus pneumoniae* mutant with a disrupted calcium efflux system was more severely inhibited at higher calcium concentrations [21]. Therefore, we inferred that the Mt-0865 strain may either have a higher rate of Mn^{2+} uptake or is unable to efficiently remove excess Mn^{2+}.

The effect of H_2O_2, UV and gamma irradiation on the survival of Mt-0865

The response of the Fur regulator to oxidative stress is very complicated in some microorganisms [5]. Because the Mt-0865 strain exhibits a growth defect and is sensitive to Mn stress, we further investigated the sensitivity of Mt-0865 to H_2O_2, UV and gamma irradiation. First, the survival of these strains was measured under oxidative stress. The results showed an increased sensitivity of Mt-0865 to H_2O_2, whereas the C-0865 strain exhibited a similar survival rate as the wild-type R1 strain (Figure 5 and Figure S4 in File S1). Furthermore, the survival rate was measured under UV and gamma irradiation. The D_{10} value, which represents the irradiating dose required to reduce the population by 90%, was used to assess the resistance of the wild-type R1 and Mt-0865 strains to gamma and UV irradiation. As shown in Figure 6A and B, the wild-type R1 strain showed higher resistance to gamma irradiation and UV radiation than the Mt-0865 strain. Similarly, the mutant strains show higher sensitivity to hydrogen peroxide, as shown in the Figure 6C

Loss of *dr0865* results in an accumulation of intracellular Mn

It has been previously reported that a high intracellular Mn^{2+}/Fe^{2+} ratio in *D. radiodurans* helps to protect proteins from oxidative damage, and contributes to its extreme resistance [15,22]. Because the Mt-0865 is sensitive to Mn stress, H_2O_2 stress, as well as UV and gamma irradiation, inductively coupled plasma mass spectrometry (ICP-MS) analyses were performed to show whether the Mt-0865 strain had lost its ability to maintain homeostasis of manganese and other ions.

As expected, even on TGY medium, the Mn (II) level in the Mt-0865 strain was almost three-fold higher than in the wild-type R1 strain (Figure 7A). Similar results were obtained when the two strains were grown on TGY medium supplemented with Mn^{2+}. Furthermore, we also found that the Mn^{2+} level was increased when the wild-type R1 strain was grown on manganese-rich TGY medium, compared to normal medium (Figure 7A). In contrast, there was no significant difference in Fe^{2+} concentrations between the Mt-0865 and wild-type R1 strains. However, the Fe^{2+}

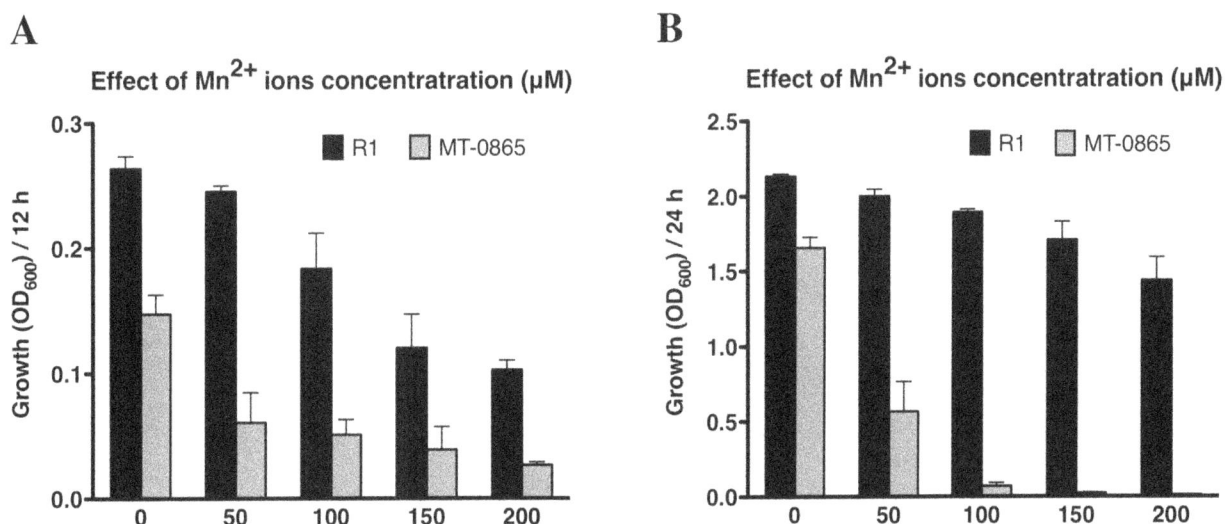

A

Effect of Mn^{2+} ions concentratration (µM)

B

Effect of Mn^{2+} ions concentratration (µM)

Figure 4. Sensitivity of the wild-type R1 strain (black bar) and mutant Mt-0865 strain (gray bar) to MnCl₂. Strains were cultured in TGY, supplemented with 0, 50,100 or 150 µM of $MnCl_2$. The OD₆₀₀ was measured after 12 and 24 h. Data represent the means ± standard deviation of three independent experiments.

Disc diffusion assay (H$_2$O$_2$)

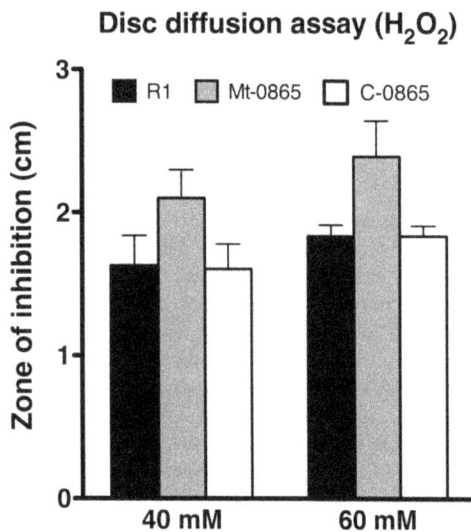

Figure 5. Hydrogen peroxide sensitivity assay for the wild-type R1, Mt-0865 and C-0865 strains, The wild-type R1 strain (black bar), Mt-0865 strain (gray bar) and C-0865 strain (white bar) were cultured in TGY plates, overlaid with filter discs saturated with 4 µl and 6 µl of 1 M solution of H$_2$O$_2$.

concentrations in both wild type and mutant strains increased under Mn stress, indicating that *D. radiodurans* has a system to regulate Mn/Fe homeostasis (Figure 7B). Collectively, these results verified that the Mt-0865 strain is sensitive to Mn^{2+} stress and that the sensitivity of the mutant to damaging agents may be caused by excess manganese.

Transcriptome changes in the *dr0865* mutant under manganese stress

Because DR0865 is homologous to transcriptional regulators and its disruption resulted in sensitivity to excess manganese, RNA sequencing (RNA-seq) was used to assess changes in transcripts of the wild-type R1 and Mt-0865 mutant strains when cultured in the presence of high (20 mM) levels of MnCl$_2$. In total, 12.7 million (M) and 13.3 M pair-end reads were obtained for the wild-type R1 and Mt-0865 strains, respectively. After cleaning the reads, 9.3 M

and 9.8 M reads mapped to the genome, and, after removing rRNA sequences, the unique sequence reads were 4.3 M and 4.4 M (Table 1).

A total of 3,098 of the 3,167 open reading frames (ORFs) of the wild-type R1 strain, and 3,060 of the 3,167 ORFs of Mt-0865 mutant strain were detected by RNA sequence data. In the Mt-0865 strain, 246 genes were up-regulated more than two-fold (Table S1 in File S1) and 317 genes were down-regulated more than two-fold in comparison to the wild-type R1 strain (Table S2 in File S1). The significantly expressed genes were classified in accordance with the Cluster of Orthologous Groups (COG) of proteins database (Table 2). The top three categories were transcription (22%), inorganic ion transport and metabolism (19.2%), and nucleotide transport and metabolism (18.5%).

In this study, our analysis focused on the genes involved in (i) Mn/Fe metabolism, (ii) ROS production, (iii) DNA damage response genes, and (vi) cell cleaning genes (Table 3).

(i) Proteins involved in Mn^{2+} and Fe^{2+} metabolism

The expression of five genes involved in Mn/Fe metabolism was significantly changed under MnCl$_2$ adaptation conditions (Table 2). *dr1236* (*mntE*), which encodes a putative manganese efflux family protein that controls the removal of excess Mn^{2+}, was repressed (Table 2). However, all ATP-dependent Mn^{2+} transporter genes, including *dr2283*, *dr2284*, and *dr2523*, were induced, The Mn^{2+} transporter gene expression pattern suggested that Mn^{2+} concentration was increased in the mutant, which is in agreement with our ICP-MS data. Furthermore, a previous DNA binding assay also supported that ATP-dependent Mn^{2+} transporter genes and *mntE* are regulated by DR0865 [23].

(ii) Proteins associated with the production of ROS

Manganese is among the essential enzyme cofactors because it protects cells from oxidative damage. However, it can be toxic at high concentrations and, therefore, its level should be strictly regulated [17,23]. Previous research showed that cytochromes, flavoproteins, iron-sulfur proteins, NADPH and NADH-dependent enzymes are regarded as the major generators of ROS [1,2]. Under Mn^{2+} stress, nine cytochrome-related genes and 12 NADH-dependent enzymes were repressed in the Mt-8065 mutant compared to the wild-type R1 strain. This indicates that the mutant strain compensates for its Mn sensitivity by dampening the production of ROS.

A   **B**   **C**

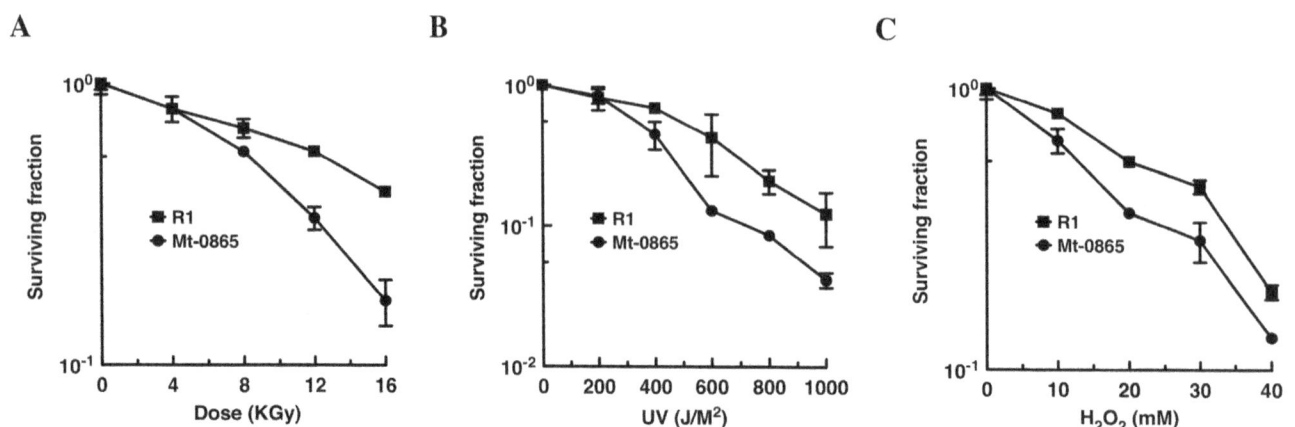

Figure 6. Survival curves of *D. radiodurans* strains exposed to (A) ionizing irradiation (B) UV irradiation and (C) Hydrogen peroxide. Wild-type *D. radiodurans* R1 (black square) was compared with the Mt-0865 strain (black circle). Error bars represent standard deviations from three replicate experiments.

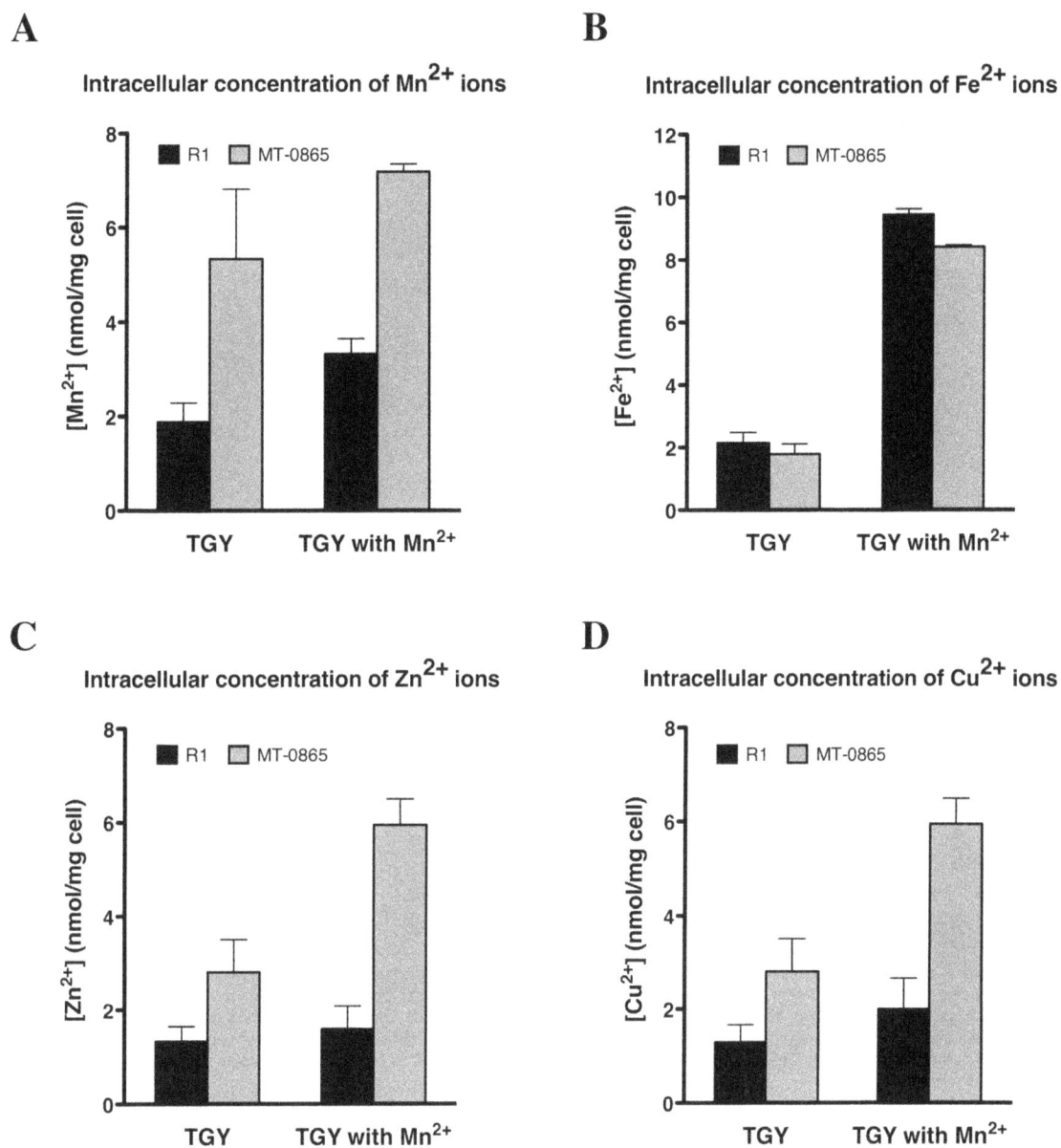

Figure 7. Analysis of the intracellular ion content of the wild-type R1 and Mt-0865 strains, cultured in a medium supplemented with or without 50 μM manganese; (A) manganese, (B) iron, (C) zinc, (D) copper. The data represent the mean ± standard deviation of three independent experiments.

Table 1. Summary of sequence reads statistics resulting from Illumina deep sequencing of mutant and wild type strain.

	Mutant Strain	Wild type strain
Raw data (read number)	13289865	12798354
Raw data (length, bp)	200	200
Clean data (read number)	10260088	9998560
Clean data (length, bp)	176	178
Mapping to genome (read number)	9831262	9299761
Mapping to rRNA	5289176 (53.8%)	4728594 (50.85%)
Unique mapping	4441705 (45.18%)	4387406 (47.18%)

Table 2. Classification of the genes with different levels of expression according to the Cluster of Orthologous Groups of proteins (COG) database.

COG_type[1]	Total genes	Induced genes	Repressed genes	Total rate[2]
Information storage and processing:				
J	165	10	6	9.6%
K	149	13	21	22.8%
L	138	5	12	12.3%
B	1	0	0	0
Cellular processes and signaling:				
D	26	0	4	15.4%
V	50	1	7	16%
T	119	5	9	11.2%
M	110	6	9	13.6%
N	21	0	2	9.5%
U	40	0	3	7.5%
O	108	6	12	16.7%
Metabolism:				
C	140	7	28	25%
G	114	6	10	14%
E	251	13	15	11.2%
F	92	10	7	18.5%
H	110	8	9	15.5%
I	99	5	3	8%
P	130	14	11	19.2%
Q	53	4	2	11.3%
Poorly characterized:				
S	251	19	23	16.7%
R	333	19	23	12.6%

[1]. J: Translation, ribosomal structure and biogenesis; K: Transcription; L: Replication, recombination and repair; B: Chromatin structure and dynamics; D: Cell cycle control, cell division, chromosome partitioning; V: Defense mechanisms; T: Signal transduction mechanisms; M: Cell wall/membrane/envelope biogenesis; N: Cell motility; U: Intracellular trafficking, secretion, and vesicular transport; O: Posttranslational modification, protein turnover, chaperones; C: Energy production and conversion; G: Carbohydrate transport and metabolism; E: Amino acid transport and metabolism; F: Nucleotide transport and metabolism; H: Coenzyme transport and metabolism; Lipid transport and metabolism; P: Inorganic ion transport and metabolism; Q: Secondary metabolites biosynthesis, transport and catabolism; S: Function unknown; R: General function prediction only.
[2]. the total number of significant genes/Number of total genes in this COG

(iii) Proteins associated with the DNA damage response

There were 15 induced genes associated with stress responses, but these did not include *katE* or *recA*. However, three antioxidant proteins, *bcp* (*dr1208*), *ahpC* (*dr1209*), and *grxA* (*dr1209*, *dra0072*), were up-regulated in the Mt-0865 strain (Table 3), which indicates that the mutant strain may be under oxidative stress. In addition to the well-characterized components of stress response systems, *D. radiodurans* encodes several proteins whose specific roles are unknown but are likely to be important for the multiple stress resistance phenotypes of the bacterium. An example of a poorly studied, but potentially important, system is the "addiction module" response, which is encoded by two genes, *mazE* (*dr0417*) and *mazF* (*dr0416*). MazF is a stable protein that is toxic to bacteria, whereas MazE protects cells from the toxic effect of MazF, and is degraded by the ClpX serine protease (*dr0202*) (Table 3). When the Mt-0865 mutant was under Mn stress, all of these genes were induced, which suggests that the mutant strain activates the antidote-toxin system to reduce cell growth to avoid the production of ROS. This result is also consistent with the expression patterns of ROS generating genes.

(iv) Cell cleaning proteins

When the Mt-0865 mutant strain was under Mn stress, we found that the cellular cleansing system, including the export of damaged DNA components and sanitization of intracellular mutagenic precursors, was also induced. First, it was observed that six ABC transporter permease genes, which may control oligonucleotide export, were activated (Table 3) [24]. The export of damaged nucleotides outside the cell might protect the organism from elevated levels of mutagenesis by preventing the reincorporation of damaged bases during DNA synthesis [25]. Second, 15 of 20 *mutT/nudix* family genes were induced, five (*dr0092*, *dr0192*, *dr0261*, *dr0274*, dr0784) of which were up-regulated significantly (Table 3). The MutT protein has an 8-oxo-dGTPase activity, which can limit mutation of DNA by hydrolyzing the oxidized products of nucleotide metabolism. The remaining intracellular mutagenic precursors could be sanitized via this superfamily [26]. Finally, it was also found that Lon protease (DR1974) and ClpX protease (DR0202) were induced approximately twofold (Table 3). These ATP-dependent proteases help with cellular sanitization by degrading damaged proteins [27].

Table 3. The significant genes were classified into three classes, Mn/Fe metabolism, ROS production genes, and Damage response genes.

ORF	Name	Description	M[1]
Mn/Fe metabolism:			
DR2283	dr2283	Mn ABC transporter permease	1.99
DR2284	dr2284	Mn ABC transporter permease	2.39
DR2523	fimA	Mn/Fe transport system substrate-binding protein	3.79
DR1236	mntE	manganese efflux protein	−1.69
DR1220	feoA	ferrous iron transport protein A	1.35
ROS production genes			
DR0342	dr0342	cytochrome complex iron-sulfur subunit	−1.45
DR0344	ccmH	cytochrome c-type biogenesis protein	−1.09
DR0346	ccmF	cytochrome c-type biogenesis protein	−1.09
DR0347	ccmE	cytochrome c-type biogenesis protein	−1.30
DR0348	dr0348	cytochrome c-type biogenesis heme exporter protein C	−1.46
DR2095	dr2095	c-type cytochrome	−1.39
DR2617	ctaA	cytochrome AA3-controlling protein	−1.11
DRC0001	drc0001	cytochrome P450-related protein	$-\infty$[2]
DRC0041	drc0041	Cytochrome P450	$-\infty$
DR1492	dr1492	NADH dehydrogenase I subunit N	−1.03
DR1493	dr1493	NADH dehydrogenase I subunit M	−1.68
DR1494	dr1494	NADH dehydrogenase I subunit L	−1.55
DR1497	dr1497	NADH dehydrogenase I subunit I	−2.54
DR1498	dr1498	NADH dehydrogenase I subunit H	−2.28
DR1499	dr1499	NADH dehydrogenase I subunit G	−1.57
DR1500	dr1500	NADH dehydrogenase I subunit F	−1.99
DR1501	dr1501	NADH dehydrogenase I subunit E	−1.76
DR1503	dr1503	NADH dehydrogenase I subunit D	−2.59
DR1504	dr1504	NADH dehydrogenase I subunit C	−2.32
DR1505	dr1505	NADH dehydrogenase subunit B	−1.79
DR1506	dr1506	NADH dehydrogenase I subunit A	−1.44
DRA0243	Hmp	Haemoglobin-like flavoprotein	−1.33
Damage response genes:			
DR1208	Bcp	Antioxidant type thioredoxin fold protein	∞
DR1209	ahpC	Thiol-alkyl hydroperoxide reductases	1.87
DRA0072	grxA	Glutaredoxin	3.38
DR2056	hslJ	Related to heat shock protein	1.18
DR0194	htpX	Predicted Zn-dependent proteases	1.69
DR0416	mazE	Regulatory protein, MazF antagonist	1.66
DR0417	mazF	ppGpp-regulated growth inhibitor	1.32
DR_B0088	kdpD	Osmosensitive K1 channel histidine kinase sensor domain	1.39
DR1667	trkH	Potassium uptake system component	−1.92
DR1678	trkG	Potassium uptake system component	−2.61
DRA0123	arsC	Arsenate oxidoreductase	−1.31
DR0455	strA	Streptomycin resistance protein	−1.62
DR2234	dr2234	involved in multidrug resistance	−2.41
DR1695	gloA	Lactoylgluthation lyase, fosphomicin resistance protein	1.49
DR0599	BS_yokD	amino glycoside N3-acetyltransferase	1.52
Cell cleaning genes:			
DR0092	dr0092	MutT/nudix family protein	2.69
DR0192	dr0192	MutT/nudix family protein	1.27

Table 3. Cont.

ORF	Name	Description	M[1]
DR0261	dr0261	MutT/nudix family protein	3.33
DR0274	dr0274	MutT/nudix family protein	3.02
DR0784	dr0784	MutT/nudix family protein	1.18
DR0202	clpX	ATPase subunit of Clp protease	1.20
DR1974	Lon	ATP-dependent Lon serine protease	1.40
DR0958	dr0958	peptide ABC transporter permease	1.65
DR0959	dr0989	peptide ABC transporter permease	1.44
DR1358	dr1358	outer membrane protein	1.18
DRA0168	dra0168	ABC transporter permease	1.54
DRA0268	dra0268	adenine deaminase-like protein	1.21
DRA0323	dra0323	urea/short-chain amide ABC transporter ATP-binding protein	1.07

[1]. M value means $\log_2$Ratio, Ratio = $FPKM_{(M-0865)}/FPKM_{(R1)}$
[2]. ∞ means gene's expression level is not detected in one sample, but detected in another sample.

qRT-PCR analysis

To confirm the transcriptome assay results, gene expression in the Mt-0865 mutant and in the wild-type R1 strains was analyzed using quantitative real time PCR (qRT-PCR) analysis. Eight genes (dr2523, dr2283, dr1709, dr1236, dr1998, dr1506, dr0348, and dr0828) were quantified under normal growth conditions and after treatment with Mn^{2+}. Four of these genes are Mn^{2+} transport genes (dr2523, dr2283, dr1709 and dr1236). The DR2523 and DR2283 proteins are ATP-dependent transporters, the DR1709 protein belongs to the Nramp family of transporters, and DR1236 is a Mn^{2+} efflux gene. The dr1998 gene encodes a major catalase (KatE), which plays an important role in the protection of D. radiodurans from oxidative stress and ionizing radiation [1]. The DR1506 protein is a NADH dehydrogenase and DR0348 is the cytochrome c-type biogenesis heme exporter protein C. It was previously shown that dr1506 and dr0348 are associated with the production of ROS [1]. The dr0828 gene encodes an isocitrate lyase, which is an enzyme in the glyoxylate cycle that catalyzes the cleavage of isocitrate to succinate and glyoxylate. Previous research has shown that when irradiated, D. radiodurans represses the tricarboxylic acid (TCA) cycle and activates the glyoxylate bypass [28].

It was observed that under normal growth conditions, the expression of dr2523, dr2283, and dr1709 increased 1.96-fold, 4.24-fold and 4.41-fold, respectively, in the Mt-0865 mutant strain compared to the wild-type R1 strain (Table S3 in File S1), while the dr1236 gene was significantly repressed 24.32-fold in the mutant strain. In addition, the transcript levels of dr1506 and dr0348 decreased, whereas the level of the dr0828 transcript increased, in the mutant strain (Figure 8A and Table S3 in File S1). Gene expression levels were also measured under Mn^{2+} stress. The expression of the Mn^{2+} transporter gene dr2539 increased 20.14-fold, while expression of dr1236 decreased 53.94-fold in the Mt-0865 mutant (Figure 8A). This suggests that under Mn^{2+} stress, the wild-type strain attempts to stop Mn^{2+} uptake and opens the Mn^{2+} efflux system, whereas this does not occur in the mutant strain. These results provide further evidence that Mn^{2+} transporter genes are not properly expressed in the mutant. In addition, the dr1506 and dr0348 genes were repressed and the dr0828 gene was induced. This suggests that, under Mn (II) stress, the wild-type strain lowers its metabolic rate to reduce ROS production and activates the glyoxylate bypass to provide energy,

whereas these adaptations are defective in the mutant strain. Overall, the pattern of gene expression indicates that the mutant strain is likely subject to more damage than the wild-type strain under Mn^{2+} stress.

Discussion

Manganese is a trace element that is essential for many cellular functions in all organisms. For example, Mn^{2+} is required as a co-factor for super-oxide dismutase, which is critical for preventing cellular oxidative stress [29]. However, high manganese levels inhibit calcium influx and promote the exchange of accumulated Ca^{2+}, and inhibit RNA and protein synthesis [3]. Thus, maintaining metal ion homeostasis is necessary for all organisms. D. radiodurans is well known for its extreme resistance to radiation and oxidants and its high intracellular Mn/Fe ratio is an important factor that contributes to this resistance. In this study, we identified a unique Mur homolog that is encoded by dr0865, and data showed that it is Mn^{2+}-specific regulator.

Sequence analyses showed that DR0865 contains three metal-binding domains that are present in the H. pylori Fur homolog. Previous data showed that the $C_{92}XXC_{95}$ motif is necessary for the construction of the ZnS_2 $(N/O)_2$ domain, while the $C_{133}XXXXC_{138}$ motif is not important [20,30]. Further research is needed to discern the function of the $C_{82}XXC_{85}$ and $C_{112}XXC_{115}$ motifs in D. radiodurans. Because the phenotype of the Mt-0865 mutant strain showed that DR0865 is a novel Mur protein, we compared the DR0865 amino acid sequence to the R. leguminosarum Mur protein [12]. The results showed that the R. leguminosarum Mur protein does not have domain I, while it contains domains II and III (data not shown). This indicates that domains II and III are important for Mn^{2+} ion regulation.

It has been suggested that the accumulation of Mn^{2+} or a higher Mn/Fe ratio benefit the radio-resistance of D. radiodurans. However, the excess of Mn^{2+} is toxic to the cell. Although the precise mechanism of Mn^{2+} toxicity is poorly understood, three mechanisms have been suggested previously. In the first mechanism, Mn^{2+} cell toxicity may be associated with its interaction with other essential trace elements, such as Fe^{2+}, Zn^{2+} and Cu^{2+} [31]. Human studies have shown that chronic exposure to Mn^{2+} appears to be associated with similar increases in cellular Fe^{2+} uptake, which consequently produces cellular oxidative stress and

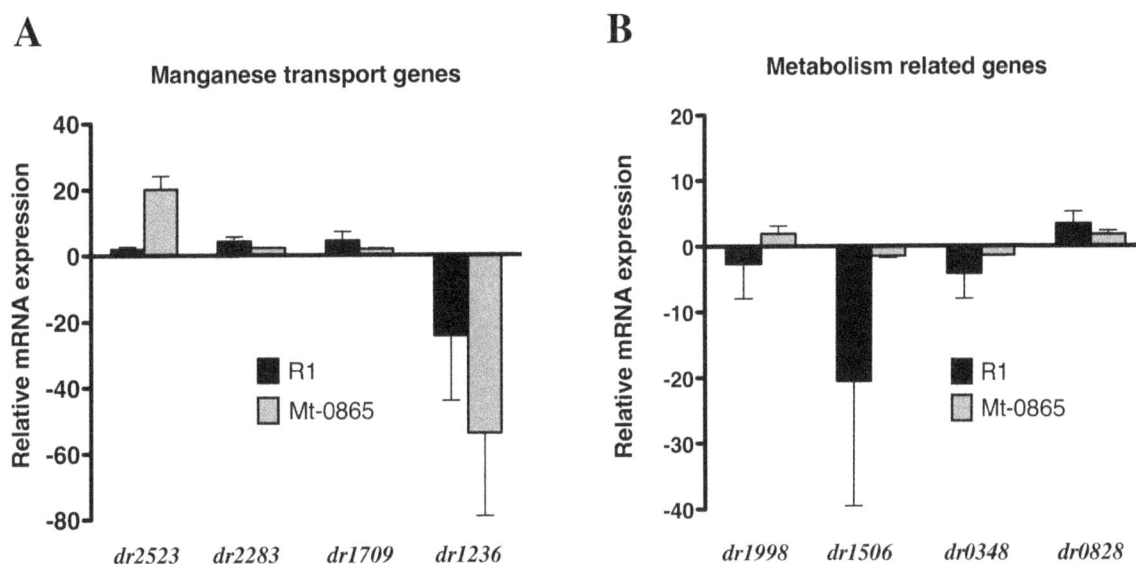

Figure 8. The expression of potential DR0865-dependent genes in wild-type *D. radiodurans* **compared to the mutant strain, under normal conditions (black bar), and wild-type R1 under manganese stress compared with the mutant strain under manganese stress (grey bar).** Error bars represent standard deviations from three replicate experiments. (A) Four manganese transport genes, (B) metabolism related genes.

also increases the concentration of Cu^{2+} and Zn^{2+} [31]. When the wild-type R1 strain was under Mn^{2+} stress, the intracellular concentration of Mn^{2+} and Fe^{2+} increased significantly, whereas the concentrations of Zn^{2+} and Cu^{2+} increased slightly (Figure 7C and D). Under normal growth conditions, the Mt-0865 strain had higher Mn^{2+}, Cu^{2+} and Zn^{2+} contents than the wild-type R1 strain. These data further confirmed that mutation of the *dr0865* gene causes a defect in the control of Mn^{2+} metabolism, which also results in changes in the concentrations of Cu^{2+} and Zn^{2+}. Interestingly, Fe^{2+} concentration was not significantly different between the wild-type and mutant strains under Mn^{2+} stress (Figure 7B), which may be due to the distinct *D. radiodurans* Fe^{2+} regulation system, which utilizes OxyR and DtxR regulators.

The second mechanism is that high intracellular levels of Mn^{2+} inhibit RNA and protein synthesis, and manganese may exert a toxic effect through such inhibition [3]. When *Bacillus stearothermophilus* was grown in media containing excess Mn^{2+}, its doubling time increased more than two-fold. A similar effect on growth was also observed in the Mt-0865 mutant under normal growth conditions. The third mechanism suggests that Mn^{2+} can participate in reactions that potentially increase ROS, which subsequently causes oxidative damage [32]. These three mechanisms may explain why the mutant was sensitive to different DNA damaging agents.

The RNA-seq data identified 562 genes (approximately 17% of the genome) that showed at least a twofold change in expression between the Mt-0865 mutant and the wild-type R1 strain, which indicates that these genes were regulated by *dr0865* either through direct or indirect mechanisms. Using the Kyoto Encyclopedia of Genes and Genomes (KEGG) database, we found that genes involved in metabolic pathways, the biosynthesis of secondary metabolites, oxidative phosphorylation and nitrogen metabolism were significantly repressed in the mutant strain. This indicates that the Mt-0865 mutant is likely to suffer more cellular damage under Mn^{2+} stress than the wild-type strain. This phenomenon may be caused by higher Mn^{2+} levels in the mutant, which would increase ROS levels and lead to DNA damage [33]. Five *mutT/*

nudix family genes were also activated but no major DNA repair genes (such as *recA* or *pprA*) were induced, which further confirms our hypothesis.

Interestingly, we found that the heme biosynthesis pathway (HemA, HemE and HemN) was slightly repressed in the Mt-0865 mutant (Table S2 in File S1). Hemes are biosynthesized from protoporphyrin and free ferrous iron [34] and are cofactors for cytochromes, catalases and peroxidases. This may explain why nine cytochrome genes were down-regulated under Mn^{2+} stress in the mutant. In addition, two vitamin B12 biosynthesis proteins, *drb0010* (cobalamin biosynthesis protein) and *drb0012* (cobyric acid synthase), were also repressed. Vitamin B12 is a water-soluble vitamin that is normally involved in DNA synthesis and regulation, as well as in fatty acid biosynthesis and energy production. The *dr0910* and *dr1076* genes, which encode cell wall protein and cell wall synthesis proteins, respectively, were also down-regulated. The reduction of vitamin B12 and cell wall proteins may be caused by high Mn (II) levels, which consequently results in the inhibition of cell growth.

Five ribonucleases (*dr0020, dr0859, dr1949, dr2374* and *drb0107*) were induced at least two-fold in the Mt-0865 mutant (Table S1 in File S1). Ribonucleases in prokaryotic toxin-antitoxin systems are proposed to function as stress-response elements. The degradation of RNA within a cell leads to fragments of RNA that are no longer needed and can be cleaned up as part of the cellular protection system. Five cation transporter genes (*dr0748a, dr0816, dr0883, dra0168, dra0361*) were also activated, which explains why the mutant had higher ion concentrations.

Overall, our work presents a biochemical mechanism for Mn (II) sensing by the Mur homolog gene in *D. radiodurans*. Using qRT-PCR and global transcriptome analysis, we provided evidence that DR0865 functions as a positive (*dr1236*) and a negative regulator (*dr2283, dr2284, dr2523*) of different classes of Mn^{2+} transporter genes. More research is needed to establish the detailed mechanism of Mur regulation of these important genes. The potential communication between OxyR and other regulators, such as DtxR (*dr2539*), should be explored to determine whether it is required for the intricate coordination of oxygen radical detoxification.

Experimental Procedures

Strains, media and primers

All the primers used in this study are listed in Table 4. The *E. coli* strains were grown in Luria-Bertani (LB) broth medium (1% tryptone, 0.5% yeast extract and 1% sodium chloride) with aeration or on LB agar plates (1.2% Bacto-agar) at 37°C supplemented with the appropriate antibiotics. All *D. radiodurans* R1(ATCC 13939) strains used in this work were grown at 30°C in TGY medium (0.5% tryptone, 0.1% glucose and 0.3% yeast extract) with aeration or on TGY plates supplemented with 1.5% Bacto-agar.

Sequence alignment

The protein sequence of previously characterized Fur proteins found in *A. ferrooxidans*, *P. aeroginosae*, *E. coli*, *B. subtilis*, *M. marinum*, *D. radiodurans*, *H. pylori* were obtained from the NCBI database. The protein sequence alignment of selected Fur proteins was generated using ClustalW.

Disruption of the DR0865 gene in *D. radiodurans*

The mutant strain was constructed as described previously [18]. Primer ME1 and ME2 were used to amplify a *Bam*HI fragment upstream of targeted genes, and primers ME3 and ME4 were used to obtain a *Hind*III fragment downstream of targeted genes respectively (Table 5). The kanamycin resistance cassette containing the *gro*EL promoter was obtained from a shuttle plasmid, pRADK. After this three DNA fragments were digested and ligated. The ligation products were used as template for PCR to amplify the resulting PCR fragment (ME1 and ME5 used as primers), which was then transformed into exponential-phase cells by $CaCl_2$ treatment [35]. The mutant strains were selected on TGY agar plates supplemented with 30 μg/ml kanamycin. Null mutants were confirmed by PCR product sizes, enzyme-digested electrophoresis, and DNA sequencing and the resulting mutant was designated Mt-0865.

Complementation of DR0865 mutant

Complementation strain was constructed as described [18,36]. Briefly, genome DNA was isolated from wild-type R1 strain. A 2500-bp region containing the *dr0865* gene was amplified by ME5 and ME6 (Table 5), and ligated to pMD-18 T-Easy vector (Takara, JP), designed as pMD-dr0865. After digested by *Nde*I and *Bam*HI, the target gene *dr0865* was ligated to *Nde*I and *Bam*HI-pre-digested pRADK, which named as pRKR. The complementation plasmid were confirmed by PCR and DNA sequence analysis, and transformed into Mt-0865, resulting in functional complementation strains. Selection for *D. radiodurans* complement strain was achieved on TGY plates, supplemented with kanamycin (30 μg/ml) and chloramphenicol (3 μg/ml).

Growth curve assay

To examine bacterial growth in vitro as described previously was little modified [37], the single clone of the wild-type R1, Mt-0865 and C-0865 strains were transferred into 5 ml liquid TGY medium. When the OD_{600} of the cultures reached 1.0, 1 mL of each culture was added to 100 mL fresh TGY medium. Three repeats were performed for each strain. The nine cultures were incubated with shaking at 30°C and samples were taken every two hours to measure the OD_{600} value. The cultures were incubated with 250 rpm at 30°C and samples were taken to measure the OD_{600} value at different time. All experiments were repeated in triplicate.

Cation sensitivity assays

Cation sensitivity assays were carried out as described previously [17]. Solutions (1 M) of manganese chloride, manganese sulfate, zinc chloride, copper chloride, copper sulfate, cobalt (II) chloride, nickel chloride, cadmium chloride, ferrous sulfate, ferrous chloride, ferric chloride, magnesium chloride, calcium chloride (sigma) were prepared in milli-Q water and filter-sterilized by passing through 0.22-μm filters. Newly fresh clone was taken from the wild-type R1, Mt-0865 and C-0865 TGY plates, into 5 ml TGY fresh media, when the cells grown up to stationary phase. Then, the cells were plated on TGY plates and overlaid with 5-mm sterile discs containing 1 M various cation solutions. The plates were incubated for three days, and the inhibition zone of each disc was measured. All the data provided here represent the mean and standard deviation of at least three independent experiments (mean ± SD of three experiments).

Similarly, to ascertain the effect of Mn^{2+} on growth of Mt-0865 and wild-type R1, 1×10^5 CFU ml^{-1}, were grown in TGY supplemented with increasing concentration of $MnCl_2$. The

Table 4. Bacterial strains and plasmids used in this study.

Strain or plasmid	Relevant marker	Source
Strains		
E. coli DH5α	Propagation for plasmid	Invitrogen
E. coli BL21(pLysS)	DR0865 expression strain	Invitrogen
D. radiodurans R1	ATCC13939	This lab
Mt-0865	As R1, but *dr0865::kan*	This study
C-0865	Mt-0865 complemented with *pRKR*	This study
Plasmids		
pMD18-T	TA cloning vector	Takara
pMR	pET-28a derivative recombinant expressing and dr0865	This study
pRADK	*E. coli-D.* radiodurans shuttle vector carrying *D.radiodurans groEL* promoter	[36]
pRKR	pRADK derivative expressing *D.radiodurans* dr0865	This study

Table 5. Primers used in this study.

Primer	Sequence (5′ → 3′)
Mutation primers	
0865upF(ME1)	CGAAGAAGTCGCCAACAACC
0865upR(ME2)	GGATCCGGAGGCAGGGTAGCAAAGCG
0865downF(ME3)	AAGCTTGGCGGGAAGTTTTTACTGCGTG
0865downR(ME4)	ACACTAACCGTTTTTCGCCATTGCC
Complement primers	
0865F (ME5)	CATATGACCGCCCGCCGCAGCAC
0865R (ME6)	GGATCCTTAGTGGGCCCCGGTCTTC
Real time PCR primers	
RT-dr1506F	GCGGGAAAGGCTGGAGTCAGGAGG
RT-dr1506R	CTTGGTGCGGGTCGCCTTTTTGGG
RT-dr0348F	CCTCGGGTACTTCATCATCCGTGGC
RT-dr0348R	TTGACGGTGGCGGTCTGGTGAATG
RT-dr1236F	CATCAATCTGGTGTGGGCGAAC
RT-dr1236R	CAAGCAGCGGGTCAAGGATGTG
RT-dr0828F	GACACCATGACCCCCACCCCCAAAA
RT-dr0828R	GGTGTACTCGATGGGCAGGCTG
RT-dr2523F	CGACGCCCATACCTTTCAGC
RT-dr2523R	GTCAGCTCCTTCACCGGCAC
RT-dr2283F	GGAGCCTGCGGACCATGA
RT-dr2283R	GCGAGCGCCAGCAGAAAA
RT-dr1998F	GGGCGTGGACAAGCGTATTC
RT-dr1998R	GTAGACGGGGGCTTCCTGCT
RT-dr1709F	GCGATGGTGATTCAGAACCT
RT-dr1709R	GTTCGGCCTGAATCCAGTAA

Note: straight line represents restriction site.

OD_{600} value was measured after 12 h and 24 h post incubation. All the data provided here represent the mean and standard deviation of at least three independent experiments (mean ± SD of three experiments).

H_2O_2 sensitivity assays (Oxidative stress assays)

The discs diffusion assay to test H_2O_2 sensitivity, was performed as described previously with a little modification [38,39]. The strain was cultured up to log phase and 130 μl aliquots were spread on TGY plates. A sterile 5 mm-diameter filter discs, containing 4 μl and 6 μl of 1 M H_2O_2 was placed on the surface of the TGY plate. After incubation at 30°C for three days, the size of the area cleared of bacteria (zone of inhibition) was measured. For the curve H_2O_2 treatment, the cultures were treated with different concentrations of H_2O_2 for 30 min and then plated on TGY plates, as prescribed previously [40]. All the data provided here represent the mean and standard deviation of at least three independent experiments (mean ± SD of three experiments).

Gamma irradiation and UV sensitivity assays

Survival curves of the wild-type R1 and Mt-0865 cells were cultured in TGY broth to OD_{600} ~1.0. For the Gamma radiation treatment, the 100 ml cultured was irradiated with different doses of ^{60}Co gamma at room temperature, which correspond to doses from 0 to 16 kGy, as previously published [41,42]. After the irradiation treatment, the culture centrifuged and then re-suspended in phosphate buffer (1XPBS Buffer, pH 7.5). The cells were plated on TGY plates and incubated at 30°C for at least three days. The colonies were counted. All the data provided here represent the mean and standard deviation of at least three independent experiments (mean ± SD of three experiments).

For the UV treatment, the cells were cultured in TGY broth to OD_{600}~1.0, as described previously [17,43]. The cells were re-suspended in 1XPBS buffer (pH 7.5), then plated on TGY plates and exposed to different doses of UV radiation at 254 nm. All the data provided here represent the mean and standard deviation of at least three independent experiments (mean ± SD of three experiments).

Assay of intracellular Mn, Fe, Zn and Cu ion concentration

The protocol for determining intracellular concentration of metals ions was identical to previously reports [17]. D. radiodurans R1 and Mt-0865 were cultured in 5 ml TGY broth and re-inoculated in 500 ml TGY broth which had been pretreated with Chelex to remove any cat-ion, and then supplemented with 50 μM manganese chloride. The cells were grown up to OD_{600} ~0.6–0.8 and harvested. After centrifugation at 10000 g, 4°C for 10 min, the pellets were washed three times with 1xPBS (pH 7.5), containing 1 mM EDTA and rinsed three times with 1xPBS, without EDTA. Cells (1/10 of the total volume) were withdrawn to measure the dry weight. For ion analysis, 1 ml of Ultrex II nitric acid (Fluka AG., Buchs, Switzerland) was added to the rest cells and incubated at 100°C for 1 h. After centrifugation at 20,000 g for 20 minutes, the supernatant was filtered against 0.45 μM membrane. The concentration of samples was analyzed for ion content by inductive coupled plasma mass spectrometry (ICP-MS, Model Agilent 7500a, Hewlet-Packard, Yokogawa Analytical System, Tokyo, Japan). A control prepared in the same manner but without 50 μM manganese chloride. All the data provided here represent the mean and standard deviation of at least two independent experiments (mean ± SD of twice experiments).

Total RNA isolation

To see the effect of Mn^{2+} on the genome level, the total RNA was extracted from the three biological replicates of wild-type R1 and Mt-0865 under Mn^{2+} stress. Briefly, the wild-type R1 and Mt-0865 strains were cultured in a 5 ml TGY broth and re-inoculated in a 500 ml TGY broth. When the cells grow to OD_{600}~0.4–0.45, 20 mM $MnCl_2$ was added to the broth and further cultured at 30°C for half an hour. The pellets were washed three times with 1XPBS buffer (pH 7.5), and total RNA was extracted from cell cultures using TRIzol reagent (Invitrogen US, ice) as the kit protocol.

Bacterial RNA sequence library construction

Total RNA from three wild-type R1 and mutant strain (Mt-0865) were pooled, respectively, and rRNA (include 16S and 23S) was removed from 4 μg total RNA by MicrobexpressTM (Ambion AM1905), and the left RNA was chemically fragmented. The sequence library construction is according to ScriptSeq mRNA-Seq Library Preparation Kit (Illumina-compatible). Briefly, the fragmented RNA is reverse-transcribed into cDNA using the SuperScript double-stranded cDNA synthesis kit (Invitrogen) with the addition of SuperScript III reverse transcriptase (Invitrogen), and random primers containing a tagging sequence at their 3'ends. This was followed by RNase A (Roche, Germany) treatment, phenol-chloroform extraction, and ethanol precipitation. The resulting cDNAs were ligated to 5' DNA/DNA adaptor, and the di-tagged cDNAs was purified by PAGE gel, the insert fragment size is 150 bp~250 bp. The purification products were PCR amplified in 18 cycles using a high-fidelity DNA polymerase. PCR products were purified using the PAGE gel. Both direct cDNAs were sequenced simultaneously using a single flow cell of the Illumina Hiseq2000. All the sequence assays were performed in Zhejiang TianKe Company.

Transcriptome analysis

The images generated by the sequencers were converted into nucleotide sequences by a base-calling pipeline. The raw reads were saved in the fastq format. Three criteria were used to filter out the raw reads according to previously published [44], (i) Remove reads with sequence adaptors; (ii) remove reads with more than 20% 'N' bases; (ii) remove low-quality reads, which have more than 40% QA ≤20 bases. All subsequent analyses were based on clean reads. Only reads with high quality value were selected and used in the mapping using Tophat [45]. No more than 2-mismatches were allowed in the alignment for each read, and only the unique mapping reads used in the latter analysis. Cufflink and Cuff-diff were used to calculate Fragments Per Kilo base of transcript per Million mapped reads (FPKM), and find significantly expressed genes, respectively. The annotation of the D. radiodurans genome obtained from NCBI.

Reverse transcription-PCR (RT-PCR) analysis of expression of genes

QRT-PCR assay utilized RNA samples obtained from different condition and first-strain cDNA synthesis was carried out in 20 μl of reaction containing 1 μg of RNA sample combined with 3 μg of random hexamers using SuperScript III Reverse Transcriptase kit (Invitrogen). Each measurement was obtained for three replicate. Then Quant SYBER Premix FX TaqTM (TaKaRa Biotechnology (Dalian) Co. Ltd, China) was used to amplification following the manufactures instruction. As an internal control, dr0089 was used as a house-keeping, encoding the glycosyl transferase [18]. All primers used in QRT-PCR are shown in Table 5. All assays were performed using the STRAGENE Mx300PTM Real-time detection. Data analysis was carried out with iCycler software (Bio-rade Laboratories). The ratio of the copy number for the treatment to the control copy number was calculated. Differences in relative transcript abundance level were calculated using $2^{-\Delta\Delta T}$ [18].

Statistical analysis

All data are presented as mean ± standard error of the mean (SEM). Statistical analysis was performed on the raw data using paired student's t-test; p values <0.05 were considered significant.

Author Contributions

Conceived and designed the experiments: YH HC. Performed the experiments: AMUHS YZ YW GY QZ LW BT. Analyzed the data: AMUHS YZ HC. Contributed reagents/materials/analysis tools: YH HC. Wrote the paper: YH HC AMUHS.

References

1. Ghosal D, Omelchenko MV, Gaidamakova EK, Matrosova VY, Vasilenko A, et al. (2005) How radiation kills cells: survival of Deinococcus radiodurans and Shewanella oneidensis under oxidative stress. FEMS Microbiol Rev 29: 361–375.

2. Cabiscol E, Tamarit J, Ros J (2000) Oxidative stress in bacteria and protein damage by reactive oxygen species. Int Microbiol 3: 3–8.

3. Cheung HY, Vitkovic L, Brown MRW (1982) Toxic effect of manganese on growth and sporulation of Bacillus stearothermophilus. Journal of General Microbiology 128: 2395–2402.

4. Faulkner MJ, Helmann JD (2011) Peroxide stress elicits adaptive changes in bacterial metal ion homeostasis. Antioxid Redox Signal 15: 175–189.

5. Andrews SC, Robinson AK, Rodriguez-Quinones F (2003) Bacterial iron homeostasis. FEMS Microbiol Rev 27: 215–237.

6. Escolar L, Perez-Martin J, de Lorenzo V (1999) Opening the iron box: transcriptional metalloregulation by the Fur protein. Journal of Bacteriology 181: 6223–6229.

7. Gaballa A, Wang T, Ye RW, Helmann JD (2002) Functional analysis of the Bacillus subtilis Zur regulon. Journal of Bacteriology 184: 6508–6514.

8. Zhang T, Ding Y, Li T, Wan Y, Li W, et al. (2012) A Fur-like protein PerR regulates two oxidative stress response related operons dpr and metQIN in Streptococcus suis. Bmc Microbiology 12: 85.

9. Gaballa A, Wang T, Ye RW, Helmann JD (2002) Functional analysis of the Bacillus subtilis Zur regulon. J Bacteriol 184: 6508–6514.

10. Hamza I, Chauhan S, Hassett R, O'Brian MR (1998) The bacterial irr protein is required for coordination of heme biosynthesis with iron availability. Journal of Biological Chemistry 273: 21669–21674.

11. Qi ZH, O'Brian MR (2002) Interaction between the bacterial iron response regulator and ferrochelatase mediates genetic control of heme biosynthesis. Molecular Cell 9: 155–162.

12. Diaz-Mireles E, Wexler M, Sawers G, Bellini D, Todd JD, et al. (2004) The Fur-like protein Mur of Rhizobium leguminosarum is a Mn(2+)-responsive transcriptional regulator. Microbiology 150: 1447–1456.

13. Battista JR, Earl AM, Park MJ (1999) Why is Deinococcus radiodurans so resistant to ionizing radiation? Trends Microbiol 7: 362–365.

14. Repine JE, Pfenninger OW, Talmage DW, Berger EM, Pettijohn DE (1981) Dimethyl sulfoxide prevents DNA nicking mediated by ionizing radiation or iron/hydrogen peroxide-generated hydroxyl radical. Proc Natl Acad Sci U S A 78: 1001–1003.

15. Daly MJ, Gaidamakova EK, Matrosova VY, Vasilenko A, Zhai M, et al. (2004) Accumulation of Mn(II) in Deinococcus radiodurans facilitates gamma-radiation resistance. Science 306: 1025–1028.

16. Kamble VA, Misra HS (2010) The SbcCD complex of Deinococcus radiodurans contributes to radioresistance and DNA strand break repair in vivo and exhibits Mre11-Rad50 type activity in vitro. DNA Repair (Amst) 9: 488–494.

17. Sun H, Xu G, Zhan H, Chen H, Sun Z, et al. (2010) Identification and evaluation of the role of the manganese efflux protein in Deinococcus radiodurans. BMC Microbiol 10: 319.

18. Chen H, Xu G, Zhao Y, Tian B, Lu H, et al. (2008) A novel OxyR sensor and regulator of hydrogen peroxide stress with one cysteine residue in Deinococcus radiodurans. PLoS One 3: e1602.

19. Chen H, Wu R, Xu G, Fang X, Qiu X, et al. (2010) DR2539 is a novel DtxR-like regulator of Mn/Fe ion homeostasis and antioxidant enzyme in Deinococcus radiodurans. Biochem Biophys Res Commun 396: 413–418.

20. Dian C, Vitale S, Leonard GA, Bahlawane C, Fauquant C, et al. (2011) The structure of the Helicobacter pylori ferric uptake regulator Fur reveals three functional metal binding sites. Mol Microbiol 79: 1260–1275.

21. Rosch JW, Sublett J, Gao G, Wang YD, Tuomanen EI (2008) Calcium efflux is essential for bacterial survival in the eukaryotic host. Mol Microbiol 70: 435–444.

22. Daly MJ, Gaidamakova EK, Matrosova VY, Vasilenko A, Zhai M, et al. (2007) Protein oxidation implicated as the primary determinant of bacterial radioresistance. Plos Biology 5: e92.

23. Sun H, Li M, Xu G, Chen H, Jiao J, et al. (2012) Regulation of MntH by a dual Mn(II)- and Fe(II)-dependent transcriptional repressor (DR2539) in Deinococcus radiodurans. PLoS One 7: e35057.

24. Zhu XN, Long F, Chen YH, Knochel S, She QX, et al. (2008) A Putative ABC Transporter Is Involved in Negative Regulation of Biofilm Formation by Listeria monocytogenes. Appl Environ Microbiol 74: 7675–7683.

25. Battista JR (1997) Against all odds: the survival strategies of Deinococcus radiodurans. Annu Rev Microbiol 51: 203–224.

26. White O, Eisen JA, Heidelberg JF, Hickey EK, Peterson JD, et al. (1999) Genome sequence of the radioresistant bacterium Deinococcus radiodurans R1. Science 286: 1571–1577.

27. Slade D, Radman M (2011) Oxidative stress resistance in Deinococcus radiodurans. Microbiol Mol Biol Rev 75: 133–191.

28. Liu Y, Zhou J, Omelchenko MV, Beliaev AS, Venkateswaran A, et al. (2003) Transcriptome dynamics of Deinococcus radiodurans recovering from ionizing radiation. Proc Natl Acad Sci U S A 100: 4191–4196.

29. Aguirre JD, Clark HM, McIlvin M, Vazquez C, Palmere SL, et al. (2013) A manganese-rich environment supports superoxide dismutase activity in a Lyme disease pathogen, Borrelia burgdorferi. J Biol Chem 288: 8468–8478.

30. Gilbreath JJ, Pich OQ, Benoit SL, Besold AN, Cha JH, et al. (2013) Random and site-specific mutagenesis of the Helicobacter pylori ferric uptake regulator provides insight into Fur structure-function relationships. Mol Microbiol 89: 304–323.

31. Banh A, Chavez V, Doi J, Nguyen A, Hernandez S, et al. (2013) Manganese (Mn) oxidation increases intracellular Mn in Pseudomonas putida GB-1. PLoS One 8: e77835.

32. Gutteridge JM, Bannister JV (1986) Copper + zinc and manganese superoxide dismutases inhibit deoxyribose degradation by the superoxide-driven Fenton reaction at two different stages. Implications for the redox states of copper and manganese. Biochem J 234: 225–228.

33. Anjem A, Imlay JA (2012) Mononuclear iron enzymes are primary targets of hydrogen peroxide stress. J Biol Chem 287: 15544–15556.

34. Moody MD, Dailey HA (1985) Iron transport and its relation to heme biosynthesis in Rhodopseudomonas sphaeroides. J Bacteriol 161: 1074–1079.

35. Funayama T, Narumi I, Kikuchi M, Kitayama S, Watanabe H, et al. (1999) Identification and disruption analysis of the recN gene in the extremely radioresistant bacterium Deinococcus radiodurans. Mutat Res 435: 151–161.

36. Gao GJ, Lu HM, Huang LF, Hua YJ (2005) Construction of DNA damage response gene pprI function-deficient and function-complementary mutants in Deinococcus radiodurans. Chinese Science Bulletin 50: 311–316.

37. Jiao J, Wang L, Xia W, Li M, Sun H, et al. (2012) Function and biochemical characterization of RecJ in Deinococcus radiodurans. DNA Repair (Amst) 11: 349–356.

38. Srinivasan VB, Vaidyanathan V, Mondal A, Venkataramaiah M, Rajamohan G (2012) Functional characterization of a novel Mn2+ dependent protein serine/threonine kinase KpnK, produced by Klebsiella pneumoniae strain MGH78578. FEBS Lett 586: 3778–3786.

39. King KY, Horenstein JA, Caparon MG (2000) Aerotolerance and peroxide resistance in peroxidase and PerR mutants of Streptococcus pyogenes. J Bacteriol 182: 5290–5299.

40. Sun H, Xu G, Zhan H, Chen H, Sun Z, et al. (2010) Identification and evaluation of the role of the manganese efflux protein in Deinococcus radiodurans. Bmc Microbiology 10: 319.

41. Hua Y, Narumi I, Gao G, Tian B, Satoh K, et al. (2003) PprI: a general switch responsible for extreme radioresistance of Deinococcus radiodurans. Biochem Biophys Res Commun 306: 354–360.

42. Wang L, Xu G, Chen H, Zhao Y, Xu N, et al. (2008) DrRRA: a novel response regulator essential for the extreme radioresistance of Deinococcus radiodurans. Mol Microbiol 67: 1211–1222.

43. Jiao J, Wang L, Xia W, Li M, Sun H, et al. (2012) Function and biochemical characterization of RecJ in Deinococcus radiodurans. DNA Repair (Amst) 11: 349–356.

44. Ren S, Peng Z, Mao JH, Yu Y, Yin C, et al. (2012) RNA-seq analysis of prostate cancer in the Chinese population identifies recurrent gene fusions, cancer-associated long noncoding RNAs and aberrant alternative splicings. Cell Res 22: 806–821.

45. Trapnell C, Roberts A, Goff L, Pertea G, Kim D, et al. (2012) Differential gene and transcript expression analysis of RNA-seq experiments with TopHat and Cufflinks. Nat Protoc 7: 562–578.

14

Comparative Transcriptomic Characterization of the Early Development in Pacific White Shrimp *Litopenaeus vannamei*

Jiankai Wei[1,2], Xiaojun Zhang[1], Yang Yu[1,2], Hao Huang[3], Fuhua Li[1], Jianhai Xiang[1]*

1 Key Laboratory of Experimental Marine Biology, Institute of Oceanology, Chinese Academy of Sciences, Qingdao, China, **2** University of Chinese Academy of Sciences, Beijing, China, **3** Hainan Guandtop Ocean Breeding Co. Ltd, Haikou, China

Abstract

Penaeid shrimp has a distinctive metamorphosis stage during early development. Although morphological and biochemical studies about this ontogeny have been developed for decades, researches on gene expression level are still scarce. In this study, we have investigated the transcriptomes of five continuous developmental stages in Pacific white shrimp (*Litopenaeus vannamei*) with high throughput Illumina sequencing technology. The reads were assembled and clustered into 66,815 unigenes, of which 32,398 have putative homologues in nr database, 14,981 have been classified into diverse functional categories by Gene Ontology (GO) annotation and 26,257 have been associated with 255 pathways by KEGG pathway mapping. Meanwhile, the differentially expressed genes (DEGs) between adjacent developmental stages were identified and gene expression patterns were clustered. By GO term enrichment analysis, KEGG pathway enrichment analysis and functional gene profiling, the physiological changes during shrimp metamorphosis could be better understood, especially histogenesis, diet transition, muscle development and exoskeleton reconstruction. In conclusion, this is the first study that characterized the integrated transcriptomic profiles during early development of penaeid shrimp, and these findings will serve as significant references for shrimp developmental biology and aquaculture research.

Editor: Sebastian D. Fugmann, Chang Gung University, Taiwan

Funding: This work was supported by National High-Tech Research and Development Program of China (863 program) (2012AA10A404 and 2012AA092205) and National Natural Science Foundation of China (31172396 and 31072203). Co-author Hao Huang is employed by Hainan Guandtop Ocean Breeding Co. Ltd. Hainan Guandtop Ocean Breeding Co. Ltd. provided support in the form of salary for author Hao Huang and shrimp samples, but did not have any additional role in the study design, data collection and analysis, decision to publish, or preparation of the manuscript. The specific role of this author is articulated in the 'author contributions' section. The funders had no role in study design, data collection and analysis, decision to publish, or preparation of the manuscript.

Competing Interests: The authors have the following interests: Co-author Hao Huang is employed by Hainan Guandtop Ocean Breeding Co. Ltd. Hainan Guandtop Ocean Breeding Co. Ltd provided shrimp samples used in this study. There are no patents, products in development or marketed products to declare.

* Email: jhxiang@qdio.ac.cn

Introduction

Pacific white shrimp (*Litopenaeus vannamei*) is one of the most economically important marine aquaculture species and farmed widespread over the world [1]. As a member of Crustacea, it has a distinctive pattern for early development by passing through embryo, nauplius, zoea, mysis and postlarvae [2]. In embryo stage, it gets through the journey from zygote to 2 cell, 4 cell, blastula, gastrula, limb bud embryo and larva in membrane. After hatching from membrane, it also experienced six nauplius stages, three zoea stages, three mysis stages and postlarvae stages before it becomes a juvenile shrimp (Figure 1). This pattern linked by metamorphosis is an important evolutionary and developmental transition and is a remarkable example of modularity in life cycles [3]. Both its morphological and physiological features change dramatically in this period, also leading to a high uncontrollability in larval rearing [4]. So the researches about early development of *L. vannamei* are of considerable significance for both developmental biology and aquaculture in penaeid shrimp.

During shrimp metamorphosis, physiological experiments are difficult to conduct due to the small size and rapid development. Up to now, the researches about early development of shrimp are mainly from morphological observation and biochemical analysis. Some studies focused on the impact of environmental factors on metamorphosis [4–6] in order to improve larvae survival rates in aquaculture [7]. Studies about the salinity and temperature optima for penaeid larvae have been performed in many species [4,8,9]. Some studies characterized enzyme activities which can be used for evaluating their physiological status. The activities of phenoloxidase, superoxide dismutase and peroxidase were measured as immunological parameters during ontology of *L. vannamei* [10]. Digestive enzyme activities which were closely connected with feeding habits also have been extensively studied such as trypsin and chymotrypsin [11,12]. Nevertheless, functional genes related to early development are rarely reported. The molecular mechanisms of many important physiological changes in metamorphosis also have not been fully understood, and little is

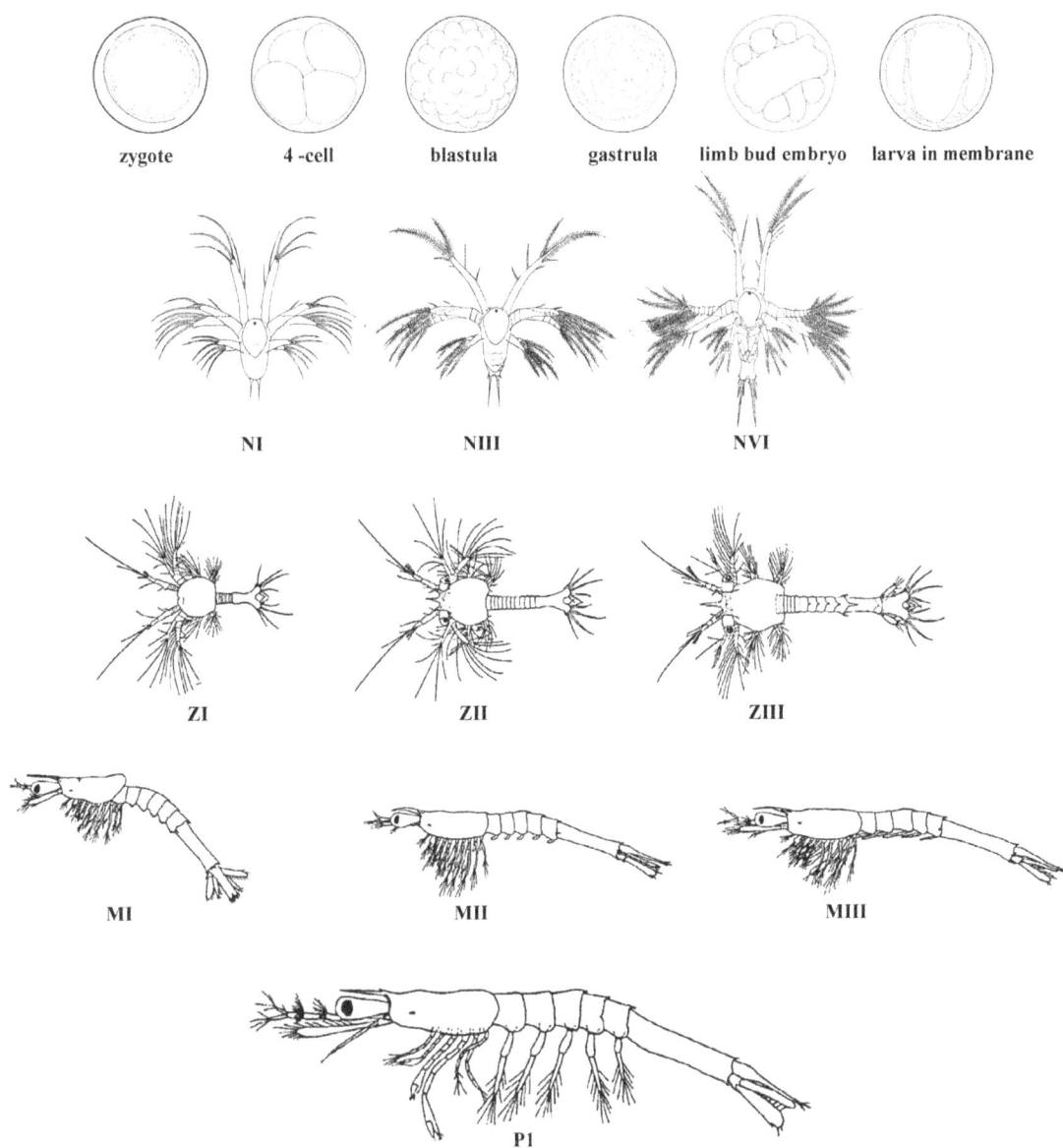

Figure 1. Embryonic and larval stages during early development of *L. vannamei.* (Z, M and P are modified from Hertzler PL, 2009). The developmental stages drawn in this figure include zygote, 4-cell, blastula, gastrula, limb bud embryo, larva in membrane, NI, NIII, NVI, ZI, ZII, ZIII, MI, MII, MIII and P1.

known about the dynamic change on gene expression level during early development.

Recently, the next generation sequencing (NGS) technology has flourished tremendously and is becoming an important method for measuring gene expression levels [13]. The advent of RNA-Seq provides a far more high-throughput and precise measurement of levels of transcripts and their isoforms than other methods [14]. It gives a general view of gene expression especially in these species lack of a fully sequenced and assembled genome such as *L. vannamei*. RNA-Seq has been reported in adult shrimp to identify immune related genes under WSSV or TSV infection [15–18]. The transcriptome of *L. vannamei* postlarvae (20 days post spawning) were also sequenced and annotated [19]. However, the transcriptional profiles across the early development for comparative analysis are still absent. The transition from comparing a few genes to whole transcriptomes is a vital approach for enhancing our understanding about this ontology [3].

In this study, we analyzed the transcriptomic characterization of *L. vannamei* during five different early development stages through Illumina high-throughput sequencing data. Results obtained from this study will contribute to further studies about molecular mechanisms for early development of *L. vannamei* and can be used for evolutionary analysis, developmental biology and functional gene research in penaeid shrimp.

Materials and Methods

Embryos and larvae sampling

The *L. vannamei* samples of different development stages were collected from Guangtai shrimp farm in Wenchang, Hainan, China. No specific permissions were required for the sampling locations and activities, and the studies did not involve endangered or protected species and locations. A total of 15 samples were collected based on their development stages: zygote, blastula,

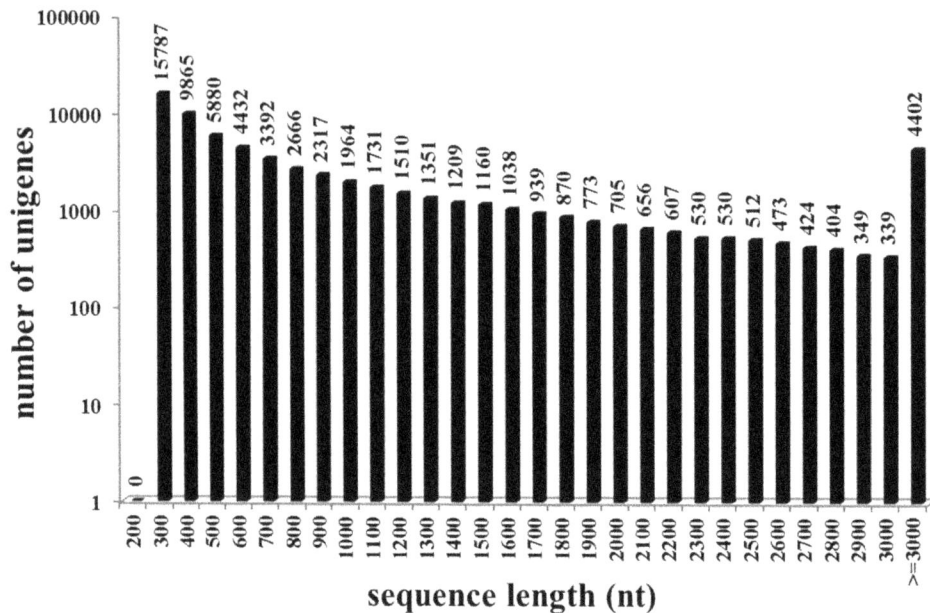

Figure 2. Length distribution of all-unigenes after clustering unigenes in each group. X axis represents sequence length intervals. Y axis represents the number of unigenes in each interval.

gastrula, limb bud embryo, larva in membrane, nauplius I (NI), nauplius III (NIII), nauplius VI (NVI), zoea I (ZI), zoea II (ZII), zoea III (ZIII), mysis I (MI), mysis II (MII), mysis III (MIII) and postlarvae 1 (P1). Each stage was identified according to observation with microscope. They were reared in a 25 m³ indoor pond with seawater at 31°C, salinity of 2.5%. They were unfed during embryo and nauplius stages. At zoea stage they were fed with spirulina and multiform formulated diet, while at mysis and postlarvae stages they were fed with artemia nauplii and multiform formulated diet. Embryos and larvae were collected with screen mesh at each stage when 90% of the population had reached the objective stage. Samples were immediately preserved in liquid nitrogen and then stored in −80°C for assays.

RNA isolation and sample pooling

The total RNA of 15 samples was extracted separately by Unizol reagent (Biostar, China) following the manufacturer's instructions, RNA were assessed by electrophoresis in 1% agarose gel and quantified by NanoDrop 1000 spectrophotometer (Thermo Scientific, USA) and Agilent 2100 Bioanalyzer (Agilent Technologies, USA). Afterwards, the RNA samples of zygote, blastula, gastrula, limb bud embryo and larva in membrane were

mixed equivalently into embryo sample (E), the RNA samples of NI, NIII and NVI were mixed equivalently into nauplius sample (N), the RNA samples of ZI, ZII and ZIII were mixed equivalently into zoea sample (Z), the RNA samples of MI, MII and MIII were mixed equivalently into mysis sample (M) and the RNA samples of postlarvae 1 were considered as postlarvae sample (P). The sample mixture was based on both morphological classification and physiological characters (Figure 1). Samples of zygote, blastula, gastrula, limb bud embryo and larva in membrane were typical stages before hatching. They mixed into E sample and represented the features of embryo in membrane. Similarly, NI, NIII and NVI composed N sample which represented nauplius stage, ZI, ZII and ZIII composed Z sample which represented zoea stage and MI, MII and MIII composed M sample which represented mysis stage. Then the five mixed RNA samples were used for library construction and sequencing.

Library construction and Illumina sequencing

RNA purification, reverse transcription, library construction and sequencing were conducted by BGI (Shenzhen, China). To sum up, beads with Oligo(dT) were used to isolate and collect poly(A) mRNA from the mixed RNA. Fragmentation buffer was

Table 1. Summary of sequencing and assembly of the transcriptome from *L. vannamei*.

Samples	Raw Reads	Clean Reads	Unigene number	Unigene Average Length	Unigene N50
E	55,895,400	51,568,556	53,822	747	1404
N	58,816,588	52,824,674	58,048	735	1314
Z	59,721,370	53,430,302	64,443	703	1219
M	59,760,506	53,902,786	66,215	705	1226
P	57,889,382	51,574,056	64,528	699	1204
All	292,083,246	263,300,374	66,815	1027	1851

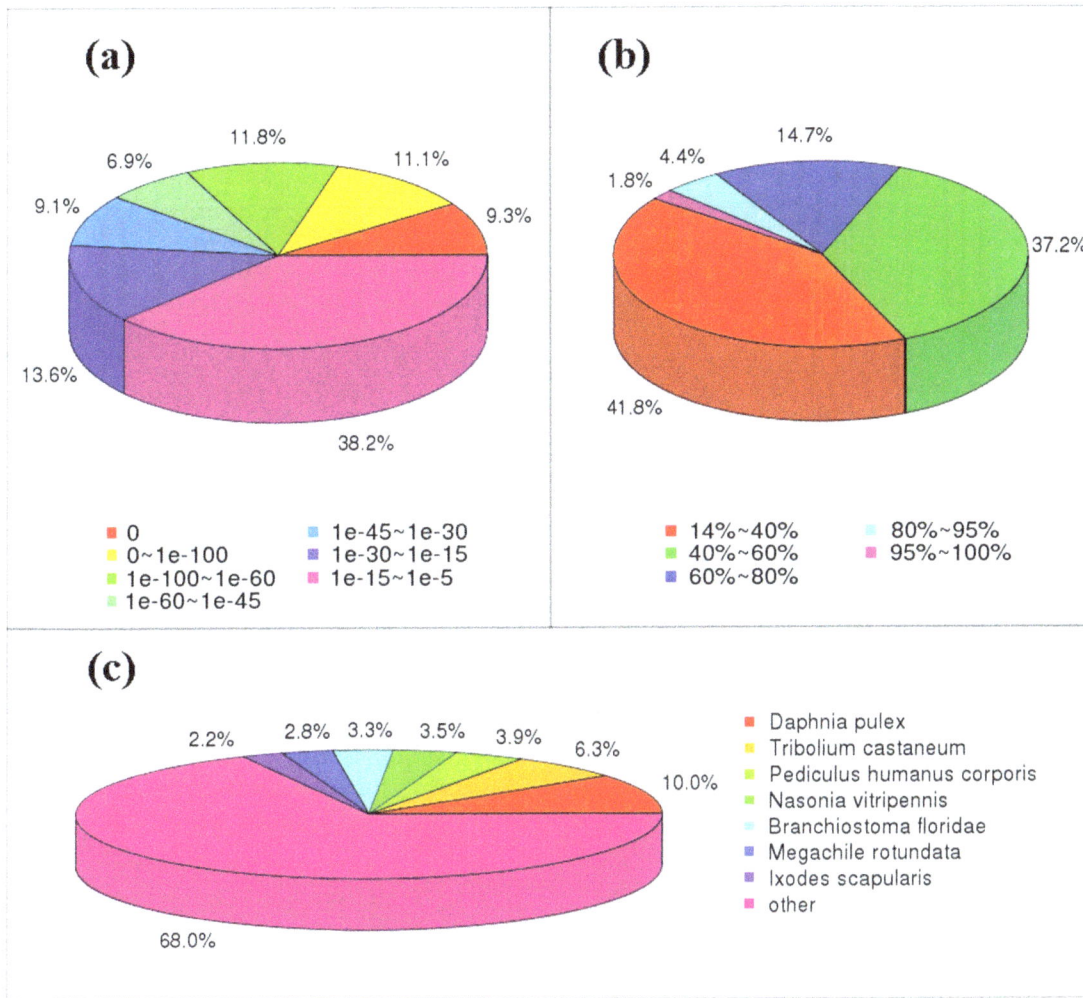

Figure 3. Summary of homology search of all-unigenes against nr database. (a) E-value distribution of the best blast hits; (b) Similarity distribution of the best blast hits; (c) Species distribution of the best blast hits.

added for interrupting mRNA to short fragments. Using these short fragments as templates, random hexamer-primer was used to synthesize the first-strand cDNA. The second-strand cDNA was synthesized using buffer, dNTPs, RNase H and DNA polymerase I. Short fragments were purified with QiaQuick PCR extraction kit (Qiagen, Germany) and resolved with EB buffer for end reparation and tailing A. After that, the short fragments were connected with sequencing adapters. And, after the agarose gel electrophoresis, the suitable 200 bp fragment were selected for the PCR amplification as templates. At last, the libraries were sequenced using HiSeq 2000 (Illumina, USA).

Sequencing data assembly and annotation

Image data from sequencing machine was transformed into raw reads by base calling, and stored in fastq format. The raw reads of all five samples were preprocessed by removing adaptors, reads with unknown nucleotides larger than 5% and low quality reads. The clean reads of each stage were then assembled into unigenes using the Trinity program [20]. Unigenes of five samples were then clustered into all-unigenes using TGICL [21]. In order to annotate all-unigenes, blast alignments [22] (E value < 1e-5) against the nr, nt, Swiss-Prot, KEGG, and COG databases were

performed. Gene ontology (GO) analysis was carried out using BLAST2GO program [23].

Analysis of differentially expressed unigenes

By means of reads mapping to all-unigenes, the FPKM (Fragments Per Kilo bases per Million fragments) value [24] of all-unigenes in each sample were obtained and used for comparing the expression difference between samples. Hierarchical clustering analysis (HCA) and principal components analysis (PCA) were performed using R [25]. We use FPKM value for comparing the expression difference between adjacent samples (E-N, N-Z, Z-M and M-P). We chose those with FDR (false discovery rate) ≤ 0.001 and absolute value of log2ratio ≥ 1 as differentially expressed genes (DEGs). Hypergeometric test was used to find significantly enriched GO terms and KEGG pathways in DEGs comparing to the whole background. After Bonferroni correction for p value, we defined corrected p value ≤ 0.05 as significantly enriched GO terms and KEGG pathways. The unigenes analyzed in this article for heat map were grouped together according to their FPKM values by Cluster 3.0 [26] and visualized by TreeView 1.6 [27].

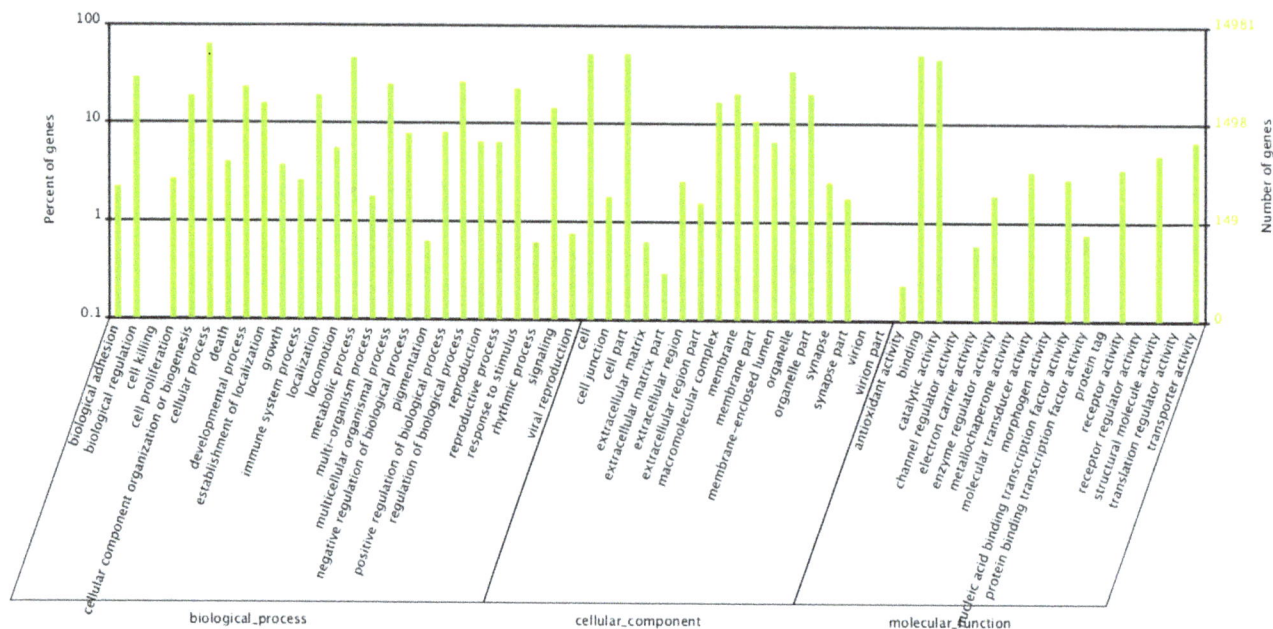

Figure 4. GO annotation of all-unigenes. Unigenes with GO annotation were divided into three major categories: biological process, cellular component and molecular function.

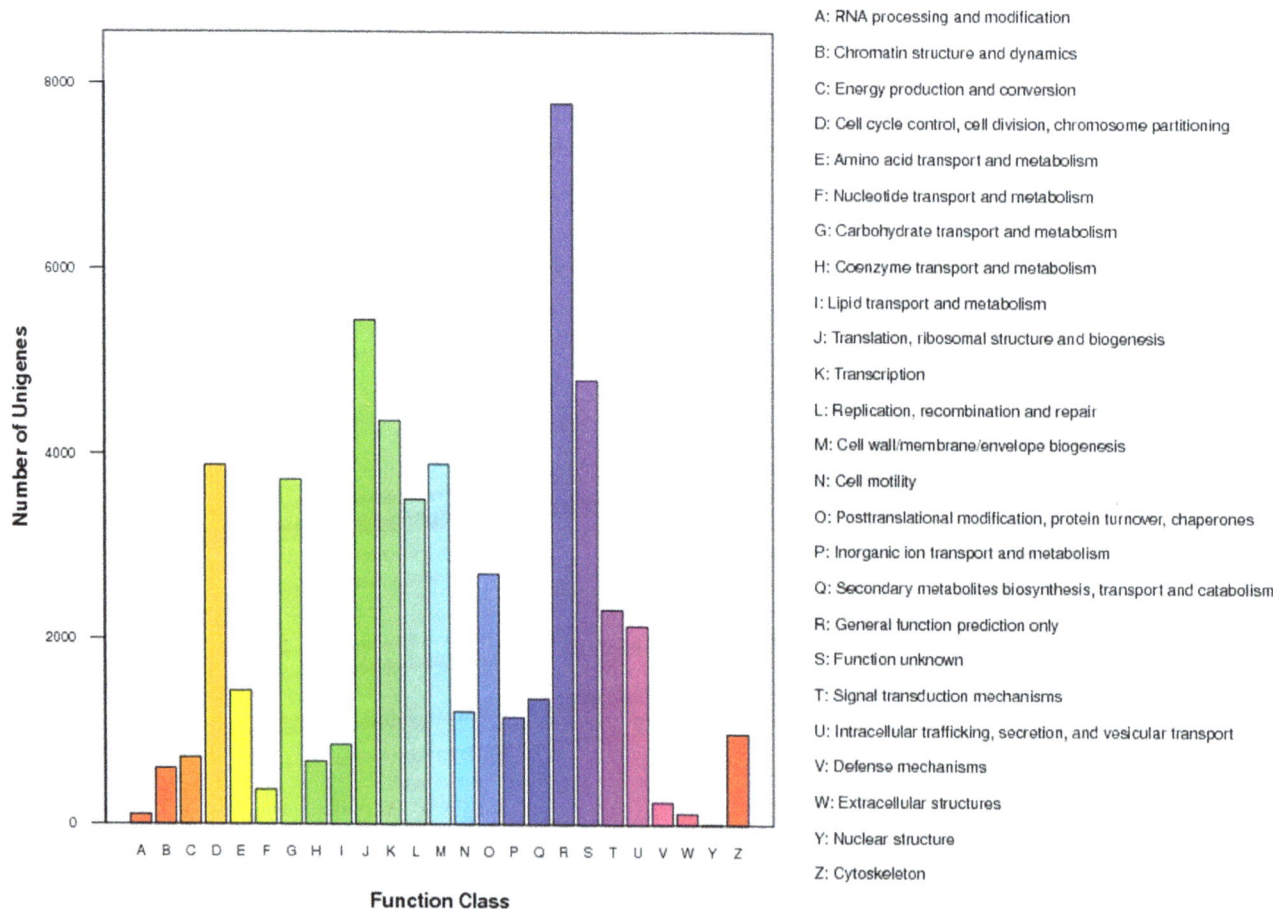

A: RNA processing and modification

B: Chromatin structure and dynamics

C: Energy production and conversion

D: Cell cycle control, cell division, chromosome partitioning

E: Amino acid transport and metabolism

F: Nucleotide transport and metabolism

G: Carbohydrate transport and metabolism

H: Coenzyme transport and metabolism

I: Lipid transport and metabolism

J: Translation, ribosomal structure and biogenesis

K: Transcription

L: Replication, recombination and repair

M: Cell wall/membrane/envelope biogenesis

N: Cell motility

O: Posttranslational modification, protein turnover, chaperones

P: Inorganic ion transport and metabolism

Q: Secondary metabolites biosynthesis, transport and catabolism

R: General function prediction only

S: Function unknown

T: Signal transduction mechanisms

U: Intracellular trafficking, secretion, and vesicular transport

V: Defense mechanisms

W: Extracellular structures

Y: Nuclear structure

Z: Cytoskeleton

Figure 5. COG classification of all-unigenes. Unigenes were classified into 25 function classes. The columns represents the number of unigenes in each class.

Table 2. Development related pathways and annotated key nodes.

KEGG pathway	Unigene number	Partial unigene annotation
MAPK signaling pathway	532	*EGF, EGFR, Grb2, Sos, Ras, NF1, p38, ERK5,*
Dorso-ventral axis formation	399	*Grk, Argos, Top, Drk, Egh*
Wnt signaling pathway	341	*wnt-1, wif-1, wnt-5, Frizzled, beta-catenin, GSK3, APC*
Hedgehog signaling pathway	262	*Hgdgehog, Gas1, ptc, smo, Fu, Ci, PKA, Slimb*
TGF-beta signaling pathway	172	*BMP, BMPR, Smad, ERK, Activin, ActivinR, SARA*
Notch signaling pathway	170	*Notch, TACE, Delta, Serrate, CSL*
VEGF signaling pathway	167	*VEGFR, Paxillin, casp9, Rac, CALN, PKC, SPK, MEK*
Jak-STAT signaling pathway	140	*JAK, STAT, CytokineR, CBP, SHP1,SHP2, PI3K, AKT*

Validation by quantitative real-time PCR

Quantitative real-time PCR (qPCR) analysis was used for validation. 18S rRNA gene was used as an internal standard and relative gene expression levels were calculated using the comparative Ct method with the formula $2^{-\Delta\Delta Ct}$ [28]. The qPCR results were then compared with transcriptome data (FPKM value) to detect their expression correlation of each gene.

Results and Discussion

Illumina sequencing and *de novo* assembly

Five cDNA libraries were constructed on the basis of five RNA samples as described in the Materials and methods section. By mix 15 samples into five groups for sequencing, it would be more comprehensive for depicting the transcriptome profiles during development and more targeted for comparison. Using Illumina HiSeq 2000, a total of 55,895,400, 58,816,588, 59,721,370, 59,760,506, 57,889,382 raw reads were obtained respectively. After removing adaptors and trimming low quality reads,

51,568,556 (92%), 52,824,674 (90%), 53,430,302 (89%), 53,902,786 (90%), 51,574056 (89%) clean reads were obtained, and these clean reads were assembled respectively and then clustered into 66,815 unigenes (Table 1). These data were deposited to Sequence Read Archive database of National Center for Biotechnology Information with accession numbers of SRR1460493, SRR1460494, SRR1460495, SRR1460504 and SRR1460505. The all-unigenes, totaling to 68 Mbp, with an average length of 1027 bp and N50 length of 1851 bp, were then used as references for annotation and expression analysis. The size distribution of all-unigenes was shown in Figure 2.

RNA-Seq is a sequencing based method that allows the entire transcripts to be surveyed in a very high-throughput and quantitative manner. It has clear advantages over other approaches and is expected to revolutionize the manner in which eukaryotic transcriptomes are analyzed [14]. By sequencing five different samples individually, we built the first gene expression profiles of *L. vannamei* during early development. The average length of unigenes in each group was around 700 bp. By clustering the

(a)

(b)

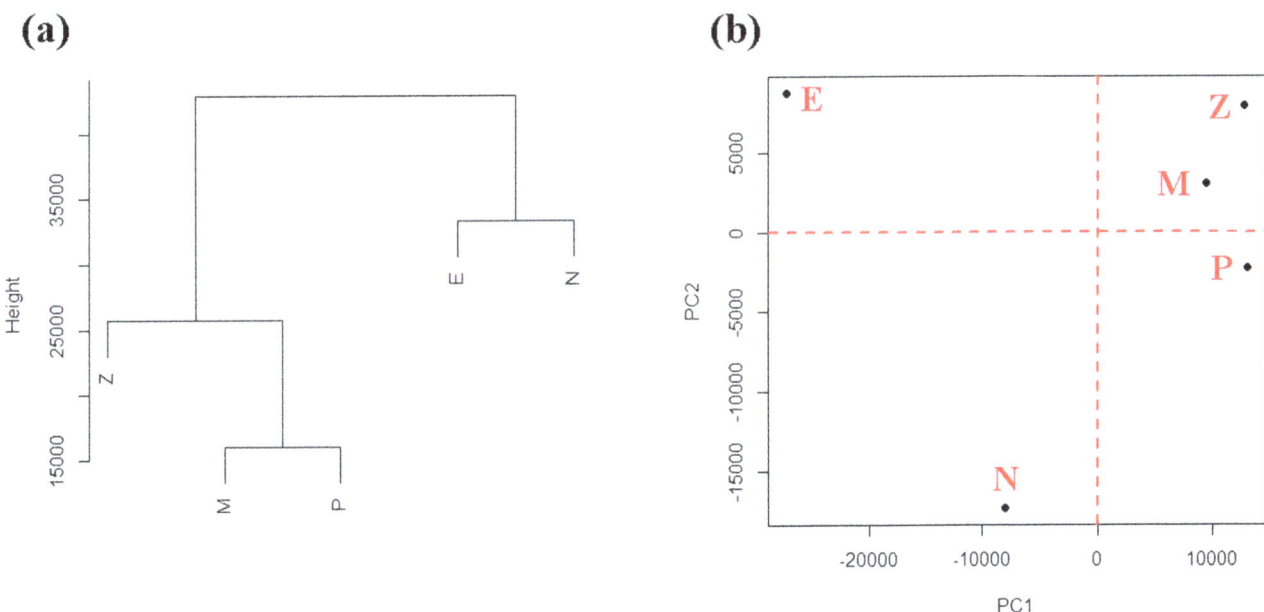

Figure 6. HCA and PCA plots for five samples using FPKM value of unigenes. (a) Hierarchical clustering analysis plot. The height represents the Euclidean distance; (b) Principal components analysis plot. X axis represents PC1 that explains 78.7% and Y axis represents PC2 that explains 11.2% of the total variability for gene expression.

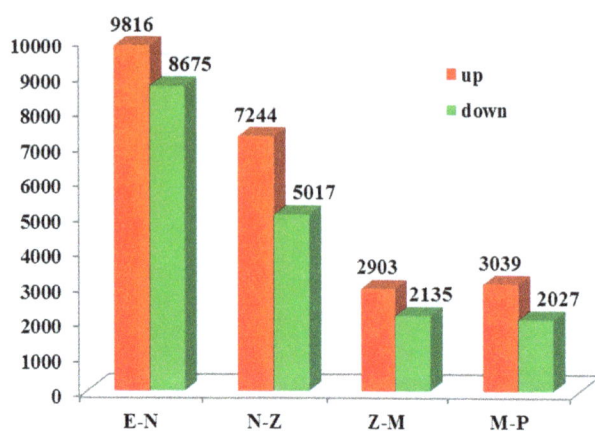

Figure 7. Statistics of differentially expressed unigenes between adjacent samples. Red columns represent the number of up-regulated genes while green columns represent the number of down-regulated genes.

unigenes together into all-unigenes, we get a significantly improved assembly result with an average length 1027 bp and total length 68 Mbp. By blast search, a large number of genes which have not been reported in penaeid shrimp before were annotated and many of them play key roles in early development. These results would contribute to both the penaeid transcriptome database and comparative analysis of gene expression profiles.

Functional annotation and classification

In order to annotate the all-unigenes, blast alignment against the nr, nt, and Swiss-Prot databases was performed. By blast searching with a cutoff E-value $< 1e-5$, 32,398 (48.5%) unigenes found putative homologues in the nr protein database from NCBI, 19,363 (29.0%) unigenes found putative homologues in the nt database and 29,022 (43.4%) unigenes found putative homologues in the Swiss-Prot database. The best aligning results are used for analysis. The E-value and similarity distribution of the best blast hits for unigenes were shown in Figure 3. The distribution of best hits over species for nr annotation was also analyzed. *Daphnia pulex* (10.0%), *Tribolium castaneum* (6.3%) and *Pediculus humanus corporis* (3.9%) possess the top three maximum unigene numbers with nr annotation.

Gene Ontology (GO) is an international standardized gene functional classification system which offers a dynamic-updated vocabulary to comprehensively describe properties of genes and their products in any organism [29]. With nr annotation, we used Blast2GO program to get GO annotation of all-unigenes. 14,981 (22.4%) were classified into diverse functional categories by GO annotation. Among them, 11,065 were mapped to biological processes including 3449 involved in development process and 551 involved in growth, 8678 were mapped to cellular components and 12,057 were mapped to molecular functions (Figure 4). Clusters of Orthologous Groups (COG) database [30] is also an important classification system for functional annotation. As for COG classification, 15,467 (23.1%) were classified into 25 functional categories (Figure 5). The largest group was "general function predicted only", followed by "translation, ribosomal structure and biogenesis" and "function unknown". "cell cycle control, cell division, chromosome partitioning", containing plenty

Table 3. The top five most significantly enriched GO terms.

Compare groups	GO – Cellular Component	GO – Molecular Funciton	GO – Biological Proecss
E-N	1. myosin filament	1. microfilament motor activity	1. dorsal closure, spreading of leading edge cells
	2. myosin complex	2. myosin light chain binding	2. ecdysone-mediated induction of salivary gland cell autophagic cell death
	3. apical cortex	3. myosin binding	3. induction of programmed cell death by ecdysone
	4. cell division site	4. structural constituent of muscle	4. induction of programmed cell death by hormones
	5. cleavage furrow	5. catalytic activity	5. anterior midgut development
N-Z	1. myosin filament	1. catalytic activity	1. striated muscle myosin thick filament assembly
	2. myosin complex	2. microfilament motor activity	2. skeletal myofibril assembly
	3. apical cortex	3. myosin light chain binding	3. skeletal muscle myosin thick filament assembly
	4. striated muscle myosin thick filament	4. structural constituent of muscle	4. myosin filament organization
	5. A band	5. myosin binding	5. myosin filament assembly
Z-M	1. ribosome	1. structural molecule activity	1. translation
	2. myosin filament	2. structural constituent of ribosome	2. skeletal myofibril assembly
	3. myosin complex	3. microfilament motor activity	3. anterior midgut development
	4. actin cytoskeleton	4. myosin binding	4. myosin II filament assembly
	5. non-membrane-bounded organelle	5. myosin light chain binding	5. myosin II filament organization
M-P	1. extracellular region	1. catalytic activity	1. metabolic process
	2. myosin filament	2. peptidase activity	2. chitin metabolic process
	3. myosin complex	3. oxidoreductase activity	3. aminoglycan metabolic process
	4. ribosome	4. chitin binding	4. polysaccharide metabolic process
	5. extracellular space	5. hydrolase activity	5. carbohydrate metabolic process

Table 4. The top ten enriched KEGG pathways.

	E-N	N-Z	Z-M	M-P
1	Vibrio cholerae infection	Vibrio cholerae infection	Vibrio cholerae infection	Amoebiasis
2	Amoebiasis	Amoebiasis	Ribosome	Vibrio cholerae infection
3	Viral myocarditis	Pancreatic secretion	Amoebiasis	Staphylococcus aureus infection
4	Cardiac muscle contraction	Protein digestion and absorption	Viral myocarditis	Glutathione metabolism
5	Complement and coagulation cascades	Influenza A	Hypertrophic cardiomyopathy (HCM)	Amino sugar and nucleotide sugar metabolism
6	mRNA surveillance pathway	Amino sugar and nucleotide sugar metabolism	Dilated cardiomyopathy	Renin-angiotensin system
7	Hypertrophic cardiomyopathy (HCM)	Tyrosine metabolism	Cardiac muscle contraction	Hematopoietic cell lineage
8	Protein digestion and absorption	Metabolic pathways	Influenza A	Linoleic acid metabolism
9	Renin-angiotensin system	Staphylococcus aureus infection	Amino sugar and nucleotide sugar metabolism	Metabolic pathways
10	Glycosaminoglycan degradation	Glutathione metabolism	Staphylococcus aureus infection	Complement and coagulation cascades

of developmental related genes, also represented a large group with 3877 unigenes.

KEGG is a database to analyze gene product during metabolism process and related gene functions in the cellular processes [31]. To identify the biological pathways involved in early development of *L. vannamei*, the KEGG pathway annotation were obtained by blast all-unigenes with KEGG database, and 26,257 (39.3%) were associated with 255 pathways including a lot of development related pathways, such as Wnt [32] (341 unigenes), Hedgehog [33] (262 unigenes), Notch [34] (170 unigenes) and so on (Table 2). The main nodes in these pathways were identified and some of them were listed in Table 2.

Clustering analysis and identification of differentially expressed genes (DEGs)

To investigate the global transcriptional differences between stages and genes during development, hierarchical clustering analysis (HCA) and principal components analysis (PCA) were performed using whole expression datasets in each sample (Figure 6). For HCA, M and P clustered together first and then clustered with Z, while E and N clustered together eventually. For PCA plot, the first two principal components (PC1 and PC2) explained 89.9% percent of the total variability in gene expression (78.7% percent and 11.3% percent respectively). PC1 divided them into two groups: one for E and N, another for Z, M and P, which is in accordance with HCA result. The bi-dimensional plot also revealed that Z, M and P shared a relatively similar expression profile, while E and N had a relatively large difference compared to Z, M and P.

To identify DEGs involved in early development, we use FPKM value for comparing the expression differences between adjacent samples (E-N, N-Z, Z-M and M-P). A large number of DEGs were screened with absolute value of log2ratio ≥ 1 and FDR ≤ 0.001 (Figure 7). Among 66,815 unigenes, 18,536 were identified as DEGs between E and N (9861 up-regulated, 8675 down-regulated) and 12,261 were identified as DEGs between N and Z (7244 up-regulated and 5017 down-regulated). The number of DEGs distinctly decreased when comparing Z with M (5038 DEGs with 2903 up-regulated, 2135 down-regulated) and M with P (5066 DEGs with 3039 up-regulated, 2027 down-regulated). The number of up-regulated genes was significantly more than that of down-regulated genes with a p-value 0.027 by paired t test.

The five samples could be clustered into three major groups (E for group1, N for group 2, Z, M and P for group3) according to HCA and PCA. In correspondence with this, a relatively high proportion of DEGs also occurred in E-N and N-Z, while a low proportion occurred in Z-M and M-P. These all indicated that more dramatic changes occurred in earlier transition. For E-N transition, this may related to the existence of maternal transcripts in E sample. Maternal gene products drive early development when the newly formed embryo is transcriptionally inactive [35,36]. Embryonic transcription is initiated and many maternal RNAs are degraded until the maternal-zygotic transition [35]. For N-Z transition, the dramatic changes may relate to the higher levels of cell differentiation with organ formation in Z stage such as the formation of midgut for digestion [37] and compound eyes for locomotion [2]. In addition, the number of up-regulated genes was more than that of down-regulated genes (p = 0.027), suggesting more genes get activated along with development in order to drive further developmental events [38].

GO term enrichment analysis and KEGG pathway enrichment analysis

GO term enrichment analysis detected significantly overrepresented GO terms in DEGs with FDR corrected p value <0.05. The top five most significantly enriched GO terms were shown in Table 3 including three ontologies: cellular component, molecular function and biological process.

The biological process ontology includes terms that represent collections of processes as well as terms that represent a specific and entire process. The enrichment of DEGs on this ontology provided a considerable perspective for understanding the biological change during early development. For E-N group, the most significant GO biological process term was "dorsal closure, spreading of leading edge cells", followed by three GO terms involved with cell death ("ecdysone-mediated induction of salivary gland cell autophagic cell death", "induction of programmed cell death by ecdysone" and "induction of programmed cell death by hormones"). For N-Z group, the top five most significant GO biological process terms all associated with muscle and skeletal development including "striated muscle myosin thick filament assembly", "skeletal myofibril assembly", "skeletal muscle myosin thick filament assembly", "myosin filament organization" and "myosin filament assembly". For Z-M group, the most significant

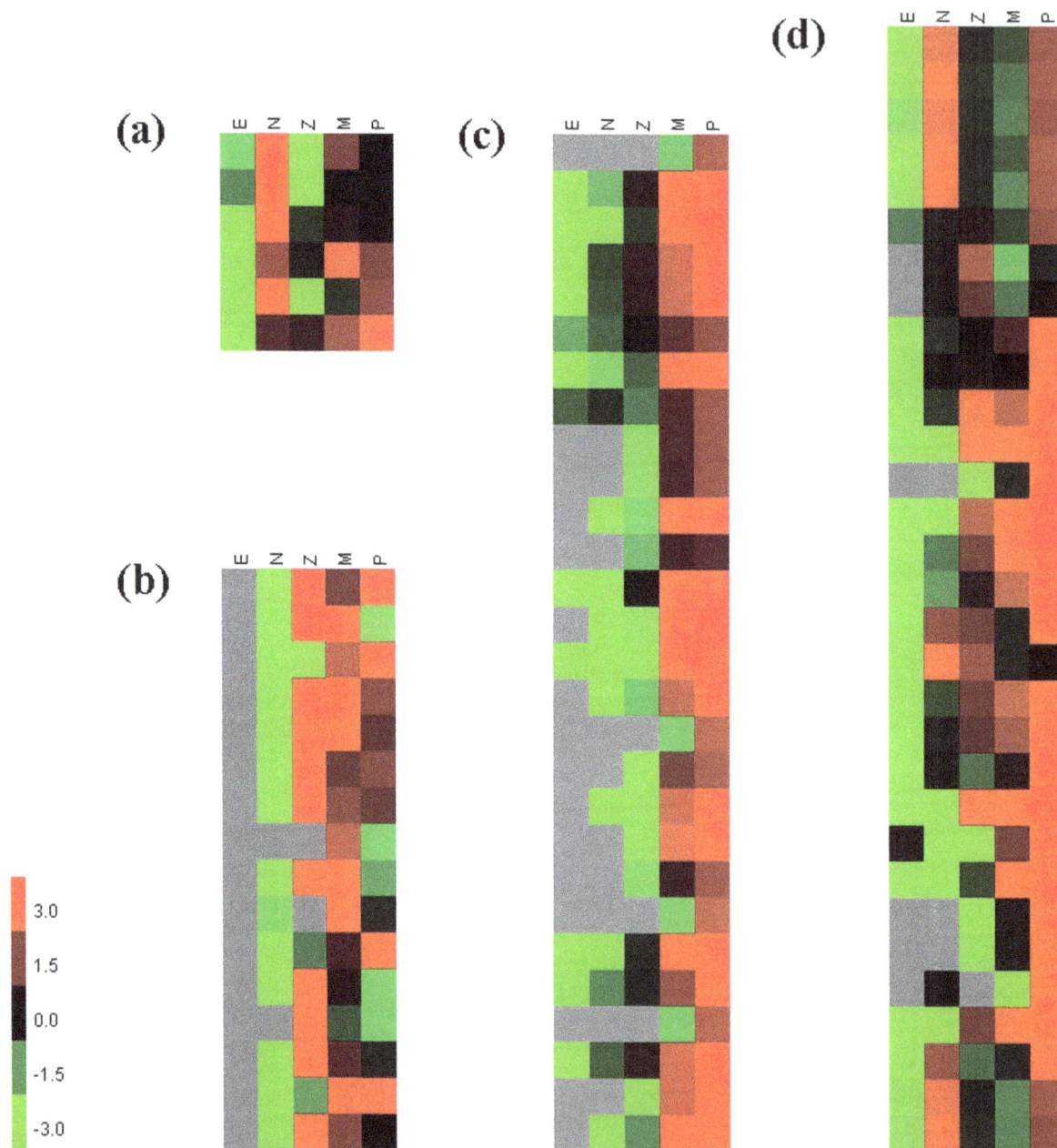

Figure 8. Expression profiles of genes involved in specific functions. The red-green heat maps were drawn according to FPKM value. The Gray units represent zero for FPKM value. (a) Heat map for nuclear hormone receptor E75; (b) Heat map for amylase; (c) Heat map for myosin heavy chain type a (partial); (d) Heat map for calcified cuticle protein.

GO molecular function term was "translation", followed by "skeletal myofibril assembly", "anterior midgut development", "myosin II filament assembly" and "myosin II filament organization". For M-P group, the top 5 most significant GO biological process terms all involved with metabolism process including "metabolic process", "chitin metabolic process", "aminoglycan metabolic process", "polysaccharide metabolic process" and "carbohydrate metabolic process".

To evaluate the pathways associated with DEGs, we conducted the KEGG pathway enrichment analysis. The top ten enriched pathways were listed in Table 4. Considering these top ten pathways, some appeared repeatedly in the four comparison groups like "Vibrio cholerae infection", "Amoebiasis" and "Staphylococcus aureus infection". At the same time, some only enriched in a specific comparison group. For example, "mRNA surveillance pathway" and "Glycosaminoglycan degradation" only appeared in E-N group, while "Hematopoietic cell lineage" and "Linoleic acid metabolism" only appeared in M-P group. For N-Z group, many metabolism related pathways were enriched including "Pancreatic secretion", "Protein digestion and absorption", "Amino sugar and nucleotide sugar metabolism", "Tyrosine metabolism", "Metabolic pathways" and "Glutathione metabolism". For Z-M group, many enriched pathways are associated with cardiac muscle like "Viral myocarditis", "Hypertrophic cardiomyopathy", "Dilated cardiomyopathy" and "Cardiac muscle contraction".

Table 5. Partial functional categories and genes involved in early development of *L. vannamei.*

Functional categories	Functional genes	Number of hits	Up-regulated hits	Up-regulated group
Ecdysone signal transduction	EcR	3	1	E-N
	RXR	2	0	
	E75	6	6	
	FTZ-F1	2	2	
digestive enzyme secretion	trypsin	29	21	N-Z
	chymotrypsin	17	9	
	amylase	16	11	
myosin II filament assembly	Distal-les	4	0	Z-M
	Ladybird	1	0	
	myosin heavy chain type a	100	28	
	myosin heavy chain type b	77	18	
exoskeleton reconstruction	cuticle protein	188	125	M-P
	calcified cuticle protein	31	27	
	early cuticle protein	24	20	
	calcification associated soluble matrix protein	11	11	
	calcification-associated peptide-1	11	10	
	crustocalcin	2	2	

Physiological changes during shrimp metamorphosis

The early development of *L. vannamei* gets through both embryonic stage and larval stage. This unique early developmental mode by metamorphosis implies dramatic changes and provides a unique opportunity to examine shrimp body reorganization [39]. Moreover, a better understanding of *L. vannamei* development would be greatly beneficial for breeding control and ensure the long-term viability of shrimp aquaculture [40]. The 15 samples we chose in this study covered the main stages across both embryonic and larval development. By sample pooling and comparative analysis, dynamic changes of gene expression can be revealed more distinctly.

GO term enrichment analysis revealed that DEGs between E and N stages were significantly enriched in hormone induced programmed cell death process which related to histolysis and reconstruction (three of five in top five most significantly enriched GO terms). Metamorphosis took place when shrimps stepped from embryo into larval development, and a series of hormones like ecdysone triggered the transition during metamorphosis and initiated programmed cell death [41]. Ecdysone signaling has been studied extensively in the larval salivary glands of Drosophila. The pulses of ecdysone regulate developmental pathways through a transcriptional hierarchy [42]: Ecdysone binds to its heterodimeric receptor which consists of ecdysone receptor (EcR) and ultraspiracle (USP). The EcR complex functions together with fushi tarazu-factor 1 (FTZ-F1) to induce transcription of the early genes including Broad Complex (BR-C), E74A, E75, and E93. The early genes then activate transcription of many late genes involved in signaling, cellular organization, apoptosis, and autophagy [43]. We identified EcR, RXR (retinoid X receptor, homolog of USP), FTZ-F1 and E75 in our dataset (Table 5) and many of them were up-regulated in E-N transition such as E75 (Figure 8), implicating the hormone signal hierarchy also existed in shrimp metamorphosis and the hormones might also act as a primary signal for programmed cell death in shrimp larval stage.

Hence we infer that the ecdysone signals might have distinctly different roles between embryo and larvae.

KEGG pathway enrichment analysis revealed that DEGs between N and Z stages were significantly enriched in digestive enzymes secretion and metabolic pathways (six of ten in top ten most significantly enriched KEGG pathways), indicating that source of nutrition was transforming greatly in this period. There was no feeding process in embryo and nauplius stage, and all of the nutrients come from the reservation of yolk. When shrimp became zoea, they started to eat unicellular algae or plant debris [44,45]. Corresponding to the initiation of feeding, the expression level of many digestive enzymes increased sharply during N-Z transition such as trypsin, chymotrypsin and amylase (Table 5). Take amylase as an example, 16 unigenes were identified as alpha amylase and their expression levels all increased sharply from nauplius stage to zoea stage, and then kept relatively stable until mysis and postlarvae stage (Figure 8). The explosion of digestive enzymes in N-Z transition is consistent to larval feeding habits and also could be a symbol of a developed digestive system [44].

DEGs between Z and M stages were significantly enriched in myosin II filament assembly and organization which related to muscle development, indicating the enhancement of motor ability in this period. The prominent morphological change in Z-M transition is the appendage formation. As larvae progress to later stages, more posterior appendages are used for locomotion [46]. Nauplius and zoea use cephalic propulsion. Mysis swim using the pereopods, while postlarvae use the pleopods [47]. Therefore, we inferred that the muscle also entered a rapid growth period along with new appendage formation. Myosin II, which composed of two myosin heavy chain (MYH) subunits and four myosin light chain subunits [48], is a major component of thick filaments in muscle. For MYH, type a and type b are primarily expressed in body-wall muscle [49]. As for appendicular myogenesis, Distal-less gene marks the initiation of appendage development [50] and ladybird genes were also essential to generate a specialized type of appendage adapted for locomotion [51]. We identified these two

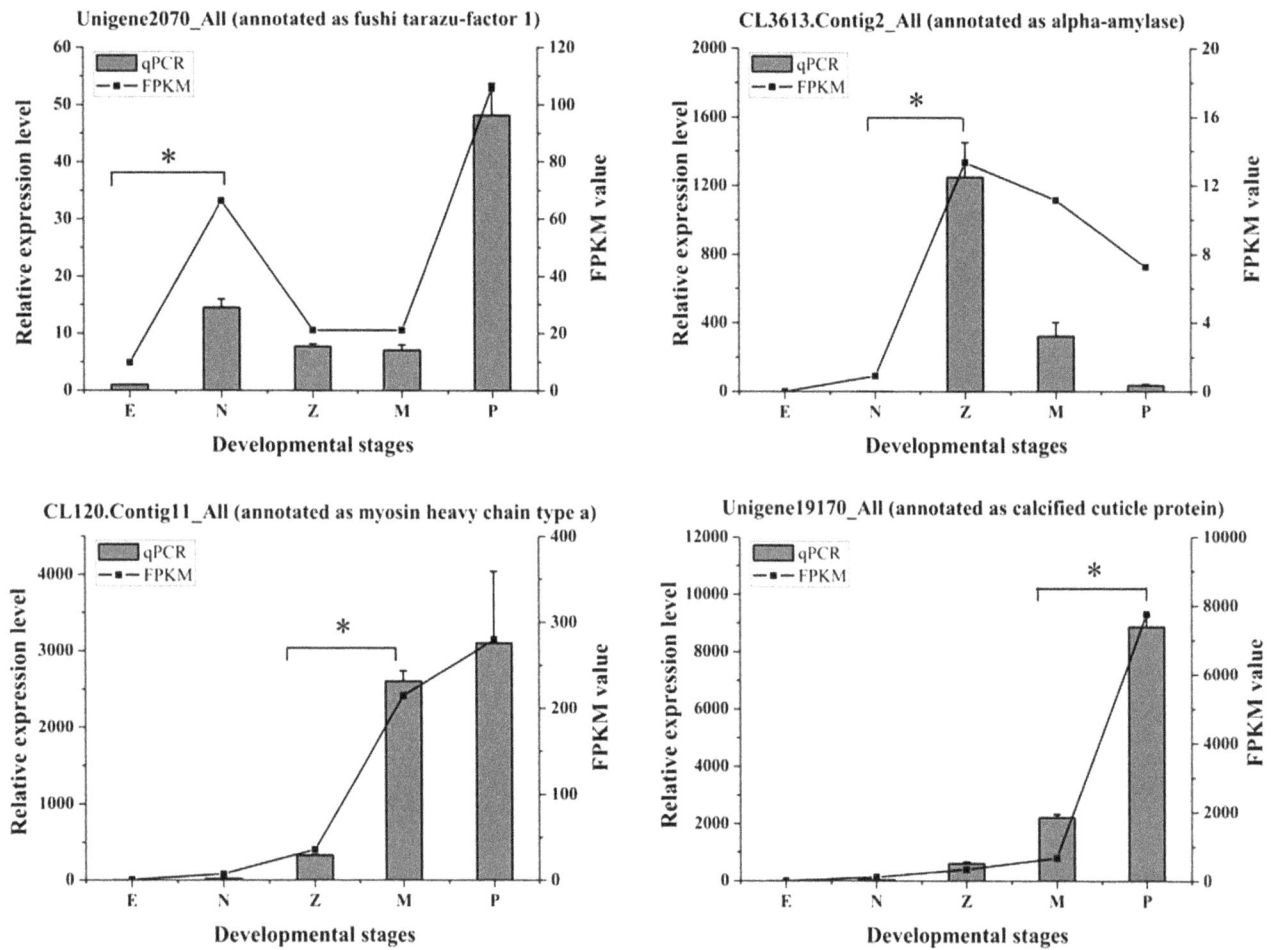

Figure 9. Expression profiles of four unigenes. X axis represents the developmental stages. Columns and bars represent the means and standard error of relative expression levels from qPCR result (Y axis at left). Lines represent the FPKM value from transcriptome result (Y axis at right). Asterisks represent the up-regulated transitions of each unigene.

key genes in our dataset (Table 5), and pathways which activate muscle development were also well-annotated including Wnt [52] and Notch [53] (Table 2). A large group for MYH type a and type b genes were identified in our dataset (Table 5), and the up-

regulated genes of MYH type a and type b in M stage implied the rapid growth for muscles (Figure 8).

DEGs between M and P stages were significantly enriched in chitin metabolism, which related to exoskeleton reconstruction. The body surface of arthropods is covered by an extracellular

Table 6. Information of primers used for real time PCR.

Amplified sequences	annotation	Sequence (5' to 3')	Tm (°C)
Unigene2070_All	fushi tarazu-factor 1	F: ATTGCCAACTCAACCGTCTCTAC	60
		R: GCTACACCAGGGAACAAACCA	
CL3613.Contig2_All	alpha-amylase	F: GCGAACACTACTGGATGATTGACA	61
		R: CGAAGACATTGAGGAAGCCG	
CL120.Contig11_All	myosin heavy chain type a	F: CGCCCTCTTTCTTCTCG	54
		R: ATCTGCGACGGTGCCTA	
Unigene19170_All	calcified cuticle protein	F: TCAGGAGCGGTGTAGGAGT	55
		R: GAAGAGTTCGTGCCAATCC	
18S rRNA	18S rRNA	F: TATACGCTAGTGGAGCTGGAA	55
		R: GGGGAGGTAGTGACGAAAAAT	

material called the exoskeleton (cuticle). The exoskeleton is an assembly of chitin and cuticle proteins. Its physical properties are determined largely by the proteins it contains, and vary widely with developmental stages and body regions [54]. Chitin is the major component for exoskeleton of penaeid shrimp [55], while cuticular proteins enhance the hardness of exoskeleton in order to protect the body from predation [56]. The variation of chitin and polysaccharide metabolism during this transition suggested the reconstruction for exoskeleton. Intriguingly, cuticle proteins involved in calcification (calcified cuticle protein and early cuticle protein) were sharply up-regulated in M-P transition (Figure 8). Meanwhile, other calcification related proteins (calcification associated soluble matrix protein, calcification-associated peptide-1 and crustocalcin) were also up-regulated (Table 5). Considering larvae moved to lower water layer during M-P transition, the reinforcement of exoskeleton by calcification might be an adaptation for the transition from a planktonic life to a benthic life. Total of 188 unigenes were annotated as cuticle proteins in our dataset. They possessed a high diversity and varied in expression patterns. The cuticular protein genes were also diversified among insects: 101 cuticular protein genes have been identified in the genome of *Drosophila melanogaster* [57] and 156 in *Anopheles gambiae* [58]. The diversity of cuticle protein in transcriptional level of shrimp might contribute to the rapid generation of exoskeletons with different physical properties in different developmental stages.

To validate our sequencing data, four analyzed differentially expressed genes (fushi tarazu-factor 1, alpha-amylase, myosin heavy chain type a and calcified cuticle protein) were selected for quantitative real-time PCR (qPCR) analysis. The information of primers was shown in Table 6. The results (figure 9) showed that the expression profiles of transcriptome data and the qPCR data were consistent. The differentially expressed genes identified by sequencing data were also obviously up-regulated in qPCR results. Unigene2070_all (annotated as fushi tarazu-factor 1) was highly expressed in nauplius and postlarvae stages, CL3613.Contig2_All (annotated as alpha-amylase) was highly expressed in zoea stage, CL120.Contig11_All (annotated as myosin heavy chain type a) was highly expressed in mysis and postlarvae stages and Unigene19170_All (annotated as calcified cuticle protein) was highly expressed in postlarvae stage.

Conclusions

Our study focused on the transcriptomes of five early developmental stages in *L. vannamei*, aiming for comparative analysis of physiological changes during shrimp metamorphosis. The RNA-Seq reads were assembled and clustered into 66,815 unigenes, of which 37,292 have been annotated. The five samples could be clustered into three major groups according to gene expression patterns and the differentially expressed genes between adjacent samples were also identified. By GO term enrichment analysis, KEGG pathway enrichment analysis and functional gene profiling, the physiological changes during shrimp metamorphosis could be better understood especially histogenesis, diet transition, muscle development and exoskeleton reconstruction. This is the first study that characterized the integrated transcriptome profiles during early development of penaeid shrimp. These findings will serve as significant references for shrimp developmental biology and aquaculture research.

Acknowledgments

We thank Mr. Junlong Zhang from Institute of Oceanology, Chinese Academy of Sciences for kindly helping us draw the Figure 1.

Author Contributions

Conceived and designed the experiments: XZ FL JX. Performed the experiments: JW XZ YY. Analyzed the data: JW XZ. Contributed reagents/materials/analysis tools: YY HH. Contributed to the writing of the manuscript: JW XZ FL JX.

References

1. FAO (2012) The state of world fisheries and aquaculture.
2. Dall W, Hill BJ, Rothlisberg PC, Staples DJ (1990) The biology of the penaeidae. Advances in Marine Biology 27: 1–461.
3. Medina M (2009) Functional genomics opens doors to understanding metamorphosis in nonmodel invertebrate organisms. Molecular Ecology 18: 763–764.
4. Zacharia S, Kakati VS (2004) Optimal salinity and temperature for early developmental stages of *Penaeus merguiensis* De man. Aquaculture 232: 373–382.
5. Brito R, Chimal ME, Gelabert R, Gaxiola G, Rosas C (2004) Effect of artificial and natural diets on energy allocation in *Litopenaeus setiferus* (Linnaeus, 1767) and *Litopenaeus vannamei* (Boone, 1931) early postlarvae. Aquaculture 237: 517–531.
6. Kiatmetha P, Siangdang W, Bunnag B, Senapin S, Withyachumnarnkul B (2011) Enhancement of survival and metamorphosis rates of *Penaeus monodon* larvae by feeding with the diatom Thalassiosira weissflogii. Aquaculture International 19: 599–609.
7. Racotta IS, Palacios E, Hernandez-Herrera R, Bonilla A, Perez-Rostro CI, et al. (2004) Criteria for assessing larval and postlarval quality of Pacific white shrimp (*Litopenaeus vannamei*, Boone, 1931). Aquaculture 233: 181–195.
8. Preston N (1985) The effects of temperature and salinity on survival and growth of larval *Penaeus plebejus*, *Metapenaeus macleayi* and *M. bennettae*. Second Australian National Prawn Seminar. 31–40.
9. Kumlu M, Eroldogan OT, Aktas M (2000) Effects of temperature and salinity on larval growth, survival and development of *Penaeus semisulcatus*. Aquaculture 188: 167–173.
10. Martin L, Castillo NM, Arenal A, Rodriguez G, Franco R, et al. (2012) Ontogenetic changes of innate immune parameters from eggs to early postlarvae of white shrimp *Litopenaeus vannamei* (Crustacea: Decapoda). Aquaculture 358: 234–239.
11. Puello-Cruz AC, Sangha RS, Jones DA, Le Vay L (2002) Trypsin enzyme activity during larval development of *Litopenaeus vannamei* (Boone) fed on live feeds. Aquaculture Research 33: 333–338.
12. Carrillo-Farnes O, Forrellat-Barrios A, Guerrero-Galvan S, Vega-Villasante F (2007) A review of digestive enzyme activity in penaeid shrimps. Crustaceana 80: 257–275.
13. Rapaport F, Khanin R, Liang Y, Pirun M, Krek A, et al. (2013) Comprehensive evaluation of differential gene expression analysis methods for RNA-seq data. Genome Biol 14: R95.
14. Wang Z, Gerstein M, Snyder M (2009) RNA-Seq: a revolutionary tool for transcriptomics. Nature Reviews Genetics 10: 57–63.
15. Chen XH, Zeng DG, Chen XL, Xie DX, Zhao YZ, et al. (2013) Transcriptome analysis of *Litopenaeus vannamei* in response to white spot syndrome virus infection. Plos One 8.
16. Xue SX, Liu YC, Zhang YC, Sun Y, Geng XY, et al. (2013) Sequencing and *de novo* analysis of the hemocytes transcriptome in *Litopenaeus vannamei* response to white spot syndrome virus infection. Plos One 8.
17. Zeng DG, Chen XL, Xie DX, Zhao YZ, Yang CL, et al. (2013) Transcriptome analysis of Pacific white shrimp (*Litopenaeus vannamei*) hepatopancreas in response to Taura syndrome virus (TSV) experimental infection. Plos One 8.
18. Sookruksawong S, Sun FY, Liu ZJ, Tassanakajon A (2013) RNA-Seq analysis reveals genes associated to Taura syndrome virus (TSV) resistance in the Pacific white shrimp *Litopenaeus vannamei*. Developmental and Comparative Immunology 41: 523–533.
19. Li CZ, Weng SP, Chen YG, Yu XQ, Lu L, et al. (2012) Analysis of *Litopenaeus vannamei* Transcriptome Using the Next-Generation DNA Sequencing Technique. Plos One 7.
20. Grabherr MG, Haas BJ, Yassour M, Levin JZ, Thompson DA, et al. (2011) Full-length transcriptome assembly from RNA-Seq data without a reference genome. Nature Biotechnology 29: 644–U130.
21. Pertea G, Huang XQ, Liang F, Antonescu V, Sultana R, et al. (2003) TIGR Gene Indices clustering tools (TGICL): a software system for fast clustering of large EST datasets. Bioinformatics 19: 651–652.
22. Altschul SF, Gish W, Miller W, Myers EW, Lipman DJ (1990) Basic local alignment search tool. Journal of Molecular Biology 215: 403–410.

23. Conesa A, Götz S, García-Gómez JM, Terol J, Talón M, et al. (2005) Blast2GO: a universal tool for annotation, visualization and analysis in functional genomics research. Bioinformatics 21: 3674–3676.

24. Mortazavi A, Williams BA, McCue K, Schaeffer L, Wold B (2008) Mapping and quantifying mammalian transcriptomes by RNA-Seq. Nature methods 5: 621–628.

25. Team RC (2005) R: A language and environment for statistical computing. R foundation for Statistical Computing.

26. de Hoon MJL, Imoto S, Nolan J, Miyano S (2004) Open source clustering software. Bioinformatics 20: 1453–1454.

27. Saldanha AJ (2004) Java Treeview-extensible visualization of microarray data. Bioinformatics 20: 3246–3248.

28. Livak KJ, Schmittgen TD (2001) Analysis of relative gene expression data using real-time quantitative PCR and the 2 (-Delta Delta Ct) method. Methods 25: 402–408.

29. Ye J, Fang L, Zheng H, Zhang Y, Chen J, et al. (2006) WEGO: a web tool for plotting GO annotations. Nucleic Acids Res 34: W293–297.

30. Tatusov RL, Koonin EV, Lipman DJ (1997) A genomic perspective on protein families. Science 278: 631–637.

31. Kanehisa M, Goto S (2000) KEGG: Kyoto Encyclopedia of Genes and Genomes. Nucleic Acids Research 28: 27–30.

32. Wodarz A, Nusse R (1998) Mechanisms of Wnt signaling in development. Annual Review of Cell and Developmental Biology 14: 59–88.

33. Ingham PW, McMahon AP (2001) Hedgehog signaling in animal development: paradigms and principles. Genes & Development 15: 3059–3087.

34. Artavanis-Tsakonas S, Rand MD, Lake RJ (1999) Notch signaling: Cell fate control and signal integration in development. Science 284: 770–776.

35. Schier AF (2007) The maternal-zygotic transition: Death and birth of RNAs. Science 316: 406–407.

36. Barckmann B, Simonelig M (2013) Control of maternal mRNA stability in germ cells and early embryos. Biochimica Et Biophysica Acta-Gene Regulatory Mechanisms 1829: 714–724.

37. Kiernan DA, Hertzler PL (2006) Muscle development in dendrobranchiate shrimp, with comparison with Artemia. Evolution & Development 8: 537–549.

38. Tan MH, Au KF, Yablonovitch AL, Wills AE, Chuang J, et al. (2013) RNA sequencing reveals a diverse and dynamic repertoire of the *Xenopus tropicalis* transcriptome over development. Genome Research 23: 201–216.

39. Truman JW, Riddiford LM (2002) Endocrine insights into the evolution of metamorphosis in insects. Annual Review of Entomology 47: 467–500.

40. Bachère E (2000) Shrimp immunity and disease control. Aquaculture 191: 3–11.

41. Lee CY, Clough EA, Yellon P, Teslovich TM, Stephan DA, et al. (2003) Genome-wide analyses of steroid- and radiation- triggered programmed cell death in Drosophila. Current Biology 13: 350–357.

42. Thummel CS (1995) From embryogenesis to metamorphosis: the regulation and function of Drosophila nuclear receptor superfamily members. Cell 83: 871–877.

43. Tracy K, Baehrecke EH (2013) The role of autophagy in drosophila metamorphosis. In: Shi YB, editor. Animal Metamorphosis. San Diego: Elsevier Academic Press Inc. pp. 101–125.

44. Muhammad F, Zhang ZF, Shao MY, Dong YP, Muhammad S (2012) Ontogenesis of digestive system in *Litopenaeus vannamei* (Boone, 1931) (Crustacea: Decapoda). Italian Journal of Zoology 79: 77–85.

45. Le Vay L, Jones DA, Puello-Cruz AC, Sangha RS, Ngamphongsai C (2001) Digestion in relation to feeding strategies exhibited by crustacean larvae. Comparative Biochemistry and Physiology a-Molecular and Integrative Physiology 128: 623–630.

46. Chu KH, Sze CC, Wong CK (1996) Swimming behaviour during the larval development of the shrimp *Metapenaeus ensis* (DeHaan, 1844) (Decapoda, Penaeidae). Crustaceana 69: 368–378.

47. Hertzler PL, Freas WR (2009) Pleonal muscle development in the shrimp *Penaeus (Litopenaeus) vannamei* (Crustacea: Malacostraca: Decapoda: Dendrobranchiata). Arthropod Structure & Development 38: 235–246.

48. Weiss A, Leinwand LA (1996) The mammalian myosin heavy chain gene family. Annu Rev Cell Dev Biol 12: 417–439.

49. Landsverk ML, Epstein HF (2005) Genetic analysis of myosin II assembly and organization in model organisms. Cellular and Molecular Life Sciences 62: 2270–2282.

50. Panganiban G (2000) Distal-less function during Drosophila appendage and sense organ development. Developmental dynamics 218: 554–562.

51. Maqbool T, Jagla K (2007) Genetic control of muscle development: learning from Drosophila. Journal of Muscle Research and Cell Motility 28: 397–407.

52. Mill C, George SJ (2012) Wnt signalling in smooth muscle cells and its role in cardiovascular disorders. Cardiovascular Research 95: 233–240.

53. Mayeuf A, Relaix F (2011) Notch pathway: from development to regeneration of skeletal muscle. M S-Medecine Sciences 27: 521–526.

54. Charles JP (2010) The regulation of expression of insect cuticle protein genes. Insect Biochemistry and Molecular Biology 40: 205–213.

55. Rocha J, Garcia-Carreno FL, Muhlia-Almazan A, Peregrino-Uriarte AB, Yepiz-Plascencia G, et al. (2012) Cuticular chitin synthase and chitinase mRNA of whiteleg shrimp *Litopenaeus vannamei* during the molting cycle. Aquaculture 330: 111–115.

56. Watanabe T, Persson P, Endo H, Fukuda I, Furukawa K, et al. (2006) Identification of a novel cuticular protein in the kuruma prawn *Penaeus japonicus*. Fisheries Science 72: 452–454.

57. Karouzou MV, Spyropoulos Y, Iconomidou VA, Cornman R, Hamodrakas SJ, et al. (2007) Drosophila cuticular proteins with the R&R Consensus: Annotation and classification with a new tool for discriminating RR-1 and RR-2 sequences. Insect biochemistry and molecular biology 37: 754–760.

58. Cornman RS, Togawa T, Dunn WA, He N, Emmons AC, et al. (2008) Annotation and analysis of a large cuticular protein family with the R&R Consensus in Anopheles gambiae. Bmc Genomics 9.

15

Quantification of Protein Copy Number in Yeast: The NAD$^+$ Metabolome

Szu-Chieh Mei, Charles Brenner*

Department of Biochemistry, Carver College of Medicine, University of Iowa, Iowa City, Iowa, United States of America

Abstract

Saccharomyces cerevisiae is calorie-restricted by lowering glucose from 2% to 0.5%. Under low glucose conditions, replicative lifespan is extended in a manner that depends on the NAD$^+$-dependent protein lysine deacetylase Sir2 and NAD$^+$ salvage enzymes. Because NAD$^+$ is required for glucose utilization and Sir2 function, it was postulated that glucose levels alter the levels of NAD$^+$ metabolites that tune Sir2 function. Though NAD$^+$ precursor vitamins, which increase the levels of all NAD$^+$ metabolites, can extend yeast replicative lifespan, glucose restriction does not significantly change the levels or ratios of intracellular NAD$^+$ metabolites. To test whether glucose restriction affects protein copy numbers, we developed a technology that combines the measurement of Urh1 specific activity and quantification of relative expression between Urh1 and any other protein. The technology was applied to obtain the protein copy numbers of enzymes involved in NAD$^+$ metabolism in rich and synthetic yeast media. Our data indicated that Sir2 and Pnc1, two enzymes that sequentially convert NAD$^+$ to nicotinamide and then to nicotinic acid, are up-regulated by glucose restriction in rich media, and that Pnc1 alone is up-regulated in synthetic media while levels of all other enzymes are unchanged. These data suggest that production or export of nicotinic acid might be a connection between NAD$^+$ and calorie restriction-mediated lifespan extension in yeast.

Editor: Mary Bryk, Texas A&M University, United States of America

Funding: Work was supported by grant MCB1322118 from the National Science Foundation to CB. The funder had no role in study design, data collection and analysis, decision to publish, or preparation of the manuscript.

Competing Interests: The authors have declared that no competing interests exist.

* Email: charles-brenner@uiowa.edu

Introduction

Calorie restriction (CR) is a powerful intervention to extend lifespan and healthspan in model organisms including yeast, flies, worms and rodents [1,2]. Although recent studies on the efficacy of CR to extend lifespan in primates were equivocal, salutary effects on biochemical parameters were observed in both studies [3,4]. Thus, the molecular mechanism by which CR promotes healthful changes remains of interest.

Saccharomyces cerevisiae is considered an excellent tool to study lifespan because it is a single-celled eukaryote with a rapid cell division cycle, compact genome, and unparalleled genetic control. In yeast, simply lowering the glucose concentration from 2% to 0.5% produces CR [5,6]. Previous studies have shown that replicative lifespan (RLS) can be extended by CR in a manner that depends on Sir2 [5], a nicotinamide adenine dinucleotide (NAD$^+$)-dependent protein lysine deacetylase [7,8], and genes encoding NAD$^+$ salvage enzymes [9]. Because NAD$^+$ is required for glucose fermentation and for Sir2 function, it was postulated that levels of glucose alter aspects of NAD$^+$ metabolism to produce a signal that increases Sir2 activity, thereby extending lifespan [10,11]. Two competing models were put forth: that CR would increase the NAD$^+$: NADH ratio [12], or that CR would increase the NAD$^+$: nicotinamide (Nam) ratio [13]. Both models posit that NADH or Nam, proposed as inhibitory metabolites, achieve levels in cells that would inhibit Sir2 activity.

Despite classical analysis of NAD$^+$ metabolism in the yeast system [14,15], the set of genes, enzymes, transporters and metabolites was incomplete at the time both models were proposed [12,13]. Surprisingly, we found that nicotinamide riboside (NR), a natural product found in milk, is capable of bypassing all known pathways to NAD$^+$ by virtue of the activity of a specific NR kinase pathway [16]. We further expanded the NAD$^+$ metabolome with biochemical and genetic characterization of NAD$^+$ biosynthetic enzymes [17–23]. Once all genes for salvage of NR to NAD$^+$ were identified, we showed that addition of NR to synthetic high glucose media increased Sir2 activity and extended lifespan so long as the NR salvage genes were present and extended lifespan correlated with the ability of NR to elevate NAD$^+$ [20]. Though these data indicate that an intervention that increases NAD$^+$ extends lifespan, it does not necessarily follow that the mechanism by which CR works is increasing NAD$^+$ or altering ratios of NAD$^+$ metabolites. We therefore developed LC-MS methods to quantify the expanded NAD$^+$ metabolome [24,25] and used these methods to determine what happens to NAD$^+$ metabolites under two conditions that extend lifespan, namely provision of NAD$^+$ precursors and glucose restriction.

Our data indicate that intracellular NAD$^+$ metabolite levels range from below 0.1 µM nicotinic acid adenine dinucleotide (NAAD) to above 500 µM (NAD$^+$) in yeast cells. Inclusion of nicotinic acid (NA) or yeast extract in media increases intracellular NAD$^+$ levels and increases concentrations of all NAD$^+$ metabolites

Table 1. Yeast strains used in this study.

Strain	Genotype
BY4741	MATa his3 Δ1 leu2Δ0 lys2Δ0 ura3Δ0
CM001	MATa urh1Δ::KanMX his3Δ1 leu2Δ0 lys2Δ0 ura3Δ0
PAB030	BY4741 NRK1::TAP-HIS3MX6
PAB054	BY4741 URH1::TAP-HIS3MX6
CM006	BY4741 NPT1::TAP-HIS3MX6
CM007	BY4741 SIR2::TAP-HIS3MX6
CM010	BY4741 QNS1::TAP-HIS3MX6
CM011	BY4741 UTR1::TAP-HIS3MX6
CM024	BY4741 NMA1::TAP-HIS3MX6
CM025	BY4741 NMA2::TAP-HIS3MX6
CM026	BY4741 PNC1::TAP-HIS3MX6
KB048	BY4741 ISN1::TAP-HIS3MX6
CM037	BY4741 POS5::TAP-HIS3MX6
CM044	BY4741 STD1::TAP-HIS3MX6
CM003	MATα BY4741 URH1::TAP-HIS3MX6
CM005	BY4741 MATa/MATα URH1::TAP-HIS3MX6 NRK1::TAP-HIS3MX6
CM018	BY4741 MATa/MATα URH1::TAP-HIS3MX6 NPT1::TAP-HIS3MX6
CM019	BY4741 MATa/MATα URH1::TAP-HIS3MX6 SIR2::TAP-HIS3MX6
CM022	BY4741 MATa/MATα URH1::TAP-HIS3MX6 QNS1::TAP-HIS3MX6
CM023	BY4741 MATa/MATα URH1::TAP-HIS3MX6 UTR1::TAP-HIS3MX6
CM034	BY4741 MATa/MATα URH1::TAP-HIS3MX6 NMA1::TAP-HIS3MX6
CM035	BY4741 MATa/MATα URH1::TAP-HIS3MX6 NMA2::TAP-HIS3MX6
CM036	BY4741 MATa/MATα URH1::TAP-HIS3MX6 PNC1::TAP-HIS3MX6
CM041	BY4741 MATa/MATα URH1::TAP-HIS3MX6 ISN1::TAP-HIS3MX6
CM043	BY4741 MATa/MATα URH1::TAP-HIS3MX6 POS5::TAP-HIS3MX6
CM046	BY4741 MATa/MATα URH1::TAP-HIS3MX6 STD1::TAP-HIS3MX6

including NADH and Nam [24]. These data are consistent with our previous observation that NR supplementation increases yeast intracellular NAD^+ levels and extends lifespan [20] and they challenge the view that NADH [26] or Nam are inhibitory metabolites.

Though intracellular NAD^+ metabolites were not altered by glucose restriction, we aimed to determine whether the abundance of NAD^+ biosynthetic proteins is altered consistent with predictions that the flux of NAD^+ metabolism may be increased by CR [13]. To answer this question, we developed a quantitative method to determine the copy number of molecules in the yeast proteome and discovered that copy numbers of NAD^+ enzymes range from 300 to 49,000 per diploid cell. Our data indicate that the levels of most NAD^+ metabolic enzymes are not altered. However, in rich media, Sir2, an enzyme that produces Nam from NAD^+, and Pnc1, the enzyme that hydrolyzes Nam to NA, are up-regulated in glucose-restricted conditions, while only Pnc1 is up-regulated in glucose-restricted synthetic media conditions. These data suggest that NA metabolism may be functionally modulated by glucose restriction in a manner that promotes increased lifespan.

Materials and Methods

Strains and media

Yeast strains used in this study are listed in Table 1. BY4741 and TAP-tagged strains except those encoding Sdt1-TAP were purchased from Open Biosystems. CM044 was obtained by

genomic integration of the TAP tag as a C-terminal fusion as shown in Figure 1A. In brief, a TAP tag fragment flanked with 3′ SDT1 sequences was first amplified from genomic DNA of PAB054 (Urh1-TAP) with primers Sdt1-TAPF (5′- TGATA-TATTGGAGTTACCACACGTTGTGTCCGACCTGTTCG-GTCGACGGATCCCCGGGTT-3′) and Sdt1-TAPR (5′- ATA-GAGGCATCTAATGCAAGTAGATTTATATACAATTATA-TCGATGAATTCGAGCTCGTT-3′). The PCR product was than transformed into BY4741 cells and plated on SDC-His media for selection. CM044 was obtained after diagnostic PCR with primers Sdt1F (5′- GACTACTCTAGGACAGATAC-3′) and Sdt1R (5′- CTAACTGCTATGATCATCAG-3′) demonstrated a 2219 bp product. CM003, a MATα Urh1-TAP strain, was obtained by introducing a GAL1-HO plasmid to PB054 and passage through galactose media to induce a mating type switch. To obtain dually tagged diploid yeast strains, all MATa, i.e. BY4741-derived TAP-tag strains were crossed with CM003. Media used for protein copy number analysis was YP (2% bacto peptone, 1% yeast extract) or synthetic complete media supplemented with filter-sterilized glucose at final concentrations of 2%, 0.5% or 0.2%.

Urh1 purification, specific activity assay and protein copy number determination

Recombinant Urh1 was expressed and purified as described [27]. To measure the NR hydrolytic activity of Urh1, we made use

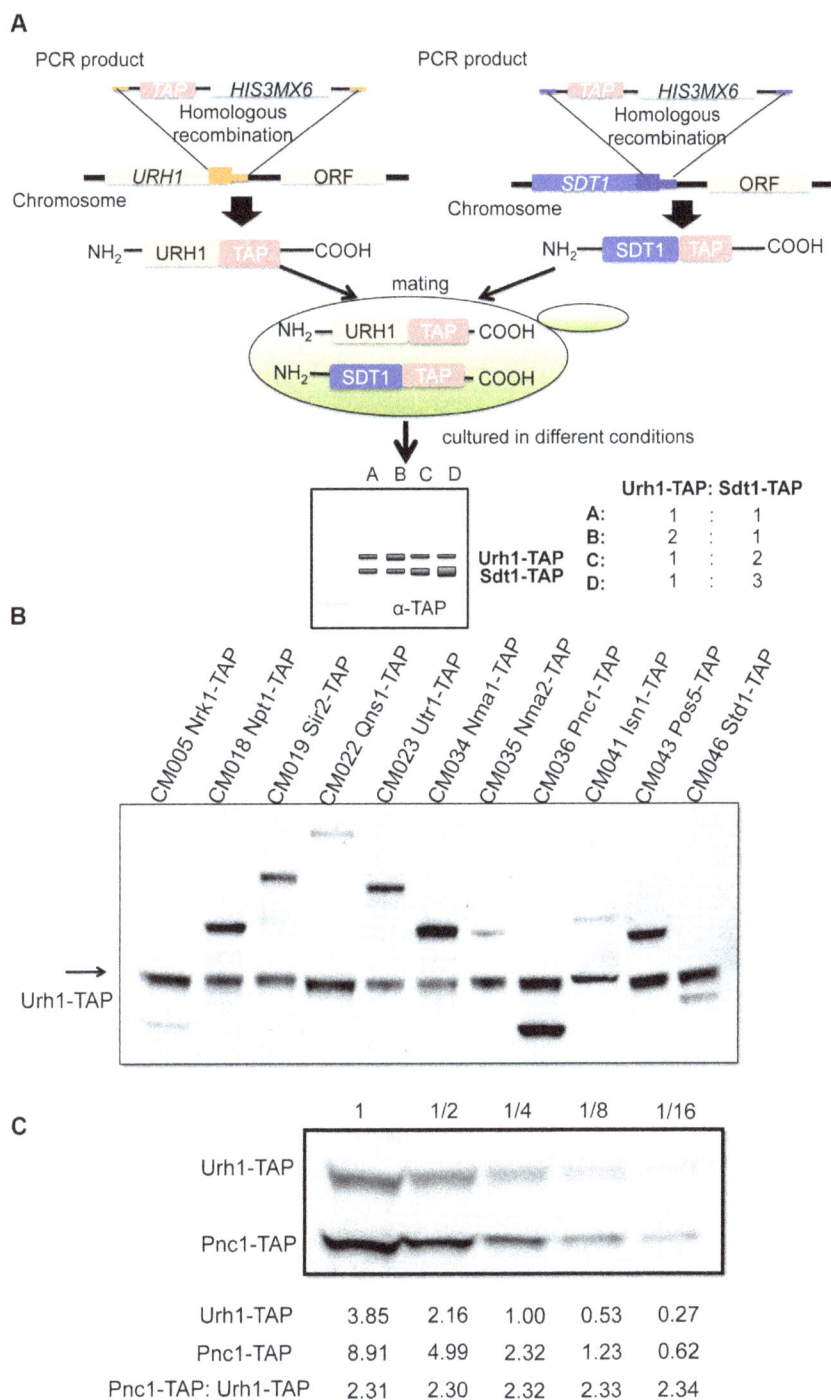

Figure 1. Dually TAP-tagged yeast strains to determine copy number relative to Urh1. (A) Strain construction to generate dually TAP-tagged strains, permitting relative protein quantification by western blot. (B) Dually tap-tagged strains at 2% glucose establish the specificity of the western assay. (C) Cell extracts from CM036 (Urh1-TAP, Pnc1-TAP) analyzed over a range of dilutions establish the linearity of TAP-tagged detection.

of the absorbance drop between NR and Nam at 269 nm, which is $2100 \ M^{-1} \ cm^{-1}$. In brief, 12 ng recombinant Urh1 or 30 μg yeast extract was incubated with 160 μM NR (10-fold higher than the K_m value) in 50 mM Tris-HCl at pH 6.8 in a 3 mm path length cuvette [27]. Reactions were followed continuously in an Ultraspec 4000 UV/Vis spectrophotometer (Amersham Pharmacia Biotech, Freiburg, Germany) at 269 nm with SWIFT II version 2.03 software and were converted to Specific Activity (SA)

in units of nmol/min/μg. To determine whether there is any NR hydrolysis activity in a *urh1* knockout strain, 0 to 12 ng of recombinant Urh1 protein plus or minus 30 μg yeast protein extract from strain CM001 was used to measure Urh1 SA. Protein copy number per cell was calculated from the ratio of the SA of Urh1 in the extract to that of pure Urh1, making use of 6 pg as the protein content per haploid yeast [28] and 37,900 Da as the molecular weight of Urh1 as in Formula 1:

A

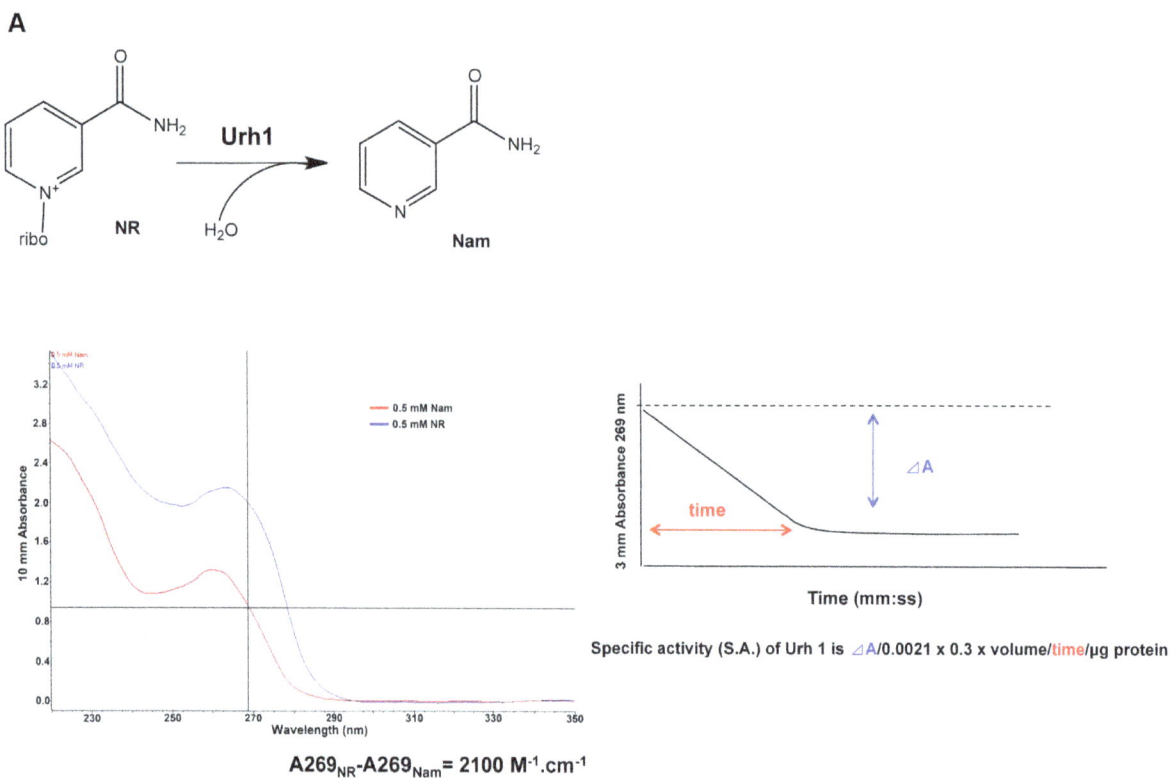

$$A269_{NR} - A269_{Nam} = 2100 \ M^{-1}.cm^{-1}$$

Specific activity (S.A.) of Urh 1 is $\triangle A/0.0021 \times 0.3 \times$ volume/time/µg protein

B

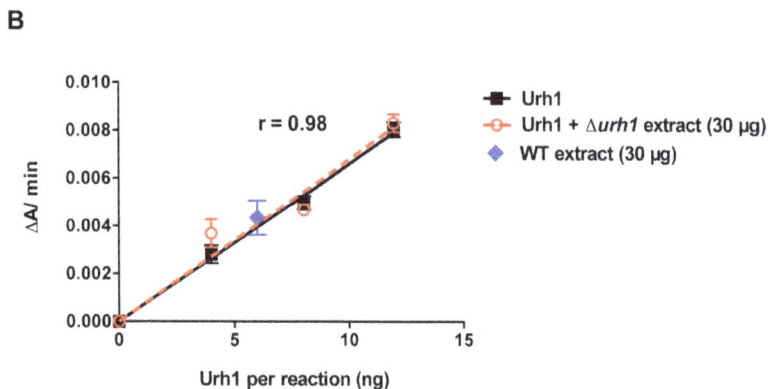

Figure 2. Urh1 enzyme activity in crude lysate can determine Urh1 absolute copy number. (A) Diagram of Urh1 SA measurements. Urh1 functions to hydrolyze NR to Nam. UV scanning result with 0.5 mM NR and Nam revealed a significant absorbance difference between NR and Nam at 269 nm, which is 2100 $M^{-1}cm^{-1}$. Urh1 SA can be calculated by monitoring the absorbance drop at 269 nm per time and protein used in the reaction. (B) Urh1 enzyme activity measurement was unaffected by addition of Urh1-free yeast extracts. Urh1 hydrolytic rate was measured with 0 to 12 ng of Urh1 in the absence or presence of 30 µg crude yeast extract from a *urh1* knockout strain. Correlation between the Urh1 and Urh1 plus *urh1*Δ extract group was 0.98.

Da as the molecular weight of Urh1-TAP, Formula 2 was applied:

$$Urh1 \ per \ cell = \frac{\dfrac{SA(crude \ extract)}{SA(pure \ Urh1)}(6 \ pg)}{37,900} \times 6.02 \times 10^{23}$$

$$Urh1 - TAP \ per \ cell = \frac{\dfrac{SA(crude \ extract)}{SA(pure \ Urh1)}(8 \ pg)}{57,900} \times 6.02 \times 10^{23}$$

Protein copy number of Urh1 in a diploid strain heterozygous for tagged Urh1 was determined after demonstration that a wild-type diploid has the same SA as the strain heterozygous for the tag. Using 8 pg as the protein content per diploid cell [28] and 57,900

Copy numbers of all the other enzymes were obtained by multiplying Urh1-TAP copy number in a given condition by the ratio of TAP-tag western signals between the other enzymes and Urh1-TAP, *i.e.*, Pnc1-TAP/Urh1-TAP.

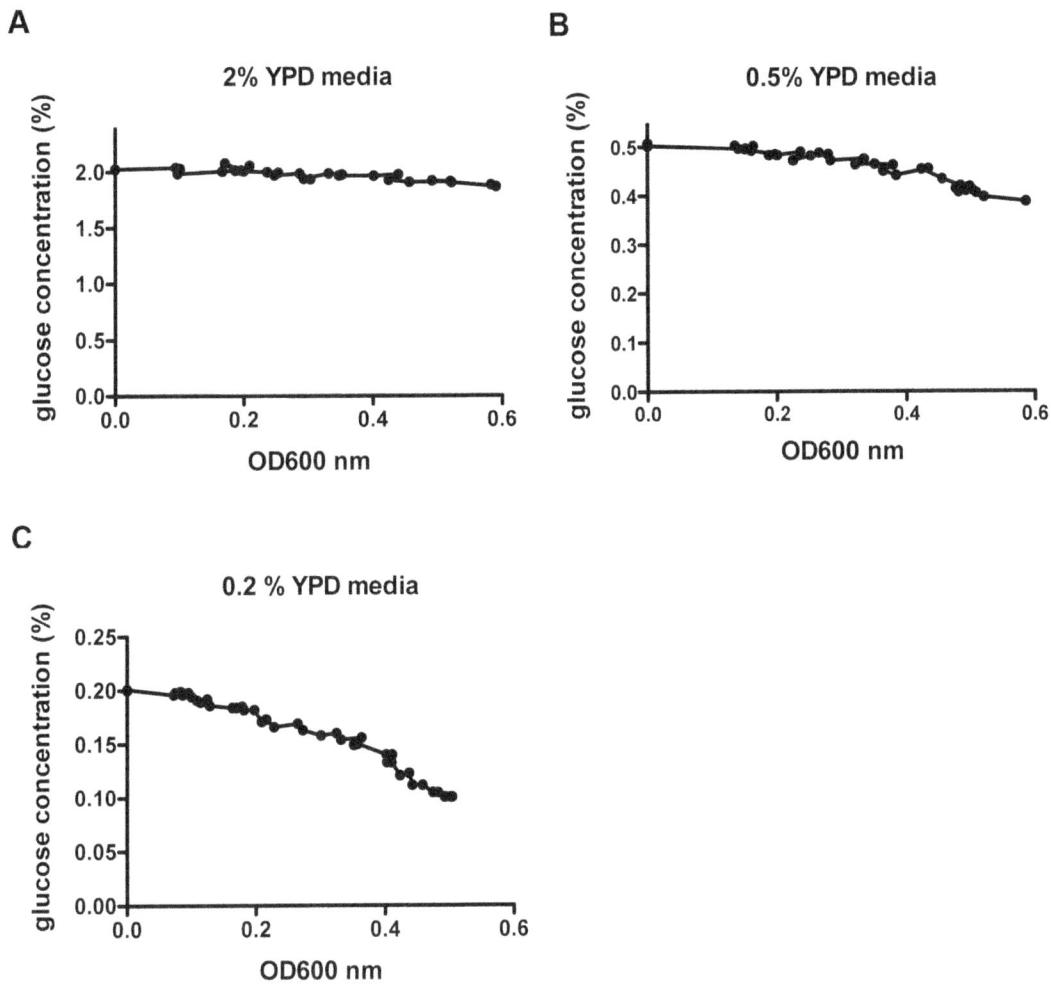

Figure 3. Glucose concentrations are maintained at acceptable levels up to $OD_{600\ nm} = 0.5$. Glucose concentrations of media from yeast strains cultured in (A) 2.0% YPD media (B) 0.5% YPD media and (C) 0.2% YPD media were measured with Glucose Assay Kits (Cayman). Data were collected from three different diploid TAP-tag yeast strains and shown as mean ± SD.

Cell extract preparation and western blotting

A single colony was inoculated in 5 ml YP 2% glucose media and allowed to divide until $OD_{600\ nm}$ reached 0.5. Cells were inoculated into 30 ml YPD cultures with 2%, 0.5% or 0.2% glucose at initial $OD_{600\ nm}$ of 0.0005. Cells were pelleted when the $OD_{600\ nm}$ reached 0.5 and lysed by glass bead beating. For western blotting, GAPDH was the loading control and 20 μg total protein extract were loaded per lane on Biorad stain-free gels. TAP-tagged proteins were detected with anti-TAP tag antibody CAB1001 (Open Biosystems) as the primary reagent and horseradish peroxidase (HRP)-conjugated goat anti-rabbit antibody as the secondary reagent (Thermo Scientific). Sir2 protein was detected with anti-Sir2 antibody sc-6666 (Santa Cruz) as the primary reagent and HRP-conjugated donkey anti-goat antibody as the secondary (Abcam). GAPDH was detected with HRP-conjugated anti-GAPDH antibody (Abcam). Signals were visualized with SuperSignal West Femto Chemiluminescent substrate (Pierce), imaged with a ChemiDoc XRS+ system (Bio-Rad), and quantified with ImageLab software. All data were collected from three individual experiments and statistical data were analyzed by one-way ANOVA.

Glucose concentration measurement

To measure the glucose concentration in yeast media, a single colony of CM018, CM019 and CM022 was inoculated in 5 ml YP 2% glucose media and allowed to grow until $OD_{600\ nm}$ reached 0.5. Cells were then inoculated in 50 ml YPD cultures with 2%, 0.5% or 0.2% glucose at an initial $OD_{600\ nm}$ of 0.0005. 0.2 ml media samples removed at indicated cell densities were centrifuged at 800 g for 10 min at 4°C to pellet yeast cells. Supernatants were then transferred to new tubes and stored at −20°C until measurement. Glucose concentration was assayed by using a Glucose Colorimetric Assay Kit (Cayman Chemical) per the manufacturer's protocol.

Protein copy number determination in SDC media

See Methods S1

Results

Novel quantification of protein copy number in yeast

At least 12 enzymes mediate the biosynthetic interconversion of NAD^+ metabolites, which include two pyridine bases, two nucleosides, two mononucleotides, and five dinucleotides. Within

Table 2. Protein copy number of enzymes in the NAD$^+$ metabolic pathway.

Protein Name	Culture Condition	Copy Number
Isn1	2.0%	2,200±700
	0.5%	2,500±300
	0.2%	2,000±200
Nma1	2.0%	10,000±1,000
	0.5%	10,000±1,000
	0.2%	9,400±1,000
Nma2	2.0%	2,000±200
	0.5%	2,300±300
	0.2%	2,000±100
Npt1	2.0%	45,000±4,000
	0.5%	42,000±7,000
	0.2%	38,000±8,000
Nrk1	2.0%	2,000±200
	0.5%	1,900±300
	0.2%	2,200±600
Pnc1	2.0%	18,000±4,000
	0.5%	36,000±3,000
	0.2%	49,000±4,000
Pos5	2.0%	9,200±1,200
	0.5%	9,500±1,700
	0.2%	11,000±2,000
Qns1	2.0%	1,500±200
	0.5%	1,800±500
	0.2%	1,800±300
Sir2	2.0%	2,000±200
	0.5%	3,200±500
	0.2%	4,900±700
Std1	2.0%	3,700±200
	0.5%	3,600±100
	0.2%	3,600±400
Urh1	2.0%	13,000±1,000
	0.5%	14,000±1,000
	0.2%	15,000±1,000
Utr1	2.0%	13,000±1,000
	0.5%	12,000±1,000
	0.2%	13,000±1,000

the set of metabolic enzymes, Urh1 is an abundant enzyme responsible for converting NR to Nam plus ribose. Though two phosphorylases have the ability to convert NR to Nam plus a ribosyl product [20,27], Urh1 is the only yeast enzyme with NR hydrolase activity, *i.e.*, the ability to convert NR to Nam in phosphate-free buffer. Taking advantage of this and with the aim of using Urh1 SA to determine its protein copy number, we developed a simple method to measure Urh1 SA by monitoring hydrolytic activity of NR. As shown in Figure 2A, NR and Nam exhibit a significant absorbance difference (2100 M^{-1} cm^{-1} at 269 nm), which was exploited to calculate Urh1 SA.

To rule out a Urh1-independent NR degradative activity in crude yeast extract, cell extracts from *urh1* knockout strain CM001 were used as control. As shown in Figure 2B, the

hydrolytic rate of recombinant Urh1 was linear in the absence or presence of Δ*urh1* yeast extract. Interpolated from linear regression of recombinant Urh1 activity, there was about 6 ng of Urh1 protein in 30 μg of crude wild-type yeast extract. Because the amount of total protein per cell can be taken as a constant [28], the ratio of crude SA of Urh1 to the SA of purified Urh1 allows one to calculate Urh1 copy number per cell.

Given a facile method to determine the protein copy number of Urh1, we developed a second tool to obtain protein copy numbers of any other yeast protein. In brief, we established 11 yeast strains in which an in-frame tandem affinity protein (TAP)-tag was integrated at the C-terminus of Urh1 coding sequences and the same TAP-tag was integrated at the C-terminus of a second enzyme in NAD$^+$ metabolism (Table 1). The resulting yeast strains

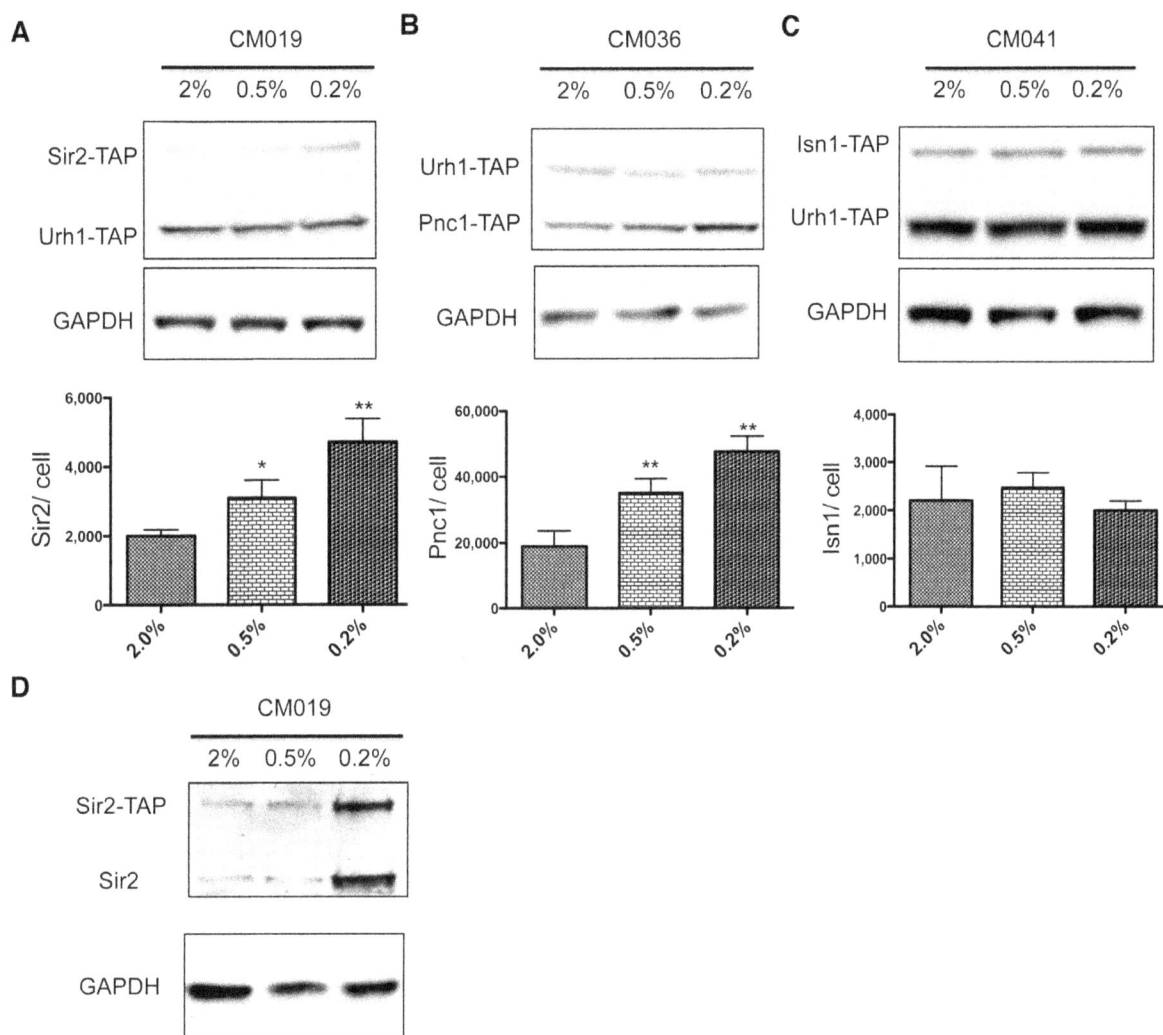

Figure 4. Pnc1 and Sir2 are induced by glucose restriction. Cell extracts from different dually TAP-tagged strains were prepared and analyzed by western blot. Urh1 expression data were converted to protein copy number using SA measurements as described in Materials and Methods. (A) Sir2 expression increases from 2,000/cell at 2% glucose to 4,900/cell at 0.2% glucose. (B) Pnc1 expression increases from 18,000/cell at 2% glucose to 49,000/cell at 0.2% glucose. (C) Isn1 expression is roughly constant at 2,000/cell under all conditions examined. (D) CM019 cell extracts probed with Sir2 antibody established no significant difference in protein expression and stability between endogenous and TAP-tagged Sir2 proteins.

were then grown in YPD at three concentrations of glucose. Extracted cellular proteins were analyzed by western blot using an anti-TAP tag antibody followed by detection with HRP-conjugated goat anti-rabbit antibody and chemiluminescent signal development. Chemiluminescent signals quantified by ImageLab software were used to determine the relative level of expression of each protein to Urh1-TAP under each experimental condition (Figures 1A, 1B and S3).

Because protein expression was expected to vary over a wide range, we aimed to determine whether this method is quantitatively sound. As shown in Figure 1C and Figure S1, the ratios between Urh1-TAP tagged protein and other tagged proteins were consistent in all serial dilutions of extract. These data indicate that the relative quantification method provides valid expression information over a wide range. Because each enzyme contains one TAP epitope and the SA of Urh1-TAP can be converted to its protein copy number per cell, the result of this analysis is a calculation of cellular protein copy number of each enzyme in each experimental condition.

Sir2 and Pnc1 copy numbers increase during glucose restriction in rich media

After establishing methods to quantify protein copy numbers, we applied this technology to determine copy number changes of enzymes in NAD$^+$ metabolism during glucose restriction under the conditions of most replicative longevity experiments, *i.e.* in rich YPD media. We aimed to maintain yeast cells in specific conditions for 10 generations to ensure evaluation of steady-state expression. However, because cell division consumes glucose, sample collection was optimized so that final glucose concentrations were maintained near the levels of initial innocula. As shown in Figure 3, with an initial OD$_{600\ nm}$ of 0.0005, a ten generation growth to an OD$_{600\ nm}$ of 0.5 produced only a drop of $\sim$ 0.1% glucose in each culture condition.

In all conditions, the ratio of crude Urh1 SA to that of homogeneous enzyme was consistently 1:6200. These data indicate that there are $\sim$1.3 fg of Urh1-TAP per cell. Using the predicted molecular weight of the Urh1-TAP tagged construct of

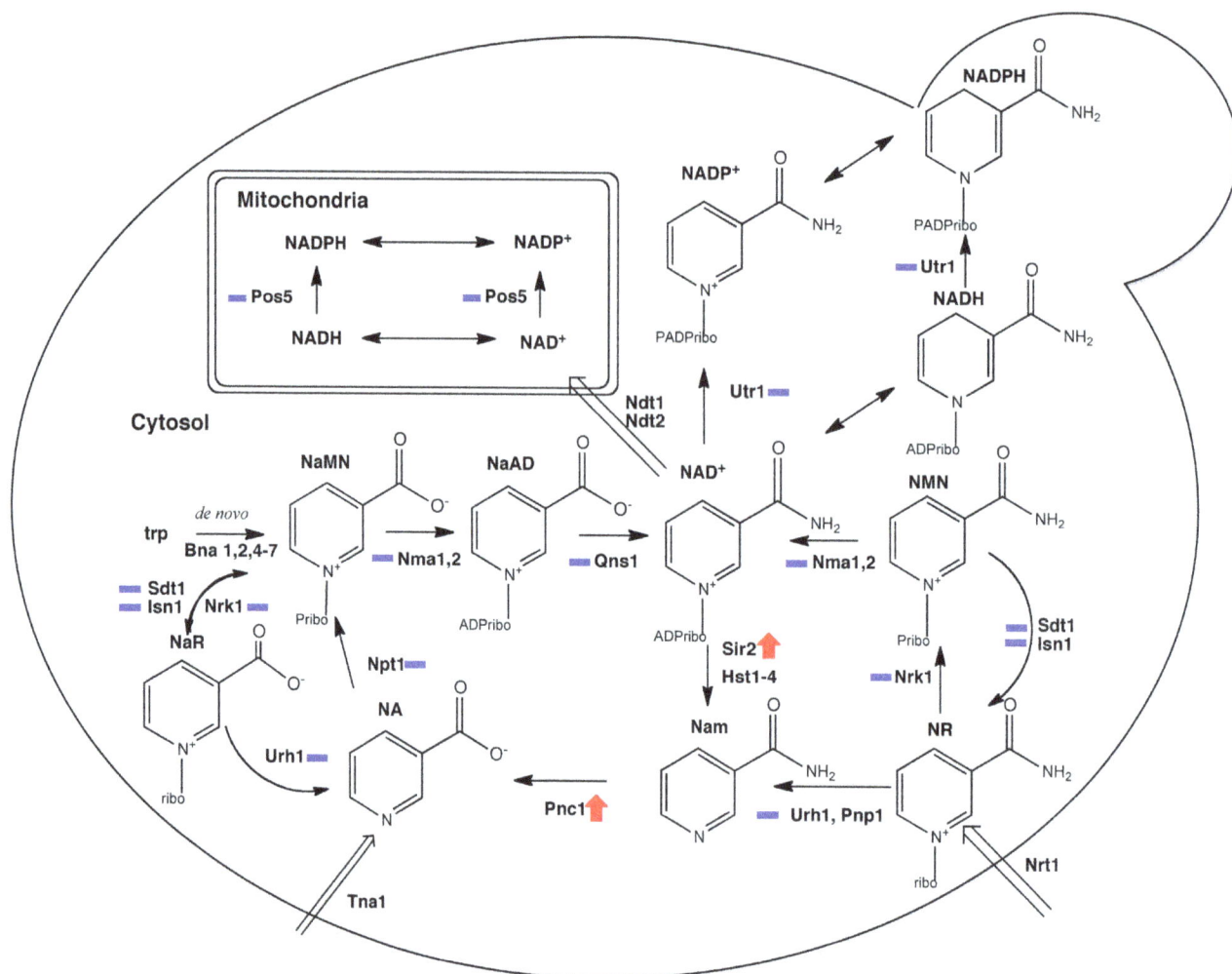

Figure 5. Summary of protein expression changes in CR. Overall, the abundance of most enzymes involved in NAD$^+$ metabolic pathways was unchanged by CR. The only two enzymes with significant alterations were Sir2 and Pnc1, which are two consecutive enzymes that convert NAD$^+$ into NA. Both are increased by CR in YPD. Pnc1 alone is increased in SDC.

57.9 kDa, this translates to an apparent Urh1 protein copy number of ~14,000 in all conditions examined (Table S1).

In rich media conditions at 2% glucose, at a range of 38,000 to 45,000 molecules per cell, the most abundant NAD$^+$ biosynthetic enzyme measured is Npt1, which converts NA to nicotinic acid mononucleotide (NAMN). Though Npt1 is crucial to maintain the level of NAD$^+$ in yeast [29] and is required for the longevity benefit of CR [5], Npt1 is neither increased nor decreased in copy number by glucose restriction. The second most abundant enzyme in rich media, high glucose conditions is Pnc1, the nicotinamidase that converts Nam to NA. In 2% glucose, there are 18,000 molecules per cell. However, its copy number rises to 36,000 when glucose is reduced to 0.5% and to 49,000 when glucose is reduced to 0.2%. Increased expression of Pnc1 in stress and low glucose conditions has been previously reported and is due to increased mRNA expression driven by 5′ stress response elements [30]. Nma1, one of two mononucleotide adenylyltransferases [31], and Pos5 and Utr1, the two major NAD$^+$/NADH kinases [18] are expressed at about 10,000 copies per cell and are unaffected by glucose concentrations. Nma2, the other mononucleotide adenylyltransferase [9] and the three enzymes that interconvert NR and nicotinic acid riboside (NAR) to nicotinamide mononucleotide

(NMN) and NAMN, namely Nrk1 [16], Isn1 and Sdt1 [22] are expressed at 2,000 copies per cell under each condition. Though discovery of NR as a vitamin that can bypass *de novo* biosynthesis of NAD$^+$ and NAD$^+$ biosynthesis from conventional niacins [16] undermined the early annotations of glutamine-dependent NAD$^+$ synthetase Qns1 as an essential gene, Qns1 can be considered to be essential unless NR is available. Despite this central role in NAD$^+$ homeostasis, Qns1 has the lowest level of expression of any NAD$^+$ metabolic enzyme examined in YPD at fewer than 2,000 copies per cell.

Sir2, a protein lysine deacetylase that converts NAD$^+$ to Nam [32], is required for longevity of long-lived Fob1-wild-type yeast strains [5,10,11]. Though much work has focused on potential mechanisms by which glucose levels might control Sir2 enzyme activity [5,6,9–13], here we show that the level of Sir2 protein is only 2,000 copies per cell at 2% glucose and increases to 3,200 and 4,900 copies per cell at 0.5% and 0.2% glucose, respectively, in YPD media. Though Sir2 expression and localization are altered by ribosomal DNA copy number [33], the level of expression of Sir2 as a function of glucose had been thought to be constant [34]. However, the lack of an earlier observation of the increase in Sir2 expression might be due to the short time of exposure to CR conditions [34]. In summary, among 12 NAD$^+$ metabolic enzymes, Sir2 and Pnc1 are the only two enzymes affected by glucose restriction (Tables 2 and S1, Figures 4, 5 and S2) To test whether the changes were confined to the expression or stability of the TAP-tagged fusion protein, we prepared cell extracts from strain CM019 and probed with a Sir2 antibody. As shown in Figure 4D, the ratio between Sir2-TAP and un-tagged Sir2 protein was consistent in all three conditions, which indicated that quantification of TAP-tagged fusion protein reflects endogenous protein expression levels.

Reduced expression of most NAD$^+$ metabolic enzymes in synthetic media

Unlike YPD media, which contains multiple salvageable NAD$^+$ metabolites, SDC media is restricted to tryptophan as a *de novo* NAD$^+$ precursor and NA as a salvageable vitamin. As shown in Tables S2 and S3, most NAD$^+$ enzymes are substantially reduced in protein accumulation in SDC media with respect to YPD. One notable difference between YPD expression and SDC expression was in accumulation of the two mononucleotide adenylyltransferases, Nma1 and Nma2. In YPD, Nma1 is the dominant enzyme at ~10,000 copies per cell while Nma2 accumulates to about 20% of this copy number. However, in SDC, the Nma1 level decreased to about 500 copies per cell while Nma2 increased to about 2,500 copies per cell, effectively becoming the dominantly expressed enzyme in SDC.

As in YPD, the concentration of glucose had little effect on accumulation of the vast majority of NAD$^+$ enzymes. Whereas Sir2 increased accumulation in glucose restricted YPD media, it did not do so in glucose restricted SDC media. However, Pnc1, which was not reduced in expression by SDC responded to glucose restriction with CR-induced protein expression.

Discussion

The connection between NAD$^+$ metabolism and lifespan extension by CR has been studied for more than a decade. Sir2 and Sir2 paralogs have been considered the major targets, which connect the requirement of NAD$^+$ salvage to CR-induced lifespan extension [7,9,10,12,13,35,36]. Besides its function as limiting the production of extra-chromosomal ribosomal DNA circles (ERCs), Sir2 is also important for asymmetrical cell division [37–39], and for deacetylation of histone H4 Lys16, which is important for maintaining telomere function in older cells [40]. It has been shown in multiple studies that NAD$^+$ levels in bulk cells are unchanged by CR [9,12,13,24,35]. In addition, having shown that intracellular NAD$^+$ metabolites are not altered by CR [24], here we developed a method to quantify the copy number of enzymes involved in NAD$^+$ metabolic pathways during normal and CR conditions. Our data indicate that levels of Sir2 and Pnc1 are increased by CR in YPD and that only Pnc1 is increased in expression by CR in SDC, while all other enzymes remain unchanged in protein expression. Though Sir2 and Pnc1 are capable of increasing degradation of NAD$^+$ to NA, previous observations of the intracellular NA level indicate that it remains below 0.5 µM in all conditioned examined [24].

Another interpretation of increased expression of Pnc1 is that it reduces the concentration of Nam, a metabolite that inhibits Sir2

at high concentrations [13]. However, because addition of NA to synthetic media elevates the concentration of Nam by 15-fold, while glucose restriction in YPD elevates Nam by a further 50% [24], there is a lack of credible evidence that Nam at intracellular concentrations of up to 50 µM shortens lifespan. Increased Pnc1 expression might either be a noncausal epiphenomenon that correlates with CR-induced lifespan extension or it could increase the rate of production of NA in a manner or pathway that does not elevate intracellular NAD$^+$ metabolites in bulk cells.

Here we showed that Urh1 copy number can be quantified in crude lysates by measurement of NR hydrolysis in phosphate-free buffer. By introducing TAP tags of Urh1 and any other protein into the same yeast strain, relative expression data obtained from western blotting were converted to protein copy number per cell. It is anticipated that this technology will be applied to diverse problems in yeast molecular and cellular biology.

Supporting Information

Figure S1 Western analysis of CM005 and CM034 strains. Cell extracts from (A) CM005 (Urh1-TAP, Nrk1-TAP) and (B) CM034 (Urh1-TAP, Nma1-TAP) analyzed over a range of dilutions further establish the linearity of TAP-tagged detection.

Figure S2 Western blot results of co-TAP-tagged strains. Equal amounts of cellular extracts isolated from dually TAP-tagged strains cultured in 2%, 0.5% or 0.2% glucose YPD media were used to perform western blot with an anti-TAP antibody. All results showed no significant difference between each culture condition. (A) CM005; (B) CM018; (C) CM022; (D) CM023; (E) CM034; (F) CM035; (G) CM043; (H) CM046.

Figure S3 GAPDH and stain-free gel images served as loading controls. (A) 20 µg of cell extracts from dually TAP-tagged strains at 2% glucose were separated in 7.5% TGX stain-free gels and imaged with ImageLab software. (B) GAPDH signal detected with specific antibody was used as loading control.

Table S1 Urh1 copy number in each dual-tag yeast strains in YPD media.

Table S2 Protein copy number of enzymes in the NAD+ metabolic pathway in SDC media.

Table S3 Urh1 copy number in each dual-tag yeast strains in SDC media.

Methods S1 Supplementary materials and methods.

Author Contributions

Conceived and designed the experiments: SCM CB. Performed the experiments: SCM. Analyzed the data: SCM CB. Contributed to the writing of the manuscript: SCM CB.

References

1. Koubova J, Guarente L (2003) How does calorie restriction work? Genes Dev 17: 313–321. doi: 10.1101/gad.1052903.

2. Fontana L, Partridge L, Longo VD (2010) Extending healthy life span–from yeast to humans. Science 328: 321 326. doi: 10.1126/science.1172539.

3. Colman RJ, Anderson RM, Johnson SC, Kastman EK, Kosmatka KJ, et al. (2009) Caloric restriction delays disease onset and mortality in rhesus monkeys. Science 325: 201–204. doi: 10.1126/science.1173635.

4. Mattison JA, Roth GS, Beasley TM, Tilmont EM, Handy AM, et al. (2012) Impact of caloric restriction on health and survival in rhesus monkeys from the NIA study. Nature 489: 318–321. doi: 10.1038/nature11432.

5. Lin SJ, Defossez PA, Guarente L (2000) Requirement of NAD and SIR2 for life-span extension by calorie restriction in Saccharomyces cerevisiae. Science 289: 2126–2128. doi: 10.1126/science.289.5487.2126.

6. Jiang JC, Jaruga E, Repnevskaya MV, Jazwinski SM (2000) An intervention resembling caloric restriction prolongs life span and retards aging in yeast. FASEB J 14: 2135–2137. doi: 10.1096/fj.00-0242fje.

7. Kaeberlein M, McVey M, Guarente L (1999) The SIR2/3/4 complex and SIR2 alone promote longevity in Saccharomyces cerevisiae by two different mechanisms. Genes Dev 13: 2570–2580. doi: 10.1101/gad.13.19.2570.

8. Imai S, Armstrong CM, Kaeberlein M, Guarente L (2000) Transcriptional silencing and longevity protein Sir2 is an NAD-dependent histone deacetylase. Nature 403: 795–800. doi: 10.1038/35001622.

9. Anderson RM, Bitterman KJ, Wood JG, Medvedik O, Cohen H, et al. (2002) Manipulation of a nuclear NAD+ salvage pathway delays aging without altering steady-state NAD+ levels. J Biol Chem 277: 18881–18890. doi: 10.1074/jbc.M111773200.

10. Lin S-J, Kaeberlein M, Andalis AA, Sturtz LA, Defossez P-A, et al. (2002) Calorie restriction extends Saccharomyces cerevisiae lifespan by increasing respiration. Nature 418: 344–348. doi: 10.1038/nature00829.

11. Jiang JC, Wawryn J, Shantha Kumara HMC, Jazwinski SM (2002) Distinct roles of processes modulated by histone deacetylases Rpd3p, Hda1p, and Sir2p in life extension by caloric restriction in yeast. Exp Gerontol 37: 1023–1030. doi: 10.1016/S0531-5565(02)00064-5.

12. Lin S-J, Ford E, Haigis M, Liszt G, Guarente L (2004) Calorie restriction extends yeast life span by lowering the level of NADH. Genes Dev 18: 12–16. doi: 10.1101/gad.1164804.

13. Anderson RM, Bitterman KJ, Wood JG, Medvedik O, Sinclair DA (2003) Nicotinamide and PNC1 govern lifespan extension by calorie restriction in Saccharomyces cerevisiae. Nature 423: 181–185. doi: 10.1038/nature01578.

14. Preiss J, Handler P (1958) Biosynthesis of diphosphopyridine nucleotide. I. Identification of intermediates. J Biol Chem 233: 488–492.

15. Preiss J, Handler P (1958) Biosynthesis of diphosphopyridine nucleotide. II. Enzymatic aspects. J Biol Chem 233: 493–500.

16. Bieganowski P, Brenner C (2004) Discoveries of Nicotinamide Riboside as a Nutrient and Conserved NRK Genes Establish a Preiss-Handler Independent Route to NAD+ in Fungi and Humans. Cell 117: 495–502. doi: 10.1016/S0092-8674(04)00416-7.

17. Bieganowski P, Pace HC, Brenner C (2003) Eukaryotic NAD+ synthetase Qns1 contains an essential, obligate intramolecular thiol glutamine amidotransferase domain related to nitrilase. J Biol Chem 278: 33049–33055. doi: 10.1074/jbc.M302257200.

18. Bieganowski P, Seidle HF, Wojcik M, Brenner C (2006) Synthetic lethal and biochemical analyses of NAD and NADH kinases in Saccharomyces cerevisiae establish separation of cellular functions. J Biol Chem 281: 22439–22445. doi: 10.1074/jbc.M513919200.

19. Wojcik M, Seidle HF, Bieganowski P, Brenner C (2006) Glutamine-dependent NAD+ synthetase. How a two-domain, three-substrate enzyme avoids waste. J Biol Chem 281: 33395–33402. doi: 10.1074/jbc.M607111200.

20. Belenky P, Racette FG, Bogan KL, McClure JM, Smith JS, et al. (2007) Nicotinamide Riboside Promotes Sir2 Silencing and Extends Lifespan via Nrk and Urh1/Pnp1/Meu1 Pathways to NAD. Cell 129: 473–484. doi: 10.1016/j.cell.2007.03.024.

21. Tempel W, Rabeh WM, Bogan KL, Belenky P, Wojcik M, et al. (2007) Nicotinamide riboside kinase structures reveal new pathways to NAD+. PLoS Biol 5: e263. doi: 10.1371/journal.pbio.0050263.

22. Bogan KL, Evans C, Belenky P, Song P, Burant CF, et al. (2009) Identification of Isn1 and Sdt1 as glucose- and vitamin-regulated nicotinamide mononucle-otide and nicotinic acid mononucleotide 5'-nucleotidases responsible for

23. Belenky P, Stebbins R, Bogan KL, Evans CR, Brenner C (2011) Nrt1 and Tna1-independent export of NAD+ precursor vitamins promotes NAD+ homeostasis and allows engineering of vitamin production. PLoS ONE 6: e19710. doi: 10.1371/journal.pone.0019710.

24. Evans C, Bogan KL, Song P, Burant CF, Kennedy RT, et al. (2010) NAD+ metabolite levels as a function of vitamins and calorie restriction: evidence for different mechanisms of longevity. BMC Chem Biol 10: 2. doi: 10.1186/1472-6769-10-2.

25. Trammell S, Brenner C (2013) Targeted, LCMS-based Metabolomics for Quantitative Measurement of NAD+ Metabolites. Comput Struct Biotechnol J 4: e201301012. doi: 10.1016/j.bmcl.2012.06.069.

26. Schmidt MT, Smith BC, Jackson MD, Denu JM (2004) Coenzyme specificity of Sir2 protein deacetylases: implications for physiological regulation. J Biol Chem 279: 40122–40129. doi: 10.1074/jbc.M407484200.

27. Belenky P, Christensen KC, Gazzaniga F, Pletnev AA, Brenner C (2009) Nicotinamide riboside and nicotinic acid riboside salvage in fungi and mammals. Quantitative basis for Urh1 and purine nucleoside phosphorylase function in NAD+ metabolism. J Biol Chem 284: 158–164. doi: 10.1074/jbc.M807976200.

28. Sherman F (2002) Getting started with yeast. Meth Enzymol 350: 3–41.

29. Smith JS, Brachmann CB, Celic I, Kenna MA, Muhammad S, et al. (2000) A phylogenetically conserved NAD+-dependent protein deacetylase activity in the Sir2 protein family. Proc Natl Acad Sci USA 97: 6658–6663. doi: 10.1073/pnas.97.12.6658.

30. Ghislain M, Talla E, François JM (2002) Identification and functional analysis of the Saccharomyces cerevisiae nicotinamidase gene, PNC1. Yeast 19: 215–224. doi: 10.1002/yea.810.

31. Winzeler EA, Shoemaker DD, Astromoff A, Liang H, Anderson K, et al. (1999) Functional characterization of the S. cerevisiae genome by gene deletion and parallel analysis. Science 285: 901–906. doi: 10.1126/science.285.5429.901.

32. Belenky P, Bogan KL, Brenner C (2007) NAD+ metabolism in health and disease. Trends Biochem Sci 32: 12–19. doi: 10.1016/j.tibs.2006.11.006.

33. Michel AH, Kornmann B, Dubrana K, Shore D (2005) Spontaneous rDNA copy number variation modulates Sir2 levels and epigenetic gene silencing. Genes Dev 19: 1199–1210. doi: 10.1101/gad.340205.

34. Medvedik O, Lamming DW, Kim KD, Sinclair DA (2007) MSN2 and MSN4 link calorie restriction and TOR to sirtuin-mediated lifespan extension in Saccharomyces cerevisiae. PLoS Biol 5: e261. doi: 10.1371/journal.pbio.0050261.

35. Anderson RM, Latorre-Esteves M, Neves AR, Lavu S, Medvedik O, et al. (2003) Yeast life-span extension by calorie restriction is independent of NAD fluctuation. Science 302: 2124–2126. doi: 10.1126/science.1088097.

36. Lamming DW, Latorre-Esteves M, Medvedik O, Wong SN, Tsang FA, et al. (2005) HST2 mediates SIR2-independent life-span extension by calorie restriction. Science 309: 1861–1864. doi: 10.1126/science.1113611.

37. Aguilaniu H, Gustafsson L, Rigoulet M, Nyström T (2003) Asymmetric inheritance of oxidatively damaged proteins during cytokinesis. Science 299: 1751–1753. doi: 10.1126/science.1080418.

38. Erjavec N, Larsson L, Grantham J, Nyström T (2007) Accelerated aging and failure to segregate damaged proteins in Sir2 mutants can be suppressed by overproducing the protein aggregation-remodeling factor Hsp104p. Genes Dev 21: 2410–2421. doi: 10.1101/gad.439307.

39. Erjavec N, Nyström T (2007) Sir2p-dependent protein segregation gives rise to a superior reactive oxygen species management in the progeny of Saccharomyces cerevisiae. Proc Natl Acad Sci USA 104: 10877–10881. doi: 10.1073/pnas.0701634104.

40. Dang W, Steffen KK, Perry R, Dorsey JA, Johnson FB, et al. (2009) Histone H4 lysine 16 acetylation regulates cellular lifespan. Nature 459: 802–807. doi: 10.1038/nature08085.

16

Dietary Inulin Supplementation Modifies Significantly the Liver Transcriptomic Profile of Broiler Chickens

Natalia Sevane, Federica Bialade, Susana Velasco, Almudena Rebolé, Maria Luisa Rodríguez, Luís T. Ortiz, Javier Cañón, Susana Dunner*

Nutrigenómica Animal, Departamento de Producción Animal, Facultad de Veterinaria, Universidad Complutense de Madrid, Madrid, Spain

Abstract

Inclusion of prebiotics in the diet is known to be advantageous, with positive influences both on health and growth. The current study investigated the differences in the hepatic transcriptome profiles between chickens supplemented with inulin (a storage carbohydrate found in many plants) and controls. Liver is a major metabolic organ and has been previously reported to be involved in the modification of the lipid metabolism in chickens fed with inulin. A nutrigenomic approach through the analysis of liver RNA hybridized to the Affymetrix GeneChip Chicken Genome Array identified 148 differentially expressed genes among both groups: 104 up-regulated ($\geq$1.4-fold) and 44 down-regulated ($\leq$0.6-fold). Quantitative real-time PCR analysis validated the microarray expression results for five out of seven genes tested. The functional annotation analyses revealed a number of genes, processes and pathways with putative involvement in chicken growth and performance, while reinforcing the immune status of animals, and fostering the production of long chain fatty acids in broilers supplemented with 5 g of inulin kg^{-1} diet. As far as we are aware, this is the first report of a microarray based gene expression study on the effect of dietary inulin supplementation, supporting further research on the use of this prebiotic on chicken diets as a useful alternative to antibiotics for improving performance and general immunity in poultry farming, along with a healthier meat lipid profile.

Editor: Marinus F.W. te Pas, Wageningen UR Livestock Research, Netherlands

Funding: This work has been supported by a grant to Research Group Nutrigenómica Animal from UCM-Santander GR35/10-A-920323. The funders had no role in study design, data collection and analysis, decision to publish, or preparation of the manuscript.

Competing Interests: The authors have declared that no competing interests exist.

* E-mail: dunner@ucm.es

Introduction

Prebiotics (e.g. fructans including inulin-type fructans [inulin and fructooligosaccharides]) are nondigestible food ingredients, whose beneficial effects on the host result from the selective stimulation of growth and/or activity of members of the gut microbiota, specifically bifidobacteria and lactobacteria [1]. Inulin, generally extracted from chicory roots (*Cichorium intybus* L.), is a prebiotic formed by a chain of fructose molecules connected by β-(2–1) glycosidic bonds, terminated by one glucose molecule, which is not decomposed by digestive enzymes due to its chemical structure [2]. However, it is a perfect carbon source for health-promoting gut bacteria. Although the inclusion of pre-biotics in the diet is known to be advantageous, their use in farm animals has been scarce [3]. Fructans supplementation is known to produce positive influences both on health and growth [4,5]: in fish, they increase intestinal growth relative to whole body weight, potentially enhancing nutrient absorption [6,7]; in broilers, a decrease in body fat deposition [8], serum cholesterol concentration and abdominal fat weight has been reported [4,5,7,9]; in rodents and, to a lesser extent in humans, inulin-type fructans can alter lipid metabolism by reducing plasma triglyceride and cholesterol concentrations [10,11]; in several animal models and in birds, these prebiotics also modify the hepatic metabolism of

lipids [5]; finally, prebiotics have also other positive effects on health, improving body functions and bone health, decreasing disease risks, reinforcing immune functions, preventing infections and intestinal diseases, and enhancing bioavailability of minerals (calcium and magnesium) [7,12,13]. However, the mechanisms through which these effects develop are not clear: it is thought to be a direct effect of the prebiotic on the host immune system by triggering receptors in the gut epithelium, which induces an immune response and activates the immune system without it becoming overactive [14]; withal, many of the desired effects are brought about by the manipulation of the gut flora, with the prebiotics providing substrates that preferentially encourage beneficial strains of bacteria to proliferate [1].

In this study, we perform a nutrigenomic approach to understand the molecular mechanisms underlying inulin supplementation effects to assess its impact in the commercial broiler. We chose to study the liver transcriptome as it is a major metabolic organ involved in many physiological processes including energy metabolism, detoxification and innate immunity. Moreover, previous results obtained in chickens by Rebolé et al. [4] and Velasco et al. [5] pointed to the modification of the hepatic metabolism of lipids by inulin. The different expression patterns from a nutrigenomic point of view help understand the mechanisms by which inulin modulates both metabolism and general immunity. Results outlined below indicate major changes in transcription of a number of genes implicated in development and maintenance of different tissues, particularly muscle and nervous system, fatty acid and protein metabolism, and immune system, gene transcription, and cell development and maintenance processes in the liver.

Material and Methods

A flow diagram of study design and results is shown in Fig. S1.

Animals

The animal protocol was approved by the Animal Care and Ethics Committee of the Universidad Complutense de Madrid (Spain) (CEA-UCM/32). Birds were handled according to the principles for the care of animals in experimentation established by the Spanish Royal Decree 1201/2005 [15].

A total of 80 one-day-old female broiler chicks (Cobb 500 genetic line) obtained from a commercial hatchery (Cobb Espanola S.A., Alcalá de Henares, Spain) were randomly allocated into 16 pens with eight replicates per treatment and five chicks per pen as described by Velasco et al. [5]. The bird groups were assigned to two dietary treatments: 1) control diet without inulin; and 2) control diet plus 5 g of inulin kg^{-1} of diet, which gave the best results on decreasing blood concentrations of triacylglycerides and increasing the capacity of sunflower oil to enhance the ratio of polyunsaturated (PUFA) to saturated (SFA) fatty acids of intramuscular fat in broilers [5]. The control basal diet (Table 1) was formulated to be adequate in all nutrients [16] and was prepared in mash form. The inulin source used in the current study was a commercial product (Prebiofeed, Qualivet, Las Rozas, Spain) obtained from chicory (C. intybus L.) roots containing 746 g kg^{-1} inulin-type fructans as determined in our laboratory [4]; therefore, the amount of this product added to the corresponding control diet at the expense of the entire diet was 6.7 g of product kg^{-1} of diet to obtain 5 g of inulin kg^{-1} of diet. Diets in mash form and water were offered ad libitum through the 34 day feeding trial. Mortality was lower than 3%. At the end of the experiment, birds were weighed and killed by cervical dislocation and liver tissue (~1 g) was placed in RNAlater (Ambion) and stored at 4°C for 24 h followed by long term storage at −20 °C prior to RNA extraction.

RNA extraction, cDNA synthesis and microarray analysis

Total RNA was extracted from 25 mg of liver tissue using the RNeasy Tissue Mini Kit (QIAGEN, Izasa, Spain). Four pools were produced consisting each in four equivalent amounts of liver samples mixed together according to the supplemented and control groups (Fig S1). Each experimental group resulted in eight samples combined in two pools which were RNA extracted for hybridization in microarray. Changes in gene expression were analyzed by microarray technology using the Affymetrix Gene-Chip Chicken Genome Array. Briefly, 200 ng of total RNA from each sample were processed, labeled, fragmented, and hybridized to the GeneChipChicken Genome Array according to the manufacturer recommendations.

The microarray normalization was carried out using functions from the Babelomics [17]. Normalized data were further analyzed using R (version 3.0.2) and the Bioconductor Limma package [18]. Differential gene expression was measured by empirical Bayes t-statistics and P-values were adjusted for false discovery rate correction [19]. Only the genes with P-value ≤0.09 and log fold change greater or equal than 1.4-fold for up-regulated genes and

lower or equal than 0.6-fold for down-regulated genes were screened out as differentially expressed genes.

Gene ontology analysis and visual pathway analysis

The Database for Annotation, Visualization and Integrated Discovery (DAVID) v6.7b [20] was used to determine pathways and processes of major biological significance and importance through the Functional Annotation Cluster (FAC) tool based on the Gene Ontology (GO) annotation function. DAVID FAC analysis was conducted on two independent gene lists containing up-regulated genes (≥1.4-fold) and down-regulated genes (≤0.6-fold) at P≤0.09. High stringency ease score parameters were selected to indicate confident enrichment scores of functional significance and importance of the given pathways and processes investigated.

Kyoto Encyclopedia of Genes and Genomes (KEGG) pathway tool was used to visually map clusters of the same chicken genes involved in common pathways and processes for both pathway-specific and molecular overview purposes. KEGG pathway tools were utilized through DAVID online tools.

Real-time PCR validation

To confirm microarray data, regulated genes in the liver tissue (ITIH5, DIO2, GIMAP5, USP18, KIAA1754, CCDC79) were selected for further validation by qRT-PCR. A total of 16 samples, eight corresponding to the inulin supplemented, and eight corresponding to the non supplemented chickens (all included in the four pools used to hybridize the microarray) were used. The total RNA was used for RT-cDNA synthesis using Superscript II First Strand cDNA Synthesis kit (Invitrogen). The resulting cDNA template was used to conduct real-time assays for all six target genes and four commonly reference genes, beta-actin (ACTB), glyceraldehyde-3-phosphate dehydrogenase (GAPDH), hypoxanthine phophoribosyl-transferase (HPRT) and glucose-6-phosphate dehydrogenase (G6PDH), by using an iCycler IQ Real-Time PCR thermocycler (Bio-Rad) and Dynamo HS SYBR Green qPCR Kit (Finnzymes, Vitro, Spain) as master mix. Primers were designed based on public available sequences (Table S1) using primer 3 (http://bioinfo.ut.ee/primer3-0.4.0/primer3/). After the selection of the most adequate annealing temperature, standard curves and the sample assays were produced in triplicate for each gene, together with the no-template controls. The following experimental run protocol was used: quantification program consisting of 42 cycles of 95°C for 30 s, 30 s at annealing temperature and 40 s at 72°C, ending with a melting program of 155 cycles of 10 s at 55°C and continuous fluorescence measurement. The results were exported into Microsoft Excel's based software Gene Expression Macro Version 1.1 (Bio-Rad Laboratories, http://www.bio-rad.com/) to calculate and normalize the expression of each gene.

The expression stability and level of the reference genes were measured using three different statistical algorithms to rank the genes by their stability values, geNorm [21], NormFinder [22] and Bestkeeper [23], being this a necessary process to guarantee that the reference genes are constitutively expressed in the tissue and treatment in question for a correct normalization. The target gene data were analyzed using PROC GLM procedure of the SAS statistical package v. 9.1.3 [24] to estimate the $2^{\Delta Ct}$ differences between treatments. Relative Expression Software (REST), which follows the Pfaffl method [25] was also used. This mathematical algorithm computes an expression ratio based on qRT-PCR efficiency and the crossing point deviation of the sample compared to a control group: $R = [(E\ target\ gene)^{\Delta Ct\ target\ gene\ (control-sample)}]/[(E\ Ref\ gene)^{\Delta Ct\ Ref\ gene\ (control-sample)}]$, where E is PCR

Table 1. Ingredients and nutrient composition of experimental control diet (g kg^{-1} as fed basis).

Ingredient	
Corn	451.8
Soybean meal (44% CP)	418.7
Sunflower oil	90.0
Calcium carbonate	10.0
Dicalcium phosphate	18.5
Sodium chloride	3.0
DL-Methionine	1.5
Antioxidant (butylated hydroxytoluene)	1.5
Vitamin and mineral premix[1]	5.0
Nutrient composition	
CP[2]	217.0
Lysine[2]	12.8
Methionine[2]	5.2
Methionine plus Cystine[3]	9.2
AME$_n$[3] (kcal kg^{-1})	3,152
Fatty acids[2,4] (g kg^{-1} of total fatty acids)	
C16:0	85.2
C18:0	36.6
C18:1n-9	299.1
C18:2n-6	548.0
SFA	121.8
MUFA	308.8
PUFA	556.6
UFA	865.4
PUFA:SFA	4.6
UFA:SFA	7.1

[1]Premix supplying (mg kg^{-1} diet): 3 retinol, 55 cholecalciferol, 25 dl-α-tocopheryl acetate, 2.5 menadione, 3 thiamine, 6 riboflavin, 7 pyridoxine, 0.2 folic acid, 0.02 cyanocobalamin, 0.2 biotin, 25 calcium pantothenate, 50 niacin, 1300 choline chloride, 60 Mn, 80 Fe, 50 Zn, 5 Cu, 0.1 Se, 0.18 I, 0.5 Co, 0.5 Mo.
[2]Determined.
[3]Calculated.
[4]SFA = saturated fatty acids; MUFA = monounsaturated fatty acids; PUFA = polyunsaturated fatty acids; UFA = unsaturated fatty acid.

efficiency of the gene transcript determined by standard curve using a serial dilution of cDNA. Normalization of the expression levels of the target genes was performed through three reference genes (*GAPDH*, *G6PDH* and *ACTB*). When differences between samples of either experiment (with or without inulin) were subjected to random, a test was carried out and the alternative hypothesis accepted for a *P* value lower than 0.08.

Results

Transcriptome profile and differential expression

A total of four Affymetrix GeneChip Chicken Genome arrays were hybridized with RNA pools resulting from mixing equal amounts of four different RNA samples from each experiment. Comparative transcriptome profiling of liver RNA samples from experimental inulin fed group versus control cDNA pools identified 112 genes over-expressed according to the elected threshold ≥1.4-fold and 46 down-regulated ≤0.6-fold ($P \leq 0.09$) (Table S2) from a total of 38,450 probes corresponding to over 28,000 chicken genes. The widespread use of arbitrary fold change cut-offs of above 2 and significance *P-values* of <0.02 in the analysis of microarray results was discarded here as it leads data

collection to look only at genes which vary wildly amongst other genes, and raises questions as to whether the biology or the statistical cutoff are more important within the interpretation [26]. In this paper we analyzed data giving priority to the biological signification of the results and set the fold change threshold at ≥ 1.4-fold for over-expressed genes and ≤0.6-fold for down-regulated genes. Among these 158 sequences, 139 shared significant homology with genes encoding proteins of known function, 9 shared homology with genes encoding proteins of unknown function, and 10 shared no significant homology with any database accession (Table S2). Out of the 148 chicken sequences with homologs, 104 were up-regulated and 44 were down-regulated in the presence of dietary inulin.

Functional annotation analyses

The expression data was analyzed using the DAVID FAC tool, obtaining enrichment scores per cluster under high stringency conditions as an indication of the biological significance of the gene groups analyzed (Table S3). From the 148 sequences with homologs, DAVID FAC analysis included 95 up-regulated and 35 down-regulated sequences in the analysis, revealing 102 enriched functional clusters with strong confident enrichment scores in the up-regulated sequences for development and maintenance of different tissues, particularly muscle and nervous system, cell processes, protein metabolism, gene transcription, response to hormones, and immune system processes, whereas the down-regulated sequences -although showing only 26 enriched functional clusters with lower enrichment scores- highlighted mainly fatty acid metabolism and intracellular organelles (Table S3, Fig. 1).

The KEGG database retrieved five pathways (Table 2, Fig. S2): Adipocytokine Signaling Pathway, Glycosphingolipid Biosynthesis, Glutathione Metabolism, Drug Metabolism - Cytochrome P450, and Metabolism of Xenobiotics by Cytochrome P450. The list of differentially expressed genes encoding proteins of known function but not included in the DAVID FAC analysis (9), along with those genes encoding proteins of unknown function (9) is shown in Table S4.

Validation of microarray data by real-time RT-PCR

In order to validate the microarray results, qRT-PCR was performed to determine the expression levels of six chicken genes - *ITIH5, DIO2, USP18, CCDC79, KIAA1754, GIMAP5*- selected from the list of sequences differentially expressed across individuals from inulin and control groups. Also, to obtain reliable qRT-PCR results, the stability of four commonly used housekeeping genes was determined (*HPRT, ACTB, GAPDH,* and *G6PDH*). The results of three programs (GeNorm, BestKeeper, and NormFinder) revealed that *ACTB, GAPDH,* and *G6PDH* were good candidate reference genes (Table 3). Despite a non significant 1.3 fold differential expression in the microarray results for *HPRT*, this gene was significantly differentially expressed when analyzed through qRT-PCR. Thereof, *HPRT* was not suitable as an endogenous control for the analysis of gene expression in the liver tissue in chickens.

The qRT-PCR expression results using REST software (Table 4) correlated with the microarray expression data for 4 out of the 6 genes tested, plus the up-regulation of *HPRT*. The qRT-PCR determination of *ITIH5, DIO2, KIAA1754, GIMAP5,* and *HPRT* mRNA levels showed a 2.2, 6.2, 2.4, 2.5, and 1.9-fold increase respectively in inulin supplemented chickens over controls, these results comparing favorably to the 2.2, 4.3, 3.3, 3.2, and 1.3–fold increase in expression determined by the microarray analysis.

Among the significant differentially expressed genes, *ITIH5, DIO2, GIMAP5,* and *HPRT* showed significant $2\Delta^{Ct}$ differences between treatments when analyzed using PROC GLM procedure (SAS) (Table 5). The activity of *ITIH5, DIO2, GIMAP5,* and *HPRT* genes explained 30%, 39%, 26% and 49% of the total variability ($P<0.08$), respectively.

However, the expression analysis of *USP18* and *CCDC79* by qRT-PCR was not significant, in contrast with the microarray data which showed both genes to be down-regulated by 0.5-fold.

Discussion

The use of inulin-type fructans in poultry feeding is known to produce positive influences both on chicken health and growth, by improving the performance [3,4], increasing the absorption of nutrients by modifications on the intestinal mucosal structure [27,28,29], stimulating the growth and/or activity of beneficial intestinal bacteria and preventing colonization by pathogenic bacteria [1]. Also, they decrease body fat deposition and improve its profile [4,5,7,8,9], along with other positive effects on health

Table 2. List of chicken genes from KEGG pathway maps differentially expressed in liver from animals supplemented with 5 g of inulin kg^{-1} diet and controls, with expression ratio, annotated gene description and KEGG ID.

Gene Symbol	Expression Ratio With vs. Without	Gene Name	KEGG ID
Adipocytokine Signaling Pathway - KEGG pathway			
TNFRSF1B	1.9	Tumor necrosis factor receptor superfamily member 1B	TNFR1
ACSL6	1.4	Acyl-CoA synthetase long-chain family member 6, transcript variant X5	FACS
PPARA	1.7	Peroxisome proliferator-activated receptor alpha	PPARα
Glycosphingolipid Biosynthesis - Ganglio Series - KEGG pathway			
ST3GAL5	1.6	ST3 beta-galactoside alpha-2,3-sialyltransferase 5	2.4.99.9
ST3GAL1	1.9	ST3 beta-galactoside alpha-2,3-sialyltransferase 1	2.4.99.4
Glutathione Metabolism - KEGG pathway			
GSTA	0.6	Glutathione S-transferase class-alpha	2.5.1.18
GSTT1	0.6	Glutathione S-transferase theta 1	2.5.1.18
RRM2B	1.5	Ribonucleotide reductase M2 B (TP53 inducible)	1.17.4.1
Drug Metabolism - Cytochrome P450 - KEGG pathway			
GSTA	0.6	Glutathione S-transferase class-alpha	2.5.1.18
GSTT1	0.6	Glutathione S-transferase theta 1	2.5.1.18
Metabolism of Xenobiotics by Cytochrome P450 - KEGG pathway			
GSTA	0.6	Glutathione S-transferase class-alpha	2.5.1.18
GSTT1	0.6	Glutathione S-transferase theta 1	2.5.1.18

Table 3. Stability of four reference genes on liver from animals supplemented with 5 g of inulin kg^{-1} diet and controls, measured through three different software: Bestkeeper, GeNorm and NormFinder.

	Bestkeeper		GeNorm		NormFinder	
	Stability value	Ranking	Stability value	Ranking	Stability value	Ranking
ACTB	0.959	1	0.659	1	0.247	3
GAPDH	0.934	2	0.690	3	0.304	2
G6PDH	0.921	3	0.727	1	0.372	1
HPRT	0.835	4	0.805	4	0.466	4

[7,12,13]. This study adds the characterization of the genetic expression patterns promoted by inulin to the evaluation on the effects of its dietary supplementation in poultry. Moreover, understanding the molecular mechanism underlying inulin effects can be a useful approach to help finding natural alternatives to the overdependence on antibiotics to enhance animal production (given that it has been directly related to the growing number of antibiotic resistances [30], which has lead to the ban of antibiotics for growth promotion by the European Union since 2006 [31] and the calls to restrict its use in other countries).

Functional analysis of the differentially expressed genes using the GO term annotations showed that the differentially expressed genes can be functionally grouped in three main classes: (i) basal processes including tissue development and maintenance (particularly muscle, nervous system processes, cell organelles processes, protein metabolism, gene transcription, and response to hormones); (ii) immune system processes; and (iii) fatty acid metabolism.

Basal processes – development and maintenance of different tissues

Although liver was the tissue explored, genes involved in the development and maintenance of different tissues, particularly nervous system and muscle, showed the highest enrichment score in the FAC analysis of up-regulated genes (Fig. 1(a)). DAVID analyses identified 46 genes with an expression range of 1.4 to 4.5 that functionally clustered into common GO terms related to nervous system, muscle, respiratory, bone, and embryonic development, or neurological, circulatory and reproductive

processes (Table S3). Thus, several differentially expressed genes may relate to other tissue-specific processes that, up to now, have not been described as expressed in liver. As an example, the high enrichment scores of neurological pathways may be due to the involvement of a neurological mechanism, e.g. through the participation of neuronal tissue in liver tissue composition. Alternatively, it could indicate that genes currently described as involved in neurological pathways, may have basic functions common to other tissues.

FAC analysis also identified protein metabolism as a significant biological process up-regulated by the addition of inulin to the chicken diets (Fig. 1(a)). DAVID analyses identified 33 genes with an expression range of 1.4 to 4.3 that functionally clustered into common GO terms related to regulation of protein metabolism, translation, peptidase activity, post-translational protein modification, protein activity, proteolysis, or protein localization (Table S3). Among them, *CAV2, TPPP, MLH1, AHCTF1, NRG1, NEFL*, and *DVL1* were also involved in the improvement of growth performance by inulin supplementation, showing the highest enrichment scores.

DAVID analyses identified 43 genes with an expression range of 1.4 to 4.3 that functionally clustered into common GO terms related to regulation of transcription and biosynthesis, chromosome organization, DNA, RNA and nucleotide binding, or transcription activity (Fig. 1(a), Table S3), showing an increased cellular activity in the presence of the prebiotic.

Finally, regarding cell organelles and cellular processes, DAVID analyses included all up-regulated genes (expression range of 1.4 to

Table 4. Differential expression results of the genes studied in the liver by Real-time PCR assay from animals supplemented with 5 g of inulin kg^{-1} diet and controls using REST software (*P*<0.08).

Gene	Type	Expression	Std. Error	95% C.I.	P(H1)	Result
ACTB	REF	0.828				
GAPDH	REF	1.093				
G6PDH	REF	1.106				
ITIH5	TRG	2.185	0.88–5.3	0.40–7.4	0.080	UP
DIO2	TRG	6.175	2.29–17.6	1.00–33.6	0.003	UP
USP18	TRG	1.082	0.44–3.0	0.16–6.4	0.880	
CCDC79	TRG	1.484	0.35–4.3	0.12–16.6	0.538	
KIAA1754	TRG	2.376	0.89–6.0	0.44–11.3	0.070	UP
GIMAP5	TRG	2.543	1.04–6.0	0.62–14.2	0.042	UP
HPRT	TRG	1.860	1.33–2.8	0.63–3.7	0.015	UP

Table 5. ANOVA results of the genes studied in the liver by Real-time PCR assay from animals supplemented with 5 g of inulin kg^{-1} diet and controls.

Dependent Variable	Mean Square	Error	R-Square	Coeff Var	Pr > F
ITIH5	48.4	11.2	0.302	71.3	0.0642
DIO2	607.5	94.7	0.390	88.5	0.0297
USP18	55.9	82.5	0.063	113.4	0.4296
CCDC79	179.5	216.0	0.077	123.4	0.3835
KIAA1754	37.2	19.1	0.178	87.8	0.1962
GIMAP5	78.6	22.7	0.257	80.1	0.0922
HPRT	6.2	0.63	0.495	34.1	0.0106

4.5) into common GO terms related to these main groups (Fig. 1, Table S3).

All these data suggested an important influence of inulin on processes and pathways that lead to the increase of growth and performance, which is in agreement with the results reported by Rebolé et al. [4] and Velasco et al. [5], who found a quadratic body weight gain in chickens supplemented with inulin, and with those of Tacchi et al. [6], who suggested an improvement in nutrient absorption as a consequence of the increase of intestinal growth caused by the addition of inulin to fish diets.

Immune system processes

The addition of inulin to the diet of chickens stimulated various immune system processes (Fig. 1(a)). DAVID analyses identified 20 up-regulated genes with an expression range of 1.4 to 4.5 that functionally clustered into common GO terms related to immune system processes, immunoglobulins, response to virus and biotic stimulus, regulation of apoptosis, immune system development and activation, immune response, or cellular response to stress and DNA damage stimulus (Table S3).

KEGG pathway visual analysis identified three genes (TNFRSF1B, ACSL6, PPARA) in the Adipocytokine Signaling Pathway that were up-regulated with inulin supplementation by 1.56 to 7.56 fold (Fig. S2(a), Table 2). However, ACSL6 and PPARA were not included in the functional clusters related to immune system processes by DAVID FAC tool (Table S3). TNFRSF1B has anti-apoptotic activity by stimulating antioxidative pathways and is considered a marker of activation of T-helper subsets regulatory T-cell (Tregs) [32]. ACSL6 encodes an enzyme that catalyzes the formation of acyl-CoA from fatty acids, ATP, and CoA, using magnesium as a cofactor, and as such was included in the mitochondrion, membrane, fatty acid metabolism, metal ion binding and carboxylic acid metabolism DAVID clusters (Table S3). Finally, PPARA, found in different clusters, is a member of peroxisome proliferator-activated receptors (PPARs), which plays a major regulatory function of genes involved in energy metabolism, affects the expression of target genes involved in cell proliferation and differentiation, and in immune and inflammation responses (see e.g. Gervois & Mansouri [33]).

Interestingly, Glutathione Metabolism is found among the pathways identified by KEGG (Fig. S2(c), Table 2). Glutathione plays important roles in antioxidant defense, nutrient metabolism, and regulation of cellular events (including gene expression, DNA and protein synthesis, cell proliferation and apoptosis, signal transduction, cytokine production and immune response, and protein glutathionylation); its deficiency contributes to oxidative stress, which plays a key role in the aging and the pathogenesis of

many diseases [34]. Three genes where included in this pathway: GSTT1 is a member of a superfamily of proteins that catalyze the conjugation of reduced glutathione to a variety of electrophilic and hydrophobic compounds identified as having an important role in human carcinogenesis; together with GSTA form the Glutathione S-transferase cluster, and both genes were down-regulated in contrast to the up-regulated RRM2B, which is found to act in Trypanosoma cruci as a mechanism to minimize the reactive oxygen species (ROS) produced by host defense [35].

These results pointed towards an effect of inulin supplementation on the reinforcement of chicken immune status by the activation of genes and pathways implicated in immune processes, while conferring a higher ability to avoid ROS.

Fatty acid metabolism

Fatty acid metabolism showed the highest enrichment score in the FAC analysis of down-regulated genes (Fig. 1(b)). DAVID analyses identified 5 genes with an expression range of 0.5 to 0.6 that functionally clustered into common GO terms related to lipid, monocarboxylic acids, carboxylic acids, oxoacids, organic acids, and ketone metabolic processes (Table S3).

Inulin supplementation seemed to foster the production of beneficial long chain fatty acid [36], as deduced from the higher expression of genes as PPARA, FASN, and ACSL6, and the down regulation of genes involved in the degradation of long branched fatty acids (ACOX2), hydrolysis of fatty acids, specifically phospholipids (JMJD7-PLA2G4B), cleavage of the ether bond of alkylglycerols (TMEM195), and in processes specifically linked with the mitochondrion and cytoplasma (CYP2J2). The down-regulation of a fatty acid elongase (ELOVL2) can be explained by its particular substrate: whereas ELOVL2 can efficiently elongate C_{20} and C_{22} polyunsaturated (PUFA) fatty acids, it cannot elongate C_{18} PUFA nor monounsaturated fatty acids (MUFA) or saturated fatty acids (SFA) [37], and its down-regulation is in concordance with the observed effect of dietary inulin addition on the increase of C18:2n-6 [5].

As supported by Rebolé et al. [4] and Velasco et al. [5], the addition of inulin-type fructans to the diet decreases body fat deposition [8], serum cholesterol concentration, and abdominal fat weight of chickens [9], which is in agreement with the gene regulation found here and explained also the increased function of the mitochondrion linked to the different fatty acid metabolism.

Validation of microarray data by real-time RT-PCR

In order to validate the microarray results, six chicken genes, selected from the list of genes differentially expressed across controls and inulin supplemented individuals, were used in qRT-

(a)

(b)

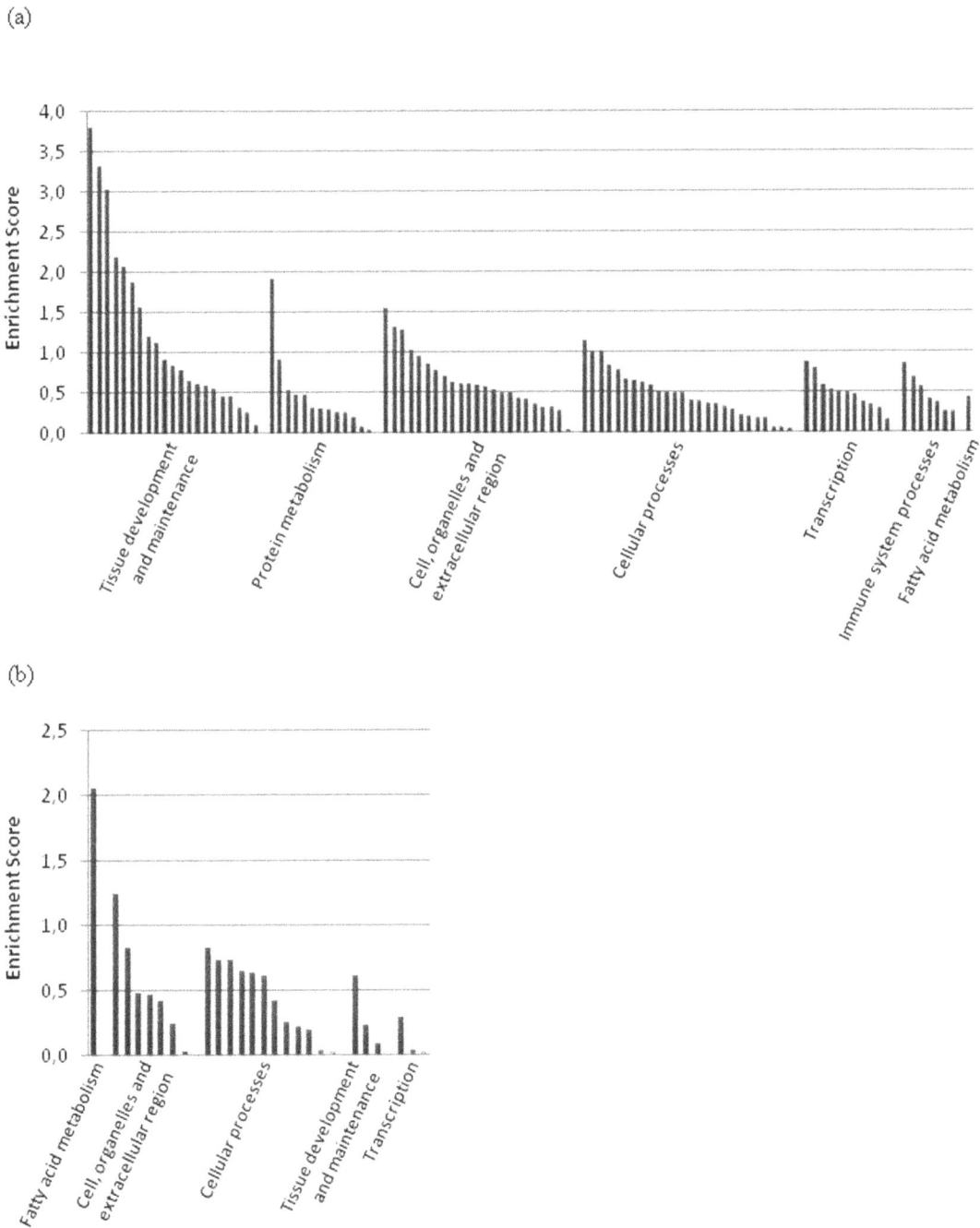

Figure 1. DAVID Functional Annotation Cluster (FAC) analysis of differentially expressed genes between chickens with and without 5 g of inulin kg^{-1} diet supplementation. DAVID FAC analysis was conducted on two independent gene lists containing 95 up-regulated genes ($\geq$1.4-fold) and 35 down-regulated genes ($\leq$0.6-fold) and P$\leq$0.09. High stringency ease score parameters were selected, to indicate confident enrichment scores of functional significance and importance of the given pathways and processes investigated. (A) Grouped major FACs for up-regulated genes ($\geq$1.4-fold). (B) Grouped major FACs for down-regulated genes ($\leq$0.6-fold). Significance is determined by corresponding enrichment scores.

PCR. To avoid bias, the tested genes were chosen from an array of different processes including biological regulation (Deiodinase iodothyronine type II, *DIO2*), biologic and metabolic process (Inter-alpha globulin inhibitor H5, *ITIH5*), immune system processes (IMAP family member 5, *GIMAP5*), protein metabolism (Ubiquitin Specific Protease 18, *USP18*), gene transcription and regulation processes (Coiled-coil domain containing protein 79 CCD79, *CCDC79*), and cell development and maintenance

processes (Inositol 1,4,5-trisphosphate receptor interacting protein, similar to *KIAA1754*-like) FAC groups. *HPRT*, initially chosen as reference gene, failed as housekeeping gene for the analysis of gene expression in liver tissue in broiler chickens due to its significant differential expression when analyzed with qRT-PCR, highlighting the importance of the correct selection of reference genes [38].

Among the genes showing differential expression, it is worth highlighting *ITIH5*, which explained 30% of the total variability found between the two feeding groups, *DIO2* -39%-, *GIMAP5* − 26%-, and *HPRT* −49%-. *ITIH5* encodes a secreted protein and is known to be highly expressed in subcutaneous adipose tissue, and associated with measures of body size and metabolism [39]. The protein encoded by *DIO2* belongs to the iodothyronine deiodinase family and activates thyroid hormone, which acts on nearly every cell in the organism -increases the basal metabolic rate, affects protein synthesis, helps regulate long bone growth and neural maturation-, being essential to proper development and differentiation of all cells types and the regulation of protein, fat, and carbohydrate metabolism [40]. KIAA1754 is an intracellular channel protein that mediates calcium (Ca^{2+}) release from the endoplasmic reticulum, and is involved in many biological processes (e.g. fertilization, muscle contraction, secretion, cell growth, differentiation, apoptosis, and synaptic plasticity) [41]. The protein encoded by *HPRT* gene plays a central role in the generation of purine nucleotides through the purine salvage pathway [42]. Thereof, the up-regulation of *ITIH5*, *DIO2*, *KIAA1754* and *HPRT* by inulin supplementation is in agreement with the higher basal activity required by an increased body weight gain reported in chickens by Rebolé *et al.* [3] and Velasco *et al.* [4].

GIMAP5 encodes a protein belonging to the GTP-binding superfamily and to the immune-associated nucleotide (IAN) subfamily of nucleotide-binding proteins, and has been implicated in autoimmune diseases, lymphocyte homeostasis and apoptosis [43].

Down-regulation of *USP18* gene, which belongs to the ubiquitin-specific proteases (UBP) family of enzymes, and *CCDC79*, one of the principal subunit oligomerization motifs in proteins, in inulin supplemented chickens was not validated by qRT-PCR. While qRT-PCR results are usually accurate and gene specific, it is possible that the microarray hybridization results can be biased for genes encoded by multigene families.

Conclusion

As far as we have notice, this is the first report of a microarray based gene expression study on the effect of inulin supplementation in any animal species. The results obtained here highlighted the functional significance and importance of inulin supplementation on processes and pathways that lead to an increase in growth and performance, while reinforcing the immune status of chickens, and fostering the production of long chain fatty acids in broilers supplemented with 5 g of inulin kg^{-1} diet. Additional information on the molecular mechanism underlying the inulin effects on gene activity and the cellular basis of feed efficiency in broilers is also provided. This nutrigenomic study supports further research on the supplementation of chicken diets with the prebiotic inulin at 5g kg^{-1} diet as a possible and useful alternative to the use of antibiotics for improving animal production and general immunity in poultry farming, along with a healthier meat lipid profile.

References

1. Roberfroid MB (2001) Functional foods: concepts and strategy. J Pharm Belg 56: 43–44.
2. Glibowski P, Bukowska A (2011) The effect of ph, temperature and heating time on inulin chemical stability. Acta Sci Pol, Technol Aliment 10: 189–196.
3. Roberfroid M, Gibson GR, Hoyles L, McCartney AL, Rastall R, et al. (2010) Prebiotic effects: metabolic and health benefits. Br J Nutr 104: S1–S63.
4. Rebolé A, Ortiz LT, Rodríguez ML, Alzueta C, Treviño J, et al. (2010) Effects of inulin and enzyme complex, individually or in combination, on growth performance, intestinal microflora, cecal fermentation characteristics, and jejunal histomorphology in broiler chickens fed a wheat- and barley-based diet. Poult Sci 89: 276–286.

Supporting Information

Table S1   Reference and selected target genes used in the Real-time PCR assay indicating name, GenBank accession number or reference (Accession), and primers used for the expression study.

Table S2   One hundred and twelve up-regulated genes (≥ 1.4-fold) and 46 down-regulated genes (≤ 0.6-fold) showing a *P-value* ≤ 0.09 when using an Affymetrix GeneChip Chicken Genome Array in RNA from chicken broilers supplemented with inulin.

Table S3   Gene list report and complete results for the 95 up-regulated and 35 down-regulated genes for which annotation information was available at DAVID Bioinformatics Resources 6.7 (http://david.abcc.ncifcrf.gov/) **in August 2013, including the corresponding DAVID scores (*P values*) and the lists of genes in every significant category.** A summary of the Functional Annotation Clusters (FAC) is also shown from page 108 to 112.

Table S4   List of differentially expressed genes encoding proteins of known function but not included in the DAVID FAC analysis, and genes encoding proteins of unknown function.

Figure S1   Flow diagram of study design and results.

Figure S2   KEGG pathway maps of chicken differentially expressed genes involved in common pathways and processes. KEGG pathway tools were utilized through DAVID online tools and the analysis were conducted on two independent gene lists containing 95 up-regulated genes (≥ 1.4-fold) and 35 down-regulated genes (≤ 0.6-fold) and P ≤ 0.09. (a) Adipocytokine Signaling Pathway; (b) Glycosphingolipid Biosynthesis - Ganglio Series; (c) Glutathione Metabolism; (d) Drug Metabolism - Cytochrome P450; (e) Metabolism of Xenobiotics by Cytochrome P450. Red boxes indicate up-regulated and down-regulated homologs. All chicken homologs identified on the KEGG maps are shown in Table 2.

Author Contributions

Contributed to the writing of the manuscript: NS SD. Designed and conceived the study: JC SD. Carried out real-time PCR validation: FB. Involved in the breeding aspects of the experiment: SV AR MLR LO. Performed the gene ontology and visual pathway analysis: NS. Contributed to the analysis of data: SD JC.

5. Velasco S, Ortiz LT, Alzueta C, Rebolé A, Treviño J, et al. (2010) Effect of inulin supplementation and dietary fat source on performance, blood serum metabolites, liver lipids, abdominal fat deposition, and tissue fatty acid composition in broiler chickens. Poult Sci 89: 1651–1662.
6. Tacchi L, Bickerdike R, Douglas A, Secombes CJ, Martin SA (2011) Transcriptomic responses to functional feeds in Atlantic salmon (*Salmo salar*). Fish Shellfish Immunol 31: 704–715.
7. Ortiz LT, Rebolé A, Velasco S, Rodríguez ML, Treviño J, et al. (2013) Effects of inulin and fructooligosaccharides on growth performance, body chemical composition and intestinal microbiota of farmed rainbow trout (*Oncorhynchus mykiss*). Aquac Nutr 19: 475–482.

8. Ammerman E, Quarkes C, Twining PV (1989) Evaluation of fructooligosaccharides on performance and carcass yield of male broilers. Poult Sci 68: 167.

9. Yusrizal Y, Chen TC (2003) Effect of adding chicory fructans in feed on broiler growth performance, serum cholesterol and intestinal length. Int J Poult Sci 2: 214–219.

10. Delzenne NM, Daubioul C, Neyrinck A (2002) Inulin and oligofructose modulate lipid metabolism in animals: Review of biochemical events and future prospects. Br J Nutr 87: S255–S259.

11. Letexier D, Diraison F, Beylot M (2003) Addition of inulin to a high carbohydrate diet reduces hepatic lipogenesis and plasma triacylglycerol concentration in humans. Am J Clin Nutr 77: 559–564.

12. Bosscher D, Loo JV, Franck A (2006) Inulin and oligofructose as functional ingredients to improve bone mineralization. Int Dairy J 16: 1092–1097.

13. Roberfroid M, Buddington R K (2007) Inulin and oligofructose: proven health benefits and claims. J Nutr 137: S2489–S2597.

14. Lomax AR, Calder PC (2009) Prebiotics, immune function, infection and inflammation: a review of the evidence from studies conducted in humans. Curr Pharm Des 15: 1428–1518.

15. Boletín Oficial del Estado (2005) Real Decreto 1201/2005 sobre protección de los animales utilizados para experimentación y otros fines científicos. BOE 252: 34367–34391.

16. NRC (1994) Nutrient Requirements of Poultry. 9th rev. ed. National Academic Press, Washington, DC.

17. Medina I, Carbonell J, Pulido L, Madeira SC, Goetz S, et al. (2010) Babelomics: an integrative platform for the analysis of transcriptomics, proteomics and genomic data with advanced functional profiling. Nucleic Acids Res 38: W210-3.

18. Smyth GK (2004) Linear models and empirical Bayes methods for assessing differential expression in microarray experiments. Stat Appl Genet Mol Biol 3: 3.

19. Hochberg Y, Benjamini Y (1990) More powerful procedures for multiple significance testing. Stat Med 9: 811–818.

20. Huang DW, Sherman BT, Lempicki RA (2009) Systematic and integrative analysis of large gene lists using DAVID Bioinformatics Resources. Nature Protoc 4: 44–57.

21. Vandesompele J, De Preter K, Pattyn F, Poppe B, Van Roy N, et al. (2002) Accurate normalization of real-time quantitative RT-PCR data by geometric averaging of multiple internal control genes. Genome Biol 3: RESEARCH0034.

22. Andersen C, Jensen J, Orntoft T (2004) Normalization of real-time quantitative reverse transcription-PCR data: a model-based variance estimation approach to identify genes suited for normalization, applied to bladder and colon cancer data sets. Cancer Res 64: 5245–5250.

23. Pfaffl M, Tichopad A, Prgomet C, Neuvians TP (2004) Determination of stable housekeeping genes, differentially regulated target genes and sample integrity: BestKeeper—Excel-based tool using pair-wise correlations. Biotechnol Lett 26: 509–515.

24. SAS: Statistical Analysis with SAS/STAT Software V9.1. SAS Institute Inc 2009.

25. Pfaffl MW, Horgan GW, Dempfle L (2002) Relative expression software tool (REST) for group-wise comparison and statistical analysis of relative expression results in real-time PCR. Nucleic Acids Res 30: 36.

26. Dalman MR1, Deeter A, Nimishakavi G, Duan ZH (2012) Fold change and p-value cutoffs significantly alter microarray interpretations. BMC Bioinformatics 13: S11.

27. Xu ZR, Hu CH, Xia MS, Zhan XA, Wang MQ (2003) Effects of dietary fructooligosaccharide on digestive enzyme activities, intestinal microflora and morphology of male broilers. Poult Sci 82: 1030–1036.

28. Pelicano ERL, Souza PA, Souza HBA, Figueiredo DF, Biago MM, et al. (2005) Intestinal mucosa development in broiler chickens fed natural growth promoters. Braz J Poult Sci 7: 221–229.

29. Rehman H, Rosenkranz C, Bölem J, Zentek J (2007) Dietary inulin affects the morphology but not the sodium dependent glucose and glutamine transport in the jejunum of broilers. Poult Sci 86: 118–122.

30. Hume ME (2011) Historic perspective: prebiotics, probiotics, and other alternatives to antibiotics. Poult Sci 90: 2663–2669.

31. Cogliani C, Goossens H, Greko C (2011) Restricting antimicrobial use in food animals: lessons from Europe. Microbe 6: 274–279.

32. Wammes LJ, Wiria AE, Toenhake CG, Hamid F, Liu KY, et al. (2013) Asymptomatic plasmodial infection is associated with increased tumor necrosis factor receptor II-expressing regulatory T cells and suppressed type 2 immune responses. J Infect Dis 207: 1590–1599.

33. Gervois P, Mansouri RM (2012) PPARα as a therapeutic target in inflammation-associated diseases. Expert Opin Ther Targets 16: 1113–1125.

34. Wu G, Fang YZ, Yang S, Lupton JR, Turner ND (2004) Glutathione metabolism and its implications for health. J Nutr 134: 489–492.

35. Mateo H (2007) Estudio de la superóxido dismutasa en Trypanosoma crucy. Tesis doctoral. Instituto de Biotecnología. Universidad de Granada.

36. Simopoulos AP (1999) Essential fatty acids in health and chronic disease. Am J Clin Nutr 70: 560S–569S.

37. Leonard AE, Pereira SL, Sprecher H, Huang YS (2004) Elongation of long-chain fatty acids. Prog Lipid Res 43: 36–54.

38. Pérez P, Tupac-Yupanqui I, Dunner S (2008) Evaluation of suitable reference genes for gene expression studies in bovine muscular tissue. BMC Mol Biol 9: 79.

39. Anveden Å, Sjöholm K, Jacobson P, Palsdottir V, Walley AJ, et al. (2012) ITIH-5 expression in human adipose tissue is increased in obesity. Obesity 20: 708–714.

40. Galton VA (2005) The roles of the iodothyronine deiodinases in mammalian development. Thyroid 15: 823–834.

41. Berridge MJ (1993) Inositol trisphosphate and calcium signalling. Nature 361: 315–325.

42. Mastrangelo L, Kim JE, Miyanohara A, Kang TH, Friedmann T (2012) Purinergic signaling in human pluripotent stem cells is regulated by the housekeeping gene encoding hypoxanthine guanine phosphoribosyltransferase. Proc Natl Acad Sci USA 109: 3377–3382.

43. Barnes MJ, Aksoylar H, Krebs P, Bourdeau T, Arnold CN, et al. (2010) Loss of T cell and B cell quiescence precedes the onset of microbial flora-dependent wasting disease and intestinal inflammation in Gimap5-deficient mice. J Immunol 184: 3743–3754.

iTRAQ-Based Proteomics Reveals Novel Members Involved in Pathogen Challenge in Sea Cucumber *Apostichopus japonicus*

Pengjuan Zhang[1], Chenghua Li[1]*, Peng Zhang[1], Chunhua Jin[1], Daodong Pan[1], Yongbo Bao[2]

1 Department of aquaculture, Ningbo University, Ningbo, Zhejiang Province, P.R China, 2 Department of Aquatic Germplasm Resources, Zhejiang Wanli University, Ningbo, Zhejiang Province, P.R China

Abstract

Skin ulceration syndrome (SUS) is considered to be a major constraint for the stable development of *Apostichopus japonicus* culture industries. In this study, we investigated protein changes in the coelomocytes of *A. japonicus* challenged by *Vibrio splendidus* using isobaric tags for relative and absolute quantification (iTRAQ) over a 96 h time course. Consequently, 228 differentially expressed proteins were identified in two iTRAQs. A comparison of the protein expression profiles among different time points detected 125 proteins primarily involved in response to endogenous stimuli at 24 h. At 48 h, the number of differentially expressed proteins decreased to 67, with their primary function being oxidation reduction. At the end of pathogen infection, proteins responsive to amino acid stimuli and some metabolic processes were classified as the predominant group. Fifteen proteins were differentially expressed at all time points, among which eight proteins related to pathologies in higher animals were shown to be down-regulated after *V. splendidus* infection: paxillin, fascin-2, aggrecan, ololfactomedin-1, nesprin-3, a disintegrin-like and metallopeptidase with thrombospondin type 1 motif (Adamts7), C-type lectin domain family 4 (Clec4g) and n-myc downstream regulated gene 1 (Ndrg1). To gain more insight into two SUS-related miRNA (miR-31 and miR-2008) targets at the protein level, all 129 down-regulated proteins were further analyzed in combination with RNA-seq. Twelve and eight proteins were identified as putative targets for miR-31 and miR-2008, respectively, in which six proteins (5 for miR-31 and 1 for miR-2008) displayed higher possibilities to be regulated at the level of translation. Overall, the present work enhances our understanding of the process of *V. splendidus*-challenged sea cucumber and provides a new method for screening miRNAs targets at the translation level.

Editor: Kenneth Söderhäll, Uppsala University, Sweden

Funding: This work was financially supported by NSFC (31101919, 4127610), Zhejiang Provincial Natural Science Foundation of China, the young academic leaders in colleges and universities in Zhejiang province (pd2013099), the Natural Science foundation of Ningbo (2013C10013), and the K.C. Wong Magna Fund at Ningbo University. The funders had no role in study design, data collection and analysis, decision to publish, or preparation of the manuscript.

Competing Interests: The authors have declared that no competing interests exist.

* Email: lichenghua@nbu.edu.cn

Introduction

Sea cucumber *Apostichopus japonicus*, famous for its superior nutritive value and supposed medicinal properties, has become one of the most important aquaculture species in China, and its aquaculture has grown rapidly since the 1980s. However, the intensification and rapid expansion of *A. japonicus* farming has also led to the occurrence of various diseases [1].Skin ulceration syndrome (SUS) is one of the most common diseases with a high mortality of 90%-100% and has become a limiting factor in the sustainable development of this industry [2]. Some reports have demonstrated that the pathogens responsible for the outbreak of skin ulceration include aspherical virus [3], *Vibrio splendidus* [4] and *Pseudomonas spp* [5]. Among these pathogens, *V. splendidus* has been widely accepted as one of the major pathogens. To date, many efforts have been made to study the pathogenic progress of SUS outbreak caused by *V. splendidus*, but the intrinsic mechanism still requires further investigation.

MicroRNAs (miRNAs) are a type of approximately 22 nt-sized small noncoding RNAs that exist in diverse organisms [6]. It has been confirmed that in plants, miRNA-target interactions are often within the coding region and nearly perfectly complementary, which triggers mRNA cleavage. By contrast, animal miRNA/target duplexes generally are interrupted by gaps and mismatches and occur in the 3′UTR of mRNAs. Pioneering genetic studies in *Caenorhabditis elegans* have identified that lin-4 can inhibit the translation of lin-14 and lin-28 mRNA without affecting the cellular level of lin-14 and lin-28 [7][8]. Recently, miRNAs have emerged as key regulators of a broad spectrum of cellular activities, including immune response [9], insulin secretion [10], and viral replication [11]. In miRNA research, the identification of the targets of individual miRNAs is of utmost importance. Our understanding of the molecular mechanisms by which individual miRNAs modulate cellular functions remains incomplete until a full set of miRNA targets is identified and validated. Since miRNAs can function through partial or full complementary base pairing to their target messenger RNAs (mRNAs), resulting in translational inhibition or mRNA degradation [12], it seems to be feasible to identify the putative targets of miRNAs by detecting the cellular level of mRNAs or proteins from the same biological sample. In our previous work, miR-31 and miR-2008 were demonstrated to be involved in SUS outbreak [13], and their candidate targets were also predicted by RNA-seq analysis by miRanda toolbox based on the reverse expression pattern between miRNA and mRNA [14]. However, negative expression correlations were not detected by qRT-PCR for most of their putative targets at the mRNA level in bacteria challenged samples, indicating that the targets might be regulated at the protein level without affecting mRNA abundance. Therefore, proteomics is a suitable method to reveal the full spectrum of miRNA targets and quantify the contribution of translational repression by miRNAs because it provides a rapid and comprehensive evaluation of protein profiles in complex protein samples. Over the past few decades, many proteomic platforms have been developed for the qualitative and quantitative characterization of protein mixtures and post-translational modifications, such as 2D gel-MS [15], LC MS/MS [16]. Currently, isobaric tags for relative and absolute quantification (iTRAQ) has unique advantages over other conventional proteomics techniques because iTRAQ identifies and quantifies many proteins from specific biological environments using labeled peptides identifiable by sensitive mass spectrometers [17]. iTRAQ analysis is further strengthened by using robust bioinformatic tools and statistical analyses to support observations [18]. Using the infection model and iTRAQ approach, many researchers have made great advances in identifying proteins involved in the pathogenic process [19] [20]. Moreover, the iTRAQ approach has also been successfully employed for the identification of miRNA targets in other species [21] [22]. However, the identification of proteins by tandem mass spectrometry requires reference protein databases that are only available for model species. Because of the recent contributions to the transcriptomic characterization of the *A. japonicus* immune system, including transcriptome analysis of diseased [14] and LPS-challenged *A. japonicus* [23], a protein database for mass spectrometry-based identification of non-model organisms has been generated. Meanwhile, NMR-based metabonomics have been performed to explore the metabolic changes in the muscle tissues of pathogen-challenged and diseased *A. japonicus*, providing a comparative understanding of the metabolic profiles under different conditions [24]. To understand the intrinsic pathogenic mechanism, we analyzed protein expression patterns over a 96 h time course pathogen infection to identify proteins and peptides changes in response to *V. splendidus* challenge and reveal miRNA targets at the translational level, thus increasing our knowledge of cellular pathways important for infection and pathogenesis.

Materials and Methods

Ethics statement

The sea cucumbers (*A. japonicus*) here are commercially cultured animals, and all the experiments were conducted in accordance with the recommendations in the Guide for the Care and Use of Laboratory Animals of the National Institutes of Health. The study protocol was approved by the Experimental Animal Ethics Committee of Ningbo University, China.

Experimental animals and conditions

One hundred healthy adult sea cucumbers *A. Japonicus* (165±23 g) were obtained from Bowang Aquaculture Company (Ningbo, China) and evenly assigned to four tanks randomly. The animals were then acclimatized in aerated natural seawater

(salinity 25 psu, temperature 16°C) for three days prior to be treated. *V. splendidus* were initially isolated from a skin ulceration diseased sea cucumbers and identified by 16S rRNA. The confirmed bacteria were cultured in liquid 2216E broth (Tryptone 5 g L-1, yeast extract 1 g L-1, pH 7.6) at 28°C, 140 rpm and centrifuged at 1,000 g for 5 min to harvest the bacteria. Live *V. splendidus* were then re-suspended in filtered seawater (FSW). For the challenge experiments, one tank used as a control, and the other three tanks were immersed with high density of *V. splendidus* with a final concentration of 10^7 CFU mL^{-1}. After being challenged for 24 h, the coelomocytes were collected from control and challenged group to confirm the exist of V. splendidus by 16S rRNA PCR. The sea cucumbers were then dissected and coelomic fluids were collected from five individuals in each tank, and approximately 50 ml of coelomic fluids were gathered at 0, 24, 48 and 96 h, respectively. Of which, 45 mL was severed as sample for iTRAQ analysis, and 5 mL for RNA quantify. The coelomic fluids was then centrifuged at 1,000 g for 5 min to harvest the coelomocytes and the coelomocytes were then stored at −80°C before protein and RNA extraction extraction.

Protein extraction, quantization, digestion and iTRAQ labeling

Two biological replicates of each group were prepared for the proteomics experiments. Briefly, the total protein of each sample was grinded to powder with liquid nitrogen and dissolved in lysis solution [9 M Urea, 4% CHAPS, 1%DTT, 1%IPG buffer (GE Healthcare)]. The mix was incubated at 30°C for 1 hour and centrifuged at 15,000 g for 15 min at room temperature. The supernatant was collected and quantified by the Bradford method [25].

For each sample, 100 μg of protein was dissolved in a dissolution buffer (AB Sciex, Foster City, CA, USA). After being reduced, alkylated and trypsin-digested, the samples were labeled following the manufacturer's instructions for the iTRAQ Reagents 8-plex kit (AB Sciex). Samples taken at 0, 24, 48 and 96 h were each labeled with iTRAQ reagents with molecular masses of 113, 114, 115, and 116 Da, respectively. Additional independent biological replicates were labeled with other reagents with molecular masses of 117, 118, 119, and 121 Da. After labeling, all samples were pooled and purified using a strong cation exchange chromatography (SCX) column by Agilent 1200 HPLC (Agilent) and separated by liquid chromatography (LC) using a Eksigent nanoLC-Ultra 2D system (AB SCIEX). The LC fractions were analyzed using a Triple TOF 5600 mass spectrometer (AB SCIEX). Mass spectrometer data acquisition was performed with a Triple TOF 5600 System (AB SCIEX, USA) fitted with a Nanospray III source (AB SCIEX, USA) and a pulled quartz tip as the emitter (New Objectives, USA). Data were acquired using an ion spray voltage of 2.5 kV, curtain gas of 30 PSI, nebulizer gas of 5 PSI, and an interface heater temperature of 150°C. For information dependent acquisition (IDA), survey scans were acquired in 250 ms and as many as 35 product ion scans were collected if they exceeded a threshold of 150 counts per second (counts/s) with a 2^+ to 5^+ charge-state. The total cycle time was fixed to 2.5 s. A rolling collision energy setting was applied to all precursor ions for collision-induced dissociation (CID). Dynamic exclusion was set for ½ of peak width (18 s), and the precursor was then refreshed off the exclusion list.

Protein Identification and Quantification

The iTRAQ data were processed with Protein Pilot Software v4.0 against the *A. japonicus* database using the Paragon algorithm [26]. Protein identification was performed with the search option of emphasis on biological modifications. The database search parameters were the following: the instrument was TripleTOF 5600, iTRAQ quantification, cysteine modified with iodoaceta-mide; biological modifications were selected as ID focus, trypsin digestion. For false discovery rate (FDR) calculation, an automatic decoy database search strategy was employed to estimate FDR using the PSPEP (Proteomics System Performance Evaluation Pipeline Software, integrated in the ProteinPilot Software). The FDR was calculated as the number of false positive matches divided by the number of total matches. Then, the iTRAQ was chosen for protein quantification with unique peptides during the search, and peptides with global FDR values from fit less than 1% were considered for further analysis. Within each iTRAQ run, differentially expressed proteins were determined based on the ratios of differently labeled proteins and p-values provided by Protein Pilot; the p-values were generated by Protein Pilot using the peptides used to quantitate the respective protein. Finally, for differential expression analysis, fold change was calculated as the average ratio of 114/113 and 118/117 at 24 h, 115/113 and 119/117 at 48 h, 116/113 and 121/117 at 96 h, respectively. and proteins with a fold change of >1.5 or <0.67 and p value less than 0.05 were considered to be significantly differentially expressed.

GO and KEGG pathway enrichment analysis

For GO and KEGG pathway enrichment analysis, the homology search was first performed for all query protein matches with blastp against the *Mus musculus* protein database. The E-value was set to < 1e-10, and the top 10 best hits for each query sequence were taken. Among the 10 best hits, the hit with the best identity to the query was picked as homologous. The GO analysis was performed with different mapping steps to link all blast hits to the functional information stored in the Gene Ontology database using the DAVID toolkit. Public resources such as NCBI, PIR and GO are used to create links with protein IDs and corresponding gene ontology information. All annotations are associated to an evidence code that provides information regarding the quality of this functional assignment. For KEGG Pathway enrichment analysis, the pathway enrichment was performed using annotated proteins in the query dataset against the KEGG database, and Cytoscape was used for protein-protein interaction network construction.

miRNA target prediction

miRNA target prediction was performed only for down-regulated proteins. The corresponding mRNA sequences were acquired from RNA-seq analysis [14]. Computational identification of miR-31 and miR-2008 targets was performed using the miRanda toolbox to search complementary regions between miRNA and the 3′UTR of mRNA with default parameters of S>90 (single-residue pair scores) and $\Delta G < -17$ kal/mol.

miRNA and RNA quantification analysis

Total RNA was extracted with the RNAiso plus reagent (Takara, Japan) following the manufacturer's instructions. The SYBR green qRT-PCR assay was used for miRNA and mRNA quantification in identical samples. In brief, 500 ng RNA containing miRNAs was polyadenylated by poly(A) polymerase and converted to cDNA by reverse transcriptase using the miScript Reverse Transcription Kit (Qiagen, Germany). For miRNA analysis, the qRT-PCR was performed using the miScript SYBR Green PCR kit (Qiagen, Germany) with the manufacturer-provided miScript Universal primer and the miRNA-specific forward primers (for mRNA, the primers were gene-specific

forward and reverse primers) in a Rotor-Gene Q 6000 Real-time PCR detection system (Qiagen, Germany).

The miRNA-specific primers were designed based on the miRNA sequences obtained from [13] and the gene-specific primers were designed based on the mRNA sequences obtained from [14]. All the primers used in qRT-PCR were shown in table 1. Each reaction was performed in a final volume of 20 μl containing 2 μl of the cDNA, 10 μM of each primer, 6 μl RNase-free water and 10 μl SYBR Green PCR Master mix (Qiagen). The amplification profile was: denaturation at 94°C for 15 min, followed by 40 cycles of 94°C for 15 s, 60°C for 30 s and 70°C for 30 s, in which fluorescence was acquired. At the end of the PCR cycles, melting curve analyses were performed. Each sample was run in triplicates for analysis. The expression levels of miRNAs were normalized to RNU6B[27], and the expression levels of mRNAs were normalized to 18S rRNA[28].

The $2^{-\Delta\Delta Ct}$ method was used to analyze the expression levels of both miRNA and mRNA, and the obtained values represented the n-fold difference relative to the control (untreated samples). The data are presented as relative expression levels (means±S.D, n = 3), and all experimental data were subjected to one-way Analysis of Variance (ANOVA) followed by multiple Duncan tests to determine differences in the mean values among the controls. Significant differences between the treated and corresponding control groups at each time point are indicated with one asterisk for P<0.05 and two asterisks for P<0.01. The error bars in the graphs represent standard deviations.

Results

Overview of changes in host proteome induced by *V. splendidus* infection

Compared with the control group, the *V. splendidus* were detected in the challenged groups. And the samples were then used for iTRAQ analysis. In conclusion, we describe global proteome changes in the *A. japonicus* coelomocytes during the short-term course of *V. splendidus* infection. We identified 2793 distinct proteins, of which 2049 were identified and quantified reliably at a global false discovery rate (FDR) of 1%. Compared with the control group, 228 identified proteins had significant changes in expression at different time points, among which 15 were observed to be differentially expressed at all examined time points (Fig. 1). A total of 99 proteins displayed increased expression trends (fold change >1.5, p≤0.05), and 129 proteins displayed decreased expression levels (fold change <0.67, p≤0.05) compared with the control group (Table S1). Notably, eight proteins related to pathology in higher animals exhibited down-regulated expression after *V. splendidus* infection: paxillin, fascin-2, nesprin-3, aggrecan, ololfactomedin-1, a disintegrin-like and metallopeptidase with thrombospondin type 1 motif (Adamts7), c-type lectin domain family 4 (Clec4g) and N-myc downstream regulated gene 1 (Ndrg1).

Proteomic analysis at each time point after pathogen infection

Among these differentially expressed proteins, there were 125, 67 and 114 proteins identified in pathogen-challenged *A. japonicus* at 24, 48 and 96 h, respectively, compared with the control group. Of these, 67, 30, and 67 proteins were uniquely identified at 24, 48, 96 h post infection, respectively (Table S2). Some well-documented immune-related proteins were also included into this group. Glutathione S-transferase were uniquely expressed at 24 h and sharply increased to 3.587-fold compared to control group. However, ficolin B and guanine nucleotide binding protein were

Table 1. Primers used in qRT-PCR.

Gene name	Forward primer (5'-3')	Reverse primer (5'-3')
Spu-miR-31	AGGCAAGATGTTGGCATAGCT	Qiagen miScript universal primer
Spu-miR-2008	ATCAGCCTCGCTGTCAATACG	
Actr3	CTACCATGTTCAGGGATTTCGG	GGTGAGAGATGACCTCTGTTTCG
Clec4g	CCAACGGGAACCAAACAAT	TCGCAAACGCCAAACCTAAC
Cndp2	CTGTCGGATGGTTGGGATACTG	CATTCAGCCAGGCAAGAACG
Coro6	GCCATACGCATTCGTCATTTG	GATTGTTTCTCCTTCCTCCTCC
Gnb2	GCAAACATTTACCGGACACGAG	CAAGTAGCAAACGGCCACTCC
H1f0	CCATCGCCGACCTGAATG	AGCCTGAAGGTGCCACTCG
Hnrnpa1	GAGGTGATTATGCCAATGTCTGC	CACCACCGCCTCTTCCTTCT
Hnrnpl	CACGCAAGTCATCCACAAAGTG	AAACATCCCTCCCATTCAGCC
HSP90b1	TACTCCGTTCTGGGTTCATGC	TCTTCCTCTTCTGGTTCCTCCTC
Hspa8	GTGCCAACCCATCATCACCA	ACCTGCACCTGGAAACCCTC
Kiaa0196	GTTGATTGGCGATACCAAGCAC	GACCTGTTCCCGCAGTTGAC
Myo10	GCCGTGGGTGCTAAGATGGT	GAGTTCTGCCTCCTGTGGACAAT
Naca	AACAGTTCCGACCACAGGAAC	CTCTTCCAACCCACCATCATC
Nnt	AGGTAAGCCTATGGCGATTGAG	CATCTCTGCGATTGGATACTGG
Plekhd1	AGGCCGAGGATTCCTTGTG	CTCTGCGGGAGATTGTGATGA
Rpl12	CTGTTGGTGGAGAAGTTGGAGC	CTGCCTGTTCTGGATGGTCAAC
Slc25a5	GCAGCTTACTTCGGGTTCTATG	GATCTTCCTCCAGCAGTCCAG
Tgds	CTCAAGGCTCAATCACCAAGC	CGACCAAGGGAGGATGTGTT
Yes1	CGAAGATTCCGAATACACCGC	AGGACACCATAGGACCACACGT

down-regulated their expression to 0.026-fold and 0.171-fold at 48 h and 96 h, respectively. More importantly, some apoptosis-related proteins, such as sarcoma oncogene (Src), vitronectin and vinculin, displayed time-dependent depressed expression (Table S2). Protein-protein interaction (PPI) analysis further indicated a diversified functional network of these novel proteins at different time points (Fig. 2).

GO and KEGG analysis of the differential expression proteins

GO enrichment demonstrated that 80, 44, and 26 protein categories were enriched in the Biological process (BP), Cellular component (CC) and Molecular function (MF) categories, respectively (Table S3). Among these, actin filament depolymerization, intracellular non-membrane-bounded organelle, and actin binding were the most abundant categories in BP, CC and MF, respectively. The differentially expressed proteins were predominately binding proteins involving in cytoskeleton organization processes at 24 h post infection. At 48 h, differentially expressed proteins were also primarily binding proteins involving in protein complex assembly for extracellular structure organization, and at 96 h, differentially expressed proteins were predominately proteins related to ribosomal assembly and regulating the actin cytoskeleton (Fig. 3). Further KEGG pathway enrichment revealed that these proteins were mainly involved in Focal adhesion and cytoskeleton regulation pathways (24 h); various sugar metabolisms and related to protein complex assembly (48 h); and ribosome, amino sugar and nucleotide sugar metabolism (96 h) (Fig. 4).

Target prediction of miR-31 and miR-2008

By integrating these results with our RNA-seq and miRanda toolbox analyses, 12 and 8 proteins were identified as putative targets for miR-31 and miR-2008 from the 129 down-regulated proteins, respectively (Table S4). Heat shock protein 90 (Hsp90b1) was the only common target for both of these miRNAs.

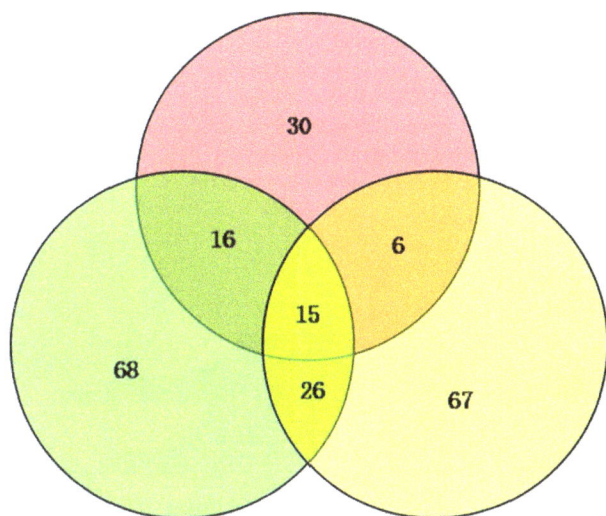

Figure 1. Changed proteome distribution between different time points, Venn diagram showing unique and shared proteins between time points.

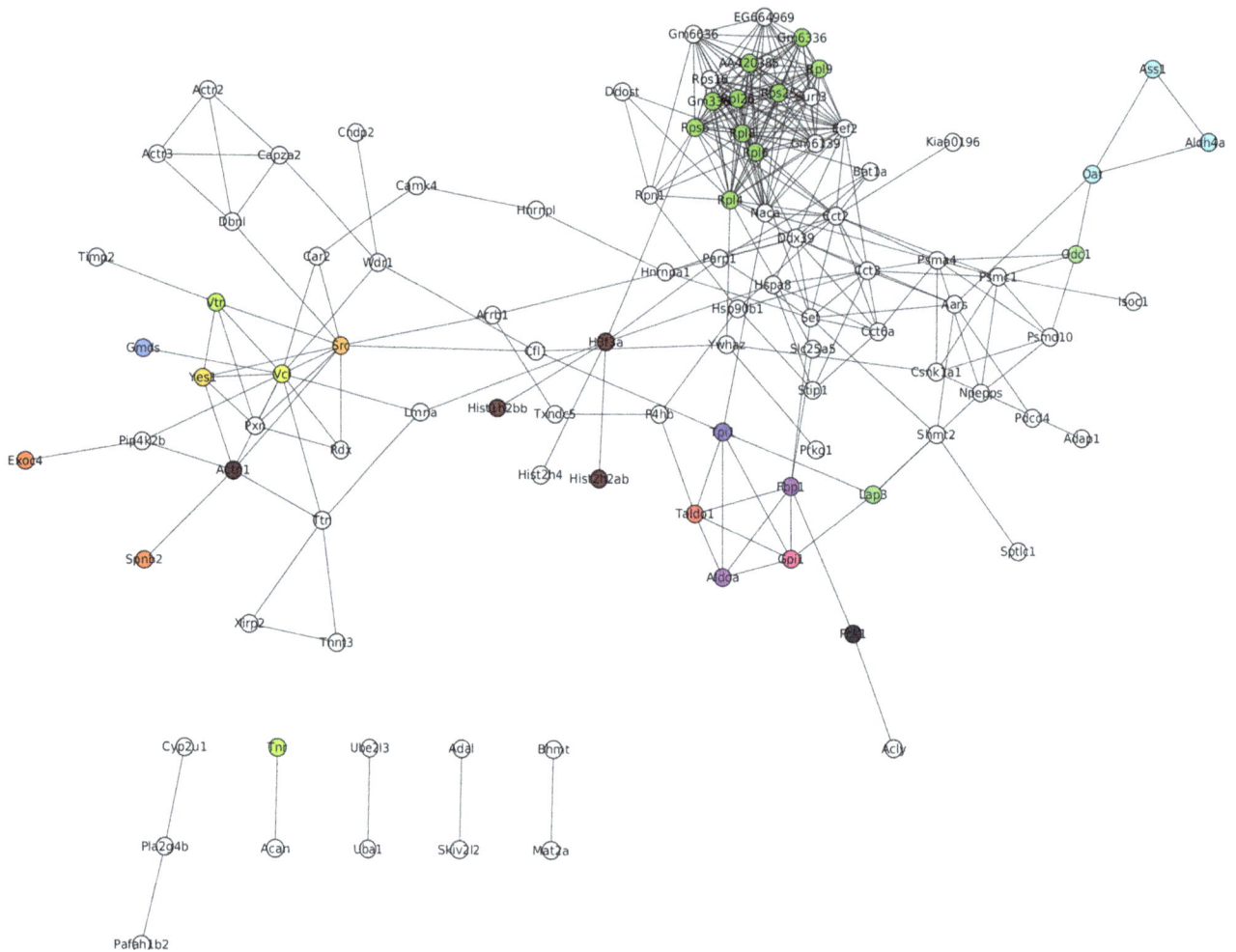

Figure 2. Protein-protein interactions were obtained by the string database. Interaction maps were created by cytoscape. The snapshot shows direct interactions found in these differentially expressed proteins from each time point.

Quantitative analysis of miRNAs and mRNAs

The expression patterns of these two miRNAs (miR-31 and miR-2008) and their 20 putative targets were examined by quantitative PCR and the results are shown in Fig. 5. miR-31 and miR-2008 were both significantly up-regulated after pathogen infection. For the targets of miR-31, five genes-WASH complex subunit strumpellin (Kiaa0196), myosin X (Myo10), TDP-glucose

4,6-dehydratase (Tgds), 60S ribosomal protein L12 (Rpl12) and H1 histone family, member 0 (H1f0)—displayed unchanged expression at the mRNA level. Guanine nucleotide binding protein beta 2 (Gnb2), nascent polypeptide-associated complex alpha (Naca), nicotinamide nucleotide transhydrogenase (Nnt) and heterogeneous nuclear ribonucleoprotein A1 (Hnrnpl) maintained constant expression of the miR-2008 targets.

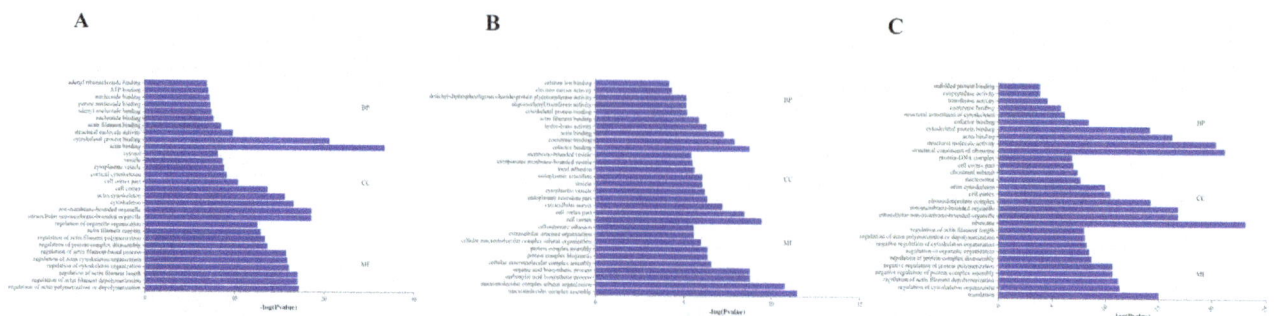

Figure 3. GO enrichment analysis of differentially expressed proteins at each time point. A: 24 h; B: 48 h; C: 96 h

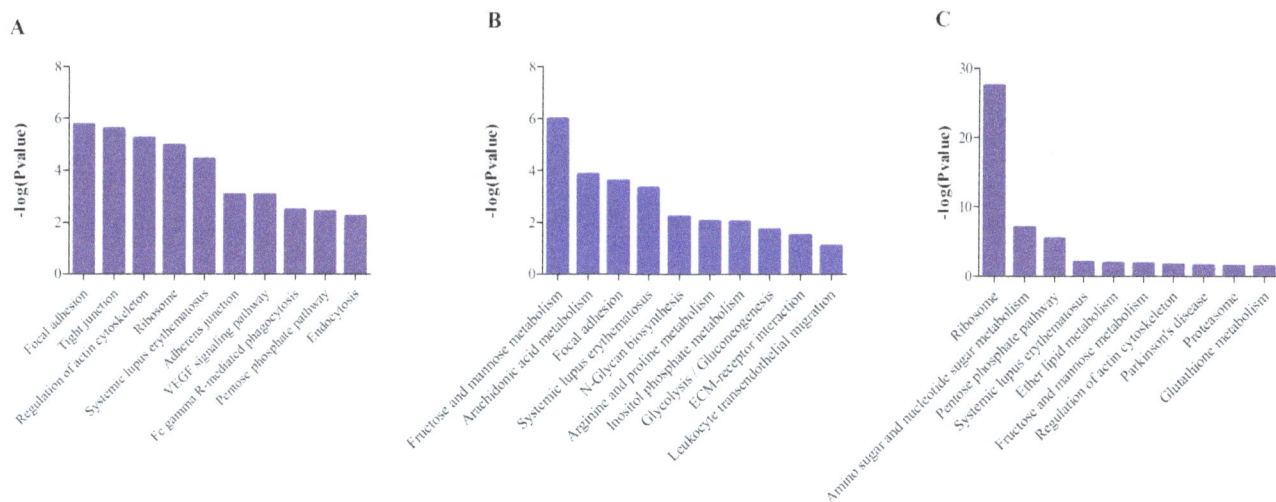

Figure 4. KEGG pathway enrichment analysis of differentially expressed proteins at each time point. A: 24 h; B: 48 h; C: 96 h.

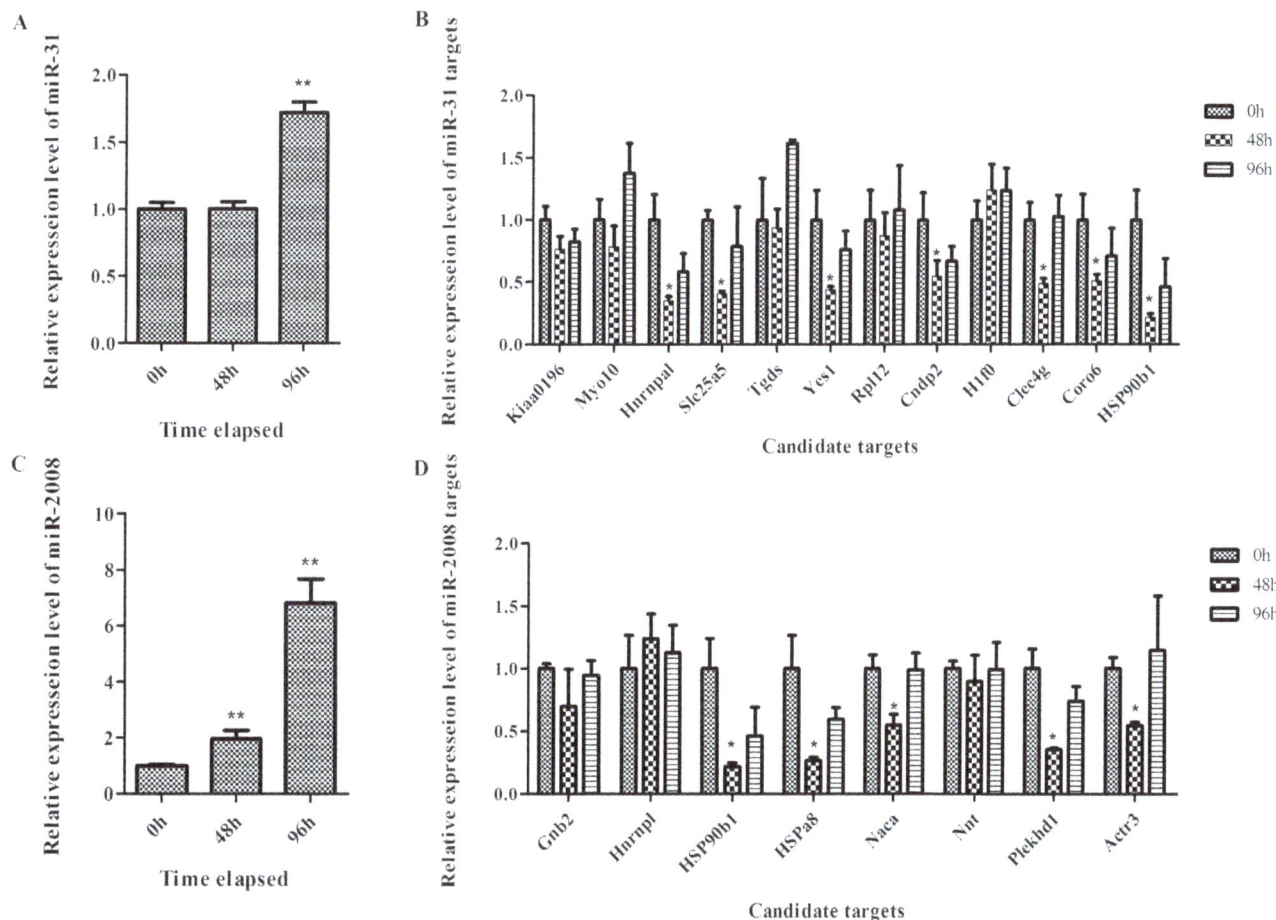

Figure 5. Time-course expression patterns of miRNAs and its putative targets of *A. japonicus* **upon** *V. splendidus* **treatment.** Hnrnpa1: Heterogeneous nuclear ribonucleoprotein L; Yes1: Yamaguchi sarcoma viral oncogene homolog 1; Cndp2: CNDP dipeptidase 2;Coro6: Coronin 6; Hspa8: Heat shock 70 kDa protein 8; Plekhd1: PH domain-containing family D member 1; Slc25a5: solute carrier family 25.

Discussion

Differentially expressed proteins between pathogen-challenged and control groups

In the current study, we used the iTRAQ method to screen proteins potentially involved in the host-pathogen interaction, and 228 proteins were identified as differentially expressed combined across all time points. However, contrary to our previous knowledge regarding immunogenic factors in invertebrates, most of these proteins have not been previously well studied. Functional annotation of these shared and unique proteins provides new insight into the pathophysiology of *V. splendidus* infection in *A. japonicus*. Of these 15 commom differential proteins in all examined time points, 13 were down-regulated (paxillin, aggrecan, Tgds, Adamts7, fascin-2, olfm-1, Clec4g, nesprin-3, Ndrg1, histone cluster 1 (Hist1h4f), histone cluster 1 (Hist1h2bb), mucin 5ac, H1 histone family) and 2 were up-regulated [titin and c-maf inducing protein (Cmip)]. Most of these down-regulated proteins are signal integration and transduction-related components and some are even involved in the activation of the immune response pathway. Given the commonalities of these shared proteins in pathogen-challenged samples, we propose that the down-regulation of these cell communication-related proteins may contribute to the immune evasion of *V. splendidus*.

Unique proteins revealed the temporal and spatial host-pathogen interaction progress

The biological roles of unique proteins identified from each comparison (Fig. 1) were further examined to understand the changes occurring at each time point. From the perspective of temporal progress of host-pathogen interaction, there was a function-biased conversion from 24 h to 96 h post-infection. In the early phase after *V. splendidus* infection (24 h), the unique proteins predominantly had actin or cytoskeletal protein binding activities and were involved in actin polymerization or depolymerization and cytoskeleton organization processes. At 48 h, the function of the uniquely differentially expressed proteins included calcium ion, vitamin or metal ion binding activities involved in oxidation-reduction and some metabolic processes. Finally, in the late phase of infection (96 h), the function of its unique proteins were transferase activities, exopeptidase activities and cofactor or coenzyme binding activities involved in translation, protein folding and some catabolic processes. The different distributions of functions among these unique proteins at each time point demonstrated a temporal process initiating from environmental information management to the final stress respondent.

On the other hand, diversified immune- and apoptosis-related proteins were also identified at each time points (Table S2), indicating that modulating immune system and apoptosis pathway were the utmost event in this host-pathogen interaction process. A number of effectors like lectin, lysozyme and Toll had been addressed to be involved into *V. splendidus* challenged sea cucumber. In the present study, diverse expression profiles of immune-related genes were detected between 24 h and other two time points. The up-regulation of Gstt1 at 24 h might result from the host immune response upon pathogen attack. As the GST family had been known as phase II detoxification enzymes which involved in response to pathogen attack, oxidative stress, and heavy-metal toxicity [29]. And the gene expression pattern of theta-class of GST in *Vibrio tapetis* and LPS challenged molluscan *Ruditapes philippinarum* also shown the similar tendency marked as the sharply raised expression level at 6 h and 24 h post infection [30]. The Ficolins were serum complement lectins which were able to recognize pathogen-associated molecular patterns (PAMPs) and trigger the activation of immune system by initiating activation of complement pathway or stimulating secretion of the inflammatory cytokines [31]. Another down-regulated immune related protein, Gnb2 belong to the G protein family, which function as G protein signaling to modulate cell motility and killing of invading pathogen. The significantly down-regulation of these two factors might represent negative modulation of the innate immune response as the insufficiency of Fcnb and Gnb2 might result in higher susceptibility to infection in individuals.

However, apoptosis related molecules like Src, vitronectin and vinculin were scarcely investigated in sea cucmber. Src is the canonical member of the non-receptor family of tyrosine kinases, which could inhibit apoptosis through the Erk1/2- dependent degradation of the death accelerator Bik [32]. Previous study also demonstrated that in the human primary macrophages, the Src tyrosine kinase activity were essential for the inflammasome activation during influenza A virus infection [33]. Vitronectin was considered to modulate neutrophil adhesion, chemotaxis, to contribute to neutrophil-associated proinflammatory processes [34], and to inhibit neutrophil apoptosis through activation of integrin-associated signaling pathways [35]. Vinculin was known to regulate extracellular signal-regulated kinase (ERK) by modulating the accessibility of paxillin for FAK interaction, thus controlling the survival and motility of cells which were critical to metastasis [36]. But recent study also shown that vinculin was emerging as a regulator of apoptosis and Shigella entry into host cells [37][38]. What is more important is that Pxn served as a "hub" protein for Src, vitronectin and vinculin, indicating a potential apoptosis related signal transportation role of paxillin under bacterial infectious conditions.

Metabolism and fuel usage during pathogen infection period

Recently, an NMR-based metabonomics study was conducted to explore the metabolic changes in muscle tissues of pathogen-challenged *A. japonicus*. This study observed that the pathogen did not induce obvious biological effects in *A. japonicus* samples 24 h after infection. Enhanced energy storage and immune responses were observed in *V. splendidus*-challenged *A. japonicus* samples at 48 h, marked by increased glucose and branched chain amino acids, respectively. Finally, infection of *V. splendidus* induced significant increases in the energy demand of *A. japonicus* samples at both 72 and 96 h, confirmed by decreased glucose and glycogen and increased ATP levels [24]. Because of technology restrictions, it is difficult to examine metabolic changes in pathogen-challenged coelomocytes. However, by reference to the metabonomics in the muscle tissues of *V. splendidus*-challenged samples and combined with our proteome analysis, we can infer metabolic changes in coelomocytes.

At 24 h post-infection, down-regulation of glucose phosphate isomerase-1 and transaldolase-1 in the pentose phosphate pathway will result in the accumulation of α-D-Glucose-6P and β-D-Fructose-6P and deficiency of D-Glyceraldehyde-3P, which may increase the concentration of pyruvate and inhibit the glycolysis pathway. Excessive pyruvate will promote the biosynthesis of leucine, isoleucine and valine, resulting in increased levels of glucose and branched amino acids. At 48 h post-infection, the up-regulation of triosephosphate isomerase-1 in the fructose and mannose metabolism pathway will increase the level of intracellular Glyceraldehyde-3P, which will accelerate the glycolysis pathway. Moreover, the up-regulation of ornithine aminotransferase and argininosuccinate synthetase-1 in arginine and proline metabolism will facilitate the urea cycle, which will further promote alanine metabolism. In addition, the typical decrease of protein synthesis-related proteins, such as Ribosomal proteins L4 and L8, that was observed at 96 h post infection may have resulted from decreased cellular amino acid levels and increased energy demand.

Coupling transcriptional and post-transcriptional miRNA regulation in host immune response

In our previous study, miR-31 and miR-2008 were identified from diseased individuals [13]. It has been widely accepted that animal miRNAs exert most of their silencing through the inhibition of translation, rather than through mRNA degradation of their targets, because of the low overall degree of sequence complementarity that animal miRNAs share with their target sites on 3′ UTRs of mRNAs[39]. Because miRNAs might result in translational inhibition without affecting the mRNA level, transcript-level approaches can miss certain targets. As a high-throughput protein platform, iTRAQ provided an easy, fast method to select certain candidate targets from the entire protein pool. Combined with expression and bioinformatic analyses, five target proteins for miR-31 and one for miR-2008 were successfully identified. Correlated miRNA and target protein expression will be investigated in our future work.

Conclusion remarkers

Overall, the resent work firstly implicated iTRAQ into addressing possible molecular events between host-pathogen interaction in *A. japonicus*. We identified several novel pathological related proteins with down-regulated expression profiles in all examined time points, which include c-maf induced proteins,

paxillin and several cytoskeleton related proteins. Also, diversified immune- and apoptosis-related proteins were also identified at each time points, revealing the temporal and spacial host-pathogen interaction progress during SUS outbreak. Importantly, six targets for our previous identified two miRNAs were also demonstrated to be regulated at translated level. The present work increase our knowledge on cellular pathways that are important for infection and pathogenesis in this non-model animals.

Supporting Information

Table S1 Summary of the total differential expressed proteins by iTRAQ.

Table S2 Common and unique expressed proteins at each times.

Table S3 GO enrichment of the total differential expressed proteins.

Table S4 Putative targets for miR-31 and miR-2008 from the down-regulated expressed proteins.

Author Contributions

Conceived and designed the experiments: CL CJ. Performed the experiments: PJZ PZ. Analyzed the data: PJZ CL. Contributed reagents/materials/analysis tools: CL DP YB. Contributed to the writing of the manuscript: PJZ CL.

References

1. Zhang C, Wang Y, Rong X, Sun H, Dong S (2004) Natural resources, culture and problems of sea cucumber worldwide. Marine Fisheries Research 25: 89–97.
2. Wang Y, Rong X, Zhang C, Sun S (2005) Main diseases of cultured *Apostichopus japonicus*: prevention and treatment. Marine Sciences 5 29: 1–7.
3. Wang P, Chang Y, Xu G, Song L (2005) Isolation and ultrastructure of an enveloped virus in cultured sea cucumber *Apostichopus japonicus* (Selenka). Journal Fishery Sciences of China 12(6): 766–770.
4. Zhang C, Wang Y, Rong X (2006) Isolation and identification of causative pathogen for skin ulcerative syndrome in *Apostichopus japonicus*. Journal of Fishery China 30: 118–123.
5. Ma Y, Xu G, Chang Y, Zhang E, Zhou W, et al. (2006) Bacterial pathogens of skin ulceration disease in cultured sea cucumber *Apostichopus japonicus* (Selenka) juveniles. Journal of Dalian Ocean University Bullitin 21(1):13–18.
6. Bartel DP (2004) MicroRNAs: genomics, biogenesis, mechanism, and function. Cell 116: 281–297.
7. Olsen PH, Ambros V (1999) The lin-4 regulatory RNA controls developmental timing in Caenorhabditis elegans by blocking LIN-14 protein synthesis after the initiation of translation. Dev Biol 15;216(2):671–680.
8. Seggerson K, Tang L, Moss EG (2002) Two genetic circuits repress the Caenorhabditis elegans heterochronic gene lin-28 after translation initiation. Dev Biol. 2002 Mar 15;243(2):215–225.
9. Chen C, Schaffert S, Fragoso R, Loh C (2013) Regulation of immune responses and tolerance: the microRNA perspective. Immunological Review 253(1):112–128.
10. Morita S, Horii T, Kimura M, Hatada I (2013) MiR-184 regulates insulin secretion through repression of Slc25a22. PeerJ 1: e162.
11. Haasnoot J, Berkhout B (2011) RNAi and cellular miRNAs in infections by mammalian viruses. Methods in Molecular Biology 721: 23–41.
12. He L, Hannon GJ (2004) MicroRNAs: small RNAs with a big role in gene regulation. Nature Review Genetics 7: 522–531.
13. Li C, Feng W, Qiu L, Xia C, Su X, et al. (2012) Characterization of skin ulceration syndrome associated microRNAs in sea cucumber *Apostichopus japonicus* by deep sequencing. Fish and Shellfish Immunology 33: 436–441.
14. Zhang P, Li C, Zhu L, Su X, Li Y, et al. (2013) De novo assembly of the sea cucumber *Apostichopus japonicus* hemocytes transcriptome to identify miRNA targets associated with skin ulceration syndrome. PLoS One 8(9):e73506.
15. Rabilloud T, Chevallet M, Luche S, Lelong C (2010) Two-dimensional gel electrophoresis in proteomics: Past, present and future. Journal of Proteomics 73(11): 2064–2077.
16. Thakur SS, Geiger T, Chatterjee B, Bandilla P, Fröhlich F, et al. (2011) Deep and highly sensitive proteomic coverage by LC-MS/MS without prefractionation. Molecular and Cell Proteomics 10(8): M110.

17. Evans C, Noirel J, Ow SY, Salim M, Pereira-Medrano AG, et al. (2012) An insight into iTRAQ: where do we stand now. Analytical and Bioanalytical Chemistry 404(4): 1011–1027.
18. Herbrich SM, Cole RN, West KP Jr, Schulze K, Yager JD, et al. (2013) Statistical inference from multiple iTRAQ experiments without using common reference standards. Journal of Proteome Research 12(2): 594–604.
19. Li Z, Lin Q, Chen J, Wu J, Lim TK, et al. (2007) Shotgun identification of the structural proteome of shrimp white spot syndrome virus and iTRAQ differentiation of envelope and nucleocapsid subproteomes. Molecular and Cell Proteomics. 6(9): 1609–1620.
20. Lü A, Hu X, Wang Y, Shen X, Li X, et al. (2014) iTRAQ analysis of gill proteins from the zebrafish (*Danio rerio*) infected with *Aeromonas hydrophila*. Fish and Shellfish Immunology 36(1): 229–239.
21. Li C, Xiong Q, Zhang J, Ge F, Bi L (2012) Quantitative proteomic strategies for the identification of microRNA targets. Expert Review of Proteomics 9(5): 549–559.
22. Ou M, Zhang X, Dai Y, Gao J, Zhu M, et al. (2014) Identification of potential microRNA-target pairs associated with osteopetrosis by deep sequencing, iTRAQ proteomics and bioinformatics. European Journal of Human Genetics doi: 10.1038/ejhg.2013.221.
23. Zhou Z, Dong Y, Sun H, Yang AF, Chen Z, et al. (2014) Transcriptome sequencing of sea cucumber (*Apostichopus japonicus*) and the identification of gene-associated markers. Molecular Ecology Resources14(1):127–138
24. Shao Y, Li C, Ou C, Zhang P, Lu Y, et al. (2013) Divergent metabolic responses of *Apostichopus japonicus* suffered from skin ulceration syndrome and pathogen challenge. Journal of Agricultural and Food Chemistry 61(45): 10766–10771.
25. Bradford MM (1976) A rapid and sensitive method for the quantitation of microgram quantities of protein utilizing the principle of protein-dye binding. Analytical Biochemistry 72: 248–254.
26. Shilov IV, Seymour SL, Patel AA, Loboda A, Tang WH, et al. (2007) The Paragon Algorithm, a next generation search engine that uses sequence temperature values and feature probabilities to identify peptides from tandem mass spectra. Molecular and Cellular Proteomics 6(9): 1638–1655.
27. Chen M, Zhang X, Liu J, Storey KB (2013) High-throughput sequencing reveals differential expression of miRNAs in intestine from sea cucumber duringaestivation. PLoS One 15;8(10):e76120.
28. Zhu L, Li C, Su X, Guo C, Wang Z, et al. (2013) Identification and assessment of differentially expressed genes involved in growth regulation in Apostichopus japonicus. Genet Mol Res 20;12(3):3028–3037.
29. Di Pietro G, Magno LA, Rios-Santos F (2010) Glutathione S-transferases: an overview in cancer research. Expert Opinion on Drug Metabolism and Toxicology 6(2):153–170.

30. Saranya Revathy K, Umasuthan N, Choi CY, Whang I, Lee J (2012) First molluscan theta-class Glutathione S-Transferase: identification, cloning, characterization and transcriptional analysis post immune challenges. Comparative Biochemistry and Physiology Part B: Biochemistry and Molecular Biology 162(1–3): 10–23.

31. Ren Y, Ding Q, Zhang X (2014) Ficolins and infectious diseases. Virological Sinica 29(1): 25–32.

32. Lopez J, Hesling C, Prudent J, Popgeorgiev N, Gadet R, et al. (2012) Src tyrosine kinase inhibits apoptosis through the Erk1/2- dependent degradation of the death accelerator Bik. Cell Death and Differentiation 19(9): 1459–1469.

33. Lietzén N, Ohman T, Rintahaka J, Julkunen I, Aittokallio T, et al. (2011) Quantitative subcellular proteome and secretome profiling of influenza A virus-infected human primary macrophages. PLoS Pathogens 7(5): e1001340.

34. Tsuruta Y, Park YJ, Siegal GP, Liu G, Abraham E (2007) Involvement of vitronectin in lipopolysaccaride-induced acute lung injury. Journal of Immunology 179(10): 7079–7086

35. Bae HB, Zmijewski JW, Deshane JS, Zhi D, Thompson LC, et al. (2012) Vitronectin inhibits neutrophil apoptosis through activation of integrin-associated signaling pathways. American Journal of Respiratory Cell and Molecular Biology 46(6): 790–796.

36. Subauste MC, Pertz O, Adamson ED, Turner CE, Junger S (2004) Vinculin modulation of paxillin-FAK interactions regulates ERK to control survival and motility. The Journal of Cell Biology 165(3): 371–381.

37. Magro AM, Magro AD, Cunningham C, Miller MR (2007) Down-regulation of vinculin upon MK886-induced apoptosis in LN18 glioblastoma cells. Neoplasma 54(6): 517–526.

38. Peng X, Nelson ES, Maiers JL, DeMali KA (2011) New insights into vinculin function and regulation. Innternational Review of Cell and Molecular Biology 287: 191–231.

39. Uhlmann S, Mannsperger H, Zhang J, Horvat EÁ, Schmidt C, et al. (2012) Global microRNA level regulation of EGFR-driven cell-cycle protein network in breast cancer. Molecular Systems Biology 8: 570.

18

Effects of Leucine Supplementation and Serum Withdrawal on Branched-Chain Amino Acid Pathway Gene and Protein Expression in Mouse Adipocytes

Abderrazak Kitsy[1,9], **Skyla Carney**[1,9], **Juan C. Vivar**[1], **Megan S. Knight**[1], **Mildred A. Pointer**[1], **Judith K. Gwathmey**[2], **Sujoy Ghosh**[1,3]*

1 Division of Cardiometabolic Disorders, Biomedical Biotechnology Research Institute, North Carolina Central University, Durham, North Carolina, United States of America, **2** Boston University School of Medicine, Boston, Massachusetts, United States of America, **3** Program in Cardiovascular and Metabolic Disorders, Duke-NUS Graduate Medical School, Singapore, Singapore

Abstract

The essential branched-chain amino acids (BCAA), leucine, valine and isoleucine, are traditionally associated with skeletal muscle growth and maintenance, energy production, and generation of neurotransmitter and gluconeogenic precursors. Recent evidence from human and animal model studies has established an additional link between BCAA levels and obesity. However, details of the mechanism of regulation of BCAA metabolism during adipogenesis are largely unknown. We interrogated whether the expression of genes and proteins involved in BCAA metabolism are sensitive to the adipocyte differentiation process, and responsive to nutrient stress from starvation or BCAA excess. Murine 3T3-L1 preadipocytes were differentiated to adipocytes under control conditions and under conditions of L-leucine supplementation or serum withdrawal. RNA and proteins were isolated at days 0, 4 and 10 of differentiation to represent pre-differentiation, early differentiation and late differentiation stages. Expression of 16 BCAA metabolism genes was quantified by quantitative real-time PCR. Expression of the protein levels of branched-chain amino acid transaminase 2 (Bcat2) and branched-chain alpha keto acid dehydrogenase (Bckdha) was quantified by immunoblotting. Under control conditions, all genes displayed induction of gene expression during early adipogenesis (Day 4) compared to Day 0. Leucine supplementation resulted in an induction of Bcat2 and Bckdha genes during early and late differentiation. Western blot analysis demonstrated condition-specific concordance between gene and protein expression. Serum withdrawal resulted in undetectable Bcat2 and Bckdha protein levels at all timepoints. These results demonstrate that the expression of genes related to BCAA metabolism are regulated during adipocyte differentiation and influenced by nutrient levels. These results provide additional insights on how BCAA metabolism is associated with adipose tissue function and extends our understanding of the transcriptomic response of this pathway to variations in nutrient availability.

Editor: Xiaoli Chen, University of Minnesota - Twin Cities, United States of America

Funding: This work was partly supported by NCMHD P20MD000175-09, NIDDK 1R21DK088319-01, American Heart Association grant AHA10SDG4230068, and Duke-NUS grant WBS R-913-200-076-263 to SG. AK was partly supported by a tuition scholarship from the Biomedical Biotechnology Research Institute, NCCU. The funders had no role in study design, data collection and analysis, decision to publish, or preparation of the manuscript.

Competing Interests: The authors have declared that no competing interests exist.

* Email: sujoy.ghosh@duke-nus.edu.sg

9 These authors contributed equally to this work.

Introduction

Branched-chain amino acids (BCAAs) make up approximately 40% of the free essential amino acids in blood and play important roles in skeletal muscle growth and maintenance, primarily as protein synthesis substrates [1]. The most widely studied BCAA, L-leucine, has also been shown to play additional roles in skeletal muscle [2], including the regulation of translation initiation [3], modulation of insulin/PI3-kinase signaling [4,5], provision of metabolic fuel [6], and donation of nitrogen for alanine and glutamine synthesis [7]. Leucine has also been implicated in muscle insulin resistance (via inhibitory phosphorylation of the insulin receptor substrate-1 by mTOR kinase) [8,9], although this finding has not been fully supported by other studies [10]. Several studies have also investigated the potentially beneficial role of BCAAs in sparing lean body mass during weight loss due to caloric restriction and muscle wastage with aging [2,11].

While the majority of studies on BCAAs have focused on their roles in skeletal muscle, obesity-associated elevations in post-absorptive plasma levels of BCAAs had been observed in some early studies [12,13], if not in all of them [14,15]. An example of earlier evidence pointing to a role of adipose tissue in obesity-associated BCAA dysregulation comes from observations of reduced expression of mitochondrial branched-chain amino acid aminotransferase (Bcat2) and branched-chain α-keto acid dehydrogenase (Bckdha) enzymes in epididymal fat from ob/ob mice and Zucker rats, compared to their lean counterparts [16]. Using gene expression pathway analysis, a more recent study also identified the adipose tissue BCAA metabolic pathway as the most significantly altered pathway in mice carrying an adipose-tissue specific knockout of the glucose transporter, Glut4 [17]. In humans, a comprehensive metabolomic profiling study [18] revealed a BCAA-related metabolite signature differentiating obese from lean human subjects. The BCAA-related signature also displayed a significant linear relationship to insulin resistance [19], correlated with HbA1c levels in weight-matched type 2 diabetic vs. non-diabetic women [20], and predicted improvements in insulin sensitivity during weight loss [21]. The association of circulating BCAA and insulin action was further found to be modified by body mass index (BMI), resting respiratory quotient, presence of type 2 diabetes, and gender [22]. A study investigating obesity-discordant monozygotic twins demonstrated significant downregulation of adipose tissue BCAA metabolism gene expression, and a negative correlation with insulin sensitivity in the obese twin [23]. In our own studies, a significant downregulation of adipose tissue BCAA metabolism gene expression was also observed in overweight, hypertriglyceridemic subjects with adipose-tissue insulin-resistance, compared to BMI-matched controls [24]. The weight of evidence from these studies suggests an association between circulating BCAA levels and insulin resistance (rather than with obesity *per se*), although evidence for whether elevated BCAA is a causal driver of insulin resistance, or a consequence thereof, is currently equivocal [25].

Nevertheless, it has become increasingly clear that adipose tissue is a significant depot for systemic BCAA homeostasis. This is based on multiple studies involving the measurement of BCAA enzyme activity levels in whole-tissue extracts, kinetics of leucine utilization in white adipose tissue (WAT), and comparative gene expression analysis of branched-chain aminotransferase (Bcat2), and branched-chain keto acid dehydrogenase (Bckdh) enzyme components in WAT [17,26–29]. However, a more comprehensive understanding of the role and significance of adipose tissue BCAA metabolism is still wanting and several important gaps remain. For example, most of these studies to date have been restricted to investigating genetic models of obesity and have focused on comparisons between adipose tissues obtained from obese and lean samples that are heterogeneous both with respect to resident cell types and stages of adipocyte differentiation. Additionally, with few exceptions [28], most studies have largely concentrated on the two common steps in BCAA catabolism, leaving the status of the downstream BCAA-specific genes largely unexamined. Since intermediates of BCAA-specific metabolism feed directly into several other metabolic processes (e.g., citrate cycle, biosynthesis of aromatic amino acids and cholesterol, metabolism of fatty acids, etc.), changes in the expression of genes that regulate BCAA metabolism intermediates are expected to have additional consequences on cellular metabolism. Furthermore, given the potentially novel role of BCAA metabolism in obesity-associated insulin resistance, there is strong impetus to carefully delineate the relationship of BCAA metabolism to adipocyte biology. Whereas where several studies have investigated the primary steps of BCAA catabolism in response to nutrient availability and starvation in liver, heart or skeletal muscle (such as the effects of whole food and protein starvation on Bckdha activity) [30–34], the role of nutrient variation on adipocyte BCAA function is relatively unknown. In the current work, we have investigated some aspects of this problem by using 3T3-L1 murine adipocytes as a model system, and interrogated changes in an expanded repertoire of BCAA metabolism gene and protein components during adipocyte differentiation under conditions of serum withdrawal or exposure to physiologically relevant increases in L-leucine. This work extends the recently reported findings from Lackey, et al. who also tested the expression of a focused subset of BCAA-associated genes

in differentiating 3T3-L1 adipocytes under conditions of PPAR-gamma agonism [28].

Materials and Methods

Materials

The Swiss mouse embryo derived preadipocyte cell line, 3T3-L1, was obtained from Dr. Howard Green (Harvard University). Dulbecco's Modified Eagle's Medium (DMEM), penicillin/streptomycin solution, phosphate-buffered saline (PBS), trypsin-EDTA, fetal bovine serum and calf serum were purchased from Invitrogen (Carlsbad, CA). Biorad Protein Assay reagent and all protein gel electrophoresis and Western blotting apparatus were purchased from Biorad (Hercules, CA). SuperSignal Femto chemiluminescence substrate was purchased from Pierce (Rockford, IL). Prestained protein and RNA molecular weight markers, gentamycin solution, agarose (molecular biology grade), and SYBR green PCR Master Mix were obtained from Fisher Scientific (Pittsburgh, PA). L-Leucine (L8912, cell culture grade) was obtained from Sigma Aldrich (St. Louis, MO). Polyclonal antibodies to mouse Bcat2 (ab95976) and Bckdha (ab90691) were obtained from Abcam (Cambridge, MA). Rabbit monoclonal antibody to mouse mitogen activated protein kinase (Mapk) and horseradish peroxidase linked anti-rabbit IgG (secondary antibody) were purchased from Cell Signaling Technologies, Inc. (Danver, MA).

Cell culture

3T3-L1 preadipocytes were cultured in growth media (89% DMEM high Glucose, 10% bovine calf serum, antibiotic/antimycotic/Pen-strep) for 2–3 days to confluency and then incubated for an additional 48 hours to induce growth arrest (Day 0). Growth arrested cells were treated with differentiation media (89% DMEM F12, 10%Fetal Bovine Serum, 1% antibiotic/antimycotic/Pen-strep, 250 uM 3-isobutyl-1-methylxanthine (IBMX), 500 uM dexamethasone, 1 uM insulin) for 2 days. At the end of 2 days (Day 2), the cells were further treated with maintenance media (89% DMEM F12, 10%Fetal Bovine Serum, 1% Antibiotic/antimycotic Pen-strep) plus 1 uM insulin for the next 4 days. Cells were thereafter continuously treated with maintenance media every 2 days until the end of the experiment at Day 10. Experiments were performed in Petri dishes or 6-well plates at a starting concentration ranging from 2.38×10^5–2.80×10^5 cells/ml. Cell treatments were as follows:

Control. Cells were cultured in regular growth, differentiation or maintenance medium as described above and harvested at Day 0, Day 4 and Day 10 for protein and RNA extraction.

Serum withdrawal. Cells were grown in growth, differentiation and maintenance medium as described above. The media were replaced with media deprived of bovine calf serum, and cells were incubated in this media for 18 hours immediately prior to Day 0, Day 4 and Day 10 of adipocyte differentiation. At the end of the treatment, cells were harvested for protein and RNA extraction.

Leucine supplementation. L-Leucine was dissolved in sterile, molecular-biology grade water to a final concentration of 6.56 g/l (50 mM). This stock solution was added to media to a final concentration of 0.5 mM. Since the media already contains 0.8 mM leucine, this treatment represents a 62.5% increase in leucine levels. This level of increase is in the range of elevations previously observed in *in vivo* leucine overfeeding studies [35] or in post-absorptive plasma of obese mice [16]. Cells were cultured in leucine containing media for 18 hours immediately prior to Day 0, Day 4 and Day 10 and then harvested for protein and RNA extraction.

The choice of exposure time was based on prior literature reports of leucine treatment or serum withdrawal of muscle and other cell types, which typically ranged from 12–24 hrs. For all treatments, the course of adipocyte differentiation was monitored by Oil Red O staining of intracytoplasmic lipid accumulation over time [36] and by quantitative polymerase chain reaction (qPCR) analysis of key differentiation related genes. Oil Red O stained cells were photographed through a Zeiss Axiovert 25 microscope attached to a Zeiss Axiocam MRC camera at 40X magnification using a red filter. Non-specific cell death was quantified by measurement of released lactate dehydrogenase activity into cell culture media [37] via absorbance at 490 nm (Cytox 96 Non-Radioactive Cytotoxicity Assay Kit, Promega, Madison, WI).

Quantitative PCR and determination of amplification efficiency

Primers for qPCR were designed via the PrimerBlast (National Center for Biotechnology Information), and Primerquest primer design softwares (Integrated DNA Technologies, Coralville, IA). Exon-spanning primers, with total GC content not exceeding 70% and PCR amplicon not exceeding 150 base pairs, were designed whenever possible. Other thermodynamic and design parameters were used at default values provided by the softwares. The sequences of the primers used in the study are provided in **Table S1**. qPCR was performed in a total volume of 20 uL using 96-well microwell plates and a Bio-Rad CFX96 Real-Time PCR Detection System. Total RNA was isolated from cells via the RNeasy mini kit (Qiagen). RNA quality was determined via Agilent Bioanalyzer and all RNA samples were found to have 260/280 ratios >2.0 and the RIN values ranged from 9.2–10 (**Table S2**). For qPCR analysis, total RNA was first converted to cDNA (Thermo Scientific Verso). cDNA synthesis reactions were carried out in a total volume of 20 uL consisting of 4 uL of 5X cDNA synthesis buffer, 2 uL dNTP Mix, 1 uL RNA Primer, 1 uL RT Enhancer, 1 uL Verso Enzyme Mix, 1 uL total RNA (100 ng) sample and 10 uL Molecular Grade Water (G-Biosciences). For qPCR, 5 microliters of cDNA, 25 uL iTaq Fast SYBR Green Supermix with Rox (Bio-Rad) and 125 nM of primers were added to each microwell, to a total volume of 20 uL. The plate was centrifuged for 4 min at 2000 rpm at 4°C prior to qPCR. The PCR was run at 95°C for 3 min, followed by 40 cycles at 95°C for 3 sec and 55°C for 30 sec. All PCRs were performed in triplicate.

To evaluate the efficiency of the amplification, a standard curve was constructed by plotting the median quantification cycle value (C_t, obtained from 3 replicates) versus six 2-fold serial dilutions of the target cDNA template, starting with 100 ng of cDNA. The slope of the standard curve was calculated by linear regression of C_t against $\log_{10}$(cDNA) and the efficiency of amplification was determined via the Standard Curve Slope to Efficiency Calculator (www.genomics.agilent.com) according to the relation: $Efficiency = 10^{(-1/slope)} -1$.

Western blotting and immunodetection

Western blotting was carried out by first transferring proteins from 10% sodium dodecyl sulfate polyacrylamide gels to PVDF membranes at 100 volts for 1 hour. The membrane was then incubated with 10–15 ml of blocking buffer (5% non-fat milk/Tris-buffered saline) at 4°C overnight. Primary antibodies of interest were added to the membrane at dilutions ranging from 1:1000–1:10,000, and further incubated at 4°C overnight. Horseradish peroxidase conjugated anti-rabbit secondary antibody was used at a dilution of 1:2000 for 1 hour. ECL detection was performed with Pierce ECL Western blotting substrate (Thermo-Scientific, Rockford, IL). Western blots were scanned on a Biorad

Molecular Imager Gel Doc XR+ scanner and quantified by ImageJ (imagej.nih.gov).

Statistical Analysis

Raw data for qPCR was captured via the CFX manager software (Biorad) and converted to cycle numbers corresponding to the threshold of detection for gene expression (Ct). Quantitation of the results was performed via the Comparative C_t method [38], using *Gapdh* as an internal calibrator gene. Summary statistics for gene and protein expression data (mean ± standard deviation) and comparison of groups via 2 way analysis of variance (ANOVA) was carried out in JMP statistical software (SAS, Cary, NC). For ANOVA, both main effects (time, treatment) and interaction effects (time * treatment) were considered. Statistical significance was set at the $p < 0.05$ level.

Results

Effects of treatments on 3T3-L1 differentiation program

The effects of the different treatments on adipocyte differentiation and adipocyte cell death were monitored separately for each treatment. We did not observe significant treatment-specific effects on 3T3-L1 adipocyte differentiation, as measured by Oil Red O staining and quantitative PCR analysis of key genes related to adipocyte differentiation (**Figure 1**). Compared to *control*, a statistically significant difference in LDH activity was noted in adipocytes exposed to *serum-withdrawal* but not *leucine* treated cells, suggesting greater cell death with serum withdrawal (p< 0.05). However the difference in the magnitude of the effect was quite small (5.86% in *serum-withdrawal* vs. 3.09% in *control*, respectively).

Gene selection for qPCR studies

Genes in the BCAA metabolism pathway, acting in the sequential metabolism of leucine, isoleucine and valine to glucogenic and ketogenic precursors (acetoacetate, acetyl CoA and succinyl CoA), were adapted from Herman et al. [17] and are shown schematically in **Figure 2**. A total of 18 BCAA metabolism pathway genes were initially tested in 3T3-L1 preadipocytes and reliable expression was observed for 16 of them (QPCR detection threshold cycles between 17–30; reference Ct for *Gapdh* = 14.2). The expression levels of the *Aldh7a1* and *Oxct2a* genes were determined to be unreliable (average Ct>35) and were therefore excluded from further analysis (**Figure S1**). Since mRNA quantification estimates are susceptible to bias if the efficiency of PCR amplification differs between the genes being tested [39], we determined the PCR amplification efficiencies of the genes of interest via serial dilutions of the target cDNA template. The amplification efficiencies ranged between 95–110% for all the genes tested (**Table S3**).

Effects of treatments on BCAA pathway gene expression at different stages of adipocyte differentiation

We first interrogated whether any of the 3 treatments changed basal expression levels of the BCAA metabolism genes. We selected 3 timepoints representing pre-differentiation (Day 0), early differentiation (Day 4) and late differentiation (Day 10) of 3T3-L1 adipocytes. Results are shown in **Figure 3**. The highest level of expression was observed for the *Hmgcs1* gene whereas very low expression levels were observed for *Pccb*, and *Mcee*. The expression levels of *Bcat2* were lower compared to *Bckdha* or *Bdk*. Moderate expression was noted for the remaining genes. Generally, the message levels of all genes were increased in the presence of *serum-withdrawal* (with the exception of *Bckdha and*

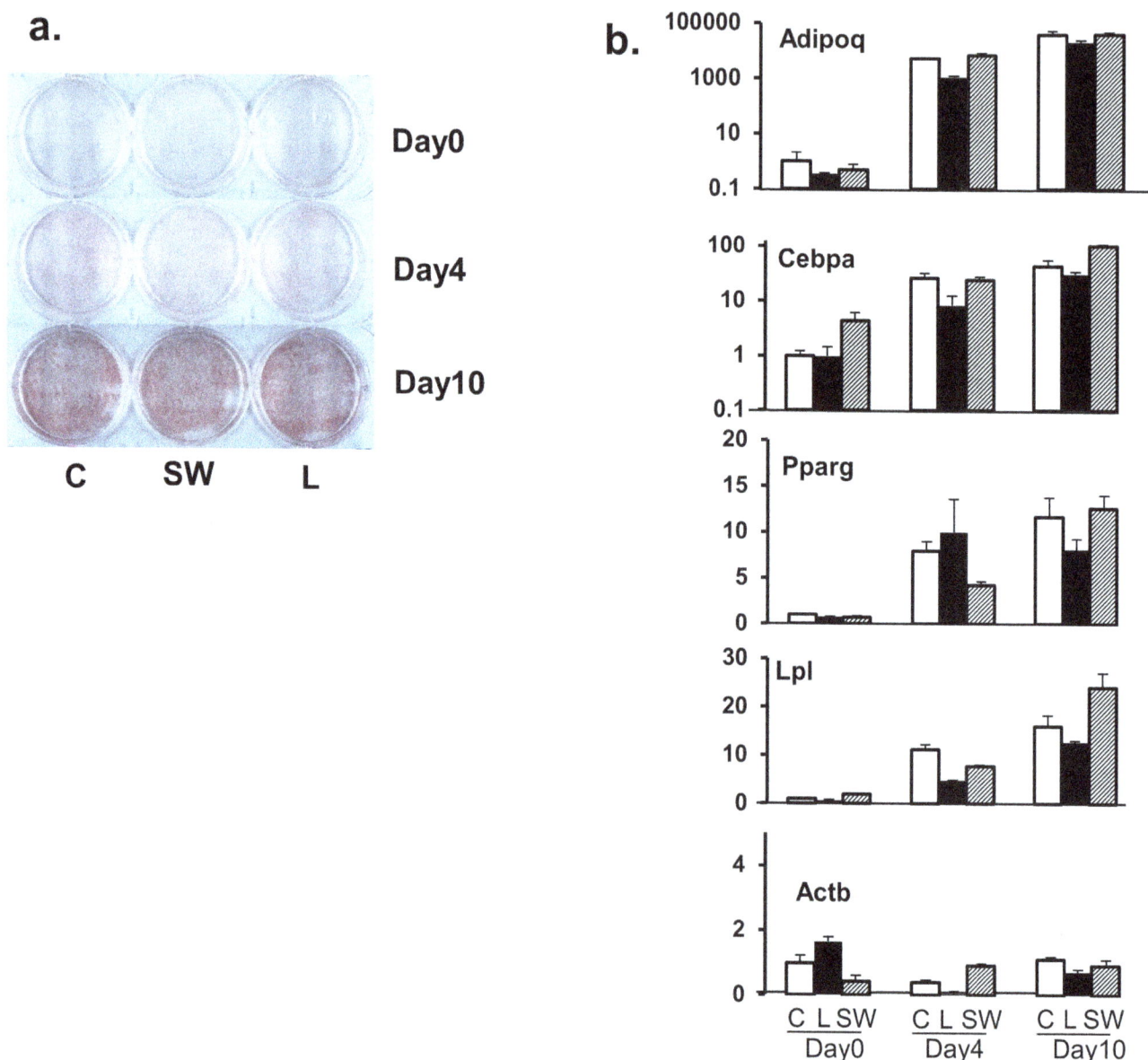

Figure 1. Effects of leucine supplementation and serum withdrawal on 3T3-L1 adipocyte differentiation. a. Oil-red O staining of cell cultures at Day 0, Day 4 and Day 10 under *control, leucine or serum withdrawal* conditions. Cells were grown under the defined conditions for the specified periods of time and stained for total neutral lipids. **b.** Quantitative PCR analysis of key genes associated with adipocyte differentiation. Gene names are indicated on the plots. Gene expression abundances were estimated by the $\Delta\Delta Q$ method after normalizing to GAPDH and referencing to Control, Day 0 samples. Due to the relatively high range of fold-change for adiponectin and CEBP-alpha, their relative abundances are expressed in the $\log_{10}$ scale. The treatments and stages of differentiation are indicated at the bottom (C,*control*; L, *leucine*; SW, *serum-withdrawal*).

Hmgcs1) compared to *control* or *leucine* supplementation, and for several genes remained higher throughout the entire time course of the experiment (e.g. *Acaa1b, Acadm, Auh*, and *Hadh*).

The results obtained in Figure 3 were then re-analyzed to determine the effects of treatment on relative changes in gene expression during the course of adipocyte differentiation. We estimated expression fold-changes based on the average expression values at each timepoint with respect to Day 0 which was set at 1.0 (**Figure S2**). In *control* and *leucine* treated samples, we observed a coordinate upregulation of a majority of the tested genes at Day 4, compared to Day 0 or Day 10. However, the expression sensitivity of *Bcat2* and *Bckdha*, the primary regulators of BCAA metabolism, differed from the rest of the genes. In differentiating cells,

both *Bcat2* and *Bckdha* mRNA expression were sensitive to leucine manipulation at both days 4 and 10 but not Day 0(fold-change between 2.5–3 fold). The fold-change pattern with serum withdrawal was more variable. For several genes, there was a monotonic increase in fold-change between Day 0 and Day 10 (e.g. *Bdk, Auh, Pccb*) whereas the opposite was observed for a subset of other genes (*Ivd, Hmgcs1, Mcee, Mut*). Serum withdrawal increased *Bcat2* levels at Day 4 (2.5-fold compared to control) and returned to control levels at Day 10. *Bckdha* levels did not change appreciably under this treatment.

To determine whether any of the observed expression changes were statistically significant, we conducted analysis of variance (ANOVA) on gene expression measures across adipocyte differ-

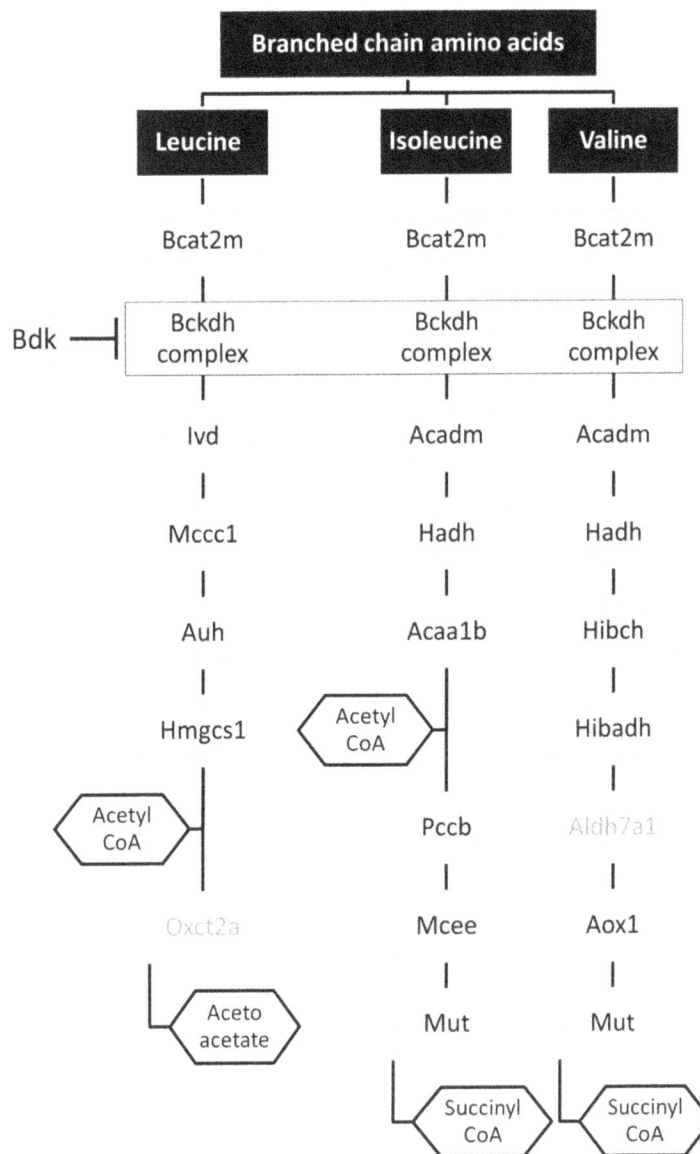

Figure 2. Organization of the BCAA metabolism pathway components studied. Genes involved in Leucine, Valine and Isoleucine metabolism are shown in gray rectangles. End-products of BCAA metabolism are listed as gray hexagons. Genes with grayed out names were not studied further due to unreliable expression levels.

entiation (three levels) and treatments (three levels). ANOVA identified statistically significant (p<0.05) changes in the expression for several BCAA metabolizing genes. For the majority of the genes, statistical significance was observed for both main effects (time and treatment), whereas interaction effects were observed for a subset of the genes. For example, a two-way analysis of variance on the *Bcat2* gene yielded a main effect for treatment (p<.005), such that the average expression was significantly higher for *leucine* than for *control* or *serum withdrawal*. The main effect of time was also significant, (p<0.0001). However, the interaction effect was not significant, (p<0.18), suggesting that the response to time was not modulated by the individual treatments. Detailed results from the ANOVA analysis are provided in **Table S4**.

Effects of treatments on key BCAA metabolism protein expression

Bcat2 and Bckdha protein expression were analyzed via Western blots on replicate cell cultures for each treatment, using mitogen activated protein kinase (Mapk) as the loading control (**Figure 4**). Bcat2 expression was undetectable in the pre-adipocyte stage (Day 0) for both *control* and *leucine* treatments. A significant increase in expression was observed during early adipogenesis (Day 4) that persisted into the differentiated state (Day 10) for *control* samples (p<0.0001 for Day 4 and Day 10 vs. Day 0). Bcat2 protein induction showed greater sensitivity to *leucine* treatment at Day 4 but was significantly reduced on Day 10, compared to *control*. For *leucine* treatments, an even greater upregulation of Bcat2 protein levels was observed at Day 4 with significant reductions at Day 10 (p<0.0001 for Day 4 and Day 10 vs. Day 0; p<0.0001 for *leucine* Day 4 vs *control* Day 4 or *leucine*

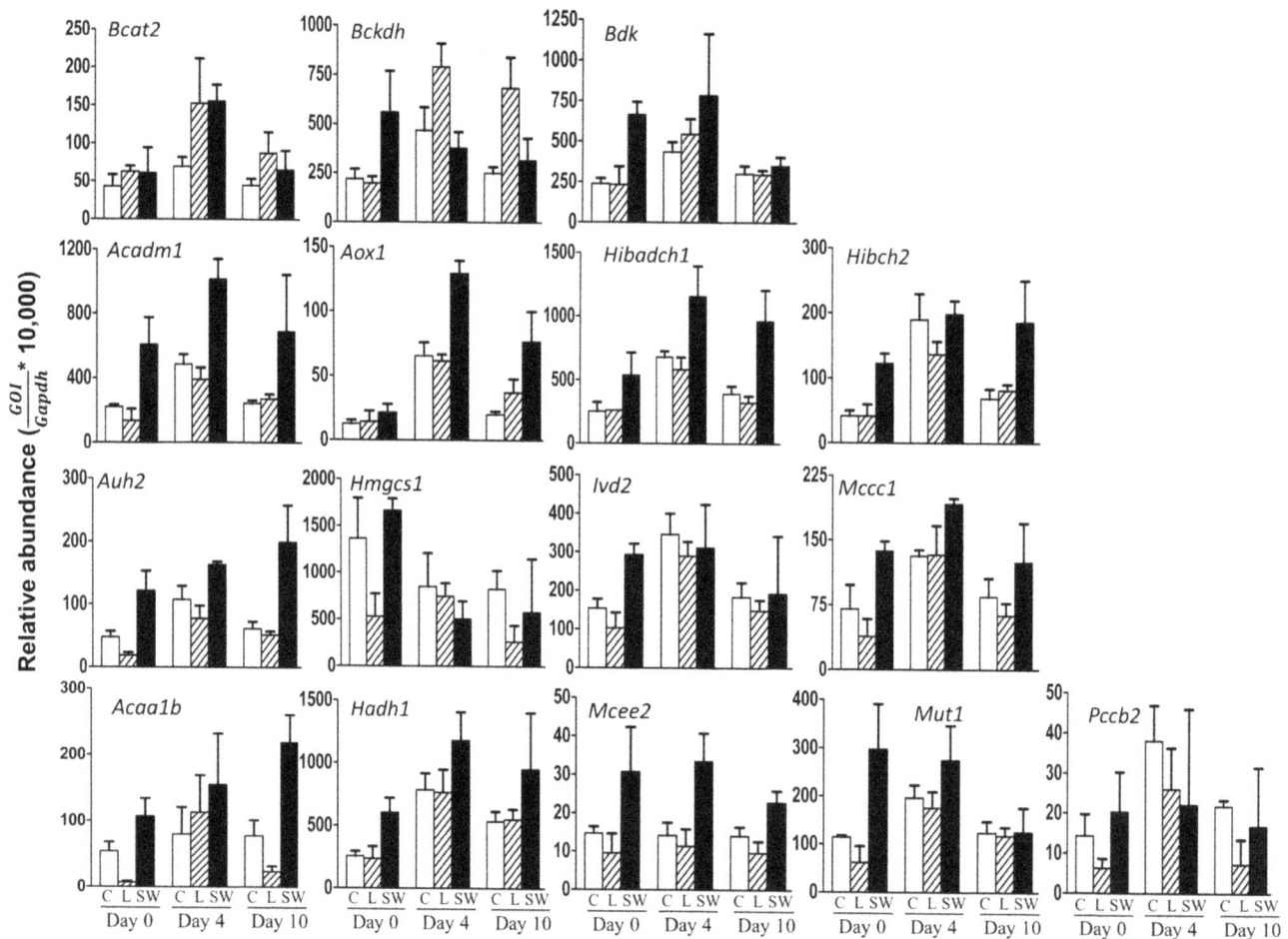

Figure 3. Effect of different treatments on expression levels of BCAA metabolism pathway genes. Quantitative PCR (qPCR) data was generated from three independent cell cultures for each treatment-timepoint combination and analyzed in triplicate. Gene abundances were normalized to *Gapdh* levels and then multiplied by a factor of 10,000 to avoid very small values. The relative abundance measures for each of the replicate cell cultures were combined to provide a summary abundance measure. Panels, from top to bottom – genes common to the metabolism of all 3 BCAAs; genes specific to valine metabolism; genes specific to leucine metabolism; genes specific to isoleucine metabolism. Gene symbols are indicated at the top of each plot. Treatments are shown as C (control, open rectangle), L (leucine, hatched rectangle), and SW (serum withdrawal, black rectangle). Results are shown for each of Day0, Day4 and Day10. *GOI*, gene of interest.

Day 10 vs. *control* Day 10). *Serum-withdrawal* led to barely detectable Bcat2 levels at all timepoints. A similar analysis for Bckdha showed the presence of 3 major bands in the Control Day 0 samples at 25, 47 and 50 kDa. The predicted molecular weight for murine Bckdha is 50 kDa, so the lower molecular weight bands could be degradation products (murine Bckdha is predicted to exist only as a single transcript). Consideration of the 50 kDa band showed the protein to be comparably expressed in the *control* and *leucine* treated samples across all timepoints. *Serum-withdrawal* again resulted in barely detectable Bckdha levels at all timepoints tested.

Discussion

Several recent studies have pointed to interplay between adipose tissue, BCAA metabolism, and insulin signaling/glucose homeostasis. For example, adipose tissue specific overexpression of the glucose transporter *Glut4* was shown to result in a coordinate downregulation of BCAA metabolism genes selectively in epididymal adipose tissue, whereas the converse was true for *Glut4* knockout mice [17]. Transplantation of adipose tissue from wild-

type mice into *Bcat2-/-* mice was further found to reduce post-absorptive plasma BCAA levels by 46%, demonstrating that adipose tissue can be a significant source of BCAA catabolism *in vivo* [17]. Other studies in humans and animal models have also largely established a negative correlation between obesity (and insulin resistance) and BCAA metabolism. For example, metabolomic analysis of blood has consistently demonstrated increases in circulating BCAA and their metabolites in obesity, suggesting altered flux through the BCAA catabolic pathway [18,20,40]. Conversely, large decreases in circulating BCAA levels have been observed in response to bariatric surgery in humans [41] along with increased expression in *Bcat2* and *BCKDHA* in subcutaneous and omental fat, suggesting a possible role of adipose tissue in regulating BCAA homeostasis following weight-loss [16]. These and other studies [18] have suggested that higher tissue and blood concentrations of BCAA may cause or exacerbate insulin resistance in obesity through leucine mediated activation of mTOR. However, the direction of causality and the specificity of the BCAA effect have been questioned in some other studies. For example, the observed drop in BCAA levels in bariatric surgery patients and the concurrent improvements in insulin

A.)

B.)

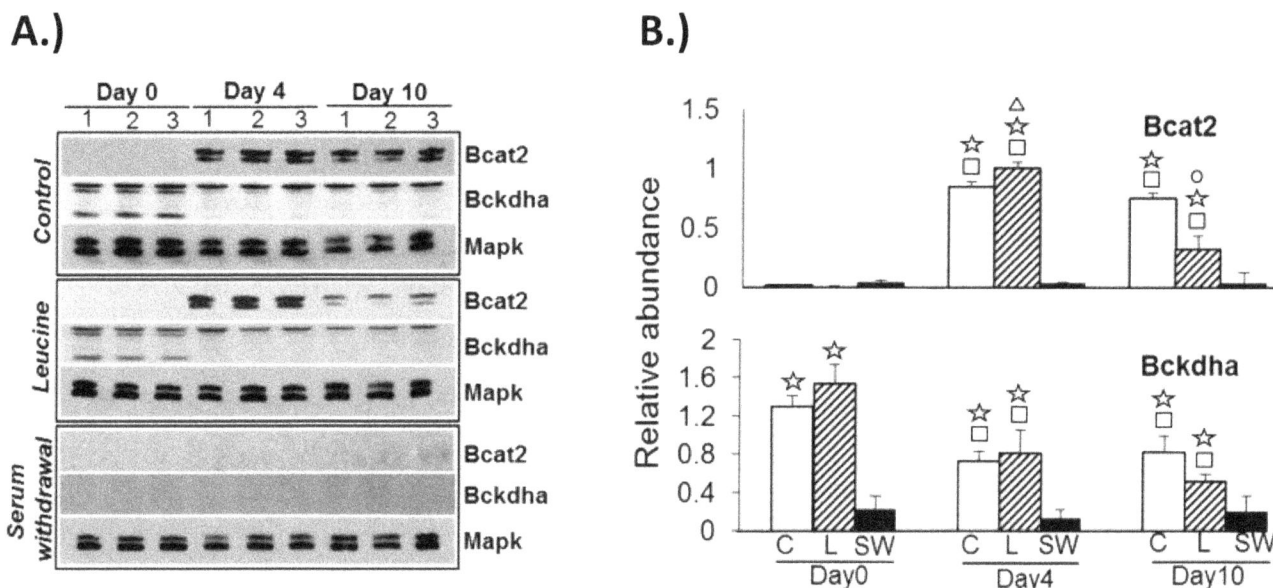

Figure 4. Western blot analysis of Bcat2 and Bckdha protein levels in treated cells over the course of adipocyte differentiation. a. Protein expression from three independent cell-culture experiments is shown for each gene. Treatments (*control, leucine or serum withdrawal*) are indicated at the left of each blot and the timepoints are indicated at the top. Mitogen activated protein kinase (Mapk) was used as loading control. **b.** Quantification of relative protein abundances by scanning densitometry on Western blots. Protein abundances are expressed relative to Mapk levels and average values from three independent cell-culture studies are shown. Significant changes in protein expression (p<0.05) are indicated as follows – comparison vs. Day 0, open squares; comparison against *serum withdrawal*, open star; comparison of Day 10 vs. Day 4, open circle; comparison of *leucine* vs. *control*, open triangle.

sensitivity is equally plausible if inhibition of BCAA is a consequence and not a cause of insulin resistance. In other studies, the reduction in insulin-stimulated glucose uptake in response to BCAA or leucine was also mimicked by other amino acids such as methionine, hisitidine, threonine and tyrosine [42]. Whether the level of increases in blood BCAA levels are adequate to activate mTOR has also not been conclusively established and even when mTOR activation is observed (for example by doubling the intake of leucine in high-fat fed mice), metabolic and inflammation-related phenotypes are often either improved or normalized [34,35]. Finally, the *Bcat2* knockout mice actually display improvements in insulin resistance rather than its exacerbation [43]. Therefore, the evidence for a contributory effect of elevated BCAA levels to obesity-associated insulin resistance is at best, equivocal [25].

Given this context, our study provides additional insights into the regulation of BCAA metabolism in adipocytes. We selected two opposing treatment conditions to exemplify nutrient excess (leucine supplementation) or nutrient restriction (serum withdrawal). We selected leucine as the BCAA of choice for supplementation studies, given its nutritional relevance as a major component of dietary proteins. Serum withdrawal was selected to mimic reduced nutrient supply such as may be encountered during starvation.

A general trend observed for the majority of the genes tested in this study was an increase in relative expression during early adipogenesis (Day 4) compared to the pre-differentiation (Day 0) or late differentiation (Day 10) stages. This trend was statistically significant for all genes (p<0.05) and observed in the *control* samples, suggesting that the behavior was intrinsic to adipocyte differentiation. The only two exceptions were for *Hmgcs1* and *Mcee* with the former showing a reduction, and the latter displaying no changes in their respective expression levels during

adipocyte differentiation. Western blot analysis showed the temporal pattern of Bcat2 protein expression to be generally consistent with that of its message for the *control* samples, (with the exception of preadipocytes where there was no detectable protein expression). Bckdha protein expression was more discordant with its gene expression in that the highest level of protein was observed in the preadipocyte samples with lower levels in the Day 4 or Day 10 samples.

Exposure of cells to increased levels of leucine induced a statistically significant increase in *Bcat2* and *Bckdha* gene expression (compared to *control*) when treatments were initiated early (Day 4) or late (Day 10) (p<0.05). The amount of exogenous leucine used in these studies was 62.5% of the basal leucine levels already present in the media (0.8 mM). As previously noted, this level of leucine increment is in the range of previous leucine overfeeding studies or post-absorptive increases in leucine levels in obese mice [16,35]. Leucine treatment did not lead to significant changes in the expression of valine-metabolizing genes, suggesting specificity in the transcriptomic response. Although several genes involved in leucine and isoleucine metabolism displayed reduced expression in response to leucine at Day 0 (compared to controls), the changes were statistically significant only for the *Acaa1b* and *Auh* genes. The significance of this finding is currently unclear and will be better ascertained when metabolic flux data becomes available. Western blot analysis showed an overall similar pattern of expression for Bcat2 and Bckdha in the leucine-treated samples as the control samples, except for Day 10 where both Bcat2 and Bckdha levels were significantly lower in the leucine-treated group. This may reflect substrate-level inhibition of BCAA metabolism, reported previously in rat heart [44], rat hepatocytes [45] and L6 myotubes [46] where, in the presence of other oxidizable substrates such as fatty acids, BCAA catabolic flux is inhibited via Bckdha inactivation [25]. We speculate that a similar mechanism may be operative in differentiating and differentiated adipocytes where the availability of fatty acids and glucose reduces the need for BCAA catabolism to tricarboxylic cycle intermediates and the excess leucine may instead be channeled to directly regulate other biological processes (e.g. translation initiation and insulin signaling via the mammalian target of rapamycin [47]).

Another key finding from the study was a statistically significant ($p<0.05$) upregulation of expression of several genes under conditions of *serum withdrawal*. Although this result may appear to be counter-intuitive, serum withdrawal is known to elicit complex and unpredictable time-dependent effects leading to diametrically opposing responses even in the same cell type [48]. Since several BCAA metabolizing genes are also involved in the catabolism of fatty acids, we hypothesize that the observed coordinated upregulation could reflect an adaptive transcriptomic response of the cells to mobilize lipid oxidation in the absence of external nutrient supply [49,50]. Exceptions to this starvation related induction were noted for *Bckdha* and *Hmgcs1* in the differentiating (Day 4) and differentiated (Day 10) samples, where the message levels were reduced compared to the *control* or *leucine* treated samples. However, both Bcat2 and Bckdha protein levels were barelydetectable under serum-withdrawal conditions at all time-points tested, suggesting a lack of correlation between protein and message levels, at least for these two proteins. We postulate this loss of Bcat2 and Bckdha will effectively block BCAA catabolism in the serum-deprived state, possibly in an effort to conserve BCAAs for protein synthesis, as previously noted in starved rats [1].

To summarize, the main findings from the study are as follows. First, the available data points to a complex relationship between mRNA and protein levels for two central BCAA metabolism genes. Thus, whereas concordance was noted in the increased message and protein levels of Bcat2 and Bckdha during early (Day 4) differentiation compared to pre- or late differentiation, serum-withdrawal led to increases in message levels of most BCAA metabolism genes whereas the protein levels for Bcat2 and Bckdha were undetectable. Second, the absence of Bcat2 protein expression, even in the presence of high Bckdha levels, in the Day 0 samples suggests a lack of direct oxidative metabolism of BCAAs in the preadipocyte. In this regard, the situation is similar to the liver where Bcat2 is nearly absent whereas Bckdha activity is maintained at high levels, possibly to enable clearance of branched-chain keto acids from portal blood. The function of Bckdha in the preadipocyte is, however, less clear. The capacity for oxidative degradation of BCAA appears to increase during the course of early adipocyte differentiation (Day 4) and is again reduced at or near completion of differentiation (Day 10), perhaps as the cellular focus shifts from lipid catabolism to lipid storage.

We should point out that the exposure periods used in our study (leucine or serum withdrawal) were comparatively short (18 hrs), implying that the findings may be more relevant for acute, and not chronic, nutrient effects. However, this experimental design was necessary since chronic exposure of adipocytes to serum deprivation led to a significant loss of cell viability. Additionally, the process of adipocyte differentiation was conducted in the presence of insulin 4 days after the start of differentiation after which insulin was withheld. The change in insulin exposure could influence the temporal gene expression. However, since the same differentiation procedure was followed for all treatments, we expect that insulin-dependent transcriptomic changes to be similar and therefore not a major confounder in our analysis. In summary, this study provides the first step in accurately quantifying the BCAA pathway transcriptome and comparing RNA and protein levels of two key regulators of BCAA metabolism (Bcat2 and Bckdha). Future

studies will further investigate the phosphorylation status of Bckdha as well as quantify the overall flux through the BCAA catabolic pathways during adipocyte differentiation under conditions of nutrient stress.

Supporting Information

Figure S1   Relative expression of BCAA pathway genes in 3T3-L1 preadipocytes. Gene expression levels were ascertained in three independent 3T3-L1 cell cultures. For each culture, quantitative PCR assays were conducted in triplicate and the average detection threshold (C_t) and the corresponding standard deviations were plotted for each gene.

Figure S2   Changes in BCAA gene expression in response to treatments and to adipocyte differentiation. Gene expression changes are represented as fold-changes compared to Day 0 for each treatment. Results were averaged over three independent cell-culture experiments. Rows from top to bottom represent the following gene categories – *top row*, genes common to the metabolism of all 3 BCAAs; *second from top row*, genes specific to valine metabolism; *third from top row*, genes specific to leucine metabolism; *bottom row*, genes specific to isoleucine metabolism. Symbols for each gene are indicated at the top of the relevant plots. Treatments are indicated as follows – *open square*, control; *open triangle*, leucine supplementation; *open circle*, serum-withdrawal.

Table S1   qPCR primer sequences for BCAA metabolism genes. Primers are named in the format of '*genename_species_orientation*'; *m* represents mouse whereas *f* and *r* stand for forward and reverse primers, respectively.

Table S2   Assessment of the quality of total RNA used in the study. Isolated RNA was analyzed via Agilent Bioanalyzer. (**a**) RNA quality as determined from absorption spectroscopy (260/280 nm absorbance ratio) and from RNA Integrity Number (RIN) estimates. (**b,c**) Electropherograms of representative total RNA isolated from 3T3-L1 cells subjected to different treatments.

Table S3   Measurement of qPCR amplification efficiency for BCAA metabolism genes. PCR amplification efficiency was determined for each gene by 2-fold serial dilutions of cDNA reverse transcribed from the RNA. Column 1, gene name; column 2, slope of the linear fit of PCR detection threshold (Ct) vs. target cDNA concentrations; column 3, correlation coefficient of the linear fit; column 4, estimated amplification of the qPCR.

Table S4   Two-way ANOVA analysis of changes in BCAA metabolizing gene expression in response to treatments. The main effects of time and treatment and their interactions were analyzed. There were 3 levels for each factor (*Treatment*: Control, Leucine and Serum-withdrawal; *Time*: Day 0, Day 4, Day 10). Treatment means were calculated based on target gene expression, normalized to GAPDH expression levels, and then multiplied by a scaling factor of 10,000.

Acknowledgments

The authors thank Professor Robert Harris, Indiana University School of Medicine, for helpful guidance on BCAA immunoblotting procedures.

Author Contributions

Conceived and designed the experiments: AK SG. Performed the experiments: AK SC MSK. Analyzed the data: AK JCV SC SG. Contributed reagents/materials/analysis tools: MAP. Wrote the paper: AK SC SG JKG.

References

1. Harris RA, Joshi M, Jeoung NH, Obayashi M (2005) Overview of the molecular and biochemical basis of branched-chain amino acid catabolism. J Nutr 135: 1527S–1530S.
2. Layman DK (2003) The role of leucine in weight loss diets and glucose homeostasis. J Nutr 133: 261S–267S.
3. Kimball SR, Jefferson LS (2001) Regulation of protein synthesis by branched-chain amino acids. Curr Opin Clin Nutr Metab Care 4: 39–43.
4. Baum JI, O'Connor JC, Seyler JE, Anthony TG, Freund GG, et al. (2005) Leucine reduces the duration of insulin-induced PI 3-kinase activity in rat skeletal muscle. Am J Physiol Endocrinol Metab 288: E86–91.
5. Patti ME, Brambilla E, Luzi L, Landaker EJ, Kahn CR (1998) Bidirectional modulation of insulin action by amino acids. J Clin Invest 101: 1519–1529.
6. Wagenmakers AJ (1998) Muscle amino acid metabolism at rest and during exercise: role in human physiology and metabolism. Exerc Sport Sci Rev 26: 287–314.
7. Ruderman NB (1975) Muscle amino acid metabolism and gluconeogenesis. Annu Rev Med 26: 245–258.
8. Carlson CJ, White MF, Rondinone CM (2004) Mammalian target of rapamycin regulates IRS-1 serine 307 phosphorylation. Biochem Biophys Res Commun 316: 533–539.
9. Newgard C (2012) Interplay between lipids and branched chain amino acids in development of insulin resistance. Cell Metabolism 15: 606–614.
10. Redman LM, Heilbronn LK, Martin CK, de Jonge L, Williamson DA, et al. (2009) Metabolic and behavioral compensations in response to caloric restriction: implications for the maintenance of weight loss. PLoS One 4: e4377.
11. Layman DK, Walker DA (2006) Potential importance of leucine in treatment of obesity and the metabolic syndrome. J Nutr 136: 319S–323S.
12. Caballero B, Finer N, Wurtman RJ (1988) Plasma amino acids and insulin levels in obesity: response to carbohydrate intake and tryptophan supplements. Metabolism 37: 672–676.
13. Felig P, Marliss E, Cahill GF Jr (1969) Plasma amino acid levels and insulin secretion in obesity. N Engl J Med 281: 811–816.
14. Forlani G, Vannini P, Marchesini G, Zoli M, Ciavarella A, et al. (1984) Insulin-dependent metabolism of branched-chain amino acids in obesity. Metabolism 33: 147–150.
15. Heraief E, Burckhardt P, Mauron C, Wurtman JJ, Wurtman RJ (1983) The treatment of obesity by carbohydrate deprivation suppresses plasma tryptophan and its ratio to other large neutral amino acids. J Neural Transm 57: 187–195.
16. She P, Van Horn C, Reid T, Hutson SM, Cooney RN, et al. (2007) Obesity-related elevations in plasma leucine are associated with alterations in enzymes involved in branched-chain amino acid metabolism. Am J Physiol Endocrinol Metab 293: E1552–1563.
17. Herman MA, She P, Peroni OD, Lynch CJ, Kahn BB (2010) Adipose tissue branched chain amino acid (BCAA) metabolism modulates circulating BCAA levels. J Biol Chem 285: 11348–11356.
18. Newgard CB, An J, Bain JR, Muehlbauer MJ, Stevens RD, et al. (2009) A branched-chain amino acid-related metabolic signature that differentiates obese and lean humans and contributes to insulin resistance. Cell Metab 9: 311–326.
19. Matthews DR, Hosker JP, Rudenski AS, Naylor BA, Treacher DF, et al. (1985) Homeostasis model assessment: insulin resistance and beta-cell function from fasting plasma glucose and insulin concentrations in man. Diabetologia 28: 412–419.
20. Fiehn O, Garvey WT, Newman JW, Lok KH, Hoppel CL, et al. (2010) Plasma metabolomic profiles reflective of glucose homeostasis in non-diabetic and type 2 diabetic obese African-American women. PLoS One 5: e15234.
21. Shah SH, Crosslin DR, Haynes CS, Nelson S, Turer CB, et al. (2012) Branched-chain amino acid levels are associated with improvement in insulin resistance with weight loss. Diabetologia 55: 321–330.
22. Thalacker-Mercer AE, Ingram KH, Guo F, Ilkayeva O, Newgard CB, et al. (2014) BMI, RQ, diabetes, and sex affect the relationships between amino acids and clamp measures of insulin action in humans. Diabetes 63: 791–800.
23. Pietilainen KH, Naukkarinen J, Rissanen A, Saharinen J, Ellonen P, et al. (2008) Global transcript profiles of fat in monozygotic twins discordant for BMI: pathways behind acquired obesity. PLoS Med 5: e51.
24. van Greevenbroek MM, Ghosh S, van der Kallen CJ, Brouwers MC, Schalkwijk CG, et al. (2012) Up-regulation of the complement system in subcutaneous adipocytes from nonobese, hypertriglyceridemic subjects is associated with adipocyte insulin resistance. The Journal of clinical endocrinology and metabolism 97: 4742–4752.
25. Adams SH (2011) Emerging perspectives on essential amino acid metabolism in obesity and the insulin-resistant state. Adv Nutr 2: 445–456.
26. Brosnan JT, Brosnan ME (2006) Branched-chain amino acids: enzyme and substrate regulation. J Nutr 136: 207S–211S.
27. Suryawan A, Hawes JW, Harris RA, Shimomura Y, Jenkins AE, et al. (1998) A molecular model of human branched-chain amino acid metabolism. Am J Clin Nutr 68: 72–81.
28. Lackey DE, Lynch CJ, Olson KC, Mostaedi R, Ali M, et al. (2013) Regulation of adipose branched-chain amino acid catabolism enzyme expression and cross-adipose amino acid flux in human obesity. Am J Physiol Endocrinol Metab 304: E1175–1187.
29. Patterson BW, Horowitz JF, Wu G, Watford M, Coppack SW, et al. (2002) Regional muscle and adipose tissue amino acid metabolism in lean and obese women. Am J Physiol Endocrinol Metab 282: E931–936.
30. Lundholm K, Edstrom S, Ekman L, Karlberg I, Walker P, et al. (1981) Protein degradation in human skeletal muscle tissue: the effect of insulin, leucine, amino acids and ions. Clin Sci (Lond) 60: 319–326.
31. Shimomura Y, Obayashi M, Murakami T, Harris RA (2001) Regulation of branched-chain amino acid catabolism: nutritional and hormonal regulation of activity and expression of the branched-chain alpha-keto acid dehydrogenase kinase. Curr Opin Clin Nutr Metab Care 4: 419–423.
32. Harris RA, Kobayashi R, Murakami T, Shimomura Y (2001) Regulation of branched-chain alpha-keto acid dehydrogenase kinase expression in rat liver. J Nutr 131: 841S–845S.
33. Holecek M (2001) Effect of starvation on branched-chain alpha-keto acid dehydrogenase activity in rat heart and skeletal muscle. Physiol Res 50: 19–24.
34. Macotela Y, Emanuelli B, Bang AM, Espinoza DO, Boucher J, et al. (2011) Dietary leucine–an environmental modifier of insulin resistance acting on multiple levels of metabolism. PLoS One 6: e21187.
35. Zhang Y, Guo K, LeBlanc RE, Loh D, Schwartz GJ, et al. (2007) Increasing dietary leucine intake reduces diet-induced obesity and improves glucose and cholesterol metabolism in mice via multimechanisms. Diabetes 56: 1647–1654.
36. Ramirez-Zacarias JL, Castro-Munozledo F, Kuri-Harcuch W (1992) Quantitation of adipose conversion and triglycerides by staining intracytoplasmic lipids with Oil red O. Histochemistry 97: 493–497.
37. Vassault A (1983) Lactate dehydrogenase. In: Bergmeyer HO, editor. Methods of Enzymatic Analysis, Vol III, Enzymes: Oxireductases, Transferases. 3rd ed. New York: Academic Press. pp. 118–126.
38. Livak KJ, Schmittgen TD (2001) Analysis of relative gene expression data using real-time quantitative PCR and the 2(-Delta Delta C(T)) Method. Methods 25: 402–408.
39. Polz MF, Cavanaugh CM (1998) Bias in template-to-product ratios in multitemplate PCR. Appl Environ Microbiol 64: 3724–3730.
40. Newgard CB (2012) Interplay between lipids and branched-chain amino acids in development of insulin resistance. Cell Metabolism 15: 606–614.
41. Laferrere B, Reilly D, Arias S, Swerdlow N, Gorroochurn P, et al. (2011) Differential metabolic impact of gastric bypass surgery versus dietary intervention in obese diabetic subjects despite identical weight loss. Science translational medicine 3: 80re82.
42. Tremblay F, Marette A (2001) Amino acid and insulin signaling via the mTOR/p70 S6 kinase pathway. A negative feedback mechanism leading to insulin resistance in skeletal muscle cells. J Biol Chem 276: 38052–38060.
43. She P, Reid TM, Bronson SK, Vary TC, Hajnal A, et al. (2007) Disruption of BCATm in mice leads to increased energy expenditure associated with the activation of a futile protein turnover cycle. Cell Metab 6: 181–194.
44. Sans RM, Jolly WW, Harris RA (1980) Studies on the regulation of leucine catabolism. Mechanism responsible for oxidizable substrate inhibition and dichloroacetate stimulation of leucine oxidation by the heart. Arch Biochem Biophys 200: 336–345.
45. Williamson JR, Walajtys-Rode E, Coll KE (1979) Effects of branched chain alpha-ketoacids on the metabolism of isolated rat liver cells. I. Regulation of branched chain alpha-ketoacid metabolism. J Biol Chem 254: 11511–11520.
46. Koves TR, Ussher JR, Noland RC, Slentz D, Mosedale M, et al. (2008) Mitochondrial overload and incomplete fatty acid oxidation contribute to skeletal muscle insulin resistance. Cell Metab 7: 45–56.
47. Takano A, Usui I, Haruta T, Kawahara J, Uno T, et al. (2001) Mammalian target of rapamycin pathway regulates insulin signaling via subcellular redistribution of insulin receptor substrate 1 and integrates nutritional signals and metabolic signals of insulin. Molecular and cellular biology 21: 5050–5062.
48. Chan A, Newman DL, Shon AM, Schneider DH, Kuldanek S, et al. (2006) Variation in the type I interferon gene cluster on 9p21 influences susceptibility to asthma and atopy. Genes Immun 7: 169–178.
49. Randle PJ, Garland PB, Hales CN, Newsholme EA (1963) The glucose fatty-acid cycle. Its role in insulin sensitivity and the metabolic disturbances of diabetes mellitus. Lancet 1: 785–789.
50. Hue L, Taegtmeyer H (2009) The Randle cycle revisited: a new head for an old hat. American journal of physiology Endocrinology and metabolism 297: E578–591.

19

Immunopathogenesis of HIV Infection in Cocaine Users: Role of Arachidonic Acid

Thangavel Samikkannu[1], Kurapati V. K. Rao[1], Hong Ding[1], Marisela Agudelo[1], Andrea D. Raymond[1], Changwon Yoo[2], Madhavan P. N. Nair[1]*

[1] Department of Immunology, Institute of NeuroImmune Pharmacology, Herbert Wertheim College of Medicine, Florida International University, Modesto A. Maidique Campus, Miami, Florida, United States of America, [2] Department of Biostatistics, Robert Stempel College of Public Health and Social Work, Florida International University, Modesto A. Maidique Campus, Miami, Florida, United States of America

Abstract

Arachidonic acid (AA) is known to be increased in HIV infected patients and illicit drug users are linked with severity of viral replication, disease progression, and impaired immune functions. Studies have shown that cocaine accelerates HIV infection and disease progression mediated by immune cells. Dendritic cells (DC) are the first line of antigen presentation and defense against immune dysfunction. However, the role of cocaine use in HIV associated acceleration of AA secretion and its metabolites on immature dendritic cells (IDC) has not been elucidated yet. The aim of this study is to elucidate the mechanism of AA metabolites cyclooxygenase-2 (COX-2), prostaglandin E2 synthetase (PGE$_2$), thromboxane A2 receptor (TBXA$_2$R), cyclopentenone prostaglandins (CyPG), such as 15-deoxy-Δ12,14-PGJ2 (15d-PGJ2), 14-3-3 ζ/δ and 5-lipoxygenase (5-LOX) mediated induction of IDC immune dysfunctions in cocaine using HIV positive patients. The plasma levels of AA, PGE$_2$, 15d-PGJ2, 14-3-3 ζ/δ and IDC intracellular COX-2 and 5-LOX expression were assessed in cocaine users, HIV positive patients, HIV positive cocaine users and normal subjects. Results showed that plasma concentration levels of AA, PGE$_2$ and COX-2, TBXA$_2$R and 5-LOX in IDCs of HIV positive cocaine users were significantly higher whereas 15d-PGJ2 and 14-3-3 ζ/δ were significantly reduced compared to either HIV positive subjects or cocaine users alone. This report demonstrates that AA metabolites are capable of mediating the accelerative effects of cocaine on HIV infection and disease progression.

Editor: Shilpa J. Buch, University of Nebraska Medical Center, United States of America

Funding: This study was supported by grants from the National Institutes Health: DA 025576 and DA034547 to Prof. Madhavan Nair. No additional external funding was received for this study. The funders had no role in study design, data collection and analysis, decision to publish, or preparation of the manuscript.

Competing Interests: The authors have declared that no competing interests exist.

* Email: nairm@fiu.edu

Introduction

During the last decade, an intertwined epidemic of drug abuse and HIV-1 infections has emerged. Globally there were an estimated 34.2 million people living with HIV [1]. Illicit drug abuse including cocaine is a significant risk factor for HIV infection and AIDS disease progression [2,3]. Cocaine is currently being used worldwide in epidemic proportions, particularly in the U.S. The 2010 report shows that 1.5 million Americans (aged 12 or older) are cocaine users [4]. Overall, about 16 million injecting drug users are present worldwide and 3 million (18.9 per cent) of them are living with HIV [1]. Previous studies suggest that cocaine use and HIV-1 infection are independently associated with immune dysfunction which leads to neuronal impairments [5,6].

Dendritic cells (DC) play a significant role as the first line of defense against viral pathogens and illicit drug effects [7,8]. HIV-1 directly affects dendritic cells (DC) and leads to dysfunction of immune system manifested by increased levels of inflammatory cytokines, chemokines and neurotoxin such as quinolinic acid and arachidonic acid (AA) [9,10]. Increasing evidence suggests that DCs play a major role in the defense against HIV infection and illicit drug such as cocaine [11–13]. Immature dendritic cells (IDC) specialize in capturing and processing antigens and plays wide role in cell maturation, migration to CD4+ T cells, and T cell

activation [14]. Previous studies indicate that AA metabolites such as COX-2, TBXA2, 5-LOX and 15d-PGJ2 found in specific DC subsets interplay with immune regulation [15,16]. Also, AA metabolites COX-2 induce T-cell tolerance to antigenic stimuli which could affect immune functions [17]. Indeed, expression of COX-2 activation subsequently affect via TBXA2, 15d-PGJ2 and 5-LOX which are the potential markers of viral replication as well as immune and neuronal impairments [18,19]. However, the COX-2 and 5-LOX can be regulated via monocytes and dendritic cells through activation of T cells signaling during inflammatory processes [20]. Furthermore, the 5-LOX enzyme plays an important role in leukotriene B$_4$, a potent inflammatory mediator in peripheral disorders [21], and neurotoxicity [22]. The members of the PGJ2 class, 15d-PGJ2 (also called cyclopentenone PGs, CyPG), play a role in checkpoint of cytokine/chemokine synthesis and intracellular translocation of HIV viral protein and viral replication [23]. 15d-PGJ2 has anti-inflammatory properties [24], and it negatively regulates PGE$_2$ synthetase. However, increased levels of AA directly bind with 14-3-3 ζ/δ protein polymerization and affect their cellular function [25]. Furthermore, decreased 14-3-3 ζ/δ proteins subsequently affect platelet aggregation mediated by platelet activating factor (PAF), which may induce apoptosis.

Studies have consistently demonstrated that cocaine use and HIV infection accelerates viral replication, disease progression which leads susceptibility and severity of immune dysfunction [3,25] which leads to HIV-associated neurocognitive disorder (HAND) [26]. HIV positive cocaine user's exhibit accelerated disease progression compared to non- cocaine using HIV positive individuals [2, 3. 26, 27]. Our recent report demonstrated that HIV derived gene product gp120 with cocaine interaction potentiated the additive effect of AA metabolite COX-2 induction in primary astrocytes [28]. Despite mounting evidence which suggests that cocaine use may exacerbate HIV disease, mechanistic studies assessing the interactive role of cocaine and HIV infection on DC and their role remains to be determined.

In this study, we investigated the role of AA metabolites COX-2, PGE$_2$, 15d-PGJ2, 14-3-3 ζ/δ, TBXA$_2$R, and 5-LOX associated immunopathogenesis in HIV positive cocaine users. We showed that cocaine use in HIV positive subjects enhances immune dysfunction and exacerbates the neurotoxin AA metabolites gene, protein and intracellular expression in IDC.

Methods and Materials

Human subjects

Blood donors were recruited from the Borinquen Health Care Center, Inc., Miami. Written and signed consent forms were obtained from all participants in compliance with Florida International University (FIU) and the National Institutes of Health (NIH) policies. All participants consent forms are kept in a lock cabinet to maintain confidentiality. The protocol was approved by the IRB of FIU. Peripheral blood from normal, HIV positives, cocaine users and HIV-1 positive cocaine users was drawn into heparin lined tubes (20cc). Human subjects were recruited through collaborating physicians, community agencies, as well as by participant referral. After eligibility was confirmed, informed consents were obtained. Participants provided medical documentation to confirm HIV and Hepatitis B and C status. For HIV positive individuals, medical documentation confirmed CD4 and Viral loads. Individuals younger than 18 years old and with Hepatitis B and/or Hepatitis C infection were deemed ineligible. Self-report data was collected concerning drug use history, and participant's blood specimens were collected by a Registered Nurse.

HIV-infected and control populations

HIV-1 infection are defined as individuals who exhibit a faster decline in CD4 lymphocytes, displaying a declining slope of >50 CD4 T cells per cubic millimeter per semester and an overall CD4 loss >30% with a current CD4 count <500 and plasma viral load >50,000 copies/ml. However, in the present study, overall patients were taking more than two antiretroviral drugs and the patients taking ART showed increase in CD4 significantly. The number of normal subjects, cocaine user, HIV positive and HIV positive cocaine user average age and CD4 counts are given in Table 1. Cocaine use with/without HIV were confirmed by urine toxicology testing on the day of blood collection. Control or normal groups (drug free and HIV negative) were healthy volunteers who were age-, sex-, and ethnically matched with HIV-1-infected subjects and cocaine users. A clinical history was obtained from all normal donors. However, in this study only cocaine user and HIV positive cocaine users were recruited. Exclusion criteria for all groups were age <18 and >50 years, pregnancy, Hepatitis B and C (Table 1).

Isolation and generation of immature DC (IDC)

DCs were prepared from PBMC as described [2]. Blood samples from normal, cocaine users, HIV positive and HIV positive cocaine users were collected and Peripheral Blood Mononuclear Cells (PBMC) were separated on a density gradient and adhered to plastic culture plates in serum containing medium. Non adherent cells were removed after 1 h at 37°C, and adherent cells were cultured for 6 days in medium containing 100 U/ml recombinant human GM-CSF and 100 U/ml IL-4 (R&D Systems). After 6 days of culture, iDC were removed by gently swirling the plate to resuspend them for use in the experiments. IDC were washed in FACS buffer (eBioscience), incubated with nonspecific IgG (20 µg/ml) for 10 min at 4°C to block FcR, stained with specific Abs for DC surface markers, and analyzed by flow cytometry. The IDC express CD80, CD86, CD40, HLA-DR, DQ, and CD11c at different levels.

RNA Extraction and Real time quantitative PCR (qRT-PCR)

Total RNA from IDC was extracted using the Qiagen kit (Invitrogen Life Technologies, Carlsbad, CA, USA) following the manufacturer's instructions. The total RNA (3 µg) was used for the synthesis of the first strand of cDNA. The amplification of cDNA was performed and using specific primers for COX-2 (Assay ID, Hs00153133), TBXA2 R (Assay ID, Hs00169054), 5-LOX (Assay ID Hs00386528), and β-actin (Assay ID, Hs99999903) (Applied Biosystems, Foster City, CA) was used as housekeeping gene for quantifying real-time PCR. Relative abundance of each mRNA species was assessed using brilliant Q-PCR master mix from Stratagene using Mx3000P instrument which detects and plots the increase in fluorescence versus PCR cycle number to produce a continuous measure of PCR amplification. Relative mRNA species expression was quantitated and the mean fold change in expression of the target gene was calculated using the comparative CT method (Transcript Accumulation Index, TAI $= 2^{-\Delta\Delta CT}$). All data were controlled for quantity of RNA input by performing measurements on an endogenous reference gene, β-actin. In addition, results on RNA from treated samples were normalized to results obtained on RNA from the control, untreated sample.

Quantification of PGE$_2$ and 15d-PGJ2 by enzyme- linked immunosorbent assay (ELISA)

Plasma were separated from normal, cocaine users, HIV positive patients, and cocaine using HIV positive subjects as previously mentioned. Plasma was analyzed for PGE$_2$ (GenWay Biotech Inc. San Diego, CA) and 15d-PGJ2 (Enzo Life Sciences, Farmingdale, NY) using commercially available ELISA kits as per the manufacturer's instructions.

Level of neurotoxin AA in plasma by GC/MS

Blood samples were centrifuged at 2400 g for 10 min at 4°C. Aliquots (1 mL) of plasma were transferred to Eppendorf tubes and were stored at −70°C until analysis. Plasma 50 µl was hydrolyzed in the presence of 1 mol/L potassium hydroxide at 40°C for 30 min. 50 µL of distilled water, 3 µL of concentrated hydrochloric acid, 200 µL of ice-cold Folch solution (chloroform–methanol, 2:1 by volume), and 50 ng of AA-d$_8$ (internal standard) were added to 50 µL of hydrolyzed sample. After thorough vortex-mixing, the sample was centrifuged at 2400 g for 5 min. The lower organic layer was transferred to another Eppendorf tube and evaporated under nitrogen. The residue was then dissolved in 100 µL distilled water, and AA was extracted by the addition of 100 µL of hexane. The hexane extracts were dried under nitrogen and AA was analyzed by GC/MS, according to the modified method of Hadley et al [29].

Table 1. Baseline characteristics of HIV positive subjects with cocaine user.

	Normal Subjects	Cocaine User	HIV Positive	HIV Positive Cocaine user
Age	32.46±6.22	41.52±3.81	36.14±3.75	44.57±4.63
Sex	Male −9 Female – 3	Male −6 Female – 6	Male −11 Female - 8	Male −9 Female – 9
CD4	–	–	≥466/µl	≥308/µl

Analysis of intracellular expression of COX-2 and 5-LOX in IDC

Cocaine and HIV induced intracellular COX-2 and 5-LOX expressions were analyzed by flow cytometry in FACS caliber (BD Bioscience, San Jose, CA). Briefly, IDC (5×10^5) were separated from normal, cocaine users, HIV positive patients, and HIV positive with cocaine user's subjects. Cells were harvested and washed twice with PBS. The cells were fixed and permeabilized by Cytofix/Cytoperm Kit (BD Pharmingen, San Diego, CA, USA) following the manufacturer's instructions. Cells were incubated with anti-COX2 (FITC) and anti-5 LOX (FITC) (antibodies-online, Atlanta, USA) for 20 min at 4° than washing with PBS and analyzed intracellular COX-2 and 5-LOX expression.

Western blot analysis

To determine the COX-2, TBXA2 R and 5-LOX protein modification in IDC and serum 14-3-3 ζ/δ protein levels in normal, cocaine users, HIV positive and cocaine using HIV positive subjects. Equal amount of total cellular protein were resolved on a 4–15% gradient polyacrylamide gel electrophoresis, transferred to a nitrocellulose membrane and incubated with their respective primary antibodies. Immunoreactive bands were visualized using a chemiluminescence western blotting system according to the manufacturers' instructions (Amersham Piscataway, NJ, USA).

Data analysis

In general total of 60 samples were utilized for the experiments. However, some of the experiments as indicated in the figures at least six samples were utilized in each group. The values obtained were averaged and data are represented as the mean ± standard error. All the data were analyzed using GraphPad Prism- software version 5. Comparisons between groups were performed using one-way ANOVA and differences were considered significant at $p \leq 0.05$ and we performed post- test using "t" test.

Results

Demographics of Cohort

A total of 60 subjects were utilized for the study. The groups were as follows: 12-control (normal) subjects, 12- cocaine users, 19-HIV positive subjects and 18- HIV positive subjects with long-standing history of cocaine use. The groups of patients and control subjects were similar in age with male predominance.

Cocaine increases HIV-1induced AA metabolites in IDC

Studies have shown that HIV positive subjects who have used illicit drugs are more sensitive to immune-neuropathogenesis than either substance abusers or HIV positive subjects [26,30]. In this study, we have investigated the possible role of cocaine increasing HIV induced AA and its metabolites COX-2, TBXA$_2$ and 5-LOX in IDC. Data presented in Fig. 1A show IDC from cocaine users

($p<0.03$), HIV positive patients ($p<0.03$) and cocaine users with HIV positive subjects ($p<0.003$) significantly upregulated COX-2 gene expression compared to normal subjects. Fig. 1B shows a significant increase in TBXA$_2$ gene expression cocaine users ($p<0.01$), HIV positive ($p<0.007$) and cocaine user with HIV positive subjects ($p<0.03$) compared with normal subjects. To examine whether AA metabolites of 5-LOX expression are similar or different from COX-2 expression, we performed expression profiles of 5-LOX. Fig. 1C results indicate that in cocaine users ($p<0.04$), HIV positives ($p<0.02$) and cocaine users with HIV positive subjects ($p<0.002$) have significantly higher levels of 5-LOX gene expression, compared to normal subjects.

Cocaine impact of AA metabolites PGE$_2$ and 15d-PGJ2 in HIV-infected Plasma

Our previous report demonstrated that AA metabolites potentiated additive effect of gp120 with cocaine [28]. Therefore, in this study we also examined the level of AA, metabolites PGE$_2$ and 15d-PGJ2 in cocaine users, HIV positives and cocaine users with HIV positives subjects. Data presented in Fig. 2A show that cocaine users have increased levels of AA but not significant; however, HIV positive ($p<0.02$) and cocaine users with HIV positives ($p<0.002$) have significantly increased higher levels of AA compared to normal subjects. The level of PGE$_2$ in cocaine users and HIV positive cocaine users shows a significant increase when compared to controls. The data in Fig. 2B show the level of PGE$_2$ respectively in HIV positive subjects ($p<0.0001$), cocaine user ($p<0.0001$) and cocaine users with HIV positive subjects ($p<0.0001$). Cocaine and HIV infection is synergistic as evidenced by significantly increased PGE$_2$ ($p<0.0001$) when compared to either HIV positive ($p<0.0001$) or cocaine use alone ($p<0.0001$). In addition, the AA metabolite of anti-inflammatory 15d-PGJ2 is the major player in controlling the onset and resolution of acute inflammation. Since no studies have reported on the quantification of 15d-PGJ2 in HIV infected drug abusers. Fig. 2C indicated that the level of 15d-PGJ2 in HIV positives cocaine users shows a significant decrease ($P<0.02$) when compared to either cocaine or HIV positive ($P<0.02$) subjects. In the analysis of present data using in ANOVA the values were significantly less than $p<0.001$ and accordingly suggest the synergistic effects in the anti-inflammatory response which leads to increased AA metabolites in HIV positive cocaine users.

Cocaine accelerates intracellular expression of COX-2 and 5-LOX

Since increased AA levels might results in modulation of intracellular levels of COX-2 and 5-LOX, and these two important enzymes are metabolized polyunsaturated fatty acids that function as parallel pathways on AA metabolism, we analyzed intra cellular expression levels of COX-2 and 5-LOX by flow cytometry. Data presented in Fig. 3 shows COX-2 expression in cocaine user (12%), HIV positive (14%) and HIV positive with

A

B

C

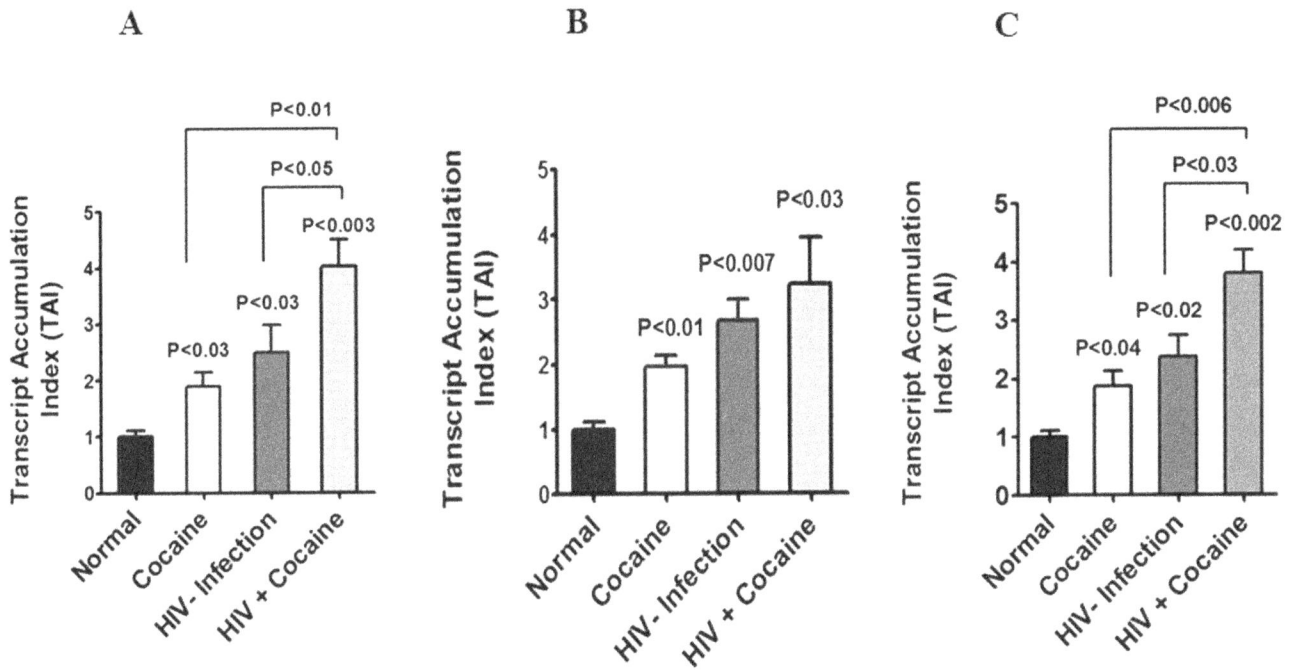

Figure 1. Effect of arachidonic acid metabolites COX-2, TBXA2 R and 5-LOX gene expression. IDC (3×10^6 cells/ml) were isolated from normal, cocaine users, HIV positives and HIV positive cocaine users. RNA was extracted and reverse transcribed followed by quantitative real time PCR for COX-2 (A), TBXA2 R (B), 5-LOX (C), and housekeeping β-actin specific primers. Data are expressed as mean $\pm$ SD of TAI values of six independent human samples.

cocaine users (27%) compared with control. Fig. 4 demonstrated that 5-LOX expression significantly increased in HIV positive cocaine users (41%) in comparison with either cocaine users (13%) or HIV positive subjects (17%) alone. These results suggest that HIV positive cocaine users have higher levels of AA and metabolites of COX-2 and 5-LOX expression.

Cocaine exacerbates HIV-induced protein modification in IDC

In order to understand protein expression mimics similar to gene expression in AA metabolites in HIV positive cocaine users. Therefore, we also analyzed COX-2, TBXA2 R, 5-LOX and 14-3-3 ζ/δ protein expression in terms of cocaine users, HIV infected

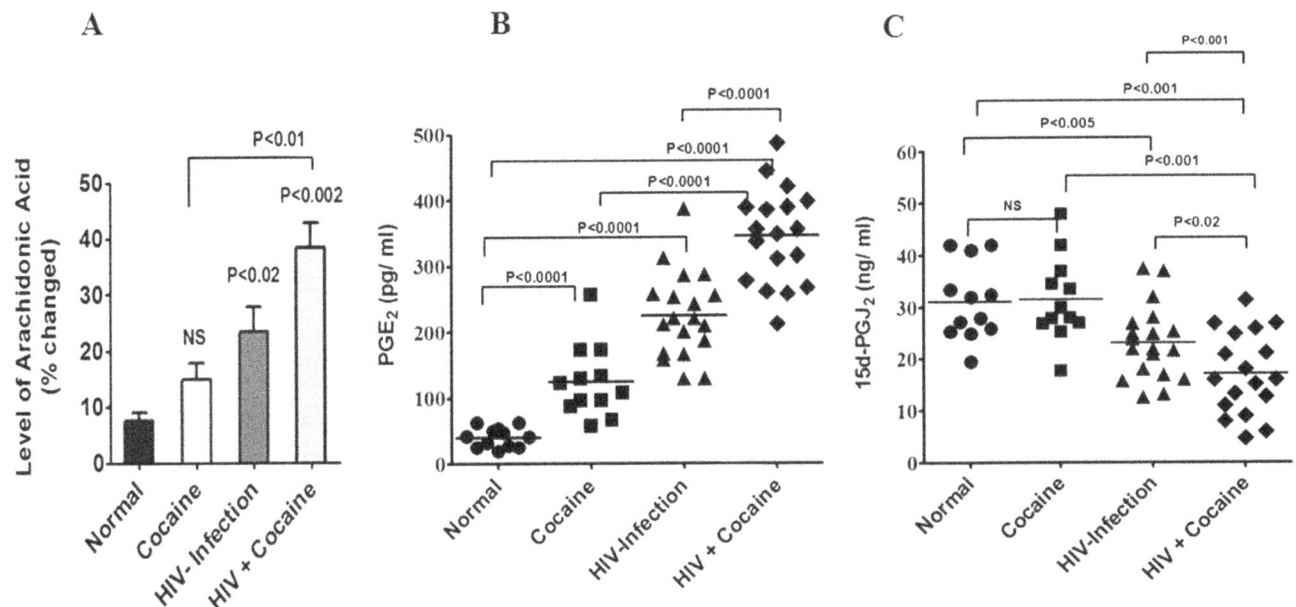

A

B

C

Figure 2. The level of AA, PGE2 and 15d-PGJ2 in cocaine users and HIV infected subjects. The plasma was separated and analyzed AA by GC/MS, PGE2 and 15d-PGJ2 level by ELISA in normal, cocaine users, HIV positives, and HIV positive cocaine users. Data are expressed as % fold change of AA (A), and ng/ml plasma (B and C).

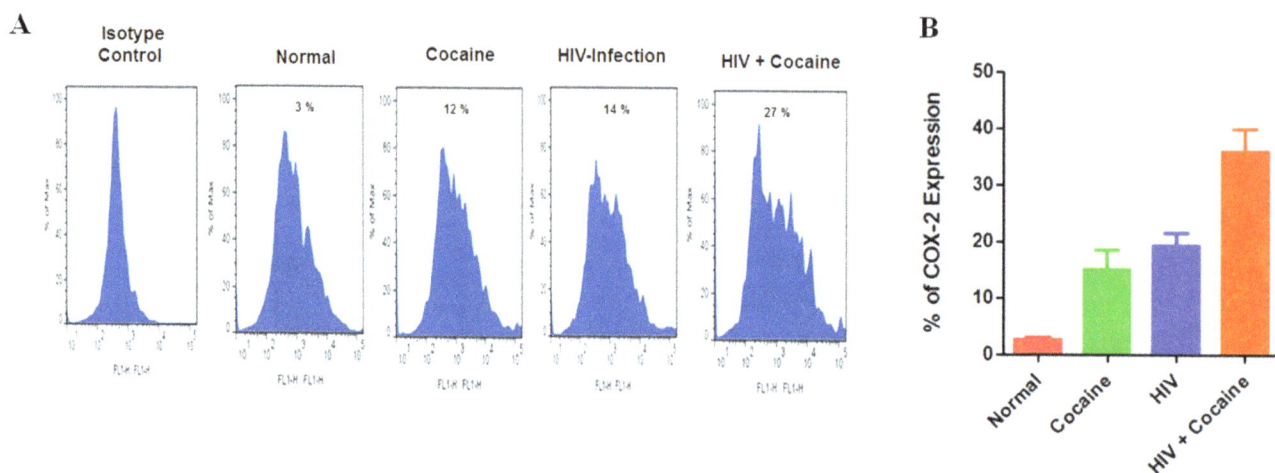

Figure 3. Effect of intracellular COX-2 expression in cocaine users and HIV infected patients. IDC (5×10^5 cells/ml) were isolated from normal, cocaine users, HIV positive and HIV positive cocaine user intracellular expression of COX-2 (A) and % COX-2 positive cells (B) was analyzed by flow cytometry. Data are expressed and represented as mean $\pm$ SD of six independent experiments.

subjects and cocaine using HIV infected patients in IDC. Fig. 5 demonstrates that COX-2 (A), TBXA$_2$ R (B) and 5-LOX (C) protein is significantly up regulated whereas 14-3-3 ζ/δ (D) significantly reduced by cocaine users, HIV positive subjects and HIV positive with cocaine users. These observed results are consistent with cocaine using HIV infected patients levels of AA and metabolites PGE$_2$ and 15d-PGJ2. Data presented in Fig. 5 E, F, G and H show the densitometry evaluation respectively COX-2 (p<0.009), TBXA$_2$ R (p<0.009), 5-LOX (p<0.004) and 14-3-3 ζ/δ (p<0.007). These results confirm increased level of AA, PGE$_2$ and COX-2 expression subsequently decreased anti-inflammatory level of 15d-PGJ2 and 14-3-3 ζ/δ protein inhibition in cocaine users, HIV positive subjects and HIV positive cocaine users when compared with normal subjects.

Discussion

Previous studies have shown that AA and its metabolites play a wide role in immune dysfunction, behavioral impairments as well as viral replication and disease progression in HIV-infection and substance abuse [9,18,23]. The increasing AA metabolites COX-2

and 5-LOX are associated with HAD or HAND. COX-2 enzyme is an important player in the regulation of immune functions (e.g. immune tolerance) of antigen-presenting cells such as macrophages or DC [31,32]. An increase in AA secretion by HIV infection and their metabolites COX-2, PGE$_2$, TBXA$_2$ and 5-LOX are found in cerebrospinal fluid (CSF) of HAD-patients [33–35]. However, overstimulation of AA leads to increase in its metabolites COX-2 and PGE$_2$ [36,37], and subsequently decreased the level of 15d-PGJ2 and 14-3-3 ζ/δ, which may play a vital role in immune dysfunction and disease progression [38,39] in HAD patients [26]. The DC differentiation, maturation, migration, and antigen presentation function are modulated by COX-2 induced prostaglandins (PGs) [15,40,41]. However, there are no reports on the impact of cocaine on AA metabolites in HIV positive cocaine users. The present study provides new insights into the functional role of COX-2 in AA metabolites TBXA2 which subsequently affects 5-LOX in HIV infection and cocaine use. Our previous *in vitro* study has shown that the HIV-1 gp120 protein induces the COX-2 mRNA expression and protein modification implicated in neuro-AIDS [42].

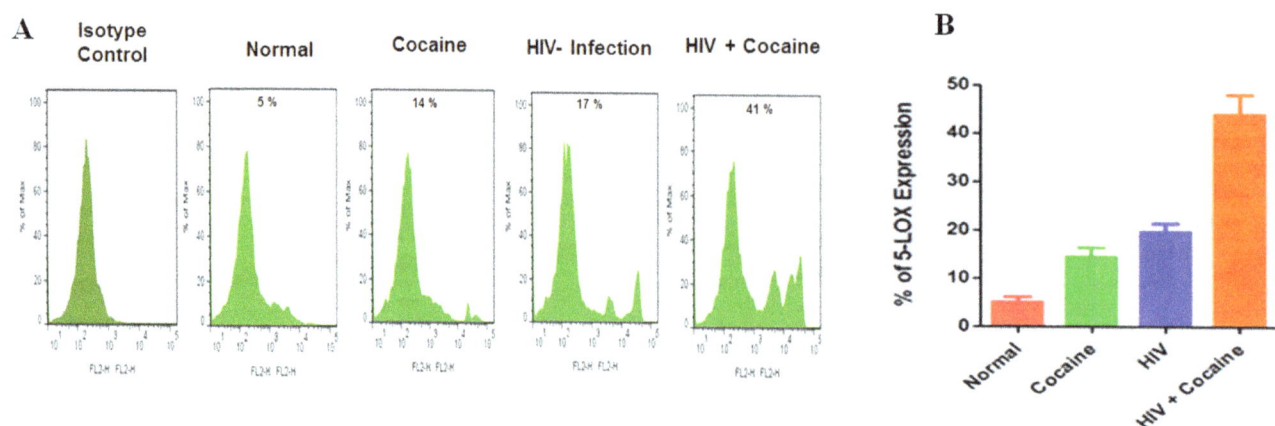

Figure 4. Effect of intracellular 5-LOX expression in cocaine users and HIV infected patients. IDC (5×10^5 cells/ml) were isolated from normal, cocaine users, HIV positives and HIV positive cocaine users intracellular expression of 5-LOX (A) and % 5-LOX positive cells (B) was analyzed by flow cytometry. Data are expressed and represented as mean $\pm$ SD of six independent experiments.

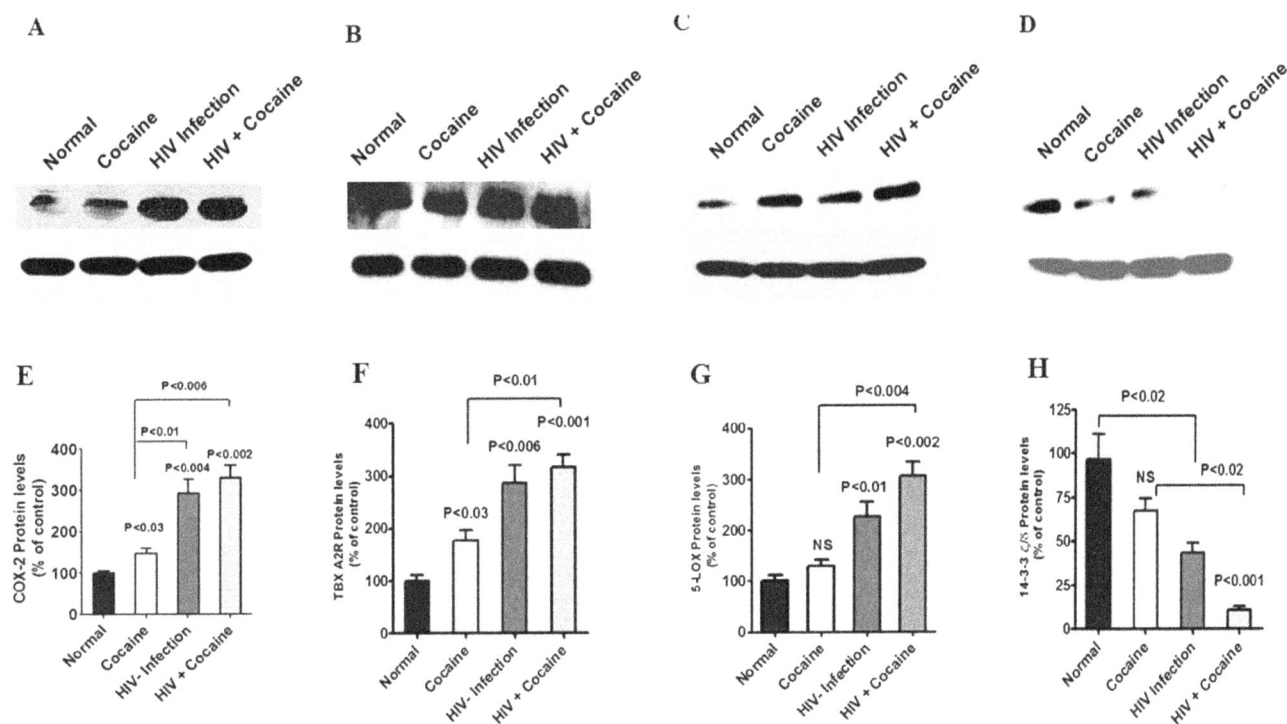

Figure 5. Effect of arachidonic acid metabolites COX-2, TBXA2 R and 5-LOX protein expression in IDC and 14-3-3 δ/ζ in plasma were isolated from normal, cocaine user, HIV positives and HIV positive cocaine users. Equal amount of IDC cell lysate and plasma protein were resolved by 4–15% SDS-PAGE and protein expression were analyzed by Western blot showing COX-2 (A), TBXA2 R (B), 5-LOX (C) and 14-3-3 δ/ζ (D). Figure E, F, G and H represented % densitometric values of COX-2, TBXA2 R, 5-LOX and 14-3-3 δ/ζ protein levels (% control). Data are expressed as mean ± SD of three independent experiments.

In the present study, we have demonstrated for the first time that cocaine users, HIV positive and HIV positive cocaine users have increased levels of AA and mRNA expression of metabolites COX-2, TBXA2, 5-LOX (Fig. 1), and the levels of AA and PGE_2 (Fig. 2) are associated with reduction in 15d-PGJ2 compared to normal subjects. It is known that AA metabolites PGE_2, COX-2, $TBXA_2$ and 5-LOX are the major players in immune dysfunction [43–45], and reduced level of 15d-PGJ2 and 14-3-3 ζ/δ, may enhance viral replication and disease progression. These studies suggest that cocaine abusing HIV positive subjects may have an enhanced role of COX-2 and AA metabolites compared to normal subjects. This is consistent with earlier reports of gp120 induced neuroblastoma cells and HIV infected pulmonary hypertension, where activation of the COX-2 and 5-LOX pathways has been observed [36,46]. Also, studies have shown that suicidal behavioral impairments are associated with 5-LOX increased in cerebral cortex brain regions [47]. These studies further confirm that the downstream effect of TBXA2 and 5-LOX are upregulated in cocaine using HIV positive subjects leading to depression and suicidal behavior. However, TBXA2 is an unstable product which initiates the PAF, blood aggregation induction and subsequently increases TBX B, which may be mediated by 14-3-3 ζ/δ leading to immune and neuronal impairments. Supporting our hypothesis, recent studies have demonstrated that 14-3-3 ζ/δ protein play a wide role in cytoskeletal translocation and platelet dysfunctions [48]. These results confirm that increased AA metabolites exacerbate the immune dysfunction in cocaine abusing HIV infected subjects.

Further, our results show that in cocaine users, HIV infected subjects, and cocaine using HIV infected subjects induction of intracellular COX-2 and 5-LOX expression (Fig. 3 and 4) is

associated with a concomitant activation of COX-2 and 5-LOX protein (Fig. 5). The main observation in this report is that HIV positive and cocaine using HIV positive subjects have higher levels of PGE_2 due to secretion of AA and COX-2 activation and subsequently reducing the level of 15d-PGJ2 and 14-3-3 ζ/δ. However, HIV positive cocaine users have higher levels of AA and metabolites. This suggests that HIV positive cocaine synergistically potentiate disease progressive effect when compared to either cocaine use or HIV positive alone.

Previous studies indicate that AA metabolites affect monocytes in HIV positive drug users [33] and HIV-1 envelop protein gp120 induced viral replication subsequently affect the immune function in human DCs [40]. In the present study, IDC in cocaine using HIV infected subjects showed an increased level of AA metabolites COX-2, TBXA2 R and 5-LOX gene expression and protein modification with significant increase of PGE_2 levels compared to normal subjects. Also, increased functional COX-2 enzyme activity in DC could lead to an enhanced production of neurotoxin AA [49], which alters immune tolerance and causes imbalance resulting in immuno-neuropathogenesis [50]. These results suggest that IDC play a role in protective mechanisms of immune function, during HIV infection and cocaine abuse since both can alter AA levels and subsequently accelerate disease progression mediated by COX-2 and 5-LOX.

Overall, the data provide evidence of an interaction of cocaine use and HIV infection leading to an association between AA and its metabolites COX-2, TBXA2 R and 5-LOX by increasing the levels of PGE_2, and formation of neurotoxin AA subsequently reducing the levels of 15d-PGJ2 and 14-3-3 ζ/δ in HIV positive cocaine users. Based on these results, AA and its metabolites potentiate the HIV disease progression by impairing the possible immune function [3,26,27,51].

Author Contributions

Conceived and designed the experiments: TS MPNN. Performed the experiments: TS KRVK MA HD ADR. Analyzed the data: TS KRVK MPNN CY. Contributed reagents/materials/analysis tools: MPNN. Wrote the paper: TS MPNN.

References

1. UNAIDS/WHO (2011) AIDS epidemic update: November 2012. UNAIDS, UNAIDS/09.36E/JC2417E. ISBN: 978-92-9173-996-7.
2. Nair MP, Mahajan SD, Schwartz SA, Reynolds J, Whitney R, et al. (2005) Cocaine modulates dendritic cell-specific C type intercellular adhesion molecule-3-grabbing nonintegrin expression by dendritic cells in HIV-1 patients. J Immunol 174: 6617–6626.
3. Baldwin GC, Roth MD, Tashkin DP (1998) Acute and chronic effects of cocaine on the immune system and the possible link to AIDS. J Neuroimmunol 83: 133–138.
4. Substance Abuse and Mental Health Services Administration (2011) Results from the 2010 National Survey on Drug Use and Health: Summary of National Findings, NSDUH Series H-41, HHS Publication No. (SMA) 11-4658. Rockville, MD: Substance Abuse and Mental Health Services Administration, 2011.
5. Klein TW, Matsui K, Newton CA, Young J, Widen RE, et al. (1993) Cocaine suppresses proliferation of phytohemagglutininactivated human peripheral blood T-cells. Int J Immunopharmacol 15: 77–86.
6. Yao H, Kim K, Duan M, Hayashi T, Guo M, et al. (2011) Cocaine hijacks sigma1 receptor to initiate induction of activated leukocyte cell adhesion molecule: implication for increased monocyte adhesion and migration in the CNS. J Neurosci 31: 5942–5955.
7. Geijtenbeek TB, van Kooyk Y (2003) DC-SIGN: a novel HIV receptor on DCs that mediates HIV-1 transmission. Curr Top Microbiol Immunol 276: 31–54.
8. Cameron PU, Lowe MG, Crowe SM, O'Doherty U, Pope M, et al. (1994) Susceptibility of dendritic cells to HIV-1 infection in vitro. J Leukoc Biol 56: 257–265.
9. Harizi H, Limem I, Gualde N (2011) CD40 engagement on dendritic cells induces cyclooxygenase-2 and EP2 receptor via p38 and ERK MAPKs. Immunol Cell Biol 89: 275–282.
10. Basselin M, Ramadan E, Igarashi M, Chang L, Chen M, et al. (2011) Imaging upregulated brain arachidonic acid metabolism in HIV-1 transgenic rats. J Cereb Blood Flow Metab 31: 486–493.
11. Reynolds JL, Mahajan SD, Aalinkeel R, Nair B, Sykes DE, et al. (2009) Proteomic analyses of the effects of drugs of abuse on monocyte-derived mature dendritic cells. Immunol Invest 38: 526–550.
12. Donaghy H, Stebbing J, Patterson S (2004) Antigen presentation and the role of dendritic cells in HIV. Curr Opin Infect Dis 17: 1–6.
13. Nair MP, Chadha KC, Hewitt RG, Chang L, Chen M, et al. (2000) Cocaine differentially modulates chemokine production by mononuclear cells from normal donors and human immunodeficiency virus type 1-infected patients. Clin Diagn Lab Immunol 7: 96–100.
14. Banchereau J, Steinman RM (1998) Dendritic cells and the control of immunity. Nature 392: 245–252.
15. Harizi H, Grosset C, Gualde N (2003) Prostaglandin E2 modulates dendritic cell function via EP2 and EP4 receptor subtypes. J Leukoc Biol 73: 756–763.
16. Valera I, Fernández N, Trinidad AG, Alonso S, Brown GD, et al. (2008) Costimulation of dectin-1 and DC-SIGN triggers the arachidonic acid cascade in human monocyte-derived dendritic cells. J Immunol 180: 5727–5736.
17. Janelle ME, Gravel A, Gosselin J, Tremblay MJ, Flamand L (2002) Activation of monocyte cyclooxygenase-2 gene expression by human herpesvirus 6. Role for cyclic AMP-responsive element-binding protein and activator protein-1. J Biol Chem 277: 30665–30674.
18. Steer AS, Corbett JA (2003) The role and regulation of COX-2 during viral infection. Viral Immunol 16: 447–460.
19. Manev H, Manev R (2007) 5-lipoxygenase as a possible biological link between depressive symptoms and atherosclerosis. Arch Gen Psychiatry 64: 1333.
20. Tsatsanis C, Androulidaki A, Venihaki M, Margioris AN (2006) Signaling networks regulating cyclooxygenase-2. Int J Biochem Cell Biol. 38: 1654–1661.
21. Crooks SW, Stockley RA (1998) Leukotriene B4. Int J Biochem Cell Biol 30: 173–178.
22. Klegeris A, McGeer PL (2002) Cyclooxygenase and 5-lipoxygenase inhibitors protect against mononuclear phagocyte neurotoxicity. Neurobiol Aging 23: 787–794.
23. Rozera C, Carattoli A, De Marco A, Amici C, Giorgi C, et al. (1996) Inhibition of HIV-1 replication by cyclopentenone prostaglandins in acutely infected human cells. Evidence for a transcriptional block. J Clin Invest 97: 1795–1803.
24. Rossi A, Kapahi P, Natoli G, Takahashi T, Chen Y, et al. (2000) Anti-inflammatory cyclopentenone prostaglandins are direct inhibitors of IκB kinase. Nature (London) 403: 103–108.
25. Brock TG (2008) Arachidonic acid binds 14-3-3zeta, releases 14-3-3zeta from phosphorylated BAD and induces aggregation of 14-3-3zeta. Neurochem Res 33: 801–807.
26. Nath A, Maragos WF, Avison MJ, Schmitt FA, Berger JR (2001) Acceleration of HIV dementia with methamphetamine and cocaine. J NeuroVirol 7: 66–71.
27. Roth MD, Tashkin DP, Choi R, Jamieson BD, Zack JA, et al. (2002) Cocaine enhances human immunodeficiency virus replication in a model of severe combined immunodeficient mice implanted with human peripheral blood leukocytes. J Infect Dis 185: 701–705.
28. Nair MP, Samikkannu T (2012) Differential regulation of neurotoxin in HIV clades: role of cocaine and methamphetamine. Cur HIV Res 10: 429–34.
29. Hadley JS, Fradin A, Murphy RC (1998) Electron captures negative ion chemical ionization analysis of arachidonic acid. Biomed Environ Mass Spectrom 15: 175–8.
30. Samikkannu T, Rao KVK, Arias A, Kalaichezian A, Sagar V, et al. (2013) HIV Infection and Drugs of Abuse: Role of Acute Phase Proteins. J of Neuroinflam. 10: 113.
31. Harizi H, Limem I, Gualde N (2011) CD40 engagement on dendritic cells induces cyclooxygenase-2 and EP2 receptor via p38 and ERK MAPKs. Immunol Cell Biol 89: 275–282.
32. Shortman K, Shalin HN (2007) Steady-state and inflammatory dendritic-cell development. Net Rev Immunol 7: 19–30.
33. Ramis I, Roselló-Catafau J, Gómez G, Zabay JM, Fernández Cruz E, et al. (1991) Cyclooxygenase and lipoxygenase arachidonic acid metabolism by monocytes from human immune deficiency virus-infected drug users. J Chromatogr 557: 507–13.
34. Griffin DE, Wesselingh SL, McArthur JC (1994) Elevated central nervous system prostaglandins in human immunodeficiency virus-associated dementia. Ann Neurol 35: 592–97.
35. Miyamoto K, Lange M, McKinley G, Stavropoulos C, Moriya S, et al. Effects of sho-saiko-to on production of prostaglandin E2 (PGE2), leukotriene B4 (LTB4) and superoxide from peripheral monocytes and polymorphonuclear cells isolated from HIV infected individuals, Am J Chin Med 1996; 24: 1–10.
36. Maccarrone M, Navarra MT, Corasaniti G, Stavropoulos S, Moriya H, et al. (1998) Cytotoxic effect of HIV-1 coat glycoprotein gp120 on human neuroblastoma CHP100 cells involves activation of the arachidonate cascade, Biochem. J 333: 45–49.
37. Corasaniti MT, Strongoli MC, Piccirilli S, Nisticò R, Costa A, et al. (2000) Apoptosis induced by gp120 in the neocortex of rat involves enhanced expression of cyclooxygenase type 2 (COX-2) and is prevented by NMDA receptor antagonists and by the 21-aminosteroid U- 74389G. Biochem Biophys Res Commun 274: 664–669.
38. Su KP (2008) Mind-body interface: the role of n-3 fatty acids in psychoneuroimmunology, somatic presentation, and medical illness comorbidity of depression. Asia Pac J Clin Nutr 17: 147–53.
39. Basso MR, Bornstein RA (2000) Neurobehavioral consequences of substance abuse and HIV infection. J Psychopharmacol 14: 228–37.
40. Wang JH, Janas AM, Olson WJ, Wu L (2007) Functionally distinct transmission of human immunodeficiency virus type 1 mediated by immature and mature dendritic cells. J Virol 81: 8933–8943.
41. Gualde N, Harizi H (2004) Prostanoids and their receptors that modulate dendritic cell-mediated immunity. Immunol Cell Biol 82: 353–360.
42. Samikkannu T, Agudelo M, Gandhi N, Reddy PV, Saiyed ZM, et al. (2011) Human immunodeficiency virus type 1 clade B and C gp120 differentially induce neurotoxin arachidonic acid in human astrocytes: implications for neuroAIDS. J Neurovirol 17: 230–238.
43. Kabashima K, Murata T, Tanaka H, Matsuoka T, Sakata D, et al. (2003) Thromboxane A2 modulates interaction of dendritic cells and T cells and regulates acquired immunity. Nat Immunol 4: 694–701.
44. Spanbroek R, Hildner M, Steinhilber D, Fusenig N, Yoneda K, et al. (2000) 5-lipoxygenase expression in dendritic cells generated from CD34+ hematopoietic progenitors and in lymphoid organs. Blood 96: 3857–3865.
45. Hedi H, Norbert G (2004) 5-Lipoxygenase Pathway, Dendritic Cells, and Adaptive Immunity. J Biomed Biotechnol 2: 99–105.
46. Porter KM, Detorio M, Schinazi R, Sutliff RL (2008) HIV-1-Induced 5-Lipoxygenase: Potential Role in HIV-Related Pulmonary Hypertension. Atlanta, GA, A440.
47. Uz T, Dwivedi Y, Pandey GN, Roberts RC, Conley RR, et al. (2008) 5-Lipoxygenase in the Prefrontal Cortex of Suicide Victims. Open Neuropsychopharmacol J 1: 1–5.
48. Calverley DC, Kavanagh TJ, Roth GJ (1998) Human signaling protein 14-3-3zeta interacts with platelet glycoprotein Ib subunits Ibalpha and Ibbeta. Blood 91: 1295–1303.
49. Harizi H, Juzan M, Grosset C, Rashedi M, Gualde N (2001) Dendritic cells issued in vitro from bone marrow produce PGE2 that contributes to the

immunomodulation induced by antigen-presenting cells. Cell Immunol 209: 19–28.

50. Xiong H, Zeng YC, Lewis T, Zheng J, Persidsky Y, et al. (2000) HIV-1 infected mononuclear phagocyte secretory products affect neuronal physiology leading to

cellular demise: relevance for HIV-1-associated dementia. J Neurovirol Suppl 1: S14–23.

51. Tyor WR, Middaugh LD (1999) Do alcohol and cocaine abuse alter the course of HIV-associated dementia complex? J. Leukoc Biol 65: 475–81.

Permissions

List of Contributors

Giovanni Minervini, Elisabetta Panizzoni, Silvio C. E. Tosatto and Alessandro Masiero
Dept. of Biomedical Sciences, University of Padua, Padua, Italy

Manuel Giollo
Dept. of Biomedical Sciences, University of Padua, Padua, Italy
Dept. of Information Engineering, University of Padua, Padua, Italy

Carlo Ferrari
Dept. of Information Engineering, University of Padua, Padua, Italy

Marcelo Falsarella Carazzolle
Laboratório Nacional de Biociências, Centro Nacional de Pesquisa em Energia e Materiais, Campinas, São Paulo, Brazil
Laboratório de Genômica e Expressão, Departamento de Genética e Evolução, Instituto de Biologia, Unicamp, Campinas, São Paulo, Brazil

Lucas Miguel de Carvalho, Jörg Kobarg and Gabriela Vaz Meirelles
Laboratório Nacional de Biociências, Centro Nacional de Pesquisa em Energia e Materiais, Campinas, São Paulo, Brazil

Hugo Henrique Slepicka
Laboratório Nacional de Luz Síncrotron, Centro Nacional de Pesquisa em Energia e Materiais, Campinas, São Paulo, Brazil

Ramon Oliveira Vidal and Gonçalo Amarante Guimarães Pereira
Laboratório de Genômica e Expressão, Departamento de Genética e Evolução, Instituto de Biologia, Unicamp, Campinas, São Paulo, Brazil

Chaoqun Wang
College of Life Sciences, Wuhan University, Wuhan, Hubei, P. R. China
State Key Laboratory for Animal Nutrition, Institute of Animal Science, Chinese Academy of Agricultural Sciences, Beijing, P. R. China

Yefu Wang
College of Life Sciences, Wuhan University, Wuhan, Hubei, P. R. China

Shulin Yang, Ningbo Zhang, Yulian Mu, Hongyan Ren and Kui Li
State Key Laboratory for Animal Nutrition, Institute of Animal Science, Chinese Academy of Agricultural Sciences, Beijing, P. R. China

Khuram Shahzad and Juan J. Loor
Department of Animal Sciences and Division of Nutritional Sciences, University of Illinois at Urbana-Champaign, Urbana, Illinois, United States of America
Illinois Informatics Institute, University of Illinois at Urbana-Champaign, Urbana, Illinois, United States of America

Massimo Bionaz
Department of Animal and Rangeland Sciences, Oregon State University, Corvallis, Oregon, United States of America

Erminio Trevisi and Giuseppe Bertoni
Istituto di Zootecnica and Centro di ricerca sulla nutrigenomica, Universita´ Cattolica del Sacro Cuore, Piacenza, Italy

Sandra L. Rodriguez-Zas
Department of Animal Sciences and Division of Nutritional Sciences, University of Illinois at Urbana-Champaign, Urbana, Illinois, United States of America
Illinois Informatics Institute, University of Illinois at Urbana-Champaign, Urbana, Illinois, United States of America
The Institute for Genomic Biology, University of Illinois at Urbana-Champaign, Urbana, Illinois, United States of America

Michal Mikula, Adriana Strzalkowska, Monika Borowa- Chmielak, Artur Dzwonek and Marta Gajewska
Department of Genetics, Maria Sklodowska-Curie Memorial Cancer Center and Institute of Oncology, Warsaw, Poland

Ewa E. Hennig and Jerzy Ostrowski
Department of Genetics, Maria Sklodowska-Curie Memorial Cancer Center and Institute of Oncology, Warsaw, Poland
Department of Gastroenterology and Hepatology, Medical Center for Postgraduate Education, Warsaw, Poland

Heyka H. Jakobs, Bernd Clement and Antje Havemeyer
Department of Pharmaceutical and Medicinal Chemistry, Christian-Albrechts-Universität zu Kiel, Kiel, Germany

Caroline Baroukh
INRA UR050, Laboratoire des Biotechnologies de l'Environnement, Narbonne, France, 2 INRIA-BIOCORE, Sophia-Antipolis, France

Rafael Muñoz-Tamayo
INRIA-BIOCORE, Sophia-Antipolis, France

Jean-Philippe Steyer
INRA UR050, Laboratoire des Biotechnologies de l'Environnement, Narbonne, France

Olivier Bernard
INRIA-BIOCORE, Sophia-Antipolis, France
LOV-UPMC-CNRS, UMR 7093, Villefranche-sur-mer, France

Jing Geng and Nelly Henry
Laboratoire Jean Perrin (CNRS FRE 3231), UPMC, Paris, France

Christophe Beloin and Jean-Marc Ghigo
Institut Pasteur, Unité de Génétique des Biofilms, Département de Microbiologie, Paris, France

Anne Jungandreas
Department of Plant Physiology, Institute of Biology, Faculty of Biosciences, Pharmacy and Psychology, University of Leipzig, Leipzig, Germany
Department of Computational Landscape Ecology, Helmholtz Centre for Environmental Research - UFZ, Leipzig, Germany

Martin von Bergen
Department of Metabolomics, Helmholtz Centre for Environmental Research - UFZ, Leipzig, Germany
Department of Proteomics, Helmholtz Centre for Environmental Research - UFZ, Leipzig, Germany
Department of Biotechnology, Chemistry and Environmental Engineering, University of Aalborg, Aalborg, Denmark

Sven Baumann
Department of Metabolomics, Helmholtz Centre for Environmental Research - UFZ, Leipzig, Germany
Institute of Pharmacy, Faculty of Biosciences, Pharmacy and Psychology, University of Leipzig, Leipzig, Germany

Benjamin Schellenberger Costa, Torsten Jakob and Christian Wilhelm
Department of Plant Physiology, Institute of Biology, Faculty of Biosciences, Pharmacy and Psychology, University of Leipzig, Leipzig, Germany

Juan Chen, Wen-Jun Hu, Juan Chen, Zhi-Jun Shen, Xiang Liu, Wen-Hua Wang and Hai-Lei Zheng
Key Laboratory of the Coastal and Wetland Ecosystems, Ministry of Education, College of the Environment and Ecology, Xiamen University, Xiamen, Fujian, China

Chao Wang
Institute of Urban and Environment, Chinese Academy of Sciences, Xiamen, P.R. China

Ting-Wu Liu
Department of Biology, Huaiyin Normal University, Huaian, Jiangsu, P.R. China

Key Laboratory of the Coastal and Wetland Ecosystems, Ministry of Education, College of the Environment and Ecology, Xiamen University, Xiamen, Fujian, China

Annie Chalifour
Department of Biology and Chemistry, City University of Hong Kong, Kowloon, Hong Kong, SAR, China

Anastasia S. Khodakova, Renee J. Smith, Leigh Burgoyne and Adrian Linacre
School of Biological Sciences, Flinders University, Adelaide, Australia

Damien Abarno
School of Biological Sciences, Flinders University, Adelaide, Australia
Forensic Science South Australia, Adelaide, Australia

Sang Kyum Kim, Woo Hee Jung and Ja Seung Koo
Department of Pathology, Severance Hospital, Brain Korea 21 PLUS Project for Medical Science, Yonsei University College of Medicine, Seoul, South Korea

María Teresa Martínez-Larrad, Arturo Corbatón Anchuelo and Manuel Serrano-Ríos
Spanish Biomedical Research Centre in Diabetes and Associated Metabolic Disorders (CIBERDEM), Madrid, Spain
Instituto de Investigación Sanitaria del Hospital Clínico San Carlos (IdISSC), Madrid, Spain

Náyade Del Prado and José María Ibarra Rueda
Instituto de Investigación Sanitaria del Hospital Clínico San Carlos (IdISSC), Madrid, Spain

Rafael Gabriel
Clinical Epidemiology Research Unit, Hospital de La Paz, Madrid, Spain

Amir Miraj Ul Hussain Shah, Ye Zhao, Yunfei Wang, Liangyan Wang, Bing Tian and Yuejin Hua
Key Laboratory of Chinese Ministry of Agriculture for Nuclear-Agricultural Sciences, Institute of Nuclear-Agricultural Sciences, Zhejiang University, Hangzhou, China

Guoquan Yan and Qikun Zhang
Laboratory of Microbiology and Genomics, Zhejiang Institute of Microbiology, Hangzhou, China

Huan Chen
Key Laboratory of Laboratory Medicine, Ministry of Education, Zhejiang Provincial Key Laboratory of Medical Genetics, School of Laboratory Medicine & Life Science, Wenzhou Medical College, Wenzhou, China
Laboratory of Microbiology and Genomics, Zhejiang Institute of Microbiology, Hangzhou, China

Jiankai Wei and Yang Yu
Key Laboratory of Experimental Marine Biology, Institute of Oceanology, Chinese Academy of Sciences, Qingdao, China
University of Chinese Academy of Sciences, Beijing, China

Xiaojun Zhang, Fuhua Li and Jianhai Xiang
Key Laboratory of Experimental Marine Biology, Institute of Oceanology, Chinese Academy of Sciences, Qingdao, China

Hao Huang
Hainan Guandtop Ocean Breeding Co. Ltd, Haikou, China

Szu-Chieh Mei and Charles Brenner
Department of Biochemistry, Carver College of Medicine, University of Iowa, Iowa City, Iowa, United States of America

Natalia Sevane, Federica Bialade, Susana Velasco, Almudena Rebolé, Maria Luisa Rodríguez,
Luís T. Ortiz, Javier Cañón and Susana Dunner
Nutrigenómica Animal, Departamento de Producción Animal, Facultad de Veterinaria, Universidad Complutense de Madrid, Madrid, Spain

Pengjuan Zhang, Chenghua Li, Peng Zhang, Chunhua Jin and Daodong Pan
Department of aquaculture, Ningbo University, Ningbo, Zhejiang Province, P.R China

Yongbo Bao
Department of Aquatic Germplasm Resources, Zhejiang Wanli University, Ningbo, Zhejiang Province, P.R China

Abderrazak Kitsy, Skyla Carney, Juan C. Vivar, Megan S. Knight and Mildred A. Pointer
Division of Cardiometabolic Disorders, Biomedical Biotechnology Research Institute, North Carolina Central University, Durham, North Carolina, United States of America

Judith K. Gwathmey
Boston University School of Medicine, Boston, Massachusetts, United States of America

Sujoy Ghosh
Division of Cardiometabolic Disorders, Biomedical Biotechnology Research Institute, North Carolina Central University, Durham, North Carolina, United States of America
Program in Cardiovascular and Metabolic Disorders, Duke-NUS Graduate Medical School, Singapore, Singapore

Thangavel Samikkannu, Kurapati V. K. Rao, Hong Ding, Marisela Agudelo, Andrea D. Raymond and Madhavan P. N. Nair
Department of Immunology, Institute of NeuroImmune Pharmacology, Herbert Wertheim College of Medicine, Florida International University, Modesto A. Maidique Campus, Miami, Florida, United States of America

Changwon Yoo
Department of Biostatistics, Robert Stempel College of Public Health and Social Work, Florida International University, Modesto A. Maidique Campus, Miami, Florida, United States of America

Index